ARCHITECTURAL BUILDING
CONSTRUCTION

VOLUME THREE

THE GARDEN FRONT

THE ENTRANCE FRONT

VIEWS OF THE SEMI-DETACHED HOUSES

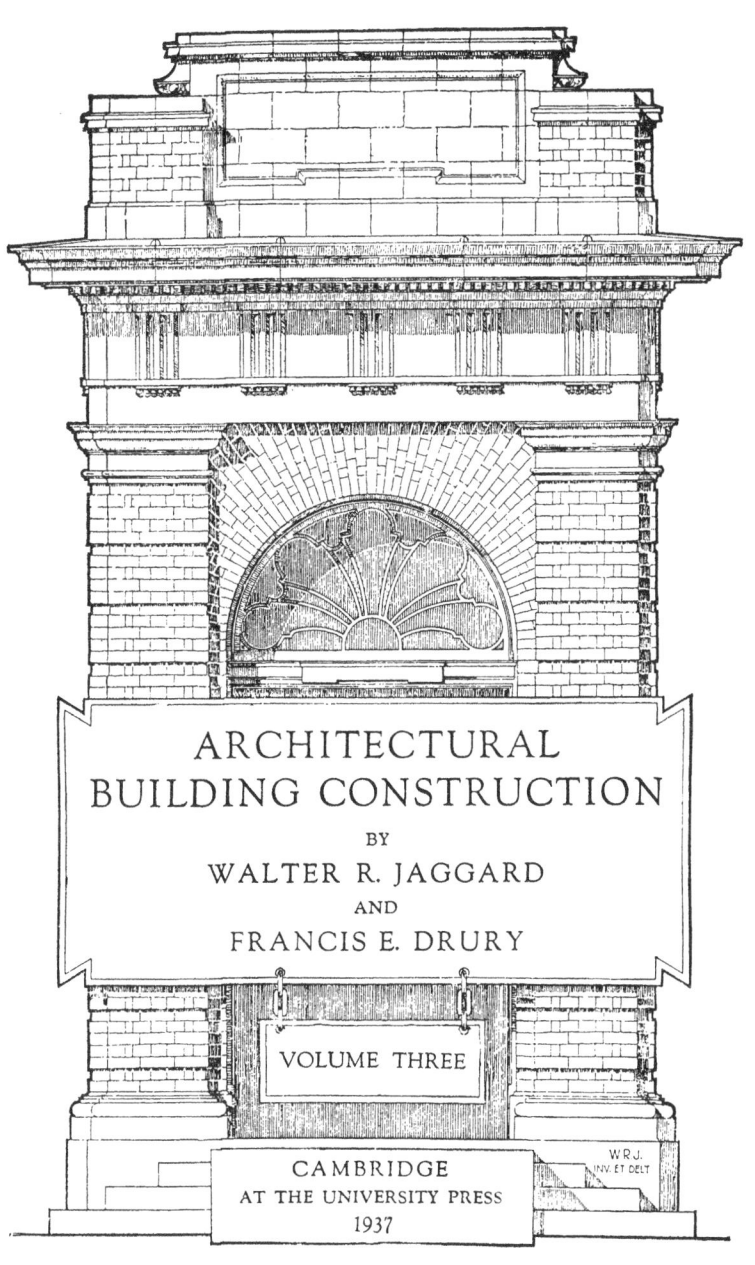

ARCHITECTURAL BUILDING CONSTRUCTION

BY

WALTER R. JAGGARD

AND

FRANCIS E. DRURY

VOLUME THREE

CAMBRIDGE
AT THE UNIVERSITY PRESS
1937

W.R.J.
INV. ET DELT

CAMBRIDGE
UNIVERSITY PRESS

University Printing House, Cambridge CB2 8BS, United Kingdom

Published in the United States of America by Cambridge University Press, New York

Cambridge University Press is part of the University of Cambridge.

It furthers the University's mission by disseminating knowledge in the pursuit of education, learning and research at the highest international levels of excellence.

www.cambridge.org
Information on this title: www.cambridge.org/9781107645363

© Cambridge University Press 1937

First edition 1923
Reprinted 1927, 1932
Second edition (Revised and Enlarged) 1937
First published 1937
First paperback edition 2014

A catalogue record for this publication is available from the British Library

ISBN 978-1-107-64536-3 Paperback

ARCHITECTURAL BUILDING CONSTRUCTION

A TEXT BOOK FOR THE ARCHITECTURAL AND BUILDING STUDENT

BY

WALTER R. JAGGARD

AND

FRANCIS E. DRURY

M.Sc.Tech., M.I.Struct.E., F.I.San.E.

Second Edition
(Revised and Enlarged by F. E. DRURY)

VOLUME THREE

CAMBRIDGE
AT THE UNIVERSITY PRESS
1937

GENERAL PREFACE

IN writing and illustrating the series of works on Architectural
Building Construction of which this is the third volume, the
authors have been actuated by the desires and objects which are
briefly set out below.

That there are many existing books on Building Construction
is a well-recognised fact; that some of them have excellent matter,
and others have good illustrations is also duly acknowledged, but
from a long experience in the practice of Architecture, both in
England and the Colonies, and many years of teaching architectural
principles and the science of building construction, the authors have
been forced to the conclusion that something more in the way of
text books is needed for the following reasons.

1. Building Construction should not be divorced from the Principles of Architectural Design. Although it is sometimes true that
we find an Architect who can design pleasing structures with little or
no knowledge of building construction, it is an undoubted fact that a
fine conception of noble architecture must be based upon an intimate
and complete knowledge of the proper use of materials, the scientific
and fit assembly of the varying units, and an honest and conscientious
co-ordination of the work of Architect, Builder and Craftsman.

It may be argued that with the present day use of steel and
reinforced concrete, together with other modern materials and
methods, we are able to construct some most extravagant fancies
in architectural design, which a few years ago would have been quite
impossible. Whilst this is quite true, and illustrates the age in which
we live, it is also true that the very great majority of our buildings
to-day are still erected with the staple materials, such as concrete,
brick, stone and timber.

2. For the creation of good architecture it is necessary to study
the work produced by our predecessors, and not only the work of
ancient civilisations and mediæval peoples, but the best work of
more modern architects must be examined. This study is rendered
comparatively easy of acquisition through the rapid and cheap
facilities offered for travel, and the many excellent books and illus-

trations which are constantly being published. Attention might here be drawn to the publication entitled, *The Development of English Building Construction*, written by Mr C. F. Innocent, and forming one of the volumes of The Cambridge Technical Series; a perusal of this book will be found both interesting and helpful. Is it possible to apply this method to the study of Building Construction? The authors have been impressed with this idea, and have endeavoured, with some success, to carry it out in their teaching. They have, however, found that in the earlier stages of education this must not be unduly pressed. For elementary students, the teacher should, to some extent, be dictatorial, and whilst selecting a well-proportioned and designed study as an example, should insist upon the construction being shown in a definite manner, although he knows that infinite variety, both in design and construction, is possible. With the more advanced student greater latitude is desirable, and in fact necessary. The authors have, therefore, impressed a certain amount of individuality in the subjects of the first two volumes, but they intend, as far as possible, in the third volume to select examples of established taste and architectural value to illustrate advanced principles of design, maintaining in some cases the constructional details given them by their designer or constructor, but in others, adapting the construction in accordance with modern methods and the more extended use of machinery.

3. Building Construction has more generally been presented to the student in the form of isolated examples, which have no relation whatever to each other, and thus the knowledge obtained cannot be applied to the actual design of a building, even of the smallest dimensions, until a very much later date. Modern methods of teaching demand a greater cohesion. The authors have endeavoured during their teaching experience to obtain or formulate one building into which all the various items comprehended in each year's work could be fitly placed, but after many attempts it was found to be impracticable, and therefore two buildings were arranged, which embody, with few exceptions, all the items necessary for an elementary knowledge of building construction, thus enabling "teaching from the structure itself to be adopted rather than the selection of isolated examples on account of their simplicity".

This method has been adopted in the first three volumes, but whilst the authors disclaim any idea of presenting great architecture, they do claim that the buildings designed fitly express their purpose, and enable them in a more or less pleasing manner to assemble the different units of the building, and at the same time to inculcate a sense of completeness in the student's work.

4. The acquisition of a knowledge of Building Construction should rightly be a plant of slow but sturdy growth, and in the majority of Architectural and Building Schools the course of instruction covers a period of from three to five years. The first volume of this series is designed to meet the needs of the first year student, the second volume will provide more than is generally required for a second year course, while the third volume will cover a large field of advanced work.

5. The authors have often felt that the ordinary orthogonal presentation of examples of building construction does not sufficiently convey the solidity of the object to an elementary student, and as it is not possible for each student to have, or to make, models of the different units for himself—although such a course would greatly make for efficiency of study—the illustrations have to a large extent been shown in perspective, isometric or pictorial projection. Photographs might, and in some cases will, be used, but the camera, whilst giving a faithful representation of the object, cannot be used to show the construction of hidden parts. On this account dissociated isometric and oblique sketches have been freely used with some slight shading to indicate differing planes, but cast shadows have generally been avoided as tending to obscure the construction, which it is desired to show in the clearest possible manner.

It is strongly recommended that in all Architectural and Building Schools correct scale models—about half full size—of the different items should be made in such a way that the parts may be dissembled, and that the student should be encouraged and advised to study and measure these carefully, and make the usual orthogonal drawings, which are, after all, the media through which Architect, Builder and Craftsman convey their ideas and wishes to one another.

It is necessary to impress strongly upon a student in the early stages of his work, that his knowledge must be presented in a clear and unmistakable way and with some architectural character and students should be encouraged carefully to complete all their drawings with full naming of parts, references, and adequate dimensions, and to ink-in and colour, or otherwise distinguish, the materials of construction. They will thus acquire the habit of thoroughness, which is of inestimable value to both draughtsman and craftsman.

In conclusion, the authors' chief endeavour has been to make these volumes of primary importance to students—architect, builder or craftsman—and since in this study, at least, they all meet upon common ground, although each with different aims and objects in life to accomplish, yet, each finding help and guidance herein, there is an augury of the future happy relations which should exist between those engaged in all the branches of the practice of architectural building.

W. R. J.
F. E. D.

June 1916

PREFACE TO VOLUMES II AND III

IN compiling this volume it was necessary for the authors to give careful consideration to the mode of treatment. At least two courses were open to them, viz. the illustration of methods of construction applied to isolated building details, by selecting examples having no definite relation to or embodiment in any one building, or, to continue the method adopted in the first volume of this series, and endeavour to relate all the items of construction which it appeared desirable to illustrate and discuss, and to apply these rationally within the boundaries of one or more buildings of a definite character.

The latter course commended itself to the authors and they decided to proceed on this basis. Two buildings of different character were selected in order to avoid the aggregation of conflicting details and doubtful combinations of materials which would, of necessity, render a single building very elaborate and complicated, if designed to contain them.

Volume II contains the treatment of such parts of the two buildings as can reasonably be placed before a good second year student in an average school of building or architecture, and the remaining study, for the complete consideration of the two buildings, is given in Volume III.

One of the buildings belongs to the class generally known under building regulations as "domestic buildings", while the other belongs to the division known as "buildings of the warehouse class".

A certain amount of overlapping in the preparation of details under the conditions of the treatment selected is inevitable, but it was found convenient and practicable to design two structures which are fairly typical of their respective classes, while embodying the features desirable for study.

These buildings are:

(1) A semi-detached suburban house.

(2) A town warehouse.

The architectural treatment of these buildings in plan and elevation has been developed with the intention of illustrating as varied a range of constructional examples as possible. It cannot be maintained, however, that all the variations of detail suggested could be embodied or economically used in one building, but each example suggested may be fitly employed as an alternative to other forms of construction in a building of the type to which it refers.

The question of cost has not been overlooked by the authors, but they have not allowed this point to interfere with the objects in view, viz. to illustrate and explain a number of sound and standard methods of construction, generally suited to the class of buildings under consideration.

First cost is not the only point to observe, because it frequently happens that an apparently costly structure is so economical in maintenance that it may ultimately become cheaper and more serviceable than a building of a more meretricious character, where a low first cost appeared to ensure economy.

Furthermore, the mere fact that a building has been erected with costly materials and supplied with much added decoration does not necessarily mean the production of a building of architectural value. Many wayside cottages have that elusive charm, which we know as architecture, and which is very often missing in some of the large town buildings. It should be the business of every student to endeavour, for himself, to find that secret by which the smallest building, by correctly expressing its functions, may, equally with the large and noble, take its place as a work of architecture.

The authors desire to record their indebtedness to all who have been kind enough to supply information, and to take an interest in the preparation of this volume.

                                                            W. R. J.
July 1922   F. E. D.

In the present edition the subjects of study have been rearranged and many revisions and additions to the text have also been made.

It is hoped that this change will be an advantage to teachers and students in meeting the needs of grouped course instruction.

Thanks are hereby accorded to Mr Norman Keep, F.R.I.B.A., and Mr R. A. Bix for their valuable assistance in preparing diagrams for this volume.

                                                            F. E. D.
April 1936

CONTENTS OF VOLUME III

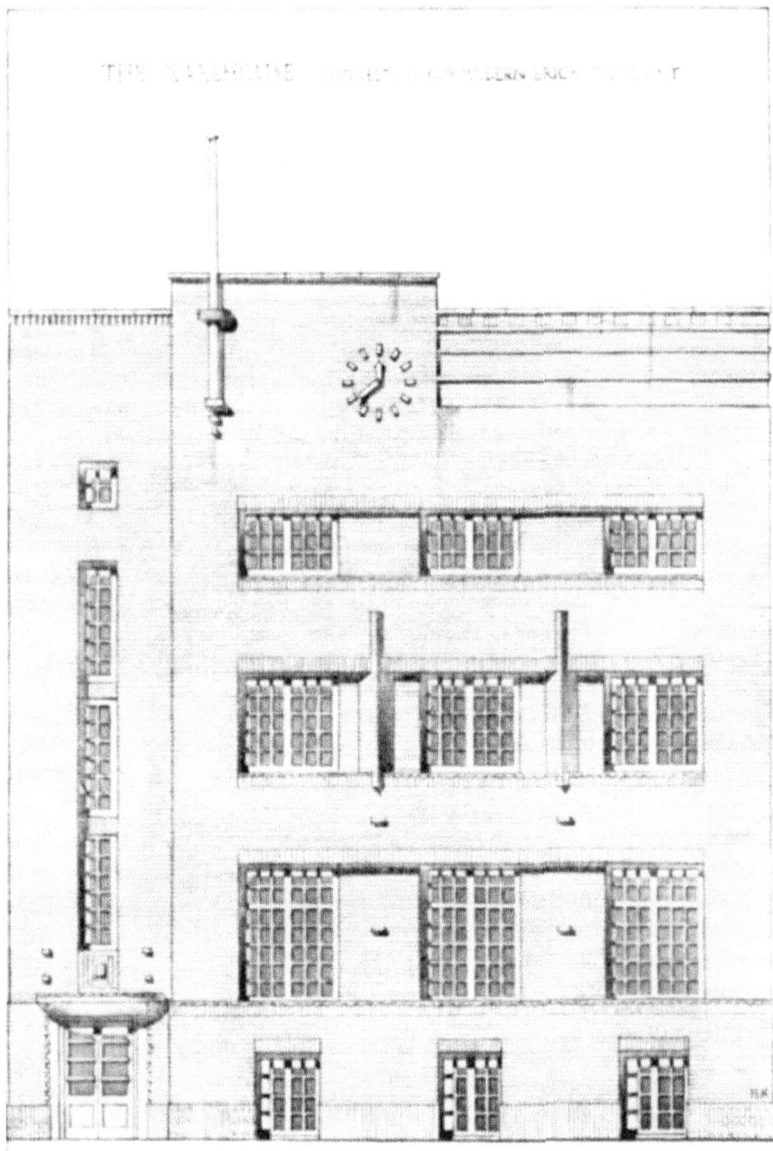

CHAPTER ONE

BRICKWORK AND MASONRY

FOUNDATIONS AND BASEMENT WALLS

Some portions of the foundations to the warehouse were treated in Vol. II.

It remains here to consider the arrangement and construction of the foundations to the long side walls which are built on sloping ground, and to the independent and attached piers which are used to support the ground floor and to increase the wall bearings for the beams of the upper floors.

1. Foundation treatment due to sloping site. Detail No. 1 shows a vertical section through the site from front to back of the warehouse.

In firm, dry ground it is unnecessary to place the foundations at a greater depth than the levels given, but in soft or unreliable ground it may be necessary to excavate to the surface of a firm underlying stratum of earth, or, if no such stratum exists at a convenient depth, to adopt a special type of foundation.

Assuming firm ground, the depth of the front foundation would be 8 ft., this being regulated by the depth of the basement and the necessary footings and foundation concrete. At the back wall the depth is 4' 3" which is sufficient to escape the normal effects of rain, frost and heavy traffic.

Taking into account the differences in ground level and excavation, there is a height of 7' 6" between front and back foundation levels.

2. Concrete to stepped foundation. If the difference in level were arranged to take place in one step, fracture of the walls immediately above the junction would be probable, because the higher part of the walls with the greater number of mortar joints and increased weight upon the base would induce more settlement at this part. To reduce or ease off the effect of this non-uniform settlement, stepped foundations are desirable; these minimise the tendency to fracture by distributing the settlement over a series of changes in depth.

An examination of the stepping in detail No. 1 shows the ground to be "benched" into three level steps. The concrete is carried horizontally and vertically in one continuous mass, with projections for the footings of piers and chimney breasts where these occur.

3. Footings to stepped foundations. In details Nos. 1 and 2 the footings are shown to butt squarely against the vertical cross face of each concrete step, but are returned at the unobstructed end to agree with their projection from the wall. This avoids any bonding difficulties and the vertical step of concrete efficiently transmits the load to the earth at each change of level.

4. Pier and foundations to ground floor pillars. In the line of the cross wall shown in detail No. 2, which forms the back enclosure to the basement at the centre of the building, two pillars are to be erected for supporting the upper floors. For this purpose brick piers are provided 2′ 7½″ square, built of *blue bricks*[1] in cement mortar and capable of supporting 12 tons per sq. ft. of horizontal area. The method of determining the size of these piers and *their foundations* is given in Chapter Five.

Should large changes of level occur, as at A in detail No. 1, the connection between the levels is made by leaving an open zig-zag joint and pouring molten asphalte into it. The joints are temporarily stopped on the face with clay, until the asphalte is firm.

The damp-proofing treatment of the ordinary main walls of the warehouse was illustrated in Vol. II.

5. Application of vertical damp proof sheeting and retention walls to basement of warehouse. Detail

[1] See Chapter on Materials, Vol. II.

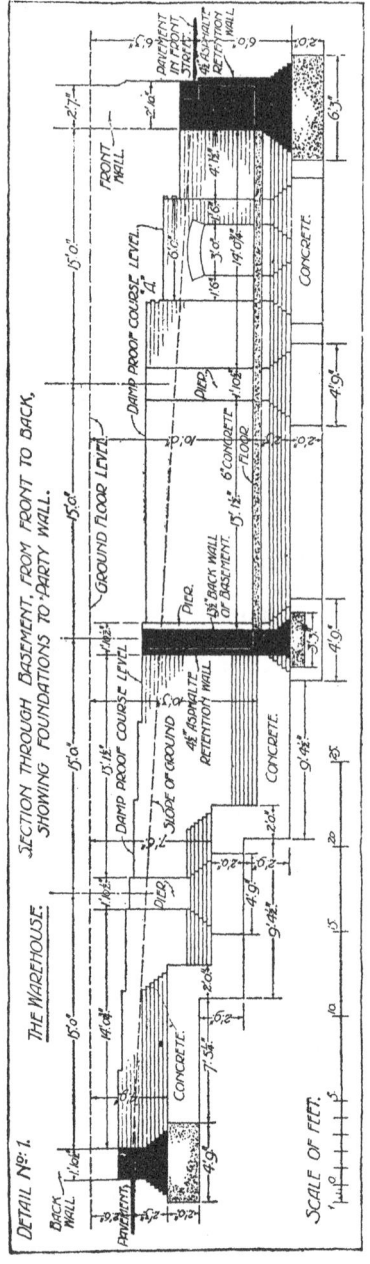

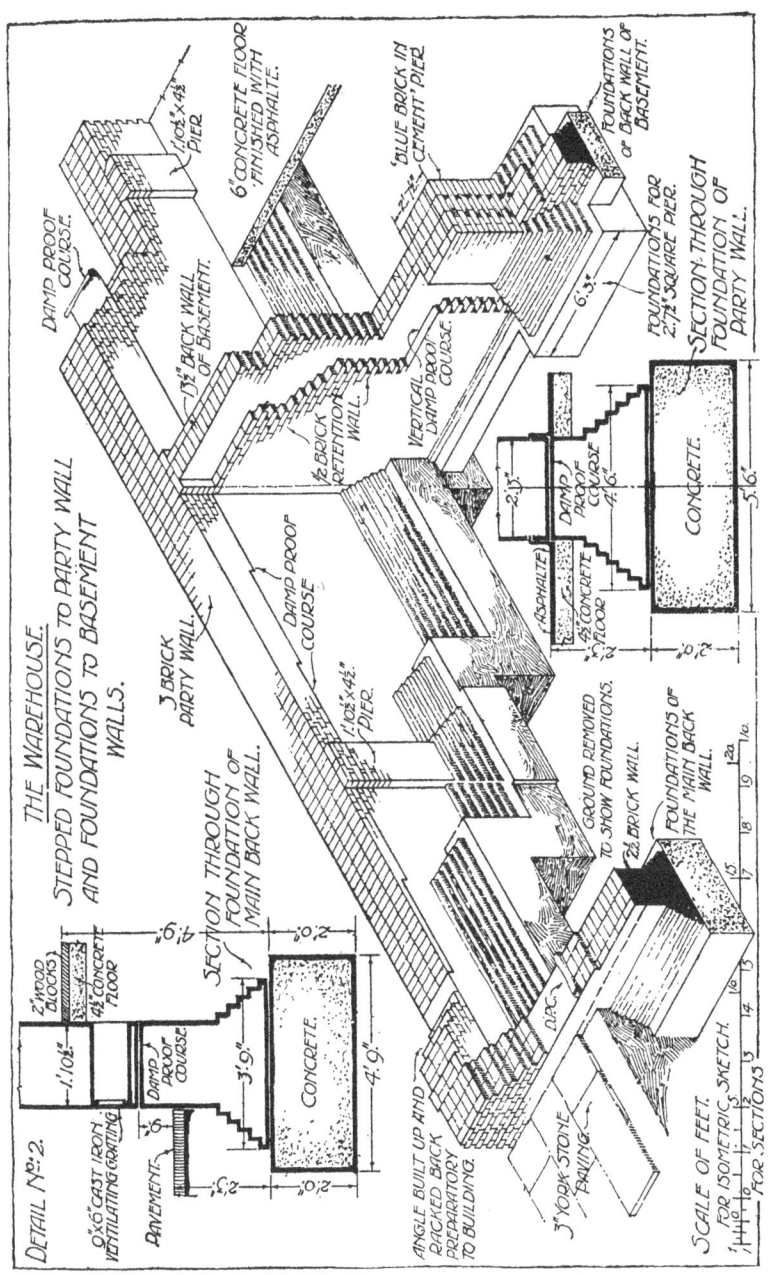

THE WAREHOUSE.
STEPPED FOUNDATIONS TO PARTY WALL
AND FOUNDATIONS TO BASEMENT
WALLS.

DETAIL Nº 2.

No. 2 shows the use of a retention wall in the cross wall at the back of the basement, while detail No. 1 illustrates the vertical sheeting at the front wall of the warehouse, taken round the base course at the pavement to keep it out of sight. Further large scale sections are given in detail No. 3 which shows the base of the external staircase wall, the division wall of the stairs, and the foundation and basement walling to the front wall of the warehouse.

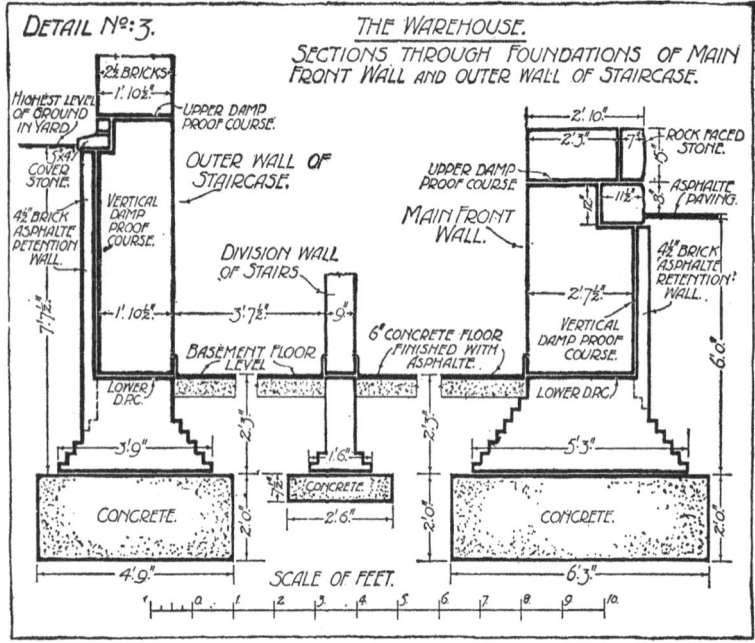

6. Pier bonding. In studying detail No. 2 the brick bonding should be noted as further applications of general principles are given.

This detail shows the bonding of the "three-brick" party wall and its attached piers, the junction of the same wall with the back wall of the warehouse, the toothing necessary for subsequent bonding to adjoining property and the bonding of the "three and a half brick" square pier to the "one and a half brick" cross wall in the basement.

7. Open area and lighting of basement to the house. Detail No. 4 gives a section across the open area taken clear of the main entrance and basement doorways, and through one of the basement windows.

The function of this area is to provide light to the basement by means of the open space in front of the windows, which are below the ground level.

8. Retaining wall. A wall becomes necessary to support or retain the earth outside this area and is known as a retaining wall. The shape and extent of the area are shown on the general plan,

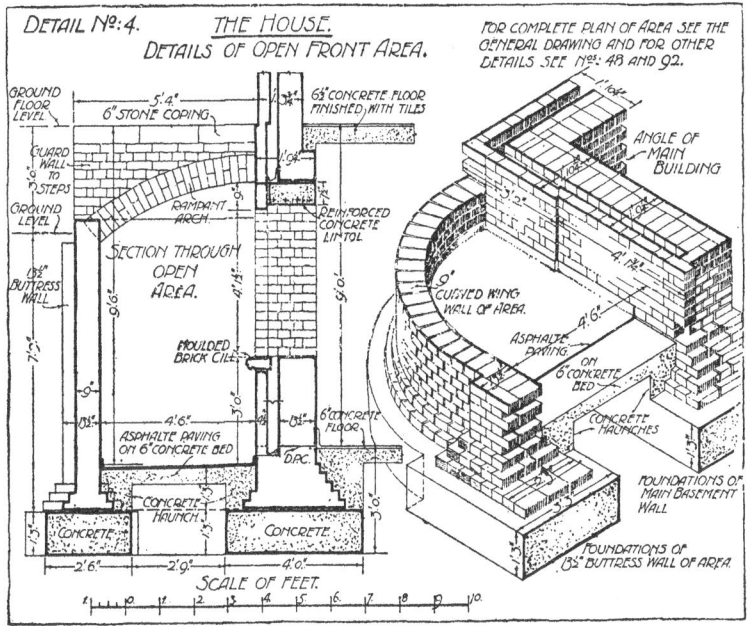

being a rectangle with two quadrant ends, the rectangle lying beneath the main entrance steps with the quadrants open to view and forming wing walls to the excavation.

The earth abutting on these walls exerts a thrust, tending to overturn them into the area. On the straight part of the retaining wall, which acts as an abutment for an arch supporting the entrance steps, the two thrusts—from arch to earth—oppose and tend to balance each other and under these forces alone a 9″ wall would probably serve the purpose; but the quadrant walls have been employed to meet the thrust of the earth towards the open parts of the area by constructing them as vertical arches of header bricks; the thrust is thus transmitted round the arch to the vertical supports, viz. the main wall of the house and the straight piece of retaining wall. Hence, the latter is made $13\frac{1}{2}''$ thick.

9. Construction of wing walls in heading bond. Heading bond has already been referred to as suitable for footings, but its logical application to walling may now be considered.

In brickwork curved on plan and constructed from standard bricks in the ordinary bonds, it is only possible to approximate to the curve except by special cutting of the exposed faces. To diminish the polygonal appearance which would result from laying stretcher bricks as shown in detail No. 5, headers may be employed to obtain a closer approximation to the curve. This arrangement causes wide back joints which can only be allowed where they are concealed as in the example. Where the curvature is too sharp for such an approximation bricks should be cut or special bricks obtained of the true curvature; ordinary bonds may then be employed.

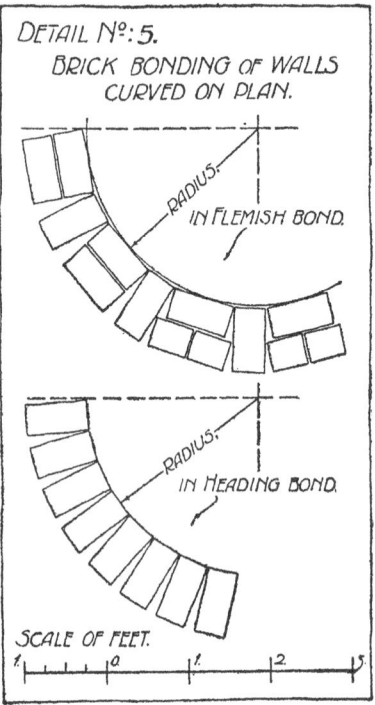

DETAIL Nº: 5.
BRICK BONDING OF WALLS CURVED ON PLAN.

IN FLEMISH BOND.

IN HEADING BOND.

SCALE OF FEET.

If the external convex surface of such a wall is to be exposed, the side joints are not permissible and the bricks must then be rough-axed to a wedge shape, or purpose made bricks obtained.

CONSTRUCTION OF WALLS

10. Thick walls. In the warehouse the brick walls vary in thickness from 9″ to 27″ and some of the thicker walls are compounded of stone facings and brick backing.

Thick walls are bonded on the usual principles, stretchers being employed on both faces of a course having an even number of half bricks in thickness and on the opposite faces of alternate courses when the thickness is an odd number of half bricks; the rest of the work is packed with headers and joints are *not* broken across the thickness of the wall.

The bond in straight portions of the walls should be studied together with the adaptations in junctions and angles; see detail No. 2.

11. Deficiencies in ordinary bonds. In Vol. I it is stated that there are deficiences of bonding in thick walls. The chief defect is caused by filling the centre of the stretching courses with header bricks, causing lack of longitudinal as compared with transverse tie in the body of the wall, the former being $2\frac{1}{4}''$ and the latter $4\frac{1}{2}''$. In high, heavily loaded walls a fault in the natural foundation which allowed settlement to occur, especially near the quoins of a building, would probably produce disturbance and withdrawal of the bricks longitudinally, because quoin settlement induces horizontal movement of part of the mass of work towards the quoin. The distribution of concentrated loads from floor girders, roof trusses, etc. in the direction of the length of thick walls is not so proportionately effective as in thin walls, because of the small side lap of the headers which form the greater bulk of the walling.

12. Methods of improving longitudinal bond in thick walls. There are several methods available for improving longitudinal tie, amongst which are reinforcement by hoop iron, or wire mesh and by placing bricks diagonally between the stretchers across the centre of the wall.

13. Hoop iron bond. Detail No. 6 shows the junction between the party wall and back wall of the warehouse, where strips of hoop iron, about $1\frac{1}{4}'' \times \frac{1}{10}''$, are laid longitudinally, one strip to each half brick of the centre headers, the series being repeated at every sixth joint. When two walls intersect at a tee junction, the terminals are single hooked over the outside strip; at an angle junction the external strips are double hooked and the others single hooked; joints in the length of a strip are welt-lapped. All joints are hammered flat and where pieces cross each other in series they may be interlaced as shown. To preserve hoop iron it may be galvanised or preferably coated with hot tar and immediately drawn through fine, dry sand which adheres and give a grip for the mortar.

Modern forms of reinforcement, including wire mesh and expanded metal and which may be used instead of hoop iron, are explained and illustrated in Vol. II.

14. Raking bonds. The oldest method of improving longitudinal tie is to fill the space between external stretchers with bricks inclined to the face of the wall, thus increasing the "length tie" with little reduction of transverse strength. The arrangement is known as "raking bond" and may be applied at every fifth course in the height of plain walls, the bricks in consecutive alternate raking courses being inclined in opposite directions. Footings are greatly improved by raking bond. There are two common forms: (a) diagonal bond, and (b) herring-bone bond.

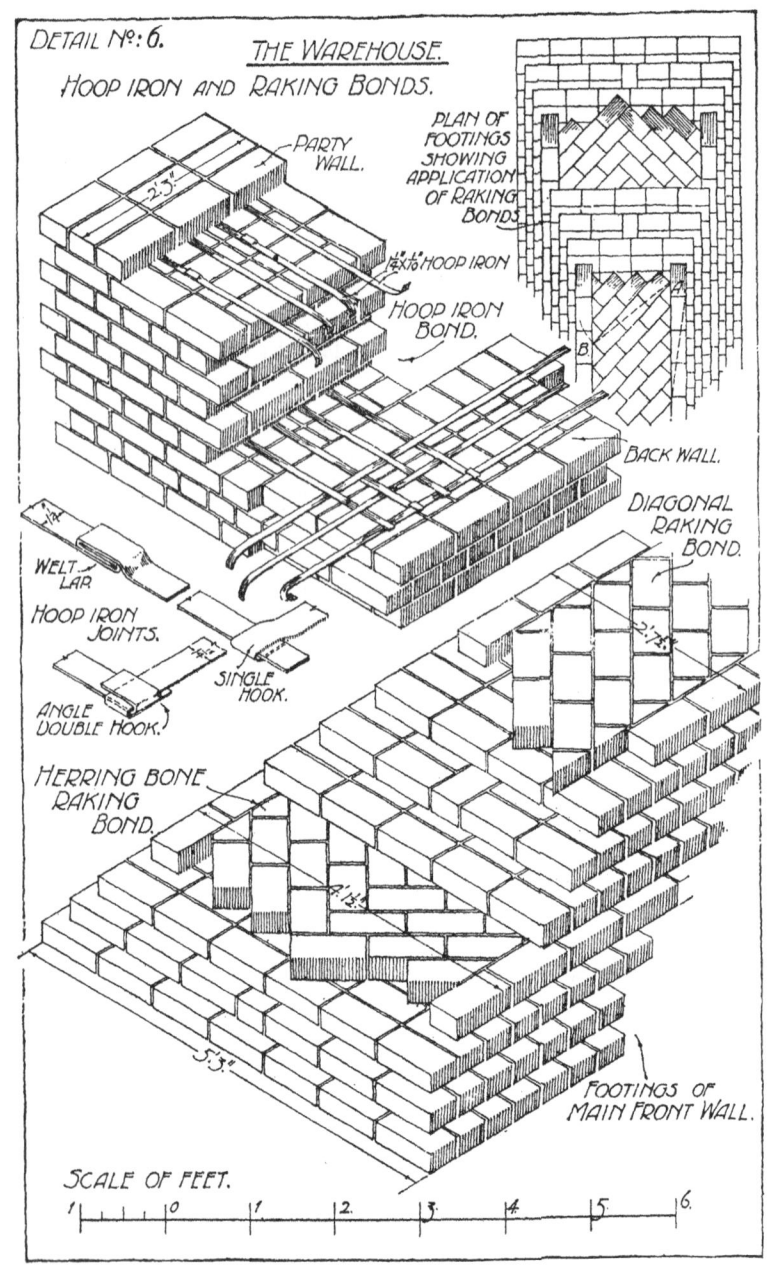

DETAIL Nº: 6.

THE WAREHOUSE.

HOOP IRON AND RAKING BONDS.

PARTY WALL.

2.3"

PLAN OF FOOTINGS SHOWING APPLICATION OF RAKING BONDS.

1½"×¹⁄₁₆" HOOP IRON

HOOP IRON BOND.

A

B

BACK WALL.

WELT LAP

HOOP IRON JOINTS.

SINGLE HOOK.

ANGLE DOUBLE HOOK.

DIAGONAL RAKING BOND.

2.1¾"

HERRING BONE RAKING BOND.

4.1½"

FOOTINGS OF MAIN FRONT WALL.

5.3"

SCALE OF FEET.

1   0   1   2   3   4   5   6.

Diagonal bond is the more effective because a number of bricks may be placed end to end as shown to the right of detail No. 6, the extreme corners of the series being in contact with the stretchers. To set out the bricks in a drawing, observe that the ends of the diagonal drawn across three bricks are here required to be in contact, hence obtain this length from the plan of three stretchers, take any point *A* on the boundary as centre, and with the diagonal as radius strike an arc to cut the opposite boundary at *B*. Upon this construct the right-angled triangles on each side, and repeat by parallel courses. This bond may be applied to any course which is three bricks or more in thickness; it is shown applied to the first course above the footings in the front warehouse wall, which is 2′ 7½″ thick at the basement level.

When applied to footings the latter should be at least four bricks thick and the *headers should be retained at the outside of the course.*

15. Herring-bone bond is used in similar positions and is shown applied in the same detail. It is more suitable for very thick walls, because the bricks are laid in courses inclined at 45° in two directions. The object of laying out the pattern should be to use the maximum number of whole bricks.

In this example the application is made to the centre course of footings, which is five and a half bricks wide and is shown with stretchers on the external faces. This arrangement illustrates a defect common to such work as the distribution of load to the stretchers is not satisfactory. Headers, if employed, would enter the walling 6¾″ and thus remedy the defect.

It should be noted that raking bonds are rapidly going out of use for large and important work. They are still used in restoration work, however, and in odd cases where metal bonding material is not available as and when required.

Arches

16. Rampant arch. When an arch is arranged with its supports at different levels, it is called a "rampant" arch. If the arch is required to rise above the higher support its outline is preferably a compound curve, but the simplest and soundest form is half of an ordinary segmental arch, and this type is suitably applied in the support provided to the steps of the main entrance to the house, as shown in detail No. 4 and to a larger scale in detail No. 10.

The arch has a span of 4′ 6″ and consists of a single rough brick ring faced at the ends with a 9″ gauged arch and bonded thereto; it springs from the 13½″ retaining wall of the area and abuts against the main wall of the building below the doorway.

17. Pointed arches. Arches formed by two arcs of curvature intersecting to produce a sharp angle at the crown are known as pointed or Gothic arches. They are commonly applied in, and are a special feature of, certain periods of medieval or Gothic architecture.

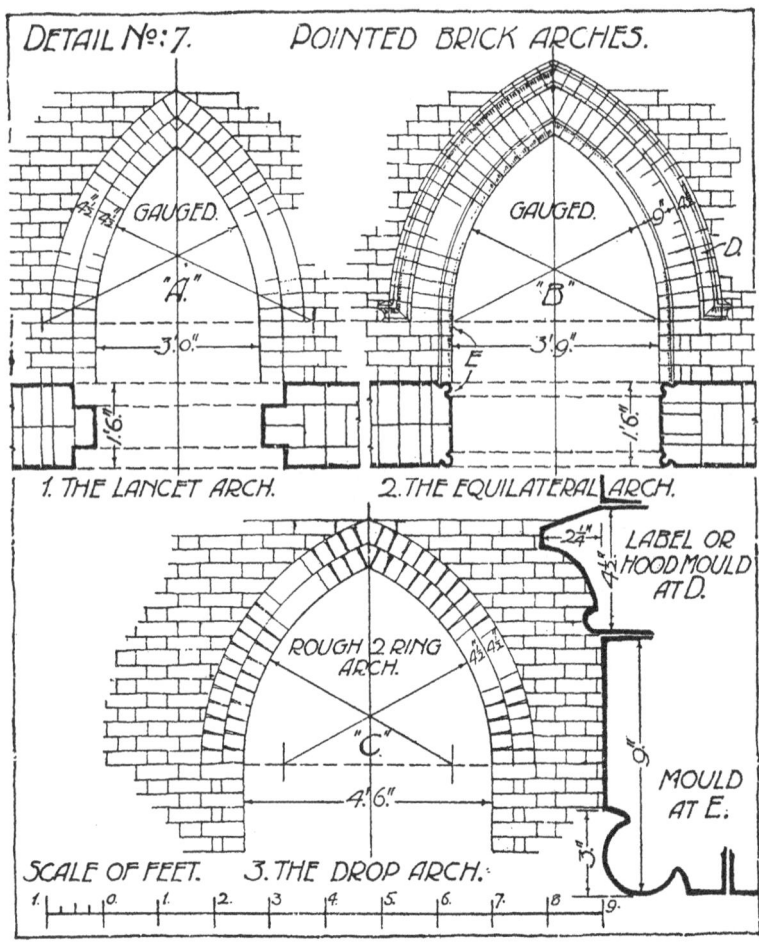

Forms of pointed arch. When the ends of the springing line and the crown of the arch lie upon the angles of an equilateral triangle, the arch is named "equilateral"; to describe the intrados curve the centres are at the ends of the springing line and the radius is equal to the span.

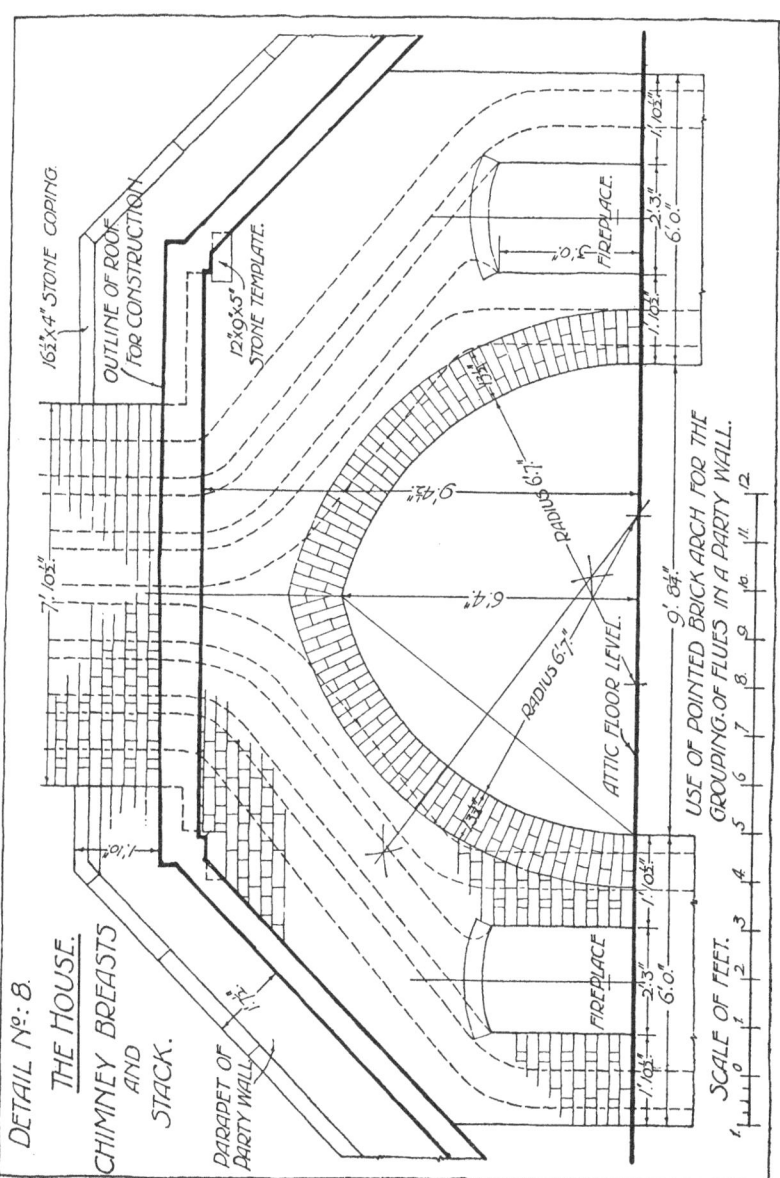

DETAIL Nᵒ. 8.

THE HOUSE.

CHIMNEY BREASTS
AND
STACK.

PARAPET OF
PARTY WALL.

16½"×4" STONE COPING.

OUTLINE OF ROOF
{ FOR CONSTRUCTION

12"×9×5" STONE TEMPLATE.

FIREPLACE.

FIREPLACE.

RADIUS 6'7".

RADIUS 6'7".

ATTIC FLOOR LEVEL.

USE OF POINTED BRICK ARCH FOR THE
GROUPING OF FLUES IN A PARTY WALL.

SCALE OF FEET.

If the crown is higher in proportion than would be given by an equilateral triangle on the span the arch is known as a "lancet" arch. The centres would then lie on the springing lines but *outside* the span.

If the crown is lower than the vertex of the equilateral triangle on the span, the form is known as a "drop" arch, and the centres of curvature would lie on the springing line but *within* the span.

These outlines are illustrated in detail No. 7 where A shows a recessed gauged lancet arch in two rings or "Orders", B a moulded gauged equilateral arch with a label mould terminating at a returned stop, and C a two-ring rough drop arch. All these examples have the same rise and varied spans to allow comparison of their general effects. This kind of arch in its full architectural application cannot be considered at present, but only in one of its simplest internal uses. Its form makes it suitable for any case where a large rise is required in comparison with the span. Suppose that it is desired to gather the flues from the dining and drawing room chimney breasts from each side to a central stack, then a pointed arch might suitably be built across the intervening space as shown in detail No. 8.

This detail shows how the flues from the two tiers of fireplaces may be gathered to one stack in the party wall of the house. There are two fireplaces on the attic floor level 2′ 3″ wide × 3′ 0″ high, with pairs of flues—one on each side of the fireplace—from the rooms below. Immediately over the openings the flues are carried, or "gathered", inwards and grouped at the central stack. Easy bends are employed at the turns of the flues and there is sufficient height to ensure a satisfactory draught at the attic fireplaces.

The arch employed is a rough-axed bonded arch, $13\frac{1}{2}″$ deep, $9′ 8\frac{1}{4}″$ span and $6′ 4″$ rise, and $13\frac{1}{2}″$ long on soffit. The face arch is only $4\frac{1}{2}″$ long and the inner portion only $4\frac{1}{2}″$ deep. The radius of the intrados curves is 6′ 7″ and the positions of the centres are determined by the construction shown on the left of the detail.

The thrust from a pointed arch is much more vertical than that from other forms, provided that it is well built with a satisfactory "key"; this latter is shown cut and mitred and set in cement mortar but in important work would be better formed in stone.

Application of tiles to Projections

18. Brick and tile cornice. Detail No. 9 shows the section of a brick and tile cornice to the back elevation of the warehouse, and illustrates a method of constructing larger projections than is possible with ordinary brick corbel courses. The bed mould consists of two brick courses, the lower of which is moulded. The corona or crowning feature is composed of four courses of tiles which overhang the bed mould 4″ and form the soffit, the lower courses exhibiting the ribs of the tiles. The top is finished with a hollow moulded brick course flaunched in cement, and in the best work is covered with 4 lbs. lead. Such cornices have a distinct architectural value in that variety of texture is obtainable by the combination of tiles and brick.

19. Brick and tile corbel to house parapet. In certain types of building it is required that the party walls shall be carried above the roof covering and where projecting wooden eaves are used they must be disconnected from the similar eaves of the adjoining property.

This may be done by allowing the party wall to project sufficiently in the form of a corbel to receive the eaves while allowing the rain water gutter to continue.

Detail No. 9 illustrates the method adopted in connection with the house and consists of a series of brick and tile corbelled courses.

The disposition of the tiles for obtaining projections of more than $2\frac{1}{4}''$ should be noticed and also the bonding of the tile lacing courses.

To terminate the projection and receive the

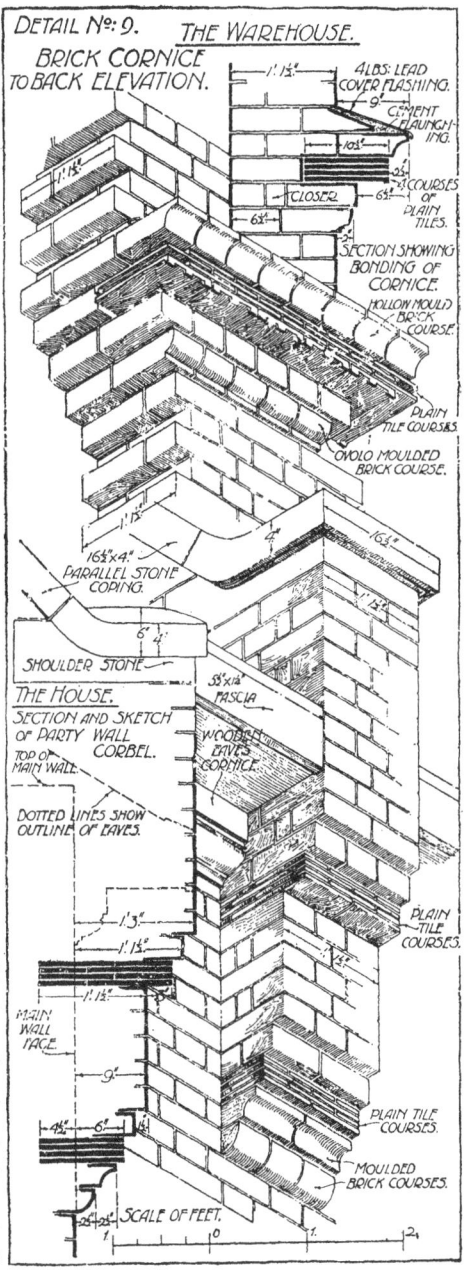

DETAIL Nº: 9.   THE WAREHOUSE.
BRICK CORNICE
TO BACK ELEVATION.

raking coping of the parapet a footstone is provided, the form of which varies somewhat from those previously detailed; see sectional elevation of the lower detail.

20. Parapets between properties. For buildings not higher than 30 ft. to be used as dwellings only, the party wall is required to be carried to the underside of the roof covering and the latter bedded solidly upon the wall in good mortar; this prevents the spread of fire from building to building especially if roofing timber is kept from entering the party wall. The latter provision is required by the Model Bye-laws, but many local authorities do not adopt it, allowing purlins to bear upon the wall $4\frac{1}{2}''$ but not in one line. The ends should be cased in $4\frac{1}{2}''$ of brickwork or other incombustible material exactly as in floor timbers except that they are covered in by brickwork to the depth of the rafter only; see construction of timber floors, Vol. II.

In other districts cast iron shoes enclosing the end of the purlin are allowed. A similar application in connection with the bearing of floor beams has been shown in Vol. II.

Domestic buildings over 30 ft. high and all commercial and public buildings, are required to have the party wall continued in the form of a parapet between the properties to a minimum height of $12''$, measured at 90° to the slope, and a thickness of at least $9''$. Such parapets must have proper copings to prevent rain from percolating into the walling below. Detail No. 8 illustrates such a parapet applied to the house, rising above the roof and intersected by the chimney stack against which the inclined and level portions terminate.

The parapet is provided with a coping, which may be of stone $4''$ thick and of rectangular cross section over the raking wall, and $5''$ thick of saddleback section on the horizontal part. The joints may be secured by lead plugs in the former case and metal cramps in the latter—see Vol. I. Alternative joints are also given·in the chapter on Masonry in Vol. II.

MAIN ENTRANCE STEPS TO HOUSE

21. Concrete groundwork for faced steps. The main entrance steps to the house are constructed by depositing concrete upon the rampant arch which spans the front area; see paragraph 16. The concrete is shaped to the desired outline and is carried to within $4\frac{1}{2}''$ of the external faces of the guard or wing walls.

The steps are to be faced with marble slabs, and the concrete is kept down at the treads $\frac{1}{4}''$ more than the slab thickness to allow for bedding and the riser faces are kept back $\frac{1}{4}''$ for the same purpose. The going of the finished steps is $11''$ and the rise $6''$.

22. Marble facings to steps. In the example of detail No. 10, the risers are faced with plain slabs of white Sicilian marble $1\frac{1}{4}''$ thick and the treads finished with nosed slabs of similar material $1\frac{1}{2}''$ thick. The slabs extend to the ends of the concrete and are thus enclosed $4\frac{1}{2}''$ within the guard walls.

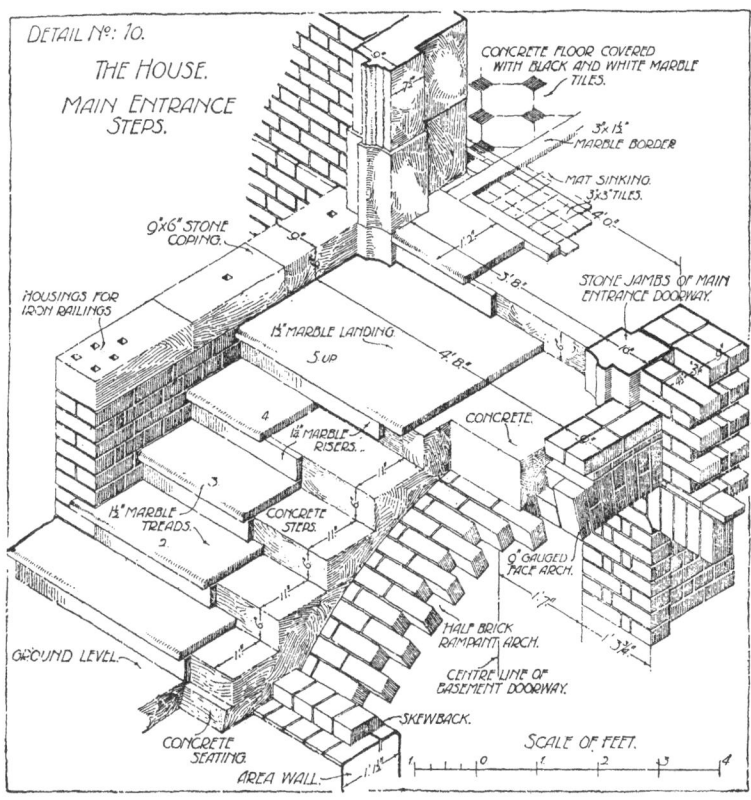

If these walls transmit considerable weight upon the slabs some danger of fracture exists unless very great care is taken to bed the slabs perfectly solid. It is often preferred to cut the slabs neatly between the walls in order to avoid danger of fracture and also to allow of easier removal for renewal or repairs. In such a case the slabs are carefully bedded in good mortar (not cement), and the joints at ends and visible edges neatly pointed in white or tinted cement mortar after the bedding is completed; for this purpose the joints should be $\frac{1}{4}''$ wide, otherwise the cement cannot enter the joint sufficiently to ensure adhesion.

23. Stone steps and facings. In stone districts the steps might be formed of solid sawn blocks with tooled faces, or for cheaper work where stone is not too easily procurable, $1\frac{1}{2}''$ stone risers and $2''$ treads may be used in the same manner as described for built-up stone steps in Vol. II, except that they would be bedded solid as proposed for marble facings.

24. Concrete steps. Concrete steps are very conveniently moulded to any form, being suitably compounded of sharp aggregates to avoid slipperiness; they are sometimes inset with dovetailed teak strips presenting end grain on the wearing surface or with lead, or rubber compounds. These steps are often reinforced with steel rods near the under surface when separately cast but if formed *in situ* expanded metal may be employed.

FAÇADE OF THE WAREHOUSE

Portions of the main façade of the warehouse have already been studied in Vol. II. The upper portion of this façade, and the main doorway now require consideration.

25. Superstructure of façade. The superstructure of the façade consists of an adaptation of what is known as an "Order of Architecture", showing three flat pilasters $3' 6''$ wide and 22 ft. high, spaced 12 ft. from centre to centre and containing windows between them lighting the first and second floors. The pilasters are surmounted by horizontal features, comprising architrave, frieze and cornice; this combination is named an "entablature".

26. The "Architectural Order". In its original sense an Order of Architecture consisted of an entablature supported by a series of isolated pillars, the features of the whole bearing a studied proportion in their relation to each other.

There are several recognised Orders which were employed by the ancient Greeks and Romans, all characterised by some distinctive features of design in "ordered" proportion, the systematised arrangement of which explains to some extent the term "Architectural Order".

In the old Orders the pillars were of circular cross section and called "columns", which in Roman usage were sometimes attached to walls instead of remaining isolated as in Grecian and early Roman examples, and were eventually merged into pilasters during the Renaissance period. In more modern architecture the same general arrangement of the members, attached to a mass of masonry, has been largely adopted, giving the appearance of the Order and having the spaces between the columns or pilasters filled with panels of stonework.

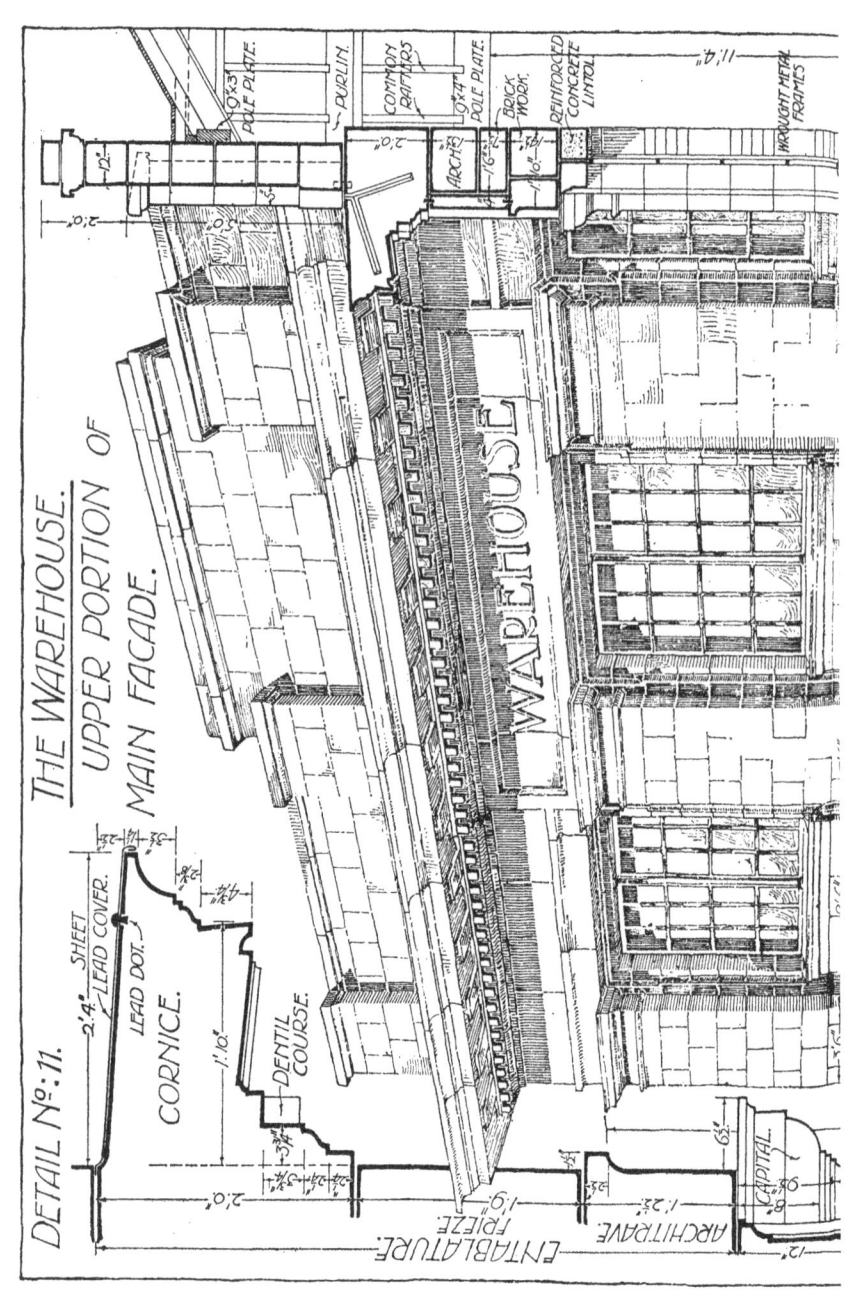

DETAIL Nº. 11.

THE WAREHOUSE.
UPPER PORTION OF
MAIN FACADE.

WAREHOUSE

CORNICE.

DENTIL COURSE.

ENTABLATURE.
FRIEZE.

ARCHITRAVE.

CAPITAL.

SHEET LEAD COVER.

LEAD DOT.

BRICK WORK.

REINFORCED CONCRETE LINTOL.

WROUGHT METAL FRAMES

COMMON RAFTERS

PURLIN.

POLE PLATE.

ARCH.

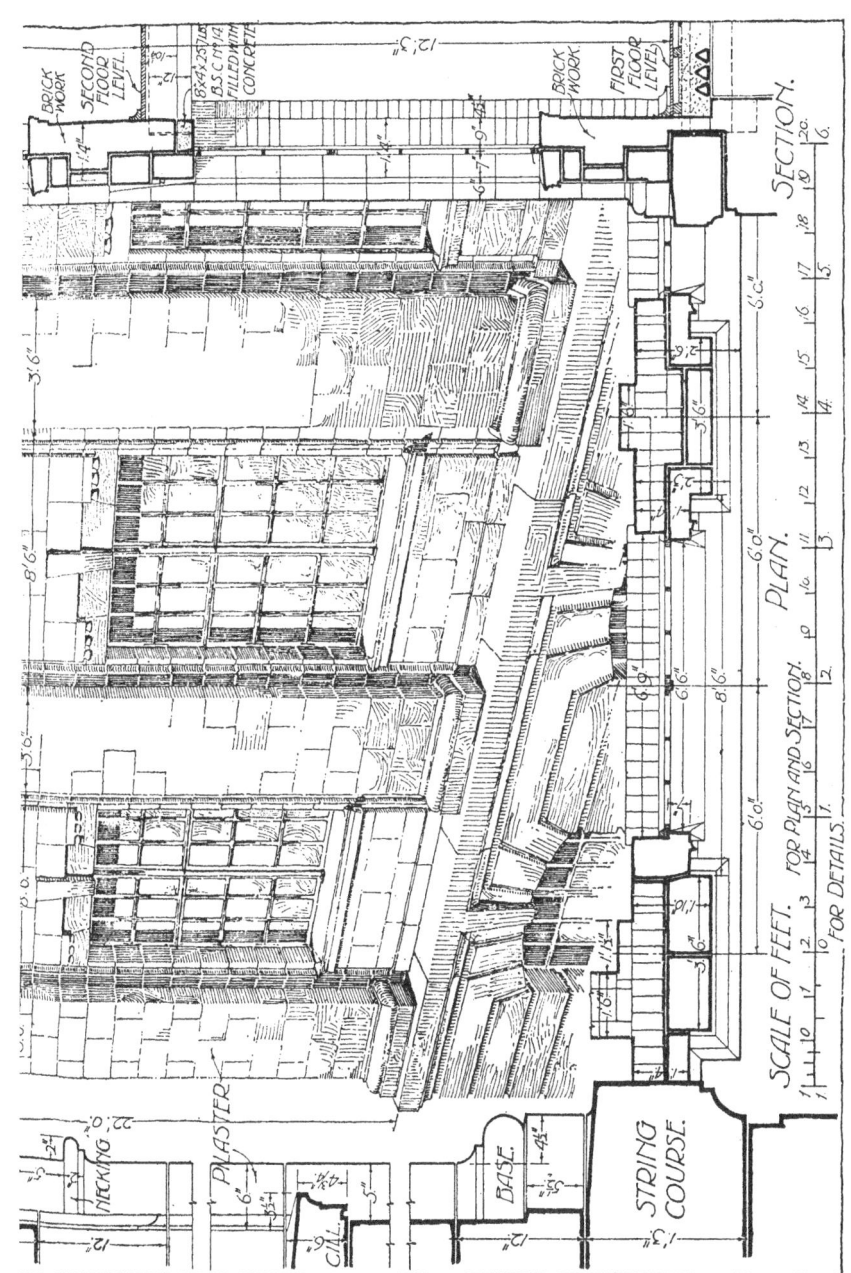

SECOND FLOOR LEVEL.

BRICK WORK

8"×4½"×5" BS CM 1:4 FILLED WITH CONCRETE

BRICK WORK

FIRST FLOOR LEVEL

SECTION.

PLAN.

SCALE OF FEET.

FOR PLAN AND SECTION.

FOR DETAILS.

NECKING

PLASTER

CILL

BASE.

STRING COURSE.

The student of architectural building should endeavour to become thoroughly acquainted with the proportions and characteristics of all the Orders of Architecture through the study of some authentic representation of notable examples and also by the actual measurement of successful applications of the Order in modern architecture. While some latitude is allowable when the proportions and features of the recognised Orders have been thoroughly assimilated, the *student* is advised to employ them in their entirety and to retain their established proportions.

27. Subdivisions of the column or pilaster. These members of the Order consist of three parts, viz. a moulded "base", a "shaft" of circular or rectangular section, and a "capital" which is the most distinctive feature in the different Orders. The capital may be a band of crowning mouldings or a carved termination to the head of the shaft.

28. Subdivisions of the entablature. The entablature is also divided into three parts, viz. architrave, frieze and cornice.

The "architrave" is a horizontal beam placed directly over and supported by the columns or pilasters, and is usually plain. The "frieze" is a plain band always broader than the architrave and divided from the latter by a projecting mould called a "tenia". The "cornice" is the upper feature of the entablature and its use, subdivisions and names of parts were given and employed in Vol. ɪ.

29. Proportions of the Order. In the present consideration of the Orders the proportions of column and entablature may be taken approximately as follows:

The height of the whole Order is divided into *five* parts of which *four* constitute the height of the column and *one* the depth of the entablature.

The entablature may also be subdivided into four parts; *one* of these is occupied by the architrave and the *remainder equally divided* between the frieze and cornice.

While these main divisions of the Order are a rough general guide to proportion, the student is urged to carry these to the greater degree of refinement peculiar to the several Orders. In all cases the diameter of the column at its base is the unit employed to proportion the whole Order.

30. Pilasters to warehouse façade. Detail No. 11 shows a part of the upper portion of the front façade. The pilasters form a part of the pier between the window openings on the first and second floors; the main dimensions of the pier in plan are 5′ 6″ wide by 1′ 4″ thick, while the stone pilaster is 3′ 6″ wide and projects 6″ from the principal surface. On the inside of the wall is an attached pier of brickwork 1' 6" wide by 4½" projection which provides for extra bearing to the ends of the main floor girders.

The capitals to the pilasters are 16½" deep, plainly moulded and finished with a 2" necking mould which divides the capital from the shaft of the pilaster. The bases are 12" deep, project 4½" from the face, and consist of a plain square block 6" deep surmounted by a large torus, fillet and hollow, the latter merging into the face of the pilaster.

31. Entablature of warehouse. The entablature of this example is 5 ft. deep, a little under one-fifth of the total height of the Order and therefore approximating to the proportions given in paragraph 29. The architrave is 1' 2½" deep, and the frieze 1' 9", these being merged into one feature to form a panel over the central pair of pilasters in which the word "warehouse" is placed in bronze letters; see also paragraph 38.

32. Cornice of warehouse entablature. The cornice is 2 ft. deep, with a weathering of 2¼" and projects 2' 4" from the principal face of the work. It consists of two parts, viz. the *corona* or upper projecting band of mouldings, two-thirds the total depth, and the *bed mould* or lower and supporting band of mouldings which occupies the remaining one-third of the depth.

In the detail the corona consists of the following mouldings, taken in order, downwards: fillet, cavetto, fillet, cyma reversa or ogee and fascia.

The soffit of the cornice falls downwards towards the front edge, which ensures a free drip and the surface is relieved and ornamented by parallel moulded slabs called "mutules". A throating is provided near the edge of the bed and in front of the mutules, which prevents rain driving and soaking inwards towards the wall face. A lead covering to the weathered top surface is employed to prevent water soaking through the vertical joints between the sections of cornice and the excessive absorption of rain from the weathered surface when left exposed, which often causes disintegration of the stone if wet weather is succeeded by frost. The lead covering is more expensive than the saddled joint formed upon the weathered surface—see Vol. I—but should be used for all important cornices having considerable projection.

The bed mould consists of two fascias divided by an ogee mould and surmounted by a fillet and small hollow which outlines the projecting mutules. The upper fascia is a plain band, notched or indented to form tooth-like projections which are named "dentils". The indents should not be wider than the dentils and commonly vary in width from one half to the full width of the latter. Spacing and dimensions of dentils are matters of importance and when employed in conjunction with mutules, as in the example given, should bear some general relation thereto, in order to allow of correct arrangement at internal and external angles where these features are returned or intersected.

33. Students' work in main outlines of façade. Having understood the general scheme of design the student should prepare a large scale detail drawing of the front elevation, with the corresponding plans and vertical section. The most convenient procedure is to outline the plans and section first and to project the elevation from these views.

For the present the detail will include only the general outline of those features of the façade already studied to which may be added at a later stage the details of windows, main doorway, etc., which immediately follow.

34. Original designs. In working out an original design the elevation would first be prepared in the form of a rough sketch, either on squared paper or approximately to scale, for the purpose of establishing satisfactory proportions for the main features of the façade. Practical dimensions may then be determined to agree closely with those established by the sketch and details of the work prepared and embodied in the working drawings with details to a much larger scale, preferably full size.

Panelled Fillings to the Façade

35. Stonework panels. The windows to the first and second floors are set within the panels of stonework which occupy the spaces between the pilasters at the two upper storeys. The four pilasters divide these upper storeys into three panels, each containing a pair of windows, one to each storey. These panels are stone faced and backed with brickwork, the full thickness being 1′ 4″, and each pair of windows therein is enclosed by a marginal mould or architrave projecting from the plain face of the panel. The architraves are broken in outline at the head of the second floor windows, a short length being stepped out at right angles and continued upwards, forming what is termed an "eared" architrave. Two additional mitres of the moulding are required by this arrangement.

The architrave commences from a plinth course underneath the first floor window opening and is continued vertically until it mitres with a similar horizontal moulding at the head of the eared portion, lying immediately beneath the entablature.

Stonework jointing should be carefully studied in this and succeeding details, particularly with regard to the correct selection of the dimensions of the stone and position of joint, in order to allow of cutting mitres and returns in the solid and to avoid interference with the horizontal coursing.

36. Heads and cills to first floor window openings. The first floor window cill is also the capping mould of a stone "dado"[1] having a projection of 1″ from the main wall surface and the same width as the stone opening. The centre portion or "die" of the dado is built of plain ashlar stonework and rests upon the plinth course referred to in the last paragraph; this is 6″ deep, projects ¾″ from the die and is returned round the angles and stopped against the base of the architrave.

The head of the same window opening consists of a flat arch, constructed with five intermediate voussoirs and abutments 12″ deep, having the latter built into the ordinary courses and oversailing into the opening. The key block, 24″ deep, passes through the stone course above the arch, and completes the structure. Terminal ornaments belonging to the upper window are worked upon the oversailing abutment voussoirs to avoid disturbing the coursing joints—see next paragraph.

37. Heads and cills to second floor windows. The second floor window cill is supported upon a stone "apron" having the same rectangular section as the dado below the first floor cill and 2′ 9″ deep. An apron in architectural construction is a plane projection below some bolder feature, such as the cill of a window opening, provided with no apparent support at its lower edge and therefore possessing the appearance of hanging from the bolder feature.

The projection must be small and the feature thus subdued, the only allowable treatment at the base being to add some simple curved outline or a pendant type of decoration such as the "guttæ" shown in the detail and worked in series near the ends of the apron. Guttæ are truncated conical drops hanging pendant-like from a soffit and in the case of small projections—as in the example—may be portions of the above, attached also to the vertical face of the wall.

The keystone of the lower arch has a greater projection than the apron and therefore penetrates it, the stones in the lower apron course being bevelled to fit the sides of the key.

Stone lintols 8″ × 9½″ are employed to span the second-floor openings and upon their faces are worked the horizontal architrave and the solid mitred return at the junction with the vertical eared portion. These lintols are comparatively small for so large a span,

---

[1] A "dado" is a supporting block or pedestal having base, plain die and cap mouldings. The term is also applied to a continuous panelling round the walls of a room.

but they are almost entirely relieved of weight by the arrangement
of the superstructure and its brick backing as described in the next
paragraph.

38. Entablature over window openings. The architrave of the
entablature is carried across the openings in three stones, and the
detail No. 12 shows the form of joint usually employed, consisting

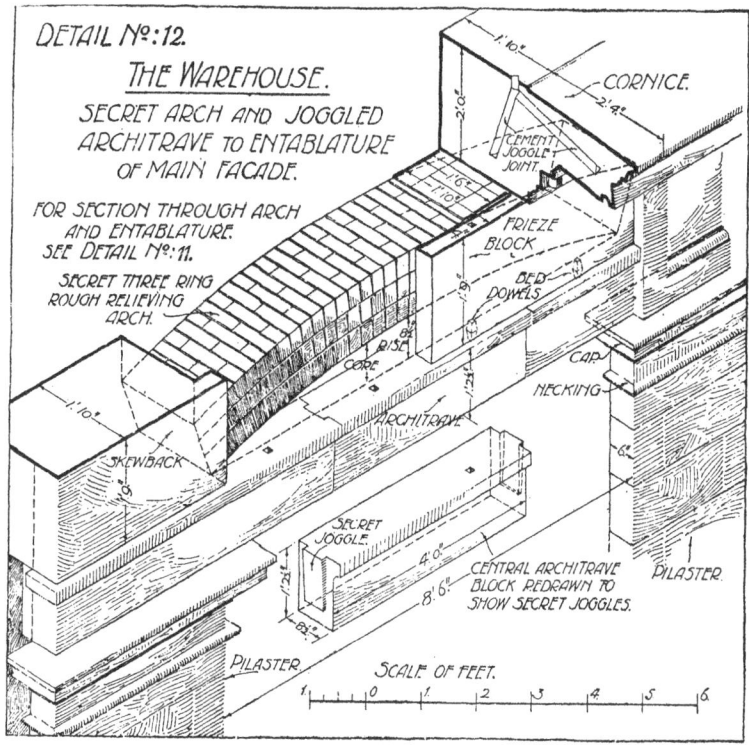

of a secret joggle to provide bearing for the middle stone and to
prevent displacement of the block in any direction. Above this
the frieze is erected, being formed by a facing of stone 4″ thick and
backed by a secret relieving arch of rough brick 18″ long and three
half bricks in depth.

The relieving arch is the principal support; it transmits the load
from the cornice, parapet wall and part of the roof, to the piers
between the windows. Care must be exercised in the construction to
prevent undue bearing of the load upon the face stones of the frieze
and architrave; for this purpose slow setting mortar with ¼″ joints

should be employed for the stonework and quick setting mortar for the arch.

39. Parapet to warehouse front. The parapet is a solid 12″ stone wall 5 ft. high from the cornice to the main level, and is finished with a 6″ moulded and feather-edged stone coping.

To maintain the general character of the superstructure the pilaster projections are continued above the cornice, with the dimensions maintained at the projecting plinth, while the body or pedestal of the parapet pilaster is diminished to 3′ 4″ wide by 6″ projection.

Above the centre panel of the superstructure, the parapet is increased in height to 7 ft. and the copings of the main portions are returned against its face. A capping mould is employed of similar section to the parapet coping, and the whole surmounted by a plain blocking course.

The parapet serves, as in previous cases, to guard the gutter behind it, to hide the hipped end of the main roof from view, and to assist in maintaining the equilibrium of the cornice.

MAIN ENTRANCE DOORWAY TO WAREHOUSE

The wing of the warehouse containing the main entrance doorway encloses the staircase and lavatories. This portion is recessed 11″ from the face of the adjacent surbase walling as seen in detail No. 13 and the front wall of the wing is 16″ thick from the ground floor level to the top floor lavatory, being constructed with a 7″ stone facing and a 9″ brick backing.

40. Main doorway. The doorway is a special feature in cut stonework, its opening being 9′ 6″ high by 5′ wide, flanked by sunk architraves and surmounted by an entablature.

The architraves are 7½″ wide and project 5″ from the face of the surrounding wall; they are recessed into plain panels, ¼″ deep, having flat margins 1″ wide. Across the head of the opening the architrave band forms the lintol.

41. Entablature over main doorway. The entablature is 2′ 0″ deep, divided into architrave, frieze and cornice and is surmounted by a small blocking course.

In the centre of the frieze is a plain projecting panel or "tablet", having the bed mould of the cornice broken around it and against which the top mould or "tenia" of the architrave is stopped. The panel is of the same form as the window apron previously described, and terminates at the bottom edge with two series of guttæ in a similar manner.

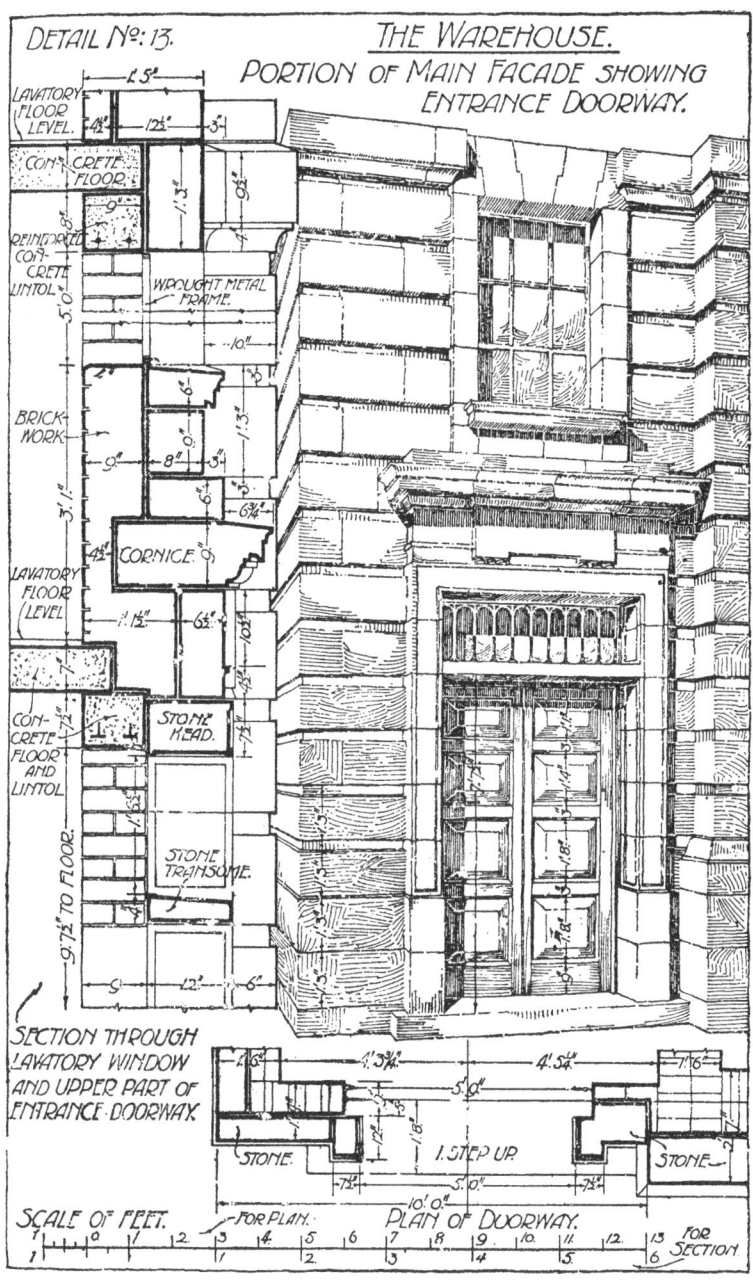

DETAIL Nº: 13.

THE WAREHOUSE.
PORTION OF MAIN FACADE SHOWING
ENTRANCE DOORWAY.

SECTION THROUGH
LAVATORY WINDOW
AND UPPER PART OF
ENTRANCE DOORWAY.

PLAN OF DOORWAY.

SCALE OF FEET.

42. Reveals, fanlight and transome. An opportunity occurs in the main entrance doorway for the use of deep reveals, which peculiarly emphasise the feature by casting broad shadows across the opening. The reveals and soffit are 12″ deep and have plain recessed panels identical in treatment with the faces of the architrave bands.

A fanlight is included in the feature, and is divided from the door opening by a weathered stone transome $12'' \times 4\frac{1}{2}''$, giving a clear height of 1′ 6″ to the fanlight and 7′ $7\frac{1}{2}''$ to the doorway.

43. Students' exercise. From the perspective view and enlarged sections of mouldings provided in detail No. 13 the student should prepare an elevation, plan and section of the stonework to the doorway to a scale of at least $\frac{3}{4}''$ to 1 ft., noting accurately the forms and general proportions of the mouldings and the jointing of the stonework. Full size details of the mouldings should also be drawn.

In tracing out the jointing and applying it to the details it should be noted that every stone is shaped from a rectangular block of the least practicable dimensions consistent with ensuring correct and sufficient bond and preserving a satisfactory architectural appearance. Further observation will show that the frieze and cornice are arranged to work in with the horizontal courses of the plain walling, these courses being 15″ deep with channelled joints 3″ broad and recessed 1″ deep along the top edges of the stones.

MAIN ENTRANCE—SEMI-DETACHED HOUSE

44. Main entrance to the house. The entrance consists of a flight of steps having five risers, leading to a broad landing immediately in front of the doorway and guarded at the sides by two stone-coped walls of 9″ brickwork; see detail No. 14.

The doorway has an opening 9′ 6″ high × 3′ 9″ wide which is flanked by a window on each side, 4′ 6″ high × 1′ 3″ wide, and divided from these by stone piers 17″ broad on the face. The whole feature is an example of the suitable combination of brickwork and masonry, where dressings of coursed stone surround the openings, each course being a multiple of a brick course in depth so that horizontal joints coincide and facilitate bonding of the two materials.

45. General treatment of design. To understand the general treatment consider the whole frontage of the projecting part of the front elevation; its width is 17′ $7\frac{1}{2}''$ and the two angles are emphasised by projecting brick quoins as previously detailed and explained in Vol. II. Between these projections, at a height of 3 ft. from the pavement level, the entrance steps, doorway and windows are centrally grouped.

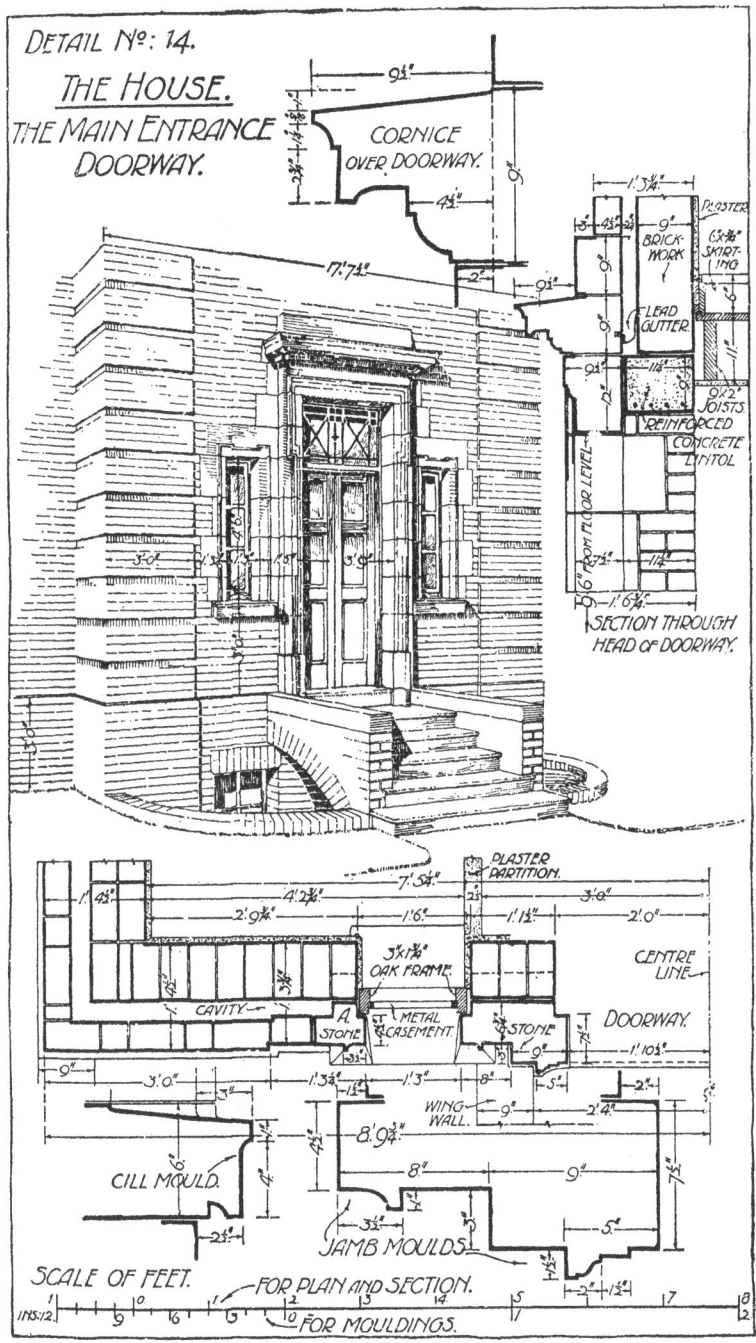

DETAIL Nº: 14.

THE HOUSE.
THE MAIN ENTRANCE
DOORWAY.

CORNICE
OVER DOORWAY.

SECTION THROUGH
HEAD OF DOORWAY.

SCALE OF FEET.

Above the doorway, on the first floor, there is a group of three windows, above which the projection terminates in a corniced gable.

46. Front doorway to house. The doorway consists of an opening intended to enclose a double margin door with a glazed fanlight over. Two plain stone piers, 9″ wide × 3″ projection, flank the opening and a 12″ deep lintol forms the head. On the faces of the piers and lintol an architrave is cut, mitred in the solid at the angles and broken in outline along the vertical length so as to form two outward projections called "ears": see also paragraph 35.

Over the horizontal architrave is a slightly projecting fascia, shown in section through the head of detail No. 14, around which the bed mould of the cornice is broken and returned.

No special reference to the cornice and blocking course is needed because their forms are now familiar and each can be obtained out of one stone.

47. Stone dressings to doorway. In order to obtain satisfactory bonding of the brick and stone work, the dressings are arranged to suit the brick courses, and the longitudinal overlap of the stones upon the brickwork is either $2\frac{1}{4}″$ or $4\frac{1}{2}″$. All stone courses are 15″ deep up to the soffit level of the windows, and above this are 12″ deep.

Because the wall is hollow, except at the window piers, the ordinary "in-bands" (or through bonders) are not employed at the outer jambs, but alternate dressings have a $2\frac{1}{4}″$ projection cut upon the back to overlap the cavity, as at A in the detail.

48. Stone dressings to piers. For the piers between the doorway and flanking windows a similar bond may be adopted for the dressings, or better, the sectional plan of detail No. 14 may be used for one course and the quoins of the next course extended $4\frac{1}{2}″$ further into the wall, as shown by the dotted line.

In all cases dressings or quoins must have adequate transverse and longitudinal tie. Fairly quick setting mortar is advisable for bedding, in order to prevent undue settlement of the brick backing, which is liable to occur because of the numerous joints therein as compared with those of the stone facing.

49. Lintols over doorway. Externally, the "architrave and fascia" course forms the lintol and enters the wall up to the line of the recess. Internally a reinforced concrete lintol is employed, $11\frac{1}{4}″$ wide and 9″ deep; this closes the cavity and bonds into three courses of brickwork. To secure the woodwork against damp, which may be caused by leakage through the external brickwork, a lead gutter is provided across the cavity as described for all such openings in Vol. II.

50. Flanking windows to doorway. The two flanking windows are intended to light the lavatory and basement stair which are separately placed on each side of the entrance.

The window openings are prepared to receive solid wood frames into which metal casements are to be inserted; for this purpose $1\frac{1}{2}''$ recesses in the jambs are sufficient.

A new feature occurs in these windows, viz. the treatment of the stone architrave. For small openings, eared architraves are somewhat difficult to arrange without destroying the original proportions of the opening; recognising this the vertical architrave is kept straight and the horizontal part is broken across the lintol.

51. Continuity of horizontal lines. An examination of the detail will show how continuity of main horizontal lines in the design is preserved; thus, the soffit of the windows coincides in level with the transome of the wood frame, while the eared architrave around the doorway breaks line at approximately the same level.

In a similar way the plinth level of the front façade is adopted as the coping level for the guard walls to the entrance steps.

The guard walls of the entrance steps and the curved wing walls of the area are provided with metal railings, which were illustrated in Vol. II.

Students' exercise. Students should prepare large scale drawings of this feature in plan, elevation and vertical section, adding full size details of the mouldings.

CHAPTER TWO

STRUCTURAL IRON AND STEEL WORK

COLUMNS AND STANCHIONS

52. Pillars. In buildings where considerable uninterrupted floor space is desired, as, for example, floors not divided into rooms, it becomes necessary for the sake of economy and efficiency to introduce pillars at intervals to support the floor girders. If the pillars are an objection, built up plate girders or compound beams may have to be employed to span the full width of the building, but this is often uneconomical and undesirable as it concentrates the whole of the roof and floor loads on the outside walls and foundations and also requires very stiff beams to prevent abnormal deflection.

A pillar is any vertical compression member, and for a floor support in modern building construction is generally of steel or reinforced concrete, though cast iron is sometimes used and even brick or plain concrete for basement pillars.

For upper floors brick and plain concrete would be too bulky and wasteful of space for the large loads to be carried, hence cast iron, steel, or reinforced concrete are the materials usually employed above the ground floor level.

In this chapter are given typical forms of pillars in cast iron and rolled steel, suitable for supporting the steel framing of the fire-resisting floors of the warehouse, and also the methods of connecting pillars and floor girders.

While cast iron columns for the support of floors are rapidly diminishing in use, they are allowed under the Building Acts and are definitely specified. In some parts of the country cast iron still finds considerable favour, hence the inclusion of examples of columns in this volume.

53. Column and stanchion. When a pillar is of circular cross section and of approximately cylindrical form it is termed a "column"; if cast into rectilinear form or built up from rolled steel sections of rectilinear outline it is termed a "stanchion".

54. Cast iron columns. These members are moulded in sand and "cast" by pouring liquid metal into the mould.

If required to support the girders of one floor only, they may be cast in one piece with flat base and cap, each being enlarged to the necessary shape and dimensions for distributing the load or for receiving it, as required.

When a cast iron column is to be continued through several storeys it is usually cast in one-storey lengths with a jointing box and cap plate, though it may be more convenient in the case of long columns to have the head-jointing box cast separately. In all cases where these divisional units are employed they should be jointed by flange and socket joints, bolted together.

55. Dimensions of cast iron columns. The thickness of column castings varies from $\frac{1}{2}''$ to $1''$ in general construction, though jointing boxes and flanges may often need to be thicker.

In the 1909 Amendment to the London Building Acts the London County Council requires that cast iron columns shall have a minimum external diameter of $5''$ and a thickness of $\frac{1}{12}$th the diameter, or $\frac{3}{4}''$, whichever is the greater. If jointed in their length, such column units must have truly turned and bored joints making a right angle with the axis of the member and must be secured with at least four bolts of a diameter not less than the thickness of metal in the column.

The method of preparing these column units for assembling and

DETAIL Nº: 15.
THE WAREHOUSE.
SKETCH OF CAST IRON
COLUMNS TO BASEMENT.
15"×6"×45LBS:
B.S.B.
BASE OF COLUMN
TO GROUND
FLOOR.
10"
SECONDARY
GIRDER.
10"×4½"×25 LBS:
B.S.B
10"
MAIN
GIRDER.
BASEMENT
COLUMN.
11"
1⅛"STIFFENERS.
CAST
IRON
BASE.
1½"STIFFENERS.
SEATED HOLE
FOR 1⅛" BOLT.

fixing will depend somewhat on the design selected and the working conditions. Some alternative arrangements are given in the details.

COLUMNS IN WAREHOUSE

In the illustrations of these columns—detail Nos. 15 to 19—perspective sketches and dimensioned details are given in separate drawings, while detail No. 20 assembles the units of the complete column and gives the relative heights.

56. Foundations and bases for cast iron columns to warehouse. Columns support considerable load when employed to collect and transmit the loads from several floors, and may require large foundation slabs when erected on poor ground in order to reduce the unit pressure on the earth to a safe value.

In the calculations relating to the dimensions of stanchions and columns—paragraph 249—it is shown that the approximate load transferred to the foundation is about 78 tons, say 80 tons to allow for weight of concrete and base.

Taking a safe pressure of 2 tons per sq. ft. on the earth, an area of $\frac{80}{2} = 40$ sq. ft is required for the foundation.

For a square foundation, the length of side $= \sqrt{40} = 6\cdot32$ ft., say 6' 6" square.

To distribute the load over the concrete an octagonal base has been selected, as shown in details Nos. 15 and 16, with a 3" inspection and grouting hole in the base plate. Let the concrete support a maximum pressure of 12 tons per sq. ft. The area of the central grouting hole is $\frac{\pi}{4} \times 3^2 = 12\cdot57$ sq. ins. or $\cdot087$ sq. ft. The area of a regular octagon $= \cdot828d^2$, where d is the diameter between parallel faces, hence the nett area $= \cdot828d^2 - \cdot087$ (sq. ft.). Let $W = $ load $= 80$ tons, $p = $ resistance of concrete $= 12$ tons per sq. ft. Then, as resistance $=$ load,

$$p\,(\cdot828d^2 - \cdot087) = W,$$
$$12\,(\cdot828d^2 - \cdot087) = 80,$$
$$\cdot828d^2 = \tfrac{80}{12} + \cdot087 = 6\cdot753,$$

and
$$d = \sqrt{\frac{6\cdot753}{\cdot828}} = \sqrt{8\cdot16} = \mathbf{2\cdot86} \text{ ft. diameter.}$$

For practical purposes, say $d = \mathbf{3}$ ft.

Note. These calculations assume axial loading and therefore uniform pressure on the base plate and earth foundation. This assumption is reliable for a stiff base when the maximum floor load exists over the entire floor, though there will be periods when the floors have loads on some parts only, which, if unsymmetrical or on single bays, will produce eccentric loading but with a *less total pressure on the base*. It may be shown that eccentricities of $\frac{1}{6}$th the width of a rectangular base may double the average pressure on one side, and entirely relieve the base from pressure on the opposite side. The *maximum* uniform load on a symmetrically planned floor with uniform panels cannot produce eccentric stresses of any consequence.

57. Cast iron base—octagonal plan. The cast iron base consists of a $1\frac{1}{2}''$ octagonal bed plate, $3'$ diameter, with a similar top plate $1\frac{1}{4}''$ thick and $1'$ $8''$ diameter, these being connected by a hollow octagonal prism $8''$ internal diameter and $1\frac{1}{2}''$ thick. At each angle is a $1\frac{1}{2}''$ bevelled stiffening bracket connecting the angles of the upper and lower plates.

Such bases are cast solid and should have all the internal angles rounded or feathered to prevent the development of cracks or flaws as the hot metal cools in the mould.[1] Holes for bolts, grouting, etc. are formed in the required positions by inserting "cores" of baked sand to stop out the liquid metal.

The top face, at least, of such bases should be planed true and at right angles to the axis of the column.

58. Concrete foundation. The concrete bed upon which the cast iron base is seated should, unless reinforced, have a thickness sufficient to produce an angle of distribution of at least 45°, extending from the outer edge of the cast iron base to the extreme base edge of the concrete; detail No. 16 shows that a thickness of $1'$ $9''$ is required in plain concrete, viz. a thickness equal to the projection.

To secure the cast iron base to the concrete and thus assist in fixing the entire base of the column, four $1\frac{1}{8}''$ bolts are enclosed in the concrete as deposited. The heads of these bolts pull upon $3'' \times 1\frac{1}{2}'' \times 5\cdot27$ lbs. B.S. channels 5 ft. long, placed with their flanges upwards.

In order to obtain a level bed and true axis to the column, the base is chipped roughly true—removing serious irregularities— packed about $\frac{3}{8}''$ above the concrete bed with wooden wedges until the top bed is level, then the base surrounded with a fillet of soft clay kept clear of the edges and the space filled with cement grout, poured through the hole in the centre. The grout should be visible along all the edges. Large bases have to be poured from both centre and edges or through intermediate pour holes; care is required to ensure that air escapes freely in order to ensure a solid bed.

As soon as the grout shows an initial set, the wedges may be withdrawn so that the weight of the base assists in bedding it to position.

59. Basement column. Detail No. 16 shows the construction of a basement column which provides support to the ground floor. It is cast in one length from the base plate to the head, extending

---

[1] See Chapter on Materials.

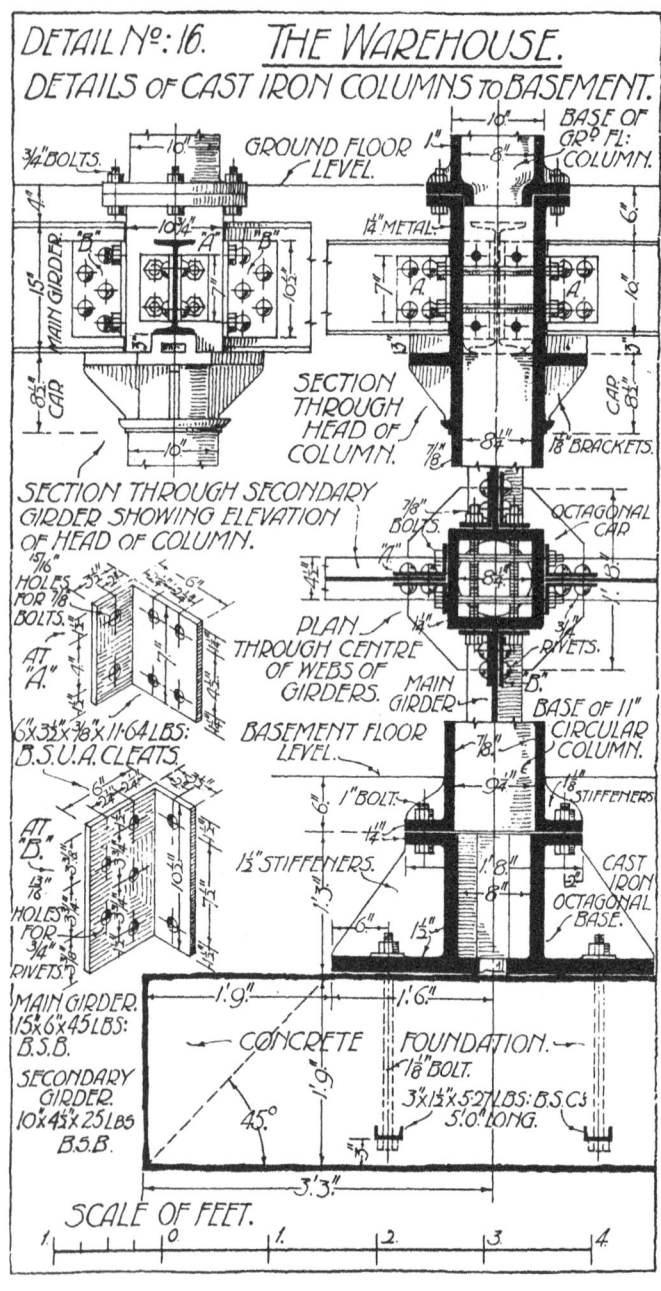

DETAIL N°: 16.   THE WAREHOUSE.

DETAILS of CAST IRON COLUMNS to BASEMENT.

above the level of the ground floor girders and providing for the connection and support of the latter. The essential provisions in such a column are: an extended and stiffened base prepared for bolting down to the cast iron base block, or directly to a stone or concrete foundation; a projecting cap plate supported by brackets, upon which the floor girders are borne; a square head against which the girders may be connected by steel angles and bolts; and a jointing flange for receiving the column above if support be required for floors at a higher level.

Where columns are of considerable length or of complicated form, the head box is often separately cast as shown in detail No. 17 where it is applied to the ground and first floor columns.

60. Bracketed base plate. The basement column is 11″ external diameter at the base and 10″ at the top and $\frac{7}{8}$″ thick throughout the shaft. Its base seating is of octagonal outline, 1′ 8″ diameter and bracketed at each angle to ensure distribution of the load and to stiffen the projection for fixing purposes. The base of the column and top of the cast iron base are machined to plane surfaces, fitted and bolted with four 1″ bolts without packing the joint.

This practice of accurate machining—turning or planing—should be followed in all cast iron pillar jointing, then packings are unnecessary; otherwise there is a temptation to place ill-fitting parts together with packed joints which would not ensure uniform distribution though they would certainly improve the transmission by easing the stress on prominent irregularities.

The base of this column should be encased in 4″ of cement concrete, up to the level of the basement floor.

61. Cap of column. The cap of the column consists of an octagonal plate $1\frac{1}{4}$″ thick projecting 5″ from the shaft and supported by $1\frac{1}{8}$″ brackets or stiffeners, all cast in the solid. There are two brackets under each main girder and one below each secondary girder; the brackets terminate upon a necking mould and the resulting appearance is a sturdy cap of strong and simple outline suitable for a position where the ironwork is to be exposed to view and where special decorative treatment would be out of place.

62. Floor beam connections to basement column. The head of the column is of square section with rounded angles and the floor beams are secured against its faces by pairs of angle cleats riveted to their webs and bolted through the head. Short bolts inserted from the inside could be used here if placed before the upper column was erected, but it is possible that the beams may require to be detached at some subsequent period without disturbing the columns, hence through bolts are employed connecting pairs of cleats on

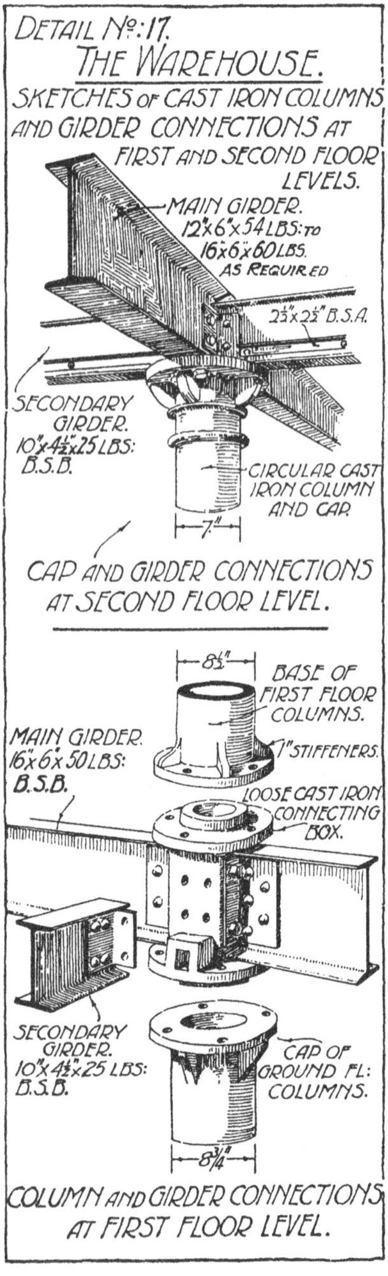

DETAIL Nº: 17.
THE WAREHOUSE.
SKETCHES of CAST IRON COLUMNS
AND GIRDER CONNECTIONS AT
FIRST AND SECOND FLOOR
LEVELS.
MAIN GIRDER.
12"x6"x54LBS: TO
16x6x60LBS.
AS REQUIRED
2½"x2½" B.S.A.
SECONDARY
GIRDER.
10"x4½"25 LBS:
B.S.B.
CIRCULAR CAST
IRON COLUMN
AND CAP.
7"
CAP AND GIRDER CONNECTIONS
AT SECOND FLOOR LEVEL.

8½"
BASE OF
FIRST FLOOR
COLUMNS.
MAIN GIRDER.
16x6x50LBS:
B.S.B.
1" STIFFENERS.
LOOSE CAST IRON
CONNECTING
BOX.
SECONDARY
GIRDER.
10"x4½"x25 LBS:
B.S.B.
CAP OF
GROUND FL:
COLUMNS.
8¾"
COLUMN AND GIRDER CONNECTIONS
AT FIRST FLOOR LEVEL.

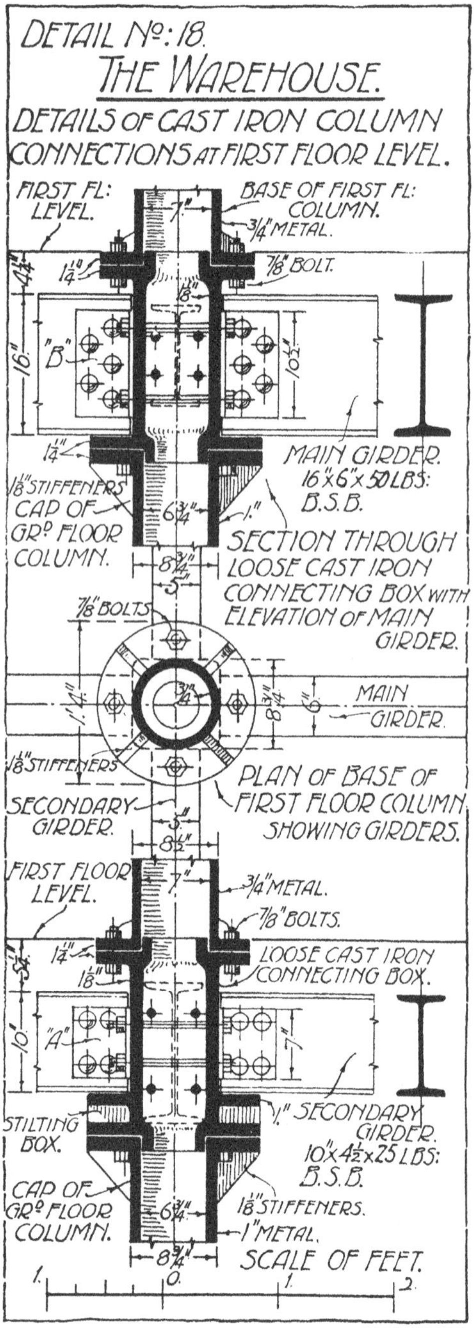

DETAIL Nº: 18.
THE WAREHOUSE.
DETAILS of CAST IRON COLUMN
CONNECTIONS AT FIRST FLOOR LEVEL.
FIRST FL:
LEVEL.
BASE OF FIRST FL:
COLUMN.
¾" METAL.
⅞" BOLT.
"B"
MAIN GIRDER.
16"x6"x50LBS:
B.S.B.
1⅛" STIFFENERS
CAP OF
GRD FLOOR
COLUMN.
SECTION THROUGH
LOOSE CAST IRON
CONNECTING BOX WITH
ELEVATION of MAIN
GIRDER.
⅞" BOLTS.
MAIN
GIRDER.
1⅛" STIFFENERS
SECONDARY
GIRDER.
PLAN OF BASE of
FIRST FLOOR COLUMN
SHOWING GIRDERS.
FIRST FLOOR
LEVEL.
¾" METAL.
⅞" BOLTS.
LOOSE CAST IRON
CONNECTING BOX.
"A"
SECONDARY
GIRDER.
10"x4½"x25 LBS:
B.S.B.
STILTING
BOX.
CAP OF
GRD FLOOR
COLUMN.
1⅛" STIFFENERS.
1" METAL.
SCALE OF FEET.

opposite sides of the head. This necessitates a difference in level for the two series of bolts to allow them to pass each other across the interior.

Sufficient clearance is necessary in the bolt holes to ensure that the girders can take solid bearing upon the column cap and stilting box, the latter becoming necessary owing to the difference in level of the girders; felt packings are advisably placed between the bearing surfaces. The bolts then secure the beams in position and produce a small measure of fixity to their ends.

63. Ground floor column. This member is 10″ diameter at the base, $8\frac{3}{4}''$ at the top and 1″ thick; it collects the load from the first and second floors and transmits it to the basement column.

Being of greater length than the basement column it is shown cast in two parts, viz. shaft, with bracketed cap and base; and a separate head to take the girders at the floor level. A sketch of the lower portion of the shaft is shown on detail No. 15 and a section on No. 16, while the upper portion of the column and the loose connecting box are shown at the bottom of detail No. 17 and in detail No. 18. The base is flanged to a diameter of 16″ with four stiffeners and a projecting rim or spigot entering a corresponding socket on the head of the basement column. This forms what is known as a thimble joint, which should be a "turned fit" and secured by bolting through the flanges.

Similar joints are made between the shaft and head and it is a matter of choice whether the spigot enters upwards or downwards.

The design of the ground floor column presumes that it is to be covered with fire-resisting plaster on expanded metal lathing and the outline of the cast iron is therefore kept as plain as the structural necessities allow, all mouldings being formed in the plaster; see Vol. II. The cap is therefore a plain circular ring 18″ diameter and $1\frac{1}{4}''$ thick with one $1\frac{1}{8}''$ bracket cast beneath each floor beam.

64. Connecting box to ground floor column. The connecting box is $1' 6\frac{1}{2}''$ long over the flanges, $8\frac{3}{4}''$ diameter and of $1\frac{1}{8}''$ metal, square on plan and with circular flanged thimble joints at top and bottom. As the girders for the first floor have different soffit levels a seating or stilting box is cast on the lower rim to receive the bottom flange of the secondary girder at each side. Where the shear at the support is not great this bearing may be dispensed with and the bolts made to transmit the whole of the vertical shear force.

Connecting bolts at flanges should be at least four in number and have a diameter at least equal to the thickness of the column shaft.

65. First floor column. This column, as shown in detail No. 19, is $8\frac{1}{2}''$ diameter at the base, 7″ at the top and $\frac{3}{4}''$ thick, and is the last of the tier, terminating at and supporting the second floor girders.

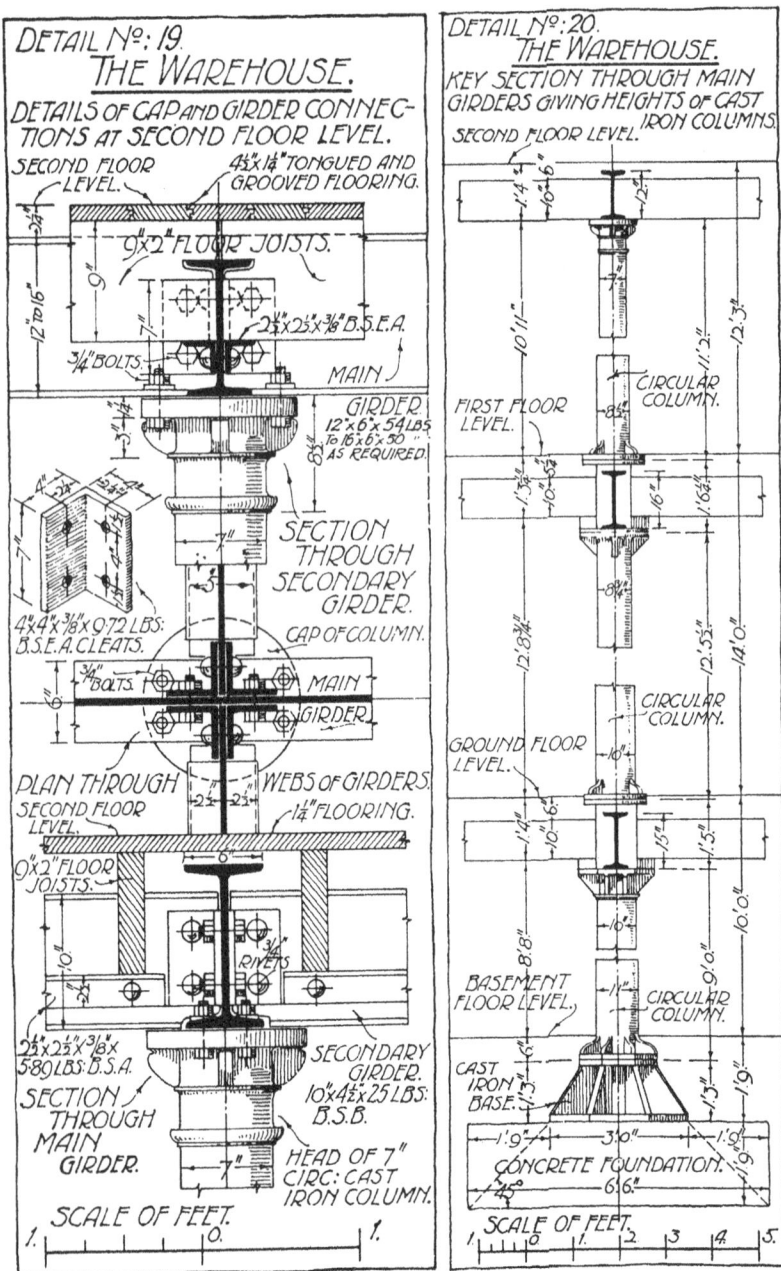

As there is no continuation beyond the cap the column is cast in one piece with a flanged and socketed joint to the first floor connecting box and a bracketed cap to receive the second floor girders which have their soffits at the same level.

The main girder might cross two spans, but it is assumed that two main and two secondary girders meet on the cap, with the former meeting at the centre and the latter notched and cleated as previously shown. The main girders should have $\frac{1}{4}''$ clearance at their ends and should be bolted through their lower flanges to the cap plate.

The floor surrounding the head of this column is constructed of $9'' \times 2''$ timber joists laid at right angles to the secondary girders and resting upon $2\frac{1}{2}'' \times 2\frac{1}{2}'' \times \frac{3}{8}'' \times 5\cdot89$ lbs. steel angles riveted thereto, one rivet below each joist. Joist ends are forked over the top flange and meet over the centre of the girder.

66. Entasis of columns. All columns of any importance should be formed with an entasis, viz. a convex curve reducing the diameter from base to cap. A straight taper gives the illusion of a concave outline and the entasis is intended to correct this, not to emphasise the convexity. The full function of the entasis and methods of setting out are matters for later study.

The first floor column would be treated in this way, the member being intended to be employed without a plaster covering.

67. Key drawing of tier of columns. Detail No. 20 is a key section showing the tier of columns employed in supporting the warehouse floors. Floor to floor dimensions are given and the several lengths of constructional parts, together with the position and depth of floor girders, and the level and dimensions of the foundation.

68. Objections to cast iron in columns and stanchions. While cast iron columns have an attractive appearance when well designed and also give the idea of great strength when correctly proportioned they are not suitable for use under all conditions. So long as the load is not so eccentric as to produce a serious bending moment, and compressive stress is produced throughout every cross section, a cast iron column is both efficient and economical, but side pressure and lack of fixity at the ends may possibly allow tension to occur on the side tending to take up a convex curve of deflection. The metal is notoriously weak in tension.

Further, if heated in a fire and suddenly quenched, cast iron tends to crack and disrupt. This is caused by the unequal contraction owing to the irregularity of the quenching process, which is accentuated if the column is in a state of tension due to buckling (see detail No. 80 and paragraph 237).

Solid steel columns are obtainable and almost exclusively used for positions where round sections of the least possible size are required.

69. Cast iron stanchions. These are sometimes used for the purpose of supporting light loads from floor beams but more often for placing flat against a wall where extra bearing is required for a girder resting on thin walls.

Shop front girders often require such assistance in cases where house property has been converted into shops and the party walls are thin. Flat backed stanchions of channel or E section may then be placed on one or both sides of a party wall where exposed to the street after forming the shop front opening, and bolted through to give them stiffness to resist buckling.

STEEL STANCHIONS

70. Because of the unreliability of cast iron and the other objections named in paragraph 68 stanchions built of rolled steel sections are in general use.

The shafts of these stanchions may be single pieces of British Standard beam, of specially broad I sections known as "broad flange beams" (B.F.B.'s), or compounded from I, ⌴,

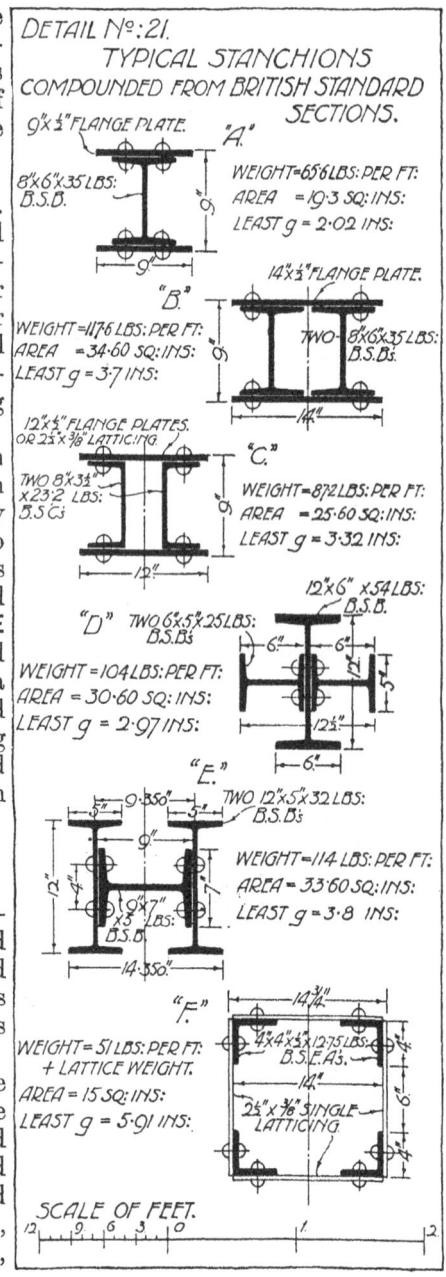

DETAIL Nº:21.

TYPICAL STANCHIONS
COMPOUNDED FROM BRITISH STANDARD
SECTIONS.

9"×½"FLANGE PLATE.   "A."

8"×6"×35 LBS:
B.S.B.

WEIGHT=65·6 LBS:PER FT:
AREA = 19·3 SQ:INS:
LEAST g = 2·02 INS:

"B."

14"×½"FLANGE PLATE.

WEIGHT=117·6 LBS:PER FT:
AREA = 34·60 SQ:INS:
LEAST g = 3·7 INS:

TWO- 8"×6"×35 LBS:
B.S.B.

12"×½"FLANGE PLATES.
OR 2½"×⅜"LATTICING

"C."

TWO 8"×3½"
×23·2 LBS:
B.S.Cs

WEIGHT=87·2 LBS:PER FT:
AREA = 25·60 SQ:INS:
LEAST g = 3·32 INS:

12"×6" ×54 LBS:
B.S.B.

"D" TWO 6"×5"×25 LBS:
B.S.B.s

WEIGHT = 104 LBS:PER FT:
AREA = 30·60 SQ: INS:
LEAST g = 2·97 INS:

"E."

TWO 12"×5"×32 LBS:
B.S.B.s

WEIGHT=114 LBS: PER FT:
AREA = 33·60 SQ:INS:
LEAST g = 3·8 INS:

"F."

WEIGHT= 51 LBS:PER FT:
+ LATTICE WEIGHT.
AREA = 15 SQ:INS:
LEAST g = 5·91 INS:

4"×4"×½"×12·75 LBS:
B.S.E.A.s.
2½"×⅜"SINGLE
LATTICING

SCALE OF FEET.

and L sections; detail No. 21 shows a few typical built-up stanchion sections, indicating at A and B the method of strengthening and compounding I sections by the use of flange plates, at C the use of channels and plates, at D and E combinations of steel I beams and at F angles and lattice bracing.

71. Steel stanchions for warehouse. In buildings such as the warehouse, loads are not very great and steel stanchions are easily obtained from British Standard beam sections. Such sections as $8'' \times 6'' \times 35$ lbs., $9'' \times 7'' \times 50$ lbs., and $10'' \times 8'' \times 55$ lbs. are often employed because of their larger ratio of breadth to depth than the average beam. They are also much heavier in the web. For light stanchions $5'' \times 4\frac{1}{2}'' \times 20$ lbs. and $6'' \times 5'' \times 25$ lbs. are also available.

Calculations made in Chapter Nine, paragraphs 250 to 252, show that $9'' \times 7'' \times 50$ lbs. B.S.B.'s are suitable for the basement stanchions and $8'' \times 6'' \times 35$ lbs. B.S.B.'s for the first floor.

Although lengths up to 40 ft. are obtainable the $9'' \times 7''$ section has been continued through the basement and ground floor storeys and the tier completed with an $8'' \times 6''$ section connected to the larger one in order to illustrate a method of longitudinal jointing.

72. Stanchion base and foundation. Commencing at the base, the stanchions are considered as an alternative to the cast iron columns previously illustrated. A sketch of this base is shown in detail No. 22 and other details at Nos. 23 and 24.

A square concrete foundation is provided $1' 9''$ thick and $6' 6''$ square as in the previous example, but reinforced with $\frac{3}{4}''$ steel rods in two series at right angles forming a $6''$ square mesh, which takes up the tension on the base of the slab due to the up-thrust from the earth, on the projecting portion beyond the cast iron base block.

It should be noted that, by reinforcement, the thickness of a base can often be reduced.

Holding down bolts $1''$ diameter are encased in the concrete and pull upon flat plates $3\frac{1}{2}'' \times \frac{1}{2}'' \times 3' 2''$ long, placed below the reinforcement.

73. Cast iron base to stanchions. Detail No. 23 shows a cast iron base 3 ft. square in plan and $12''$ deep with a top plate $1' 11''$ square, both having rounded angles. The top and base plates are $1\frac{1}{4}''$ thick and are connected by a $1\frac{1}{4}''$ vertical boxing at the centre with an intermediate diaphragm immediately below the web of the steel stanchion; four pairs of $1\frac{1}{4}''$ stiffening brackets distribute the load over the rim of the base plate and make the base rigid.

The top face is planed true for jointing the stanchion base and the latter is bedded upon it with a packing of lead or felt, as it may not be practicable to plane the thin steel plate which would be buckled slightly in riveting to the gussets and angles. It is, however, worth the labour to true the top seating of the cast iron base in order to reduce irregularities and to give a truly level bed on which to start the erection.

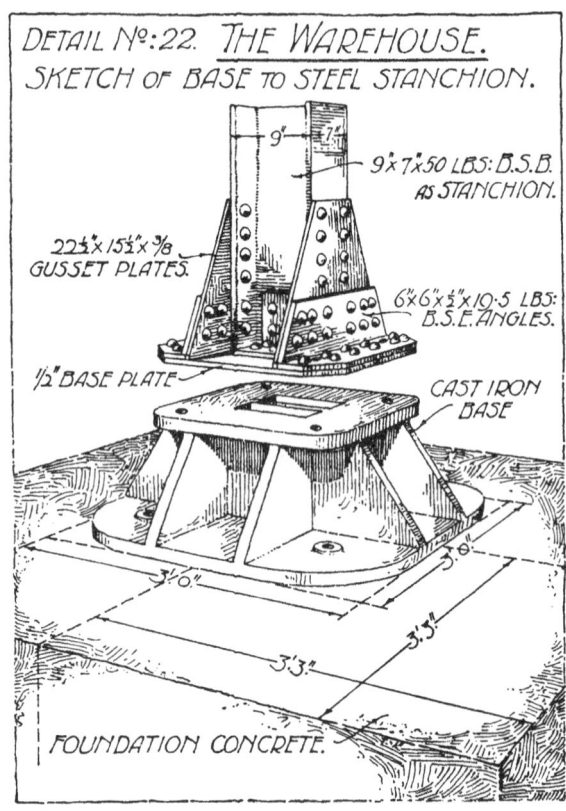

DETAIL Nº:22. THE WAREHOUSE.
SKETCH OF BASE TO STEEL STANCHION.

9"x7"x50 LBS: B.S.B.
AS STANCHION.

22½"x15½"x⅜
GUSSET PLATES.

6"x6"x½"x10·5 LBS:
B.S.E.ANGLES.

½"BASE PLATE

CAST IRON
BASE

3'·0"

3'·5"

3'·3"

FOUNDATION CONCRETE.

The base block is levelled and grouted to the concrete as explained in paragraph 58, and also bolted down thereto, the nuts bearing upon raised seatings which are machined parallel to the upper surface with little labour; they ensure uniform grip and firmness of the base.

Note. In modern work cast iron bases are little used. Portland cement concrete of high quality makes it unnecessary to spread the base to so great an extent as formerly and it is now usual to make the steel plate base to the stanchion (see next paragraph) large enough to distribute the load adequately over the concrete foundation. Details Nos. 22 and 23 are retained to illustrate the provision, if required.

The use of thick slabs or "blooms" of steel for column bases will be dealt with in a later volume.

74. Steel "stanchion base". The basement stanchion is a $9'' \times 7'' \times 50$ lbs. B.S.B.

The base to this stanchion is required to transmit its load over the broader face of the cast iron base and hence is spread by a steel

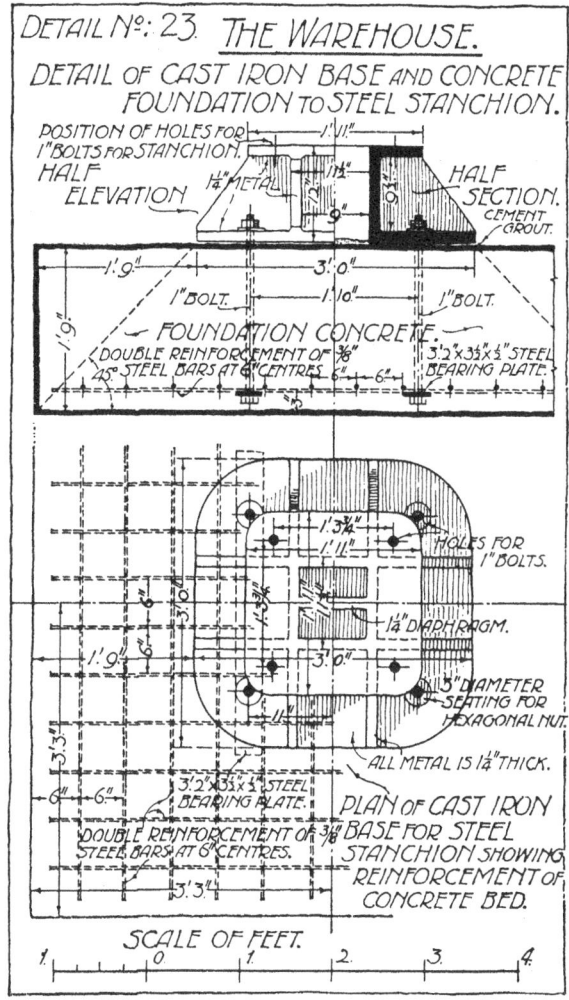

DETAIL No: 23. THE WAREHOUSE.

DETAIL OF CAST IRON BASE AND CONCRETE FOUNDATION TO STEEL STANCHION.

SCALE OF FEET.

base plate attached to the stanchion by riveted gussets and angles, as shown in detail No. 24. The base of the B.S.B. is cut dead square with a view to bearing solidly upon the $\frac{1}{2}''$ base plate, which is then attached to the flanges by first riveting $\frac{3}{8}''$ gussets of trapezoidal form thereto and connecting these at their lower edges by $6'' \times 6'' \times \frac{1}{2}''$ steel angles.

75. Design of steel bases to stanchions. In designing any steel base it is necessary to know the object to be accomplished in using it. It may serve the following purposes:

(a) Provide a sufficient and rigid projection to allow of the base being conveniently fixed by bolting to a cast iron or similar base.

(b) To spread the load uniformly over a large area, e.g. upon concrete, brickwork, masonry or metal.

Unless the transmission of load is directly through metal surfaces in assured contact the base will be called upon to resist bending moment at its projections due to up-thrust on the plate, and, as it is difficult to ensure metal to metal contact from the base of the stanchion to the steel plate—and hence to the cast iron diaphragm and vertical ribs in this example—it is necessary to stiffen the projections. Gusset plates do this efficiently if spread to the edges of the plate; they transfer some or all the flange load to the base by transmission through the rivets.

The gussets may bear upon the plate irregularly or may not bear at all when connected thereto by angle iron cleats, because the least inaccuracy in setting out and punching or drilling the holes may result in the angle projecting past the bottom edge of the gusset. In good work the base of a pillar should be machined after the assembly of the gussets, angles, and any stiffeners employed, before attaching the base plate. When this is done, the rivets employed to attach the gussets and angles to the stem of the stanchion need only be capable of transmitting 60 per cent. of the load, because the end of the main member will be in actual contact with the plates. If not machined, sufficient rivets should be employed to transmit all the load to the base plate.

76. Number of rivets in stanchion bases. Assuming that the rivets must transmit all the load the number required to connect the flanges and web of the stanchion to the base plate may be decided as follows:

Assume $\frac{3}{4}''$ rivets to be employed. They must transmit stress safely either in shear or bearing. In single shear one $\frac{3}{4}''$ rivet resists 2·65 tons (see Vol. II) and in bearing on a $\frac{3}{8}''$ gusset plate resists $\frac{3}{8}'' \times \frac{3}{4} \times 12 = 3·37$ tons.

Taking the value for shear, which is the lower, to decide the number of rivets in the flange:

Total flange load = single shear resistance of 1 rivet × No. of rivets ......(1).

$$\text{But flange load} = \frac{\text{total load} \times \text{area of flange}}{\text{area of stanchion}}$$

$$= \frac{78 \text{ (tons)} \times 5·67 \text{ (sq. inches)}}{14·71 \text{ (sq. inches)}},$$

$$\therefore \text{ flange load} = 30 \text{ tons} \dots\dots\dots\dots\dots\dots\dots\dots\dots\dots\dots\dots(2).$$

Inserting value of (2) in equation (1)

$$30 \text{ (tons)} = 2 \cdot 65 \text{ (tons)} \times n \text{ (rivets)}.$$

or
$$n = \frac{30}{2 \cdot 65} = 11 \cdot 3 \text{ (say 12 rivets)}.$$

Thus, 12 rivets are required to fix the gusset to the flange, or to fix the base angle to the gusset in the vertical plane. A rivet passing through all three pieces serves the two purposes simultaneously.

To decide the number of rivets in the connection of the web to the base plate, note that rivets are in double shear which, under the British Standard specification, is equal to twice the single shear resistance, therefore each rivet will resist $2 \times 2 \cdot 65 = 5 \cdot 3$ tons. The bearing is upon a web $\cdot 4''$ thick and $= \cdot 4 \times \frac{3}{4} \times 12 = 3 \cdot 6$ tons. Bearing resistance is now the least.

Now load in web = bearing resistance of 1 rivet × No. of rivets   ......(3),

$$\text{but load in web} = \frac{\text{total load} \times \text{area of web}}{\text{area of stanchion}} = \frac{78 \times 2 \cdot 94}{14 \cdot 71}$$

$$= 15 \cdot 6 \text{ tons.}$$

Inserting this value in equation (3)

$$15 \cdot 6 \text{ (tons)} = 3 \cdot 6 \text{ (tons)} \times n \text{ (rivets)},$$

or
$$n = \frac{15 \cdot 6}{3 \cdot 6} = 4 \cdot 3 \text{ (say 4 rivets)}.$$

(4 rivets are therefore required.)

77. Connection of angles to base plate. The angles connecting gusset to base plate have their horizontal flanges secured by a number of $\frac{3}{4}''$ rivets counter-sunk and left flush on the underside of the base. The strength of the angle flanges and of the base plate, and the number of rivets required, depend upon shear and deflection and cannot be treated here.

In the example, detail No. 24, ten $\frac{3}{4}''$ counter-sunk rivets are employed to each angle flange and the base is secured to the cast iron block by $1''$ bolts placed near the angles. The holes for rivets are placed chequerwise or zig-zag and the pitch should not generally be nearer than $3 \times$ diameter of rivet, centre to centre.

78. Connection of floor girders to stanchion. When girders are jointed into continuous stanchions the connections are made with angle cleats or with cleats and brackets. For girders supporting considerable loads brackets may become necessary as the ordinary flange cleats lack stiffness. Detail No. 25 shows how these brackets may be constructed from tee and angle sections along with packing plates, and also the arrangement of the rivets for securing them to the web and flange of the stanchion.

The parts of a bracket are separately shown at the bottom of the detail, and consist of a $4'' \times 4'' \times \frac{1}{2}''$ angle $6''$ long, a $6'' \times 3'' \times \frac{3}{8}''$ tee $9''$ long, and a packing plate $6'' \times 5\frac{1}{2}'' \times \frac{1}{2}''$; they are fixed to the stanchion by six $\frac{3}{4}''$ rivets and would resist a shear of $15 \cdot 9$ tons. Above the

girders additional flange cleats are shown which would increase the
rigidity of the connection and add to the shear resistance.

For lighter girders top and bottom flange cleats only are often
employed, especially if the stanchion has narrow flanges. Side or

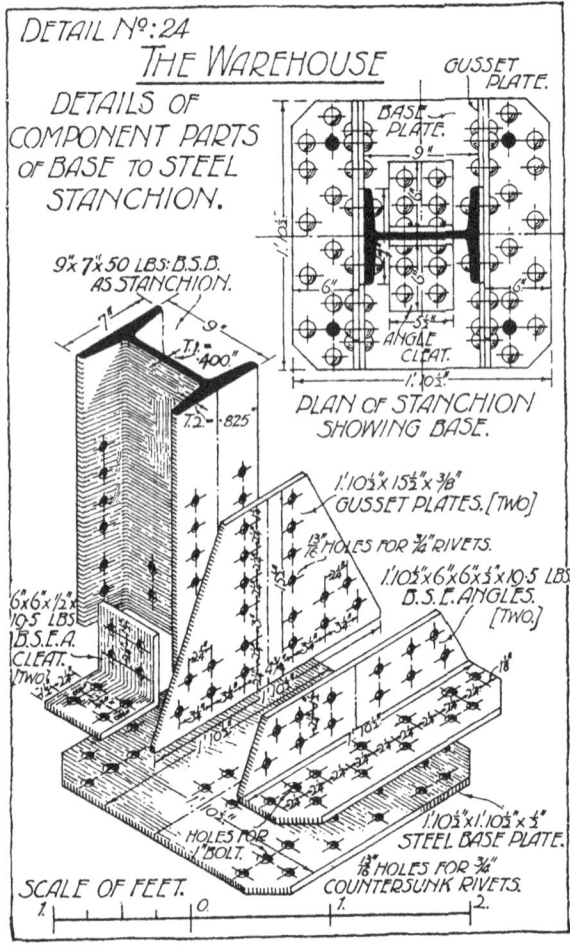

web cleats are, however, in most common use as these may be
standardised and also employed for connecting girder to girder as
shown in detail No. 31 and also shown for stanchion connections
at detail No. 27.

It should be noted that any of the methods described, or combinations of them, may be employed, the most practicable and

economical arrangement for any case depending upon the loads to be carried and the measure of fixity of the joint which is desired.

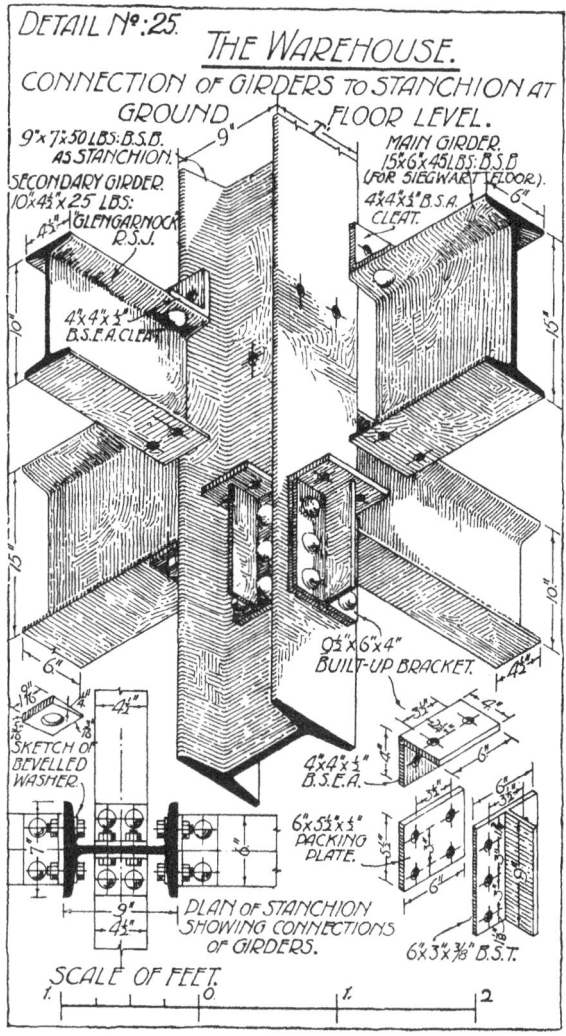

DETAIL N°.25. THE WAREHOUSE.
CONNECTION of GIRDERS to STANCHION at
GROUND FLOOR LEVEL.

79. Flange bolts and washers. When cleats have to be bolted to flanges of stanchions and beams, the bevel of the flanges prevents a solid bearing of the head or nut being obtained; bevelled washers are employed to remedy this, as shown in detail No. 25. A pair of parallel surfaces is then obtained and the grip ensured.

For the same purpose some washers are made with spherical seatings and the nuts with a corresponding bearing surface, so that for any bevel within a reasonable range the nut and washer become self-adjusting and grip effectively.

80. Longitudinal joints in stanchions. It is often necessary to join two lengths of stanchion in a tall building for one of two reasons:

(*a*) It may be impossible to obtain the required height in one length.

(*b*) It may be uneconomical to allow the size of stanchion required for the lowest storey to continue throughout the height of the building without diminution of size.

To meet the latter case it is often convenient to select a smaller section—suitable for the upper storeys—and to add flange plates as required; see detail No. 21.

If two lengths are to be jointed the connection should be made at or near a floor level. Many types of joint are in use but the one selected for illustration and shown in detail No. 26 is both dependable and ensures continuity of the stanchion without appreciably reducing the rigidity at the joint.

In the detail the joint is made just below the

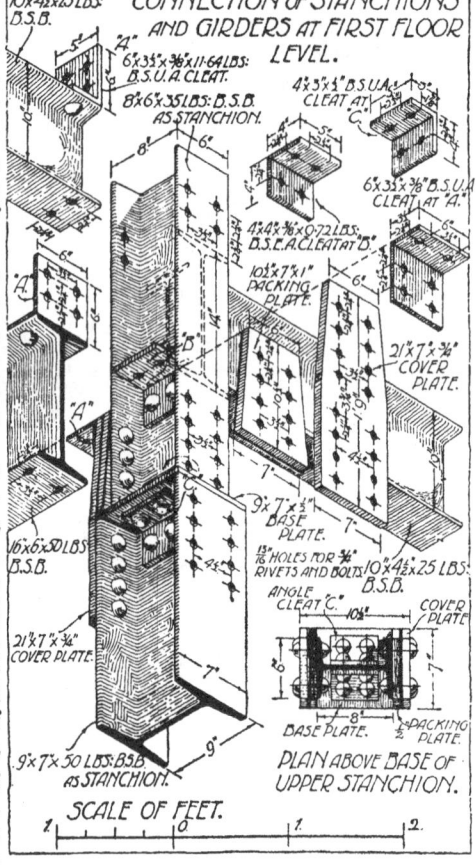

DETAIL N°: 26   THE WAREHOUSE.
CONNECTION OF STANCHIONS
AND GIRDERS AT FIRST FLOOR
LEVEL.

first floor level so that the upper section carries the weight of the second floor while the girders of the first floor transmit their weight through the joint to the larger stanchion below.

The axes of the two parts of the stanchions are in one vertical line and to transmit the pressure from the upper to the lower portion a $9'' \times 7'' \times \frac{1}{2}''$ plate is inserted between the square cut ends. The web load is provided for also by $4'' \times 3'' \times \frac{1}{2}''$ angle cleats 5'' long, having two $\frac{3}{4}''$ rivets to each flange, and the flange load of the stanchion is transmitted through flange covers $7'' \times \frac{1}{2}'' \times 21''$ long, having eight $\frac{3}{4}''$ rivets on each side of the joint. The difference in width of the two stanchion segments is made up by two $7'' \times \frac{1}{2}''$ packing plates $10\frac{1}{2}''$ long, thus getting solid bearing for the rivets.

81. Girder connections at stanchion joint. The whole of the joint is arranged so that the soffits of the main girders have the same level as the top ends of the covers and packing plates, see detail No. 26; thus, by adding an angle cleat $6'' \times 3\frac{1}{2}'' \times \frac{3}{8}'' \times 6''$ long riveted to the stanchion along with the covers and packing, an extended bearing is provided, and the girder flanges may be bolted to the angle. The top flanges of the main girders, and both top and bottom flanges of the secondary girders, are connected by $4'' \times 4'' \times \frac{1}{2}''$ angle cleats 5'' and 6'' long, with two $\frac{3}{4}''$ rivets to each flange of the angles.

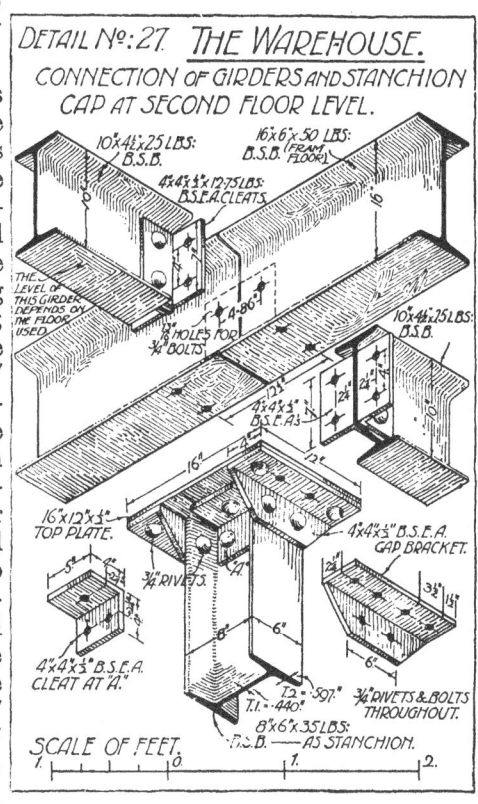

DETAIL Nº:27. THE WAREHOUSE.
CONNECTION OF GIRDERS AND STANCHION
CAP AT SECOND FLOOR LEVEL.

82. Head of stanchion to first floor storey. Detail No. 27 shows the head of the first floor stanchion, which supports two main and two secondary girders at the second floor level. These girders have their soffits at the same level and bear upon a cap plate $16'' \times 12'' \times \frac{1}{2}''$ thick projecting beyond the stanchion 4'' on each side in the direction of the main girders and 3'' in the direction of the secondary girders. The flange projections are supported by $4'' \times 4'' \times \frac{1}{2}''$ angles 12'' long, bevelled back to the flange width at the bottom edges and secured by two $\frac{3}{4}''$ rivets on each flat. Web angles $4'' \times 4'' \times \frac{1}{2}'' \times 5''$ long connect the cap plate to the web and stiffen the former.

Main girders joint centrally over the stanchion with $\frac{1}{4}''$ clearance between and the secondary girders are notched at the bottom flange to allow the end to pass forward to within $\frac{1}{8}''$ of the main girder webs; $4'' \times 4'' \times \frac{3}{8}''$ web cleats $7''$ long with two $\frac{3}{4}''$ bolts to each arm complete the joint.

The main girders are bolted, flange to cap, as shown in the detail and the arrangement forms a well-connected group of members, though not producing a rigid fixing.

83. Key drawing to tier of stanchions. Detail No. 28 shows the complete tier of stanchions, with lengths, floor levels and foundation. It should be studied as a unit, and the positions noted of the detailed portions embodied in the foregoing examples.

The whole unit is an alternative to the cast iron columns illustrated in detail No. 20.

84. Practical procedure in preparation and erection. In all steel work the preparation of the units is carried out in structural workshops to large scale dimensioned drawings, and as many parts as may be convenient assembled before sending to the erectors.

Angle cleats are riveted to the ends of girders, and lower flange cleats to the stanchions; then the connections on the site are made by bolts or, if feasible, by rivets,

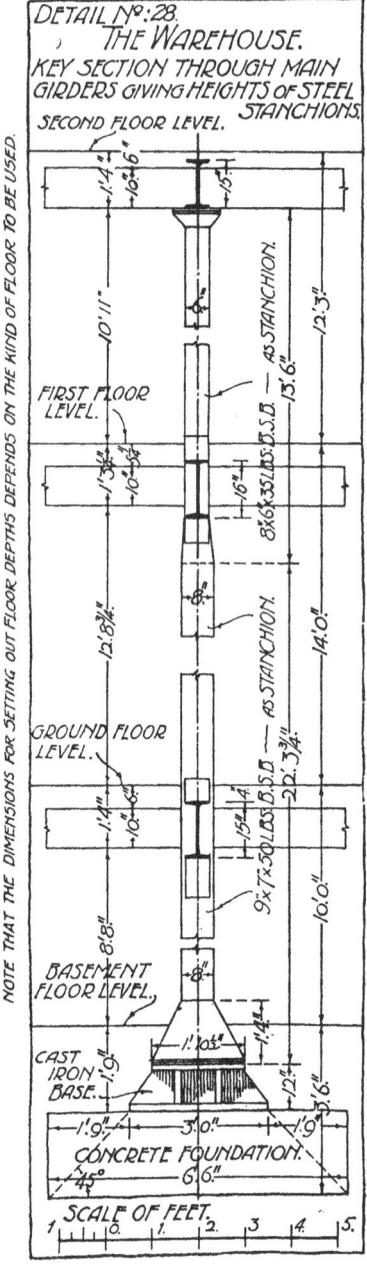

DETAIL N°:28.
THE WAREHOUSE.
KEY SECTION THROUGH MAIN GIRDERS GIVING HEIGHTS OF STEEL STANCHIONS.
SECOND FLOOR LEVEL.

except in the case of cast iron where bolts must be employed. Rivets driven on the site are called field rivets; those driven at the works, shop rivets. Riveting should be insisted upon for all important work where practicable, as the efficiency of ordinary bolts compared with rivets is only about 80 per cent., except when subjected to direct tension, when bolts are more reliable and stronger.

CHAPTER THREE

STEEL FRAMED AND FIRE-RESISTING FLOORS

PATENT FORMS OF FIRE-RESISTING FLOOR

85. There are many forms of fire-resisting floor in which steel beams are framed together, to provide support for panels consisting of reinforced concrete slabs, terra-cotta or concrete lintols, or light reinforced beams of tiles or concrete.

Most of these floor panels are patented and each system possesses some worthy features which may commend it under certain conditions.

The choice of patent floors presents a difficulty because there are so many excellent methods; it has been convenient to divide these floors into classes, and make selections from each which are typical of the group and exhibit clearly the main principles involved.

All the systems selected have been sufficiently used to become well known, but students and practitioners should give close attention to other examples and to all fresh attempts to improve existing systems and to economise material and labour.

86. Reinforced concrete slabs. Concrete slabs are very commonly formed *in situ* as described in the earlier portion of this chapter, but reinforced by steel on the tension side of the concrete. This reinforcement may consist of plain round or square bars set at 3″ to 6″ centres across the short span of the slab, crossed by lighter rods at 12″ to 18″ centres; the latter serve as distributing bars, spreading isolated loads over a greater area and also preventing localised contraction during setting.

This method of reinforcing slabs is common in monolithic structures. Its chief disadvantages are the labour involved in placing it in position and the liability to disturbance when depositing the concrete around the steel, unless the latter is wired in position. It possesses certain advantages, however, which must be considered when dealing with reinforced concrete structures.

To overcome the difficulty of placing individual rods, several patent materials have been produced, such as woven wire mesh, electrically welded steel mesh, expanded steel, Hy-rib, etc.

The Expanded Metal Co., Ltd., supply two forms of steel floor-mesh made by cutting and expanding plain or ribbed sheets of steel. The former produces a diamond-shaped mesh 3″ to 6″ wide across the short diagonal and the latter a rib-mesh of square outline and 2″

to 8″ side, with main strands running lengthwise of the sheet and lighter strands at right angles.

87. Concrete floor to entrance vestibule of warehouse. An example of the use of diamond mesh expanded metal reinforcement

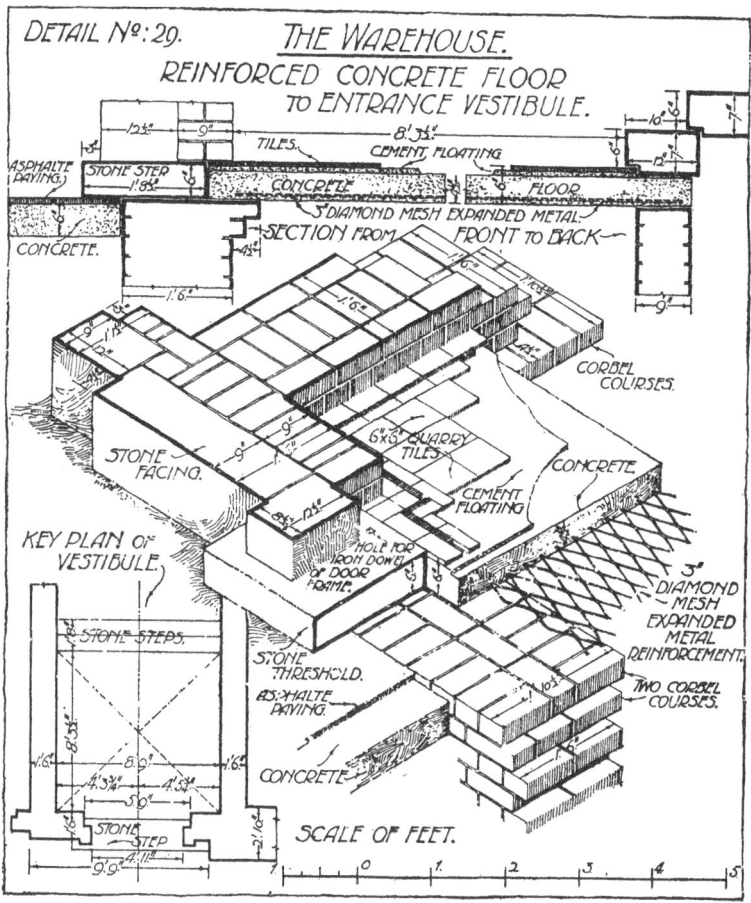

is given in the entrance vestibule of the warehouse. The key plan on detail No. 29 shows the shape of the floor, which is a rectangle 8′ 9″ × 8′ 3½″. At the sides the floor is carried by 4½″ brick offsets and at the front edge by two corbel courses; the back end rests entirely upon a 9″ wall, which receives the foot of the solid stone stairs.

The reinforcement should be laid with the long way of the mesh in the direction of the shorter span, which is important in floor bays where the length and breadth vary considerably. In this case the floor slab is practically square.

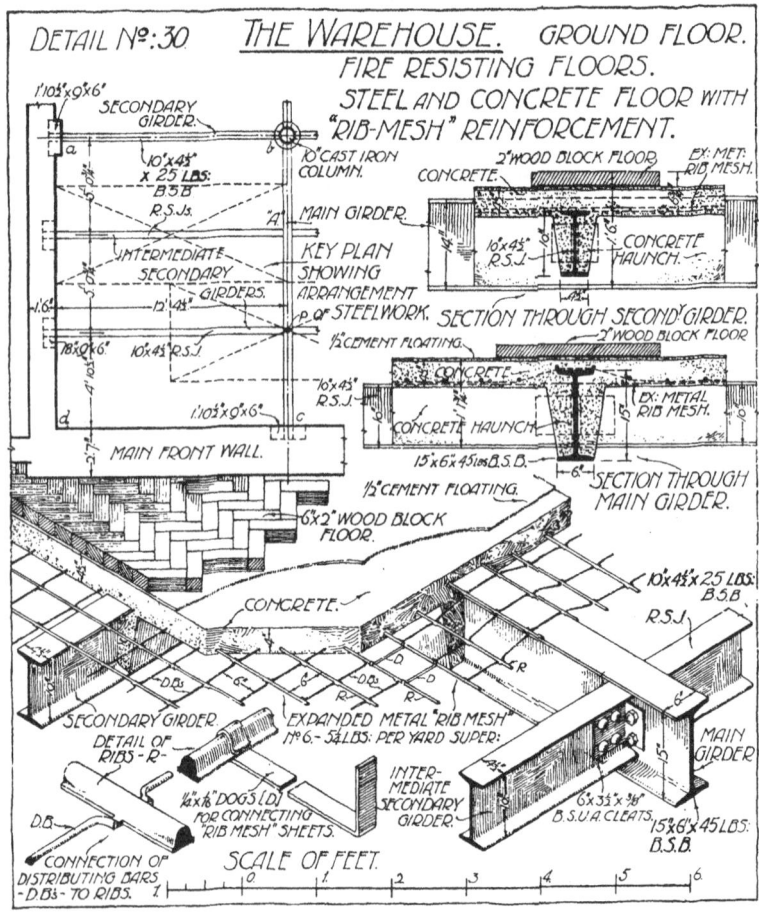

Rough shuttering is placed in a level position and supported on posts from the basement, a layer of concrete about 1″ thick screeded over it and the sheets of expanded metal laid thereon, overlapping 6″ at side joints where these are required in the width of the sheets.

The body concrete is then filled up, screeded level and left to set.

In this case the floor is finished by 6″ square quarry tiles with bands of narrow border tiles bedded in portland cement mortar.

88. Ground floor of warehouse. Rib-mesh reinforcement is applied in the ground floor of the warehouse.

This floor consists of four longitudinal bays about 15 ft. long, each divided into three transverse bays about 12 ft. wide.

To obtain the necessary floor support without using large and uneconomical cross girders, internal pillars are employed at the intersections of the bays, as shown on the general plans of the ware-

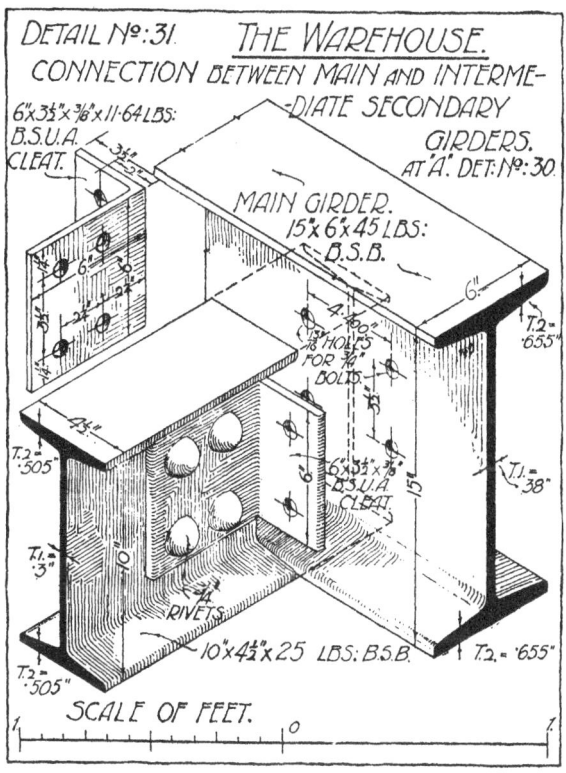

DETAIL Nº:31. THE WAREHOUSE.
CONNECTION BETWEEN MAIN AND INTERME-
-DIATE SECONDARY
GIRDERS.
AT 'A.' DET:Nº:30.

6"x3½"x⅜"x11·64LBS:
B.S.U.A.
CLEAT.

MAIN GIRDER.
15"x 6"x45 LBS:
B.S.B.

18 HOLES
FOR ⅞"
BOLTS

6"x3½"x⅞
B.S.U.A
CLEAT.

4
RIVETS.

10"x4½"x25 LBS: B.S.B.

SCALE OF FEET.

house. Main girders span from wall to pillar, and pillar to pillar, lengthwise of the building and secondary girders at right angles to these, the latter varying in size and disposition with the particular construction selected for the floor.

A key plan of a corner bay of the ground floor is given in detail No. 30.

The main girders are B.S.B.'s 15" × 6" × 45 lbs. and the secondary girders are B.S.B.'s 10" × 4½" × 25 lbs., set at approximately 5 ft. centres. Detail No. 31 shows the connection between these girders, the joint being formed by employing standard angle cleats to connect the webs.

89. Expanded metal rib-mesh floor. The concrete slab to this floor has a total thickness of 4″, is continued over the top flanges of the girders and is reinforced with No. 6 "rib-mesh," having a 6″ square mesh and weighing 5¼ lbs. per yard super. Its main ribs are marked R in the detail (No. 30) and its distributing bars DB. The whole sheet lies flat upon the secondary girders and is bent up 2″ to pass the main girders. Where the two edges of the sheets meet they may be wired together, but to prevent loss of covering power they are preferably connected by special clips or dogs marked DD, which are bent over the outer ribs of the adjacent sheets and also shown in a separate sketch on the same detail; these virtually continue the distributing bars across the junction.

Wood shuttering is placed ½″ below the metal reinforcement and concrete deposited to the required depth.

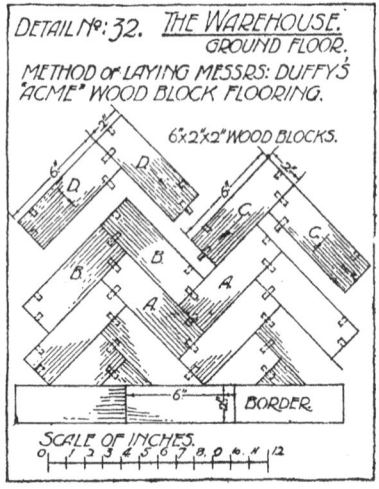

DETAIL Nº: 32. *THE WAREHOUSE.*
GROUND FLOOR.
METHOD OF LAYING MESSRS: DUFFY'S
"ACME" WOOD BLOCK FLOORING.

6×2″×2″ WOOD BLOCKS.

6″   BORDER.

SCALE OF INCHES.
0 1 2 3 4 5 6 7 8 9 10 11 12

The *effective depth* of the slab is approximately 3½″ which is measured from the top of the concrete to the centre of the wire mesh.

90. Finish of ground floor, with wood blocks. One method of finishing a concrete floor with plain wood blocks was illustrated and described in Vol. I. A further application with an improved type of block is now considered.

To level up the concrete and form a bed for wood blocks a ½″ floating of 1 to 2 cement mortar is employed; 6″ × 2″ × 2″ blocks are laid upon the dry floating in bituminous mastic as previously described, but the edges are further secured by wood dowels on Duffy's Acme system, which practically converts the flooring into one unit, ensuring a surface as permanently true as it is practicable to obtain.

The blocks are dowelled on sides and ends and must be laid in the order indicated by the lettered diagram, detail No. 32.

91. Protection of girders. The main girders of the ground floor project about 11″ and the secondary girders about 9″ below the floor slab and are partially protected against fire and oxidation by concrete haunches springing from the lower flanges. The haunches have bevelled sides and are formed by depositing the concrete into wood shuttering, upon the edges of which the horizontal shuttering is supported, the haunches being filled previous to the formation of the floor slab.

This method leaves the flange exposed; it requires periodical painting for protection and should only be employed for basements and unimportant positions, where no provision is made for the storage of inflammable goods.

92. Fireclay lintol and terra-cotta block floors. There are several patent floors in which hollow fireclay lintols or terra-cotta blocks are employed to form the floor panels; these are designed for use in one of the following ways:

(a) Placed in short lengths of $1\frac{1}{2}$ to 4 ft. between the flanges of steel joists and providing a clothing or protection to the lower flanges.

(b) Placed end to end upon shuttering, jointed in cement, and reinforced longitudinally or in both directions by steel bars grouted into the joints.

Amongst the better known systems belonging to the first division are Homan and Rodgers, Dawnay, King and Fawcett floors. All of these employ lintols of special form to span between filler joists, upon which the levelling concrete is deposited. They are known as self-centering floors because wood shuttering is dispensed with.

Detail No. 37 shows some of the forms of lintol employed.

This type of floor is somewhat heavy as compared with the reinforced slab or reinforced lintol floor but is otherwise very satisfactory and economical in the labour of erection.

In the second division are the Fram, Kleine, Cullum and Diespeker floors, which are designed to cover considerable spans without intermediate joists and are easily applied to panels 12 ft. square. These floors require shuttering in the form of horizontal platforms which are generally hung from the supporting girders, but are able to compete successfully with self-centering floors.

93. Homan and Rodgers' lintol floor. As a typical example of a lintol floor the Homan and Rodgers' method has been selected for one application to the first floor of the warehouse.

A key plan shows this application to an angle bay 15' $4\frac{1}{2}$" × 12' $6\frac{3}{4}$", which is given, along with the construction, in detail No. 33.

The main girders are B.S.B.'s 16" × 6" × 50 lbs., secondary beams (acting as stiffeners to pillars and as transverse ties to the building) are B.S.B.'s 10" × $4\frac{1}{2}$" × 25 lbs. while the joists supporting the lintols are B.S.B.'s 6" × 3" × 12 lbs. placed across the shorter span and resting upon the main girders and party wall.

Should the span of the bay be too much for an economical depth
of joist to be employed, the bay might be divided lengthwise by an
intermediate secondary girder forming two panels, and the joists
placed parallel to the side walls.

These joists are placed at 1' 6½" centres and as close as may be
convenient to the top flanges of the main girders.   Equilateral

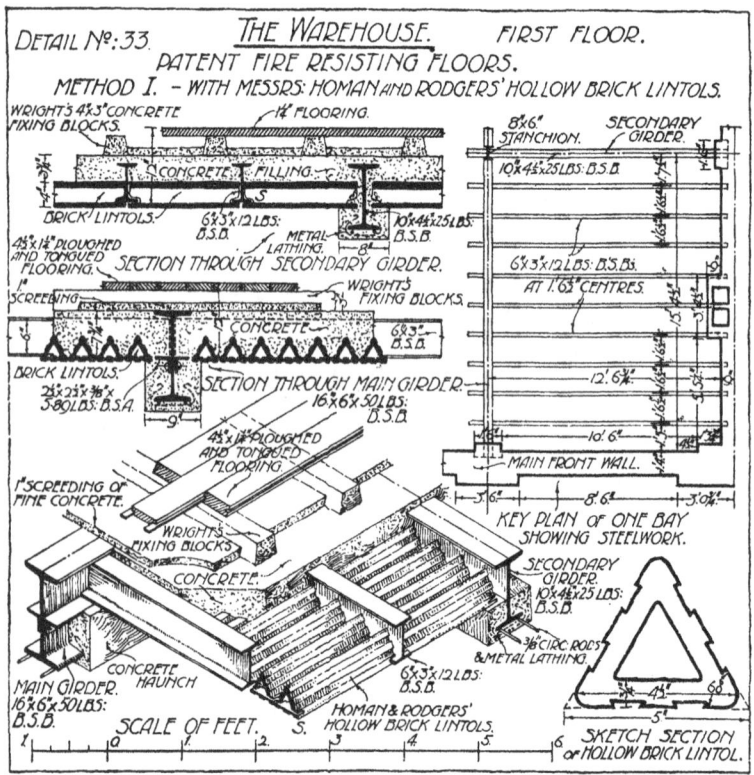

triangular lintols, 4¾" side with rounded angles, span between the
joists, resting with their slotted ends (S) upon the lower flanges, and
have their bottom flanges continued under the joists to the centre
of the latter, thus protecting the steel with ¾" of fireclay or terra-
cotta over the entire surface of the joists. To provide a key for the
ceiling plaster and concrete filling each face is traversed longitudinally by two dovetailed grooves and the surfaces are also somewhat
rough and porous.

Where the lintols bear upon secondary girders which are deeper
than the joists, steel angles may be riveted to the girder webs to

receive the notched lintols, and provide support in the same plane
as the flanges of the joists; or the lintols may rest upon concrete
haunches as shown in the detail.

Upon the lintols light concrete is filled to encase completely the
joists and girders, or it may be stopped at a convenient level for
laying the flooring.

Many kinds of floor finish are applicable, but $4\frac{1}{2}'' \times 1\frac{1}{4}''$ ploughed
and tongued deal or hard wood flooring is selected, nailed to
Wright's $4'' \times 3''$ patent fixing blocks, of dovetailed section, which
are in turn bedded level upon the concrete and secured by a $1''$
screeding of finer concrete thickened up at the edges of the bearers.
These blocks are made of concrete similar to the patent partition
blocks and are supplied in lengths of $12''$ and $24''$.

Protection of the girders is afforded by totally encasing the lower
flanges of the girders with concrete, key for the soffit being afforded
by bending expanded metal around the bottom flanges and ensuring
space for key by packing off with two $\frac{3}{8}''$ round iron rods. The concrete is filled into a temporary wooden boxing.

94. The Fram reinforced hollow-tile floor. The Fram fire-resisting
floor is applied to the first floor of the warehouse as an alternative
to the lintol floor just described, and is illustrated in detail No. 34,
which contains the key plan and details of one panel.

The principle of construction is to convert rows of terra-cotta
blocks, backed with concrete, into beams by reinforcing the under
surface with tension rods. In the simpler form of reinforcement these
rods run in one direction across the short span of the floor between
the rows of lintols and have their ends bent upwards, or hooked,
above the supporting flanges of the girders; alternate or occasional
bars may also be bent across the depth of the blocks at a short
distance from the ends and hooked over the opposite side of the top
flanges of the girders, or again, hooked continuity stirrups may be
employed to link adjacent panels together as shown in the detail.

The blocks employed by the Fram Company are $10''$ wide $\times 6''$
deep $\times 11''$ long, and may be employed with either the $6''$ or $10''$
dimension vertical or with the blocks alternately placed according
to the depth of the floor or amount of concrete covering desired,
both of which depend upon the required strength of the floor. These
blocks are $1''$ thick open at the ends and are divided into two com-
partments by a central partition or diaphragm.

In erecting the panel wooden shuttering is first suspended from
the steel work, or supported on posts and bearers, and the blocks are
placed upon it in rows across the shorter span, with their open ends
abutting on each other; these are jointed end-wise in cement mortar
to give them a solid connection for the transmission of thrust over

the upper portion, and also to prevent the grout used at a later stage from entering the cavities. These rows of blocks extend to the haunches of the girders in the direction of their span.

Between the rows of blocks, at every joint, a steel tension rod is placed, the joint being wide enough to give a clear space all round the rod for grout or fine concrete to penetrate. The bottom cover to

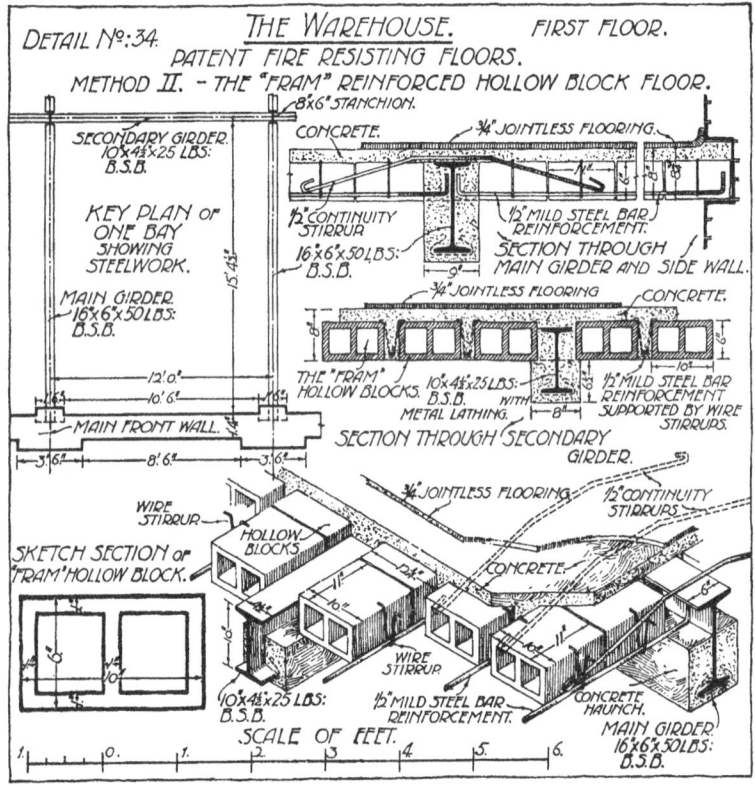

the steel should be at least ½″, which is secured by supporting the rods on wire loops or stirrups of **U** form with the top ends bent outwards to rest upon the blocks. Cement grout composed of 1 part of portland cement and 1 part fairly fine sand, is poured into the joints, adheres to the blocks, covers and effectively anchors the steel. Fine concrete may be used to fill up the joints when the steel is covered.

When complete the panel is covered with concrete to a depth of 1½″ or more and the whole left for a period of about fourteen days before removing the shuttering.

95. Principles underlying the construction of the Fram and similar floors. (See also paragraph 100 for Helicon, Cullum, Diespeker and Kleine floors.)

When the panel is set and the shuttering removed, load is supported in the following manner:

Any weight placed upon the floor—including its own weight—tends to cause vertical movement of its units, each one over the others, which is known as shear. If this is successfully resisted by the cement bedding, grout and concrete, a bending action takes place causing a compression on the upper portion and tension on the lower portion of the slab. The tiles and concrete can supply the compressive resistance but have little capacity to supply tensile resistance, hence the tension rods come into play and are enabled to exert their strength because of the grip and adhesion existing between the steel and cement, and between the cement and the blocks.

When a panel is approximately square the principle may be economically applied in both directions, but in such a case the ends of the blocks must be stopped or special blocks employed with closed vertical faces on all sides, otherwise the grout would enter the cavities. Ordinary blocks may be employed by cementing covers over their open ends, but this method cannot be so effective in transmitting and resisting shear stresses as a solidly moulded block.

When two series of reinforcing bars are employed the primary series is placed $\frac{1}{2}''$ clear of the soffit, and the secondary series immediately upon the first series without the spacing loops.

This example is further used to illustrate the application of jointless flooring laid directly upon the concrete. The covering is compounded of suitable waste materials including sawdust, asbestos, cork and rubber amalgamated with a binding composition which enables it to be laid in a semi-plastic condition and when set is sufficiently elastic to produce a tough, impermeable flooring without joints, and therefore very satisfactory from a hygienic standpoint.

It is claimed that these floorings are durable, easily laid on any rigid base, and that they are warm and resilient. They are suitable for public buildings, schools, hospitals and workrooms.

Doloment, Ebnerite, Linolite, Petropine and Wirolithic floors are amongst the better known.

96. Reinforced concrete beam or lintol floors—with pre-cast units.
Light pre-cast beams or lintols reinforced with steel are in common use as constructional units, for bridging fire-resisting floors by spanning between secondary girders. The units are placed side by side at right angles to the secondary girders and rest either (a) upon the lower flanges, or (b) upon concrete haunches, or (c) upon special

fire-resisting blocks built upon the flanges. Where girders are unnecessary they may rest upon the walls.

The Siegwart floor beams are a good example of this class.

A variation of the type occurs in the Armoured Tubular floor, in which reinforced T beams span between the secondary girders at 10″ centres and the spaces between them are bridged with hollow concrete tubes in short lengths, notched to rest upon the rebated edges of the beams.

Siegwart and Armoured Tubular floors are considered in some detail in the succeeding paragraphs.

97. Siegwart floor beams. Siegwart beams are hollow concrete castings, $\frac{7}{8}″$ to 1″ thick, of variable dimensions in cross section and reinforced with steel in the two lower angles. They are roughly a hollow rectangle in cross section, slightly rounded at the upper angles and eased by triangular fillets at the lower angles. Their form and method of support are shown in the lower part of detail No. 35.

In this example the beams rest upon concrete haunches enclosing and protecting the lower flange of the girder, with the soffit covering retained by stout metal lathing bent round the flange and packed off the steel face by two $\frac{3}{8}″$ round iron rods.

For a bay of the warehouse floor a suitable plan would be to insert a central secondary girder across the bay and to place the

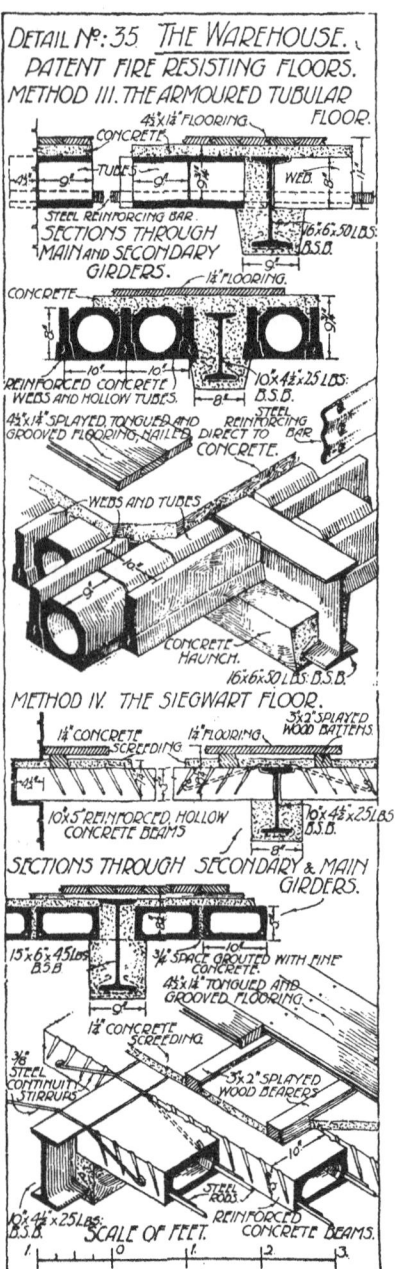

DETAIL Nº: 35   THE WAREHOUSE.
PATENT FIRE RESISTING FLOORS.
METHOD III. THE ARMOURED TUBULAR FLOOR.

Siegwart beams parallel to the main girders, thus leaving a span of about 7' 6" to be bridged by them. A suitable size of beam would then be $10'' \times 5'' \times \frac{7}{8}''$ thick in cross section, reinforced with two $\frac{3}{8}''$ and two $\frac{1}{4}''$ round steel rods to each beam; such units would support an inclusive load of about 270 lbs. per ft. sup. of floor surface.

98. Method of laying Siegwart beams. The beams are laid side by side with a space of $\frac{1}{2}''$ between them, this space being filled up on the soffit edge when laying by a fillet of cement mortar; as soon as this is sufficiently set the whole joint is grouted up solid, as described for Fram lintols.

The vertical joint faces of the beams are furrowed, as shown in the details, which allows keys of cement to form in the grouted joint and thus ensures resistance to shear and transmission of isolated loads through adjacent beams.

Some advantage is gained by fixing the ends of beams and floor slabs, because this reduces the stress at the centre of the panel, and some degree of fixity is obtained, as shown in the detail, by placing a $\frac{3}{8}''$ bent steel rod—called a continuity stirrup—over the top flange of the supporting girder and embedding it in the cement joint between pairs of beams on each side. The ends are hooked to provide a secure grip on the cement grout which is preferable to the straight bar or half cranked end.

Floor beams of this kind can be further strengthened by placing additional tension rods in the bases of the joints between beams, these joints being spaced to suit the enclosed reinforcement; the principle is exactly the same as that already described for the Fram reinforcement. The additional tension rods may be placed parallel to the soffit—suited to a free support—or bent upwards and clipped over the top flanges of the supporting girders if some fixity of the ends is desired.

The Siegwart floor—like others of this series—is light, strong and efficient and requires no shuttering except to deposit the haunch of concrete upon the girders.

The labour of assembling is economical and the structure can be speedily completed.

In the example shown the flooring selected is of $4\frac{1}{2}'' \times 1\frac{1}{4}''$ tongued and grooved boards, nailed direct to $3'' \times 2''$ splayed wood bearers, which are bedded upon the beams and secured by a $1\frac{1}{4}''$ screeding of concrete.

The total thickness of the finished slab and flooring is $8\frac{1}{4}''$, but, as the bulk of this is hollow, the dead weight is low.

99. Armoured Tubular floor. This is another unit construction which has already been referred to, and is also illustrated in detail

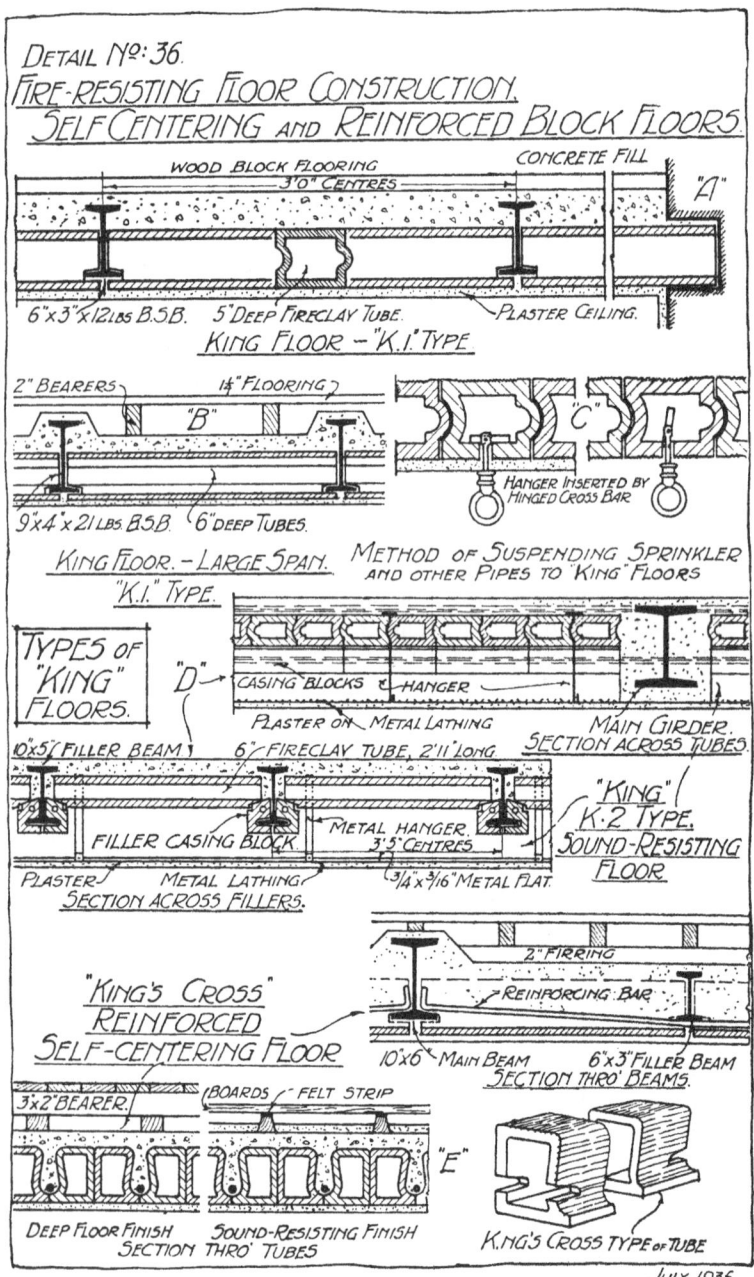

DETAIL Nº: 36.

FIRE-RESISTING FLOOR CONSTRUCTION.
SELF CENTERING AND REINFORCED BLOCK FLOORS.

WOOD BLOCK FLOORING
3'0" CENTRES
CONCRETE FILL
"A"

6"x3"x12 LBS. B.S.B.   5" DEEP FIRECLAY TUBE.   PLASTER CEILING.
KING FLOOR - "K.I." TYPE

2" BEARERS   1¼" FLOORING
"B"
"C"
HANGER INSERTED BY HINGED CROSS BAR

9"x4"x21 LBS. B.S.B.   6" DEEP TUBES.
KING FLOOR. - LARGE SPAN.
"K.I." TYPE.

METHOD OF SUSPENDING SPRINKLER
AND OTHER PIPES TO "KING" FLOORS

TYPES OF "KING" FLOORS.   "D"
CASING BLOCKS   HANGER
PLASTER ON METAL LATHING
MAIN GIRDER.
SECTION ACROSS TUBES.

10"x5" FILLER BEAM   6" FIRECLAY TUBE, 2'11" LONG
"KING" K.2 TYPE.
SOUND-RESISTING FLOOR
FILLER CASING BLOCK   METAL HANGER 3'5" CENTRES
PLASTER   METAL LATHING   ¾"x³⁄₁₆" METAL FLAT.
SECTION ACROSS FILLERS.

"KING'S CROSS" REINFORCED SELF-CENTERING FLOOR
2" FIRRING
REINFORCING BAR
10"x6" MAIN BEAM   6"x3" FILLER BEAM
SECTION THRO' BEAMS.

3"x2" BEARER.   BOARDS   FELT STRIP
"E"
DEEP FLOOR FINISH   SOUND-RESISTING FINISH
SECTION THRO' TUBES
KING'S CROSS TYPE OF TUBE

JULY. 1936.

No. 35. As detailed, it is suitable for application to a floor bay of the warehouse, and is an alternative to the constructions already considered.

The beam units are inverted tees $8'' \times 3''$, the webs being $1\frac{1}{2}''$ thick and the table $3''$ wide by $2\frac{1}{2}''$ deep; the sides of the flanges are tapered back to form seatings $1\frac{1}{2}''$ wide, which concentrates the load from the filling tubes close to the webs and facilitates the placing of the filling units.

Each beam is reinforced with a single corrugated bar of rolled steel having transverse grip studs at intervals as shown in the separate sketch; these bars resist the tensile stresses.

The tee beams are spaced at $10''$ centres and bridged by tubular lintols $8\frac{1}{2}''$ wide by $8''$ deep and $9''$ long.

Shuttering is not required, the beams being first placed across the shorter span and temporarily secured, when the lintols can be dropped successively into position and the panel quickly covered.

In the illustration the lintols and beams are shown levelled up and covered with $1''$ to $1\frac{1}{2}''$ of breeze concrete; $1\frac{1}{4}''$ secret nailed flooring is secured direct to the concrete by nailing through the edges of the tongues.

At the sides of the panel the first beam slightly overlaps the concrete haunch of the girder and the remaining space is filled with concrete when levelling the surface. This marginal space filling is necessary with most unit floors especially if the panel is of irregular dimensions; in some cases two or more beams have to be laid side by side to bridge margins too small for the lintols.

100. Further illustrations of fire-resisting floor construction are given in details Nos. 36, 36 A and 36 B. They are chiefly of the self-centering and hollow block types.

1. *King floors.* In detail No. 36, at A, is shown the King K 1 floor, with filler beams at 3 ft. centres and hollow fireclay tubes birdsmouthed upon the lower flanges of the steel beams. These tubes form a permanent center for the floor finishings. The tubes are $4''$ to $8''$ deep; shuttering is not required and a flush ceiling is produced. Where the beams are necessarily deep, and a solid floor fill unduly heavy, the floor may be constructed as shown at B. The beams are encased in concrete and wood bearers and laid in the hollow to receive the floor boards which are raised clear of the beam casing.

Pipes for sprinkler installations, etc., may be suspended from the floor by drilling $\frac{3}{4}''$ holes in the soffit of the blocks and using the hinged-bar hanger shown in the detail at C.

The King K 2 type of floor is illustrated at D. This is a sound-resisting floor. The tubes are laid on filler casing-blocks, which provide insulation as well as protection. The ceiling is suspended by metal hangers from the centering tubes, and the ceiling surface is of plaster on metal lathing. An almost unbroken void therefore occurs between the ceiling and the floor slab and adds to the sound resisting properties of the floor. The type is suitable for flats, offices and hotels, and buildings where sound-deadening is desirable.

Another type of floor (K 3)—not illustrated—is available, in which the tubes differ in shape from the ordinary flat type. The sides of the blocks are curved to an ogee form and allow of a dovetailed tongue of concrete being deposited between the tubes, which links the covering concrete solidly with the tubes. The floor is capable of resisting shock-loading owing to the keying action of the concrete tongues and to the fact that transverse distribution of load does not depend solely on shear resistance on the adhesional surfaces.

This floor may be reinforced if desired and so designed as to carry unusual loads.

The detail No. 36 at E shows another type of King floor, known as the King's Cross floor, and is definitely intended as a reinforced floor for considerable span. The tubes are somewhat similar to those described in the preceding paragraph but are flat on one face, so as to give appropriate spacing to the reinforcement. The tubes are laid in pairs giving about 12″ centres between the steel rods, which are placed near the base of the dovetailed recess between pairs of tubes and covered with concrete.

The section through a large beam and through an adjoining smaller beam shows how the reinforcing bars are placed. They pass below the smaller beams (the function of which is to support the tubes) and are cranked upward with square bent ends to rest on the lower flanges of the larger beam.

Any type of floor finish may be applied.

Note the floor finish in the example where dovetailed battens are floated in (or clipped down to) the concrete, and covered with soft felt strips, upon which the boards are laid. This is a good sound-resisting provision.

2. *Helicon floor*. This floor is another variety of the hollow block reinforced floor. The blocks are 12″ wide and of variable depth and are supplied in fireclay or in pumice concrete for lightness and sound resistance.

The section shown in detail No. 36 A is similar to those previously described, but a special provision is made by the use of bevelled blocks and metal fittings for housing conduits and suspending pipes. Reinforcement, floor finish, etc., can be arranged to suit any conditions.

3. *Cullum floor*. The Cullum floor is of the same type as the

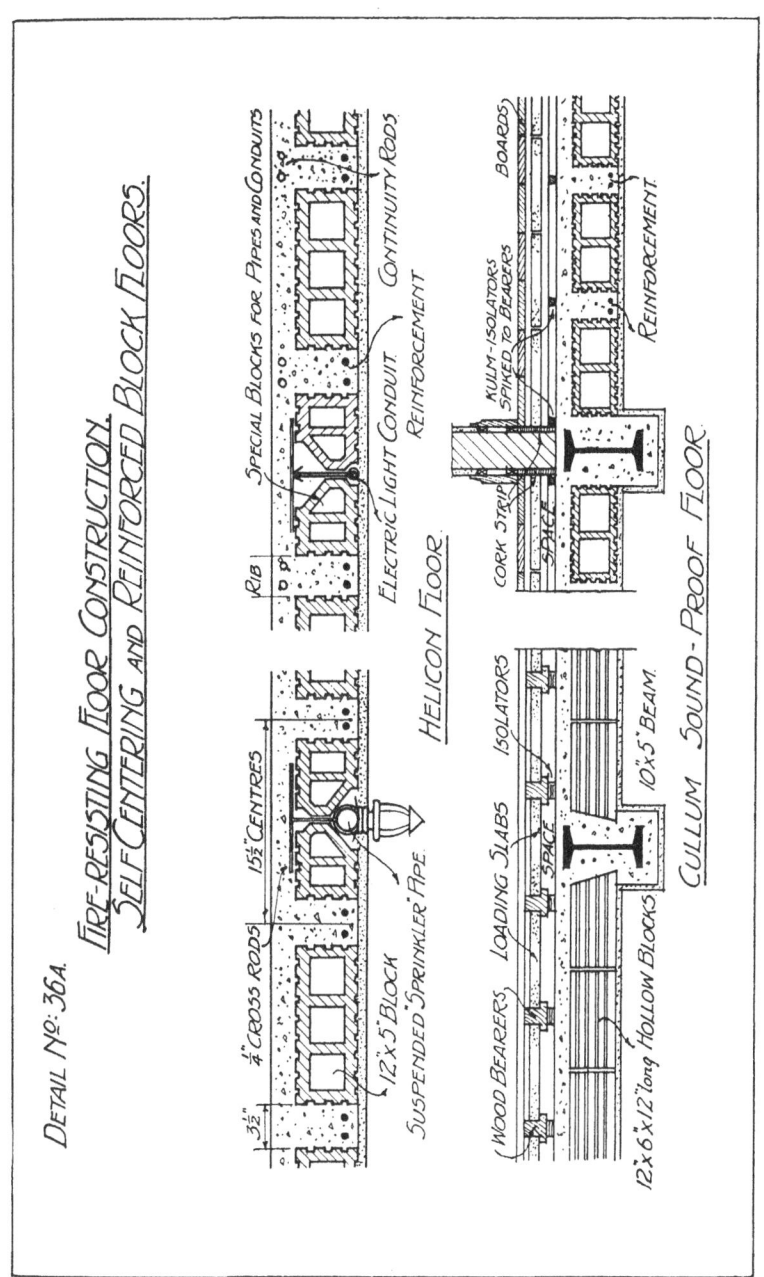

DETAIL Nº. 36A.

FIRE-RESISTING FLOOR CONSTRUCTION.
SELF CENTERING AND REINFORCED BLOCK FLOORS.

HELICON FLOOR.

CULLUM SOUND-PROOF FLOOR.

preceding examples with a specially designed sound-proof floor finish. For a boarded floor the arrangement is as follows: Rebated timber bearers are placed transversely to the lines of floor reinforcement. These bearers rest on rubber isolators to absorb sound energy. Loading slabs are placed between the bearers, resting in the rebates. They do not touch either the floor surface or floor boards. Their purpose is to provide a floating mass to reduce the effects of impacts on the flooring surface.

The floor boards are nailed direct to the rebated bearers, and a cork strip is inserted between the ends of the bearers and the adjoining wall or partition.

The floor finish is therefore a floating construction, and it depends for its efficiency on the correct proportioning of the weight of the loading slabs and the form and size of the rubber isolators.

The main features of the construction are shown in detail No. 36 A.

4. *Diespeker's "Bigspan" floor.* A transverse and a longitudinal section of this floor are given at F in detail No. 36 B. It is a reinforced floor of the hollow block type. The blocks vary from 10″ to 12″ square on plan and may be obtained from 3″ to 6″ deep. The top side of the block may be either raised slightly at the centre, or flat. The flat blocks may be used to present either the smaller dimension or the larger one in the depth of the floor as shown in the sectional details. Hence the overall depth of a floor slab, with 1½″ of concrete on top, may vary from 4½″ to 11½″ and, with deeper concrete cover (up to 4″) may be 14″ in overall depth.

The blocks are laid in line, end to end, with 3″ to 4″ spaces between. These spaces are filled with concrete, which is reinforced near the base, thus forming a series of ribs like those of a reinforced concrete tee beam.

The placing and bending of the reinforcement is done according to the established principle for continuous floor beams in reinforced concrete, and the hooking and bending is shown in the smaller sectional detail.

These floors are commonly designed of varying depth for clear spans up to 24 ft. and for superimposed loads up to 2 cwts. per sq. ft.

The method of housing conduits is shown in section, and it should be noted that the blocks are keyed all round their faces for the better adhesion and shear resistance of the concrete and for the secure keying of plaster ceilings.

The dead weight of these floors varies from 36 lbs. per sq. ft. at 4½″ thick to 110 lbs. at 14″ thick, to which must be added the extra weight of flooring material above the level concrete and the plaster ceiling, where used.

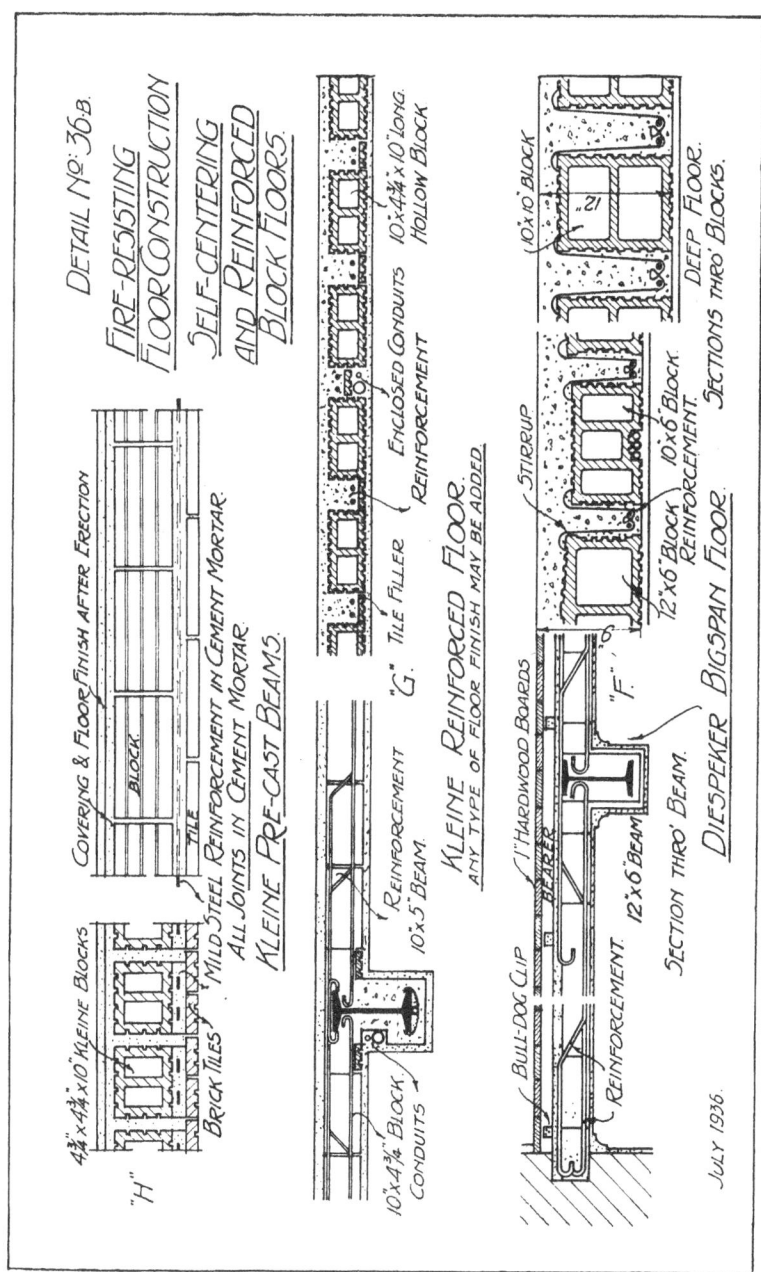

DETAIL No. 36.B.

FIRE-RESISTING FLOOR CONSTRUCTION

SELF-CENTERING AND REINFORCED BLOCK FLOORS.

COVERING & FLOOR FINISH AFTER ERECTION

BLOCK.

TILE

"H"

4¾" × 4¾" × 10" KLEINE BLOCKS

BRICK TILES

MILD STEEL REINFORCEMENT IN CEMENT MORTAR. ALL JOINTS IN CEMENT MORTAR.

KLEINE PRE-CAST BEAMS.

ENCLOSED CONDUITS 10" × 4¾" × 10" LONG. HOLLOW BLOCK

TILE FILLER

REINFORCEMENT

"G"

KLEINE REINFORCED FLOOR.
ANY TYPE OF FLOOR FINISH MAY BE ADDED.

REINFORCEMENT 10" × 5" BEAM.

10" × 4¾" BLOCK CONDUITS

10" × 10" BLOCK

"21"

DEEP FLOOR. SECTIONS THRO' BLOCKS.

STIRRUP

12" × 6" BLOCK 10" × 6" BLOCK REINFORCEMENT.

DIESPEKER BIG-SPAN FLOOR.

1" HARDWOOD BOARDS

BEARER

"F"

SECTION THRO' BEAM.

12" × 6" BEAM

REINFORCEMENT 12" × 6" BEAM

BULL-DOG CLIP

JULY 1936.

Where double (or two-way) reinforcement is required for square bays, the blocks are provided with stopped ends so that they may be laid with clear spaces between the lines in both directions.

5. *Kleine floors.* This floor was the earliest of the rectangular hollow block floors, and the many similar types of reinforced hollow block floor are variations of the Kleine system.

Originally this system incorporated a thin flat steel bar placed horizontally near the base of each joint and running throughout the length of the joint to reinforce the underside of the floor slab. In modern applications—as shown in detail No. 36 A at G—it is more usual to employ round bar reinforcement as in many other similar floors, and to make special provision for the housing of conduits.

The Kleine block is shown in detail No. 37. It is supplied with stopped ends for double (or two-way) reinforcement.

The reinforcing bars are provided with loose spacing washers which slide along the bar and can be placed to ensure clear spaces between the units and so allow the concrete filling to pass round the bars.

The blocks are usually $10'' \times 10''$ in plan and from $4\frac{3}{4}''$ to $8''$ thick. For thicker floors as required for large spans the blocks may be used on edge, either wholly or alternately.

Where pre-cast beams are required in order to avoid centering the Kleine Company provide unit beams which are constructed from their usual hollow blocks and soffit tiles.

Between the tiles and blocks for each unit is a reinforced layer of cement mortar, containing two lines of flat steel bar. The tiles and blocks are bonded in length as shown in detail No. 36 B at H—side elevation. Note that for each unit in section, the soffit tiles overlap the blocks on one side so as to seal the soffit and provide the space for the side jointing.

These floor beams are normally suitable for spans up to 12 ft. but can be specially designed and reinforced for larger spans if desired.

101. Comparison of floor beams and lintols. In detail No. 37 a number of unit floor beams, hollow tiles and lintols are assembled for comparison.

These include Siegwart and Armoured Tubular units, as representing pre-cast concrete beams; Fram, Kleine and Laing's hollow tile blocks for building up reinforced slab floors; and Homan and Rodgers', Dawnay's, King's and Fawcett's special fireclay lintols for spanning between filler joists.

The following have not been previously described:

Kleine blocks; these are similar to the Fram block, except that they can be obtained with solid ends for the application of double reinforcement.

Laing's Ferro-Brick fireproof floor blocks are made to allow of a series of blocks being assembled to form a beam before erection. The blocks are brought together end to end with cement joints and a reinforcing bar is placed in the groove on each side and secured by fine concrete or cement mortar. When set the beams can be lifted into position like pre-cast beams and centering is not required.

The Dawnay lintol is roughly approximate in sectional form to a cast iron beam. It is intended as a permanent center and concrete can be deposited upon it to any required depth. These lintols may be reinforced by passing a steel rod through a hole in the lower flange.

Fawcett's lintol is a form in which the units are prepared to stand between the joists at an inclination, such that the short diagonal is at right angles to the joists. This facilitates placing in position and also distributes the load over more floor area. The section is a flat soffit with an arched cover.

DETAIL No: 37. THE WAREHOUSE.
TYPES OF CONCRETE, TERRACOTTA and FIRECLAY UNITS FOR FIRE RESISTING FLOORS.

The King No. 1 *floor lintol* is a rectangular hollow block with a flat segmental joggle on its vertical edges. It is made 6″ wide, 3″ to 12″ deep and in lengths up to 4 ft. with the ends notched to cover the soffit of the supporting joists or to lay flush with the flanges if desired.

102. Points of importance in comparing fire-resisting floors. The necessity for clothing steelwork with a protective material has already been explained. This should be done not only as a protection against fire, but also as a protection against oxidation. Thus, all floors in which hollow blocks of concrete or terra-cotta are employed are open to oxidation of the steel supports unless the steel is solidly encased by filling the spaces between the ends of lintols and joists with concrete or grout. With many of the hollow lintols this is an impossibility, especially on the overlapping soffit where the tile must fit loosely and leave an air space.

Some patentees of hollow tile floors claim that the whole floor can be ventilated to advantage by providing air inlets through the external walls and thus provide a strong cooling current of air within the floor in case of fire. This proposal is a mistake. The air conveyed in normal conditions through the floor will often be humid and thus it assists in the oxidation of steelwork which cannot be inspected. Further, any defects in the closing of the ceiling, filling or skirting would provide assistance to combustion if the hollow spaces communicated directly with the open air.

The most perfect floors from the standpoint of fire protection and steel preservation are those of solid construction such as reinforced concrete slabs, Dawnay's solid lintol floor, etc., but in some cases they have the great disadvantage of weight; hence the introduction of the lighter hollow forms which also embody the principle of self-centering. They are also more sound-resisting than solid floors of concrete which are often troublesome in passing sound freely from floor to floor.

Where hollow lintol floors are employed they should not be ventilated but every care should be taken to clothe as much steel as possible with concrete, in the form of haunches and filling, and also to provide a hard plaster ceiling of close composition to prevent the circulation of air, as far as possible, through the hollow floor space where it would come in contact with the steel.

Reinforced unit tiles and pre-cast reinforced beams have a great advantage in exposing neither joists nor subsidiary steel beams to oxidation. The main beams are easily clothed by a haunch of concrete and the ends can be kept sufficiently clear to allow the upper parts of the girders to be packed solid; there is no objection to a little of the filling concrete entering the open ends of the units during this operation.

CHAPTER FOUR

SUBSIDIARY METAL CONSTRUCTION

METAL GRILLES

103. Metal grilles to windows and fanlights. The term "grille" in architectural work is applied to any open-work guard placed for the protection of glazing, open vents, lift openings, etc.

Such fittings are capable of good architectural treatment but they should not suffer from over-elaboration and must not unduly obstruct light.

A few simple applications are made in the warehouse and house details.

104. Iron grille to basement windows of warehouse. This is illustrated in detail No. 38 and consists of wrought iron rails $2\frac{1}{4}'' \times \frac{1}{2}''$, vertical rods $\frac{7}{8}''$ diameter and circular ring fillings made from $\frac{7}{8}'' \times \frac{5}{16}''$ metal between the upper and lower pairs of rails. The frame is prepared by housing the rods through the inner rails, pinning and riveting the upper and lower ends to the outer rails, and securing the rings to the horizontal bars by rivets.

To fix the frame, metal lugs are housed to the stone jambs and run with lead and the rail ends bolted to them.

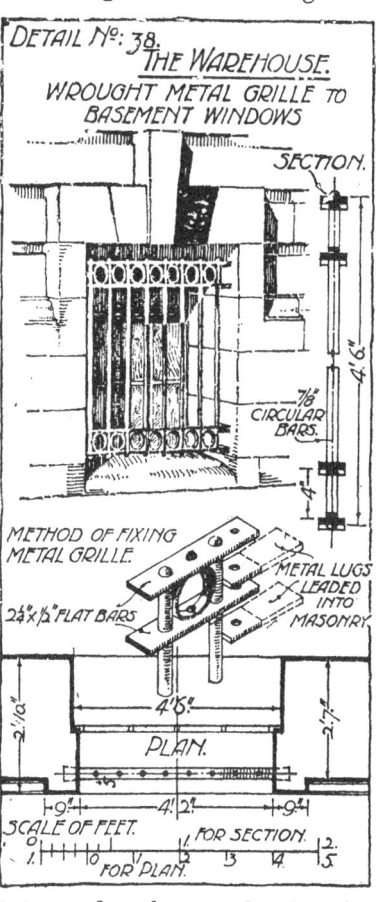

DETAIL Nº: 38.
THE WAREHOUSE.
WROUGHT METAL GRILLE TO BASEMENT WINDOWS

SECTION.

7/8" CIRCULAR BARS.

METHOD OF FIXING METAL GRILLE.

METAL LUGS LEADED INTO MASONRY

$2\frac{1}{4}'' \times \frac{1}{2}''$ FLAT BARS

4'6"

PLAN.

SCALE OF FEET.
FOR SECTION.
FOR PLAN.

The object of this grille is not to render the opening burglar proof but only to prevent the illegitimate use of the easily accessible window recess.

105. Bronze grilles to fanlights. Bronze grilles are suitably applied to fanlights over external doors and to good vestibule screens. They may be either built up from cast bars when the design is in simple lines or may be cast as an entire frame from a hand prepared pattern.

An application to the front door fanlight of the warehouse is shown in the perspective detail No. 13. The rims and bars of the grille are $\frac{1}{4}''$ thick and the margin $1''$ wide to allow for screwing to the frame. Fixing is done at the sides and top only, the bottom rim being kept clear of the transome so that rain water may drain off freely.

The grilles to the swing doors of the vestibule may be cast solid, or made from separate bars and brazed together; details Nos. 169 and 170. A similar grille of wrought iron is applied to the fanlight of the main entrance to the house and harmonises with the wrought iron railings on the entrance steps. See detail No. 14.

106. Treatment of fanlights. The design of fanlights may be such as to add considerable distinction to an entrance doorway.

Old and pleasing examples are still numerous in all parts of the country. Many are of cast lead, others of cast iron, while many are of wood cut into such ornamental forms as the one suggested on the title page of this volume.

The designs vary considerably with the period in which they were used, those of the Georgian and Queen Anne periods differing greatly from those designed by the brothers Adams.

Students should take every opportunity of noting, sketching and measuring any such examples, as an aid to their appreciation of, and development in design.

METAL CASEMENTS

107. Steel casements to ground floor front windows of warehouse. Detail No. 39 shows the application of steel casements to the front windows of the warehouse on the ground floor. These openings have segmental heads to which the casements are fitted; the general arrangement of the latter, with their opening parts, is shown in the elevation.

Each opening is divided into two parts by a cast iron or gun-metal mullion $2\frac{1}{4}''$ broad, having a moulded capital, necking and plinth, and a horizontal division, forming two heights, is provided by a double rail to which a projecting weather bar is screwed and which throws a strong shadow; it thus serves to define the transome as in an ordinary window. The primary function of the weathering piece is to discharge rain water clear of the head of the opening casement,

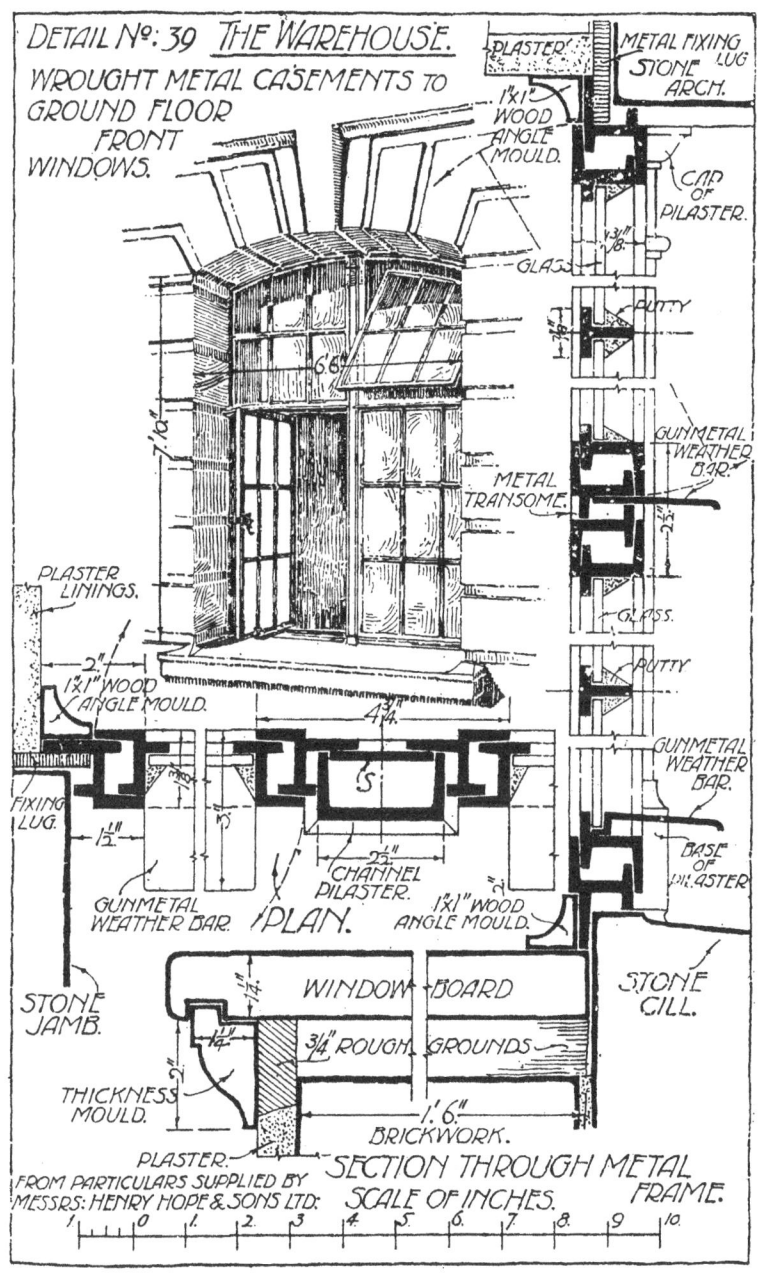

DETAIL Nº: 39 THE WAREHOUSE.
WROUGHT METAL CASEMENTS TO
GROUND FLOOR
        FRONT
WINDOWS.

PLASTER.
METAL FIXING LUG
STONE ARCH.
1"x1" WOOD ANGLE MOULD.
CAP OF PILASTER.
GLASS.
PUTTY.
GUNMETAL WEATHER BAR.
METAL TRANSOME.
GLASS.
PUTTY.
GUNMETAL WEATHER BAR.
BASE OF PILASTER.

PLASTER LININGS.
2"
1"x1" WOOD ANGLE MOULD.
FIXING LUG.
1½"
GUNMETAL WEATHER BAR.
CHANNEL PILASTER.
1"x1" WOOD ANGLE MOULD.
PLAN.
STONE CILL.

STONE JAMB.
THICKNESS MOULD.
WINDOW-BOARD
¾" ROUGH GROUNDS.
1'.6"
BRICKWORK.

PLASTER.
SECTION THROUGH METAL FRAME.
FROM PARTICULARS SUPPLIED BY
MESSRS: HENRY HOPE & SONS LTD:
SCALE OF INCHES.
1 0 1 2 3 4 5 6 7 8 9 10.

immediately below. There is a similar weathering piece to perform the same function at the cill.

Two opening parts are provided, one in the upper half which is horizontal pivot hung, and the other on the opposite side of the lower half, vertical pivot hung, but with the pivots set at about one-third the width of the light from the jamb. This latter arrangement enables the outside face of the glass to be cleaned from within when the casement is fully opened.

108. Sections of casement frames to ground floor. The general form of the framing may be gathered from the plan and section of the casement. The frame has two flanges and a web, one flange forming a glazing rebate and the other an overlap to the sash for weather resistance; the position of the overlap depends upon the purpose of the particular bar, being on the outside for covering an inward opening part and on the inside for an outward opening.

The upper sash is centre-pivoted, the lower portion moving outwards. This arrangement requires a change of position in the sections of the frame and sash above and below the pivot. A similar arrangement applies to the vertically pivoted casement which requires a change of section in the corresponding portions of the frame and sash.

It will be observed that the cill weathering piece is attached to the casement sash, being bedded into and screwed to the bottom rail and that the glass is retained by a rebate in its back edge; this arrangement provides for the weathering piece to move with the sash and the length, therefore, cannot exceed that of the space between the rebates of the stiles and requires the angles in plan to be rounded off sufficiently to pass the fixed frame.

109. Fixing of ground floor casements to front windows. These casements are virtually in two portions and may be so prepared, if desired, to allow of fixing after the stonework of the opening is complete. In such a case the two halves would be placed in position by arranging the outer stiles to clasp built-in lugs at the jambs and folding the opposite edges inwards against the centre mullion. The frames would be set-screwed to the mullion through their rebates and also to the lugs in the jambs through their back flanges.

The mullion should be dowelled to the cill and housed into the head and the whole frame bedded in oil putty and neatly cut off and smoothed at the exposed edges. If the opening is more than 5 ft. wide, or if the casement is made in two parts, plug and screw fixings are desirable at intermediate points in the cill and head.

As the mullions are required to present a fair face inside the building and it is not convenient to cast them of hollow rectangular section, they are made of channel section with stiffener ribs at intervals and either faced up with timber or by thin metal plates set-screwed to the stiffeners as indicated at S in plan.

110. Finishings to ground floor windows. The finishings to metal casements vary with the nature of the frames, but are usually severely plain. In this example the jambs are plastered and the joint between casement and plaster covered by a small cavetto mould fitted round the margin of the frame. If architraves are desired they may be of hard plaster or wood, and a wood window board and thickness mould used to complete the internal finish.

In lieu of the window board, terra-cotta cill tiles may be employed of any special section. They are usually $9'' \times 9'' \times 2\frac{1}{4}''$ to $3\frac{1}{8}''$ thick, with special left and right hand seating blocks for the reception of the jambs. The blocks possess many advantages, where cleanliness and cost of maintenance have to be considered.

Metal Panels and Window Frames to Main Façade of Warehouse

111. In the Chapter on Masonry the panelled treatment of the space between the two-storey pilasters of the main façade was referred to and also the modern alternative of employing metal frames for this purpose.

Much work of this character has been very successfully applied to commercial buildings, in some instances with great elaboration of detail and ornament. By using cast iron of sufficient thickness and paying due attention to its preservation both before and after fixing, a very durable panelled filling can be obtained with a maximum of window space.

The cast iron facing between the casements may be either backed with brickwork or concrete, or merely finished by clothing with wood panelling, but the former methods are much preferable because of their fire-resistance and prevention of loss of internal heat by conduction.

112. Metal facings and casements to front of warehouse. Detail No. 40 shows a simple architectural application of metal casements and facings to the two upper storeys in the main façade, as supplied by, amongst other makers, the Crittal Manufacturing Co., Ltd., of Braintree.

The space to be filled is $22' 0'' \times 7' 5''$ wide, and the window frames and panelling are set back so that a $9''$ brick backing to the panels finishes flush on the inside of the wall.

The latter treatment might be varied according to the thickness of the walls so that useful internal recesses may be obtained at each window.

113. Section of panels, frieze and casement. A section through the upper casement shows the construction of the frieze, casement, cill, and intermediate panelling at the second floor level. (At the first floor level the treatment is similar with the addition of a base mould.)

In this example the intermediate cast iron work is in three pieces, viz. head of lower window, panels and cill. The panels are cast in one piece with $5\frac{3}{4}'' \times 4'' \times \frac{5}{8}''$ lugs; these are connected by a $\frac{3}{4}''$ round bar for building into the $9''$ brick backing, which is supported across the opening by a reinforced concrete lintol.

The head is a shaped projection with a rebate to take the casement frame on the under side and a flange which passes behind the panelling, being previously fixed thereto and supported by set screws from behind.

The cill is supported upon the panelling by a similar overlap and the joint is guarded by the drip and throating on the soffit.

114. Junction of cast iron front with jambs. In many cases all these members are fitted between the jambs of the stonework and depend for their support entirely upon the backing. In such cases it is difficult to make the vertical joints weathertight, and it is preferable to allow all the members to obtain a bearing upon the jambs at each end, which necessitates building in, but ensures that no water will get behind the metal work. If it is thought undesirable to allow all the metal front to be so treated, the cill might desirably enter the jambs to the depth of the shallow rebate for the casements, usually $\frac{3}{4}''$ to $1''$; its junction can then be bedded and pointed to render it watertight.

The cast iron frieze and capping mould is of uniform thickness, rebated at the bottom edge to receive the casement and is itself notched into the reinforced lintol to provide bearing against the grip of the fixings. To secure the casting when not built in at the ends, tapped lewis plugs of cast iron are housed and leaded into the brickwork over the opening and the frieze casting fixed by strong set screws passing through tubular sleeves upon which they grip; the length of the sleeve allows the bottom edge to close tight against the lintol before bearing against the back of the casting.

115. Casements between cast iron frieze and cill. The casements illustrated in the detail are of a similar general type to those previously described in Vol. II.

Like the "lok'd bar" pattern, the glazing is done from within and the horizontal sash bars are continuous at the flanges, to receive which the flanges of the vertical bars are cut and expanded.

At the cast iron cill the casement is made weathertight by the channel shaped rail, forking over a projection upon the cill where it

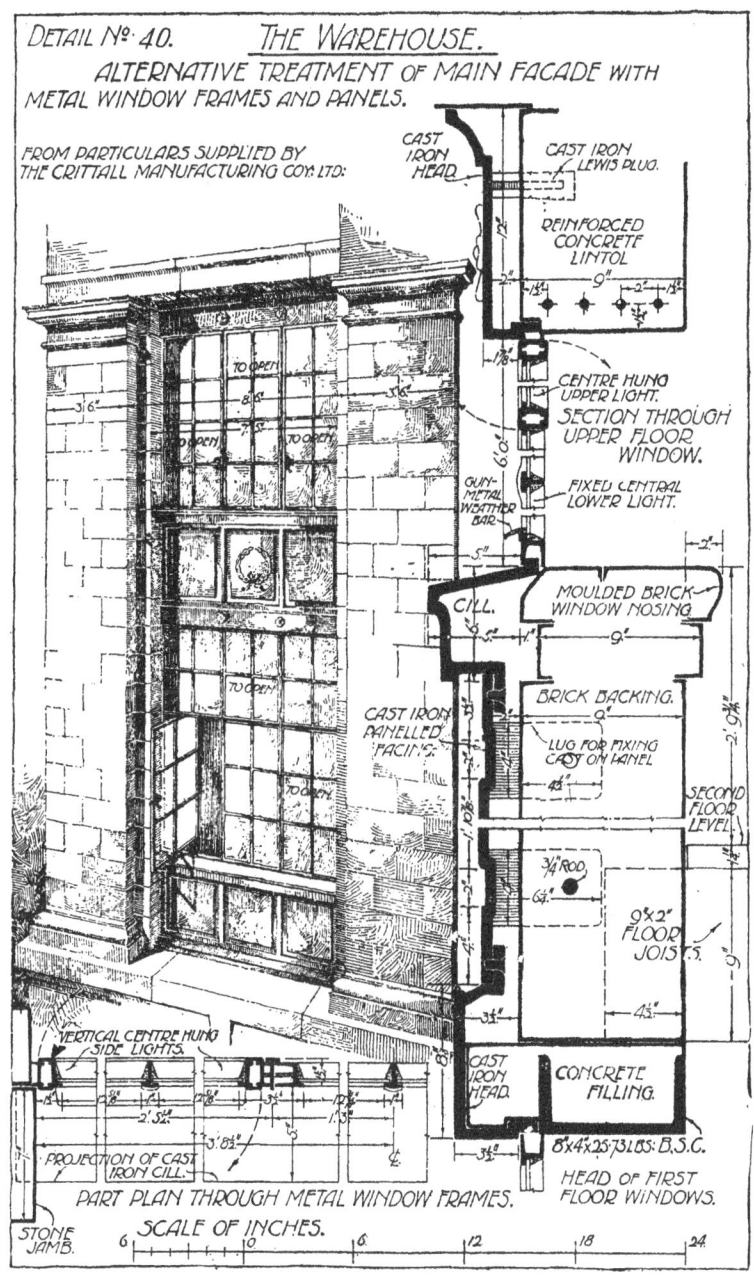

DETAIL Nº 40. THE WAREHOUSE.

ALTERNATIVE TREATMENT OF MAIN FACADE WITH METAL WINDOW FRAMES AND PANELS.

FROM PARTICULARS SUPPLIED BY THE CRITTALL MANUFACTURING COY. LTD.

CAST IRON HEAD.

CAST IRON LEWIS PLUG.

REINFORCED CONCRETE LINTOL

CENTRE HUNG UPPER LIGHT.

SECTION THROUGH UPPER FLOOR WINDOW.

FIXED CENTRAL LOWER LIGHT.

GUN-METAL WEATHER BAR.

CILL.

MOULDED BRICK WINDOW NOSING.

BRICK BACKING.

LUG FOR FIXING CAST ON PANEL

CAST IRON PANELLED FACINGS.

SECOND FLOOR LEVEL.

¾" ROD.

9" × 2" FLOOR JOISTS.

TO OPEN

VERTICAL CENTRE HUNG SIDE LIGHTS.

CAST IRON HEAD.

CONCRETE FILLING.

PROJECTION OF CAST IRON CILL.

PART PLAN THROUGH METAL WINDOW FRAMES.

8"×4"×25·731 BS: B.S.C.

HEAD OF FIRST FLOOR WINDOWS.

STONE JAMB.

SCALE OF INCHES.

6   10   6   12   18   24

should be bedded in oil putty and guarded by a curved gun-metal weather bar.

The casement is inserted after the other iron work has been completed to receive it, is bedded all round the rebate in oil putty or cement and fixed by screwing through the stiles to lead or cast iron plugs in the jambs.

The interior is finished in plain brickwork with flush joints and a two-course moulded cill serves instead of the more usual window board.

116. Sizes of casement frames. Throughout the whole of the illustrations of metal casements it should be observed that the sectional dimensions of the bars and frames are comparatively small. The average thickness over the opening parts is $1\frac{1}{2}''$ and the dimensions of any bar about $1''$ to $1\frac{1}{4}'' \times \frac{7}{8}''$.

The resulting appearance, even where opening parts are included, is that of a light sash, which ensures the admission of a maximum amount of daylight. Compared with wood sashes and frames much less light is excluded for the same size of opening and, in addition, really good metal sashes properly protected and maintained by painting are much more durable than soft-wood sashes.

The outlines of metal sashes are of necessity distinctly severe, but there is no real objection to this, because the architectural value depends upon the dimensions of the lights and the proportions of these dimensions as subdivisions of the whole frame, rather than upon elaboration of outline.

CHAPTER FIVE

STRUCTURAL DESIGN

INTRODUCTION. WALLS AND FOUNDATIONS

117. Steel framing and reinforced concrete play an important part in large modern structures though their use may not be evident on the façades of buildings where special architectural treatment has been made a feature and accomplished by facing with stone, brick or terra-cotta, which allow of the adoption of conventional types of design.

As the skeleton framing is intentionally the primary load supporting portion of these structures special attention must be paid to the principles of structural design.

While many local authorities require strict adherence to building regulations governing the erection of these special structures, a knowledge of structural mechanics is essential in order to compute the sizes of members in accordance therewith; the following chapters deal with the usual principles and problems of design in the preparation of foundations, piers, pillars of steel and cast iron and timber construction, which arise in connection with the warehouse and semi-detached house selected for study.

The application of these principles in the design of domestic buildings is not often practised, but the student will benefit by the process, because proof is afforded that the dimensions and forms determined by long practice and experience are usually in agreement with modern theory, and confidence in the use of established rules may thus be gained.

DIRECT STRESSES ON WALLS AND FOUNDATIONS

118. When any part of a structure is loaded by the application of an external force, bodily movement is only prevented by such part supplying an equal and opposite force to resist the load. This force is termed a reaction.

If a loaded member is in equilibrium under a direct thrust or pull, the material of which it is composed must be compressed or extended under the force, but for a safe condition must be capable of regaining its original form when the load is removed. The change of dimensions under load is called strain and the internal force in the material caused by the load and producing strain is called stress.

For this stress to be uniformly distributed over the cross section of a member or part of a structure, the line of action of the load and reaction must pass through the centre of area of that cross section.

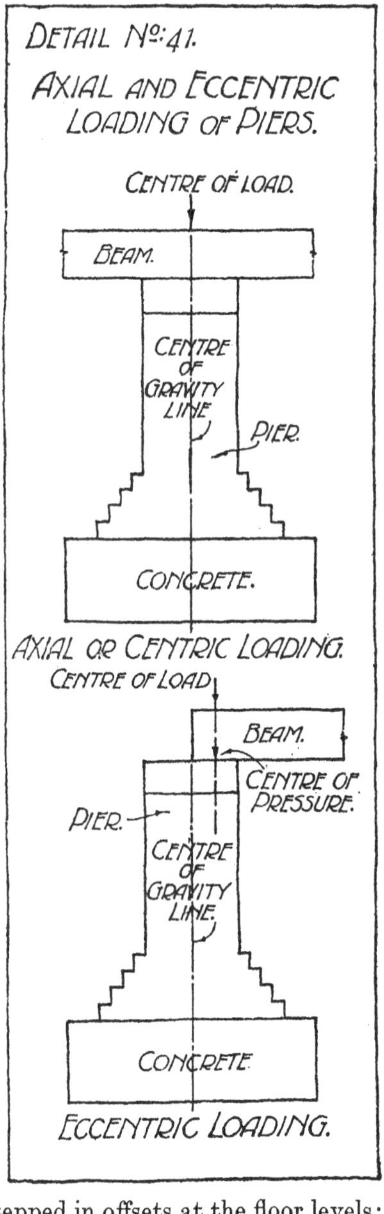

In a symmetrical cross section so loaded, it may be assumed that at a short distance from the point of application of the load the stress becomes uniform, viz. of equal intensity per unit of area over the entire section.

An original load may be applied either in one line (at a point) or spread more or less uniformly over an area, and the reaction may be supplied under similar conditions, but if the "resultant" line of the load or reaction passes through the centre of area of a section, the stress will be uniform and the member, if straight, is said to be *axially* loaded.

"Axial" or "centric" loading produces uniform stress while "non-axial" or "eccentric" loading produces non-uniform stress.

The two conditions are shown in detail No. 41.

119. Walls and pillars. While many walls and pillars are not axially loaded because the floor girders and roof trusses which they support have their bearing surfaces near one face, there are many conditions tending to correct the eccentricity thus caused.

Outside walls are commonly stepped in offsets at the floor levels;

DETAIL Nº:41.

AXIAL AND ECCENTRIC LOADING OF PIERS.

CENTRE OF LOAD.

BEAM.

CENTRE OF GRAVITY LINE

PIER.

CONCRETE.

AXIAL OR CENTRIC LOADING.

CENTRE OF LOAD

BEAM.

CENTRE OF PRESSURE.

PIER.

CENTRE OF GRAVITY LINE.

CONCRETE

ECCENTRIC LOADING.

this gives weight in a position to counteract the floor and roof loads.

The rigidity of a structure consisting of walls, pillars and stiff floor framing also assists in reducing eccentric effects. Stresses due to eccentric loading must, however, be fully taken into account in many cases, especially where deliberate side loading is adopted by the use of brackets or corbels; while out of the range of this volume the stresses arising from non-axial loading must form an important part of the study of architects and constructional engineers.

In the case of pillars supporting free beams symmetrically arranged and also of party walls between similar properties, it is quite legitimate to assume axial loading for the maximum estimated load on the structure, because the condition assumes full loading over the entire floor panels supported by the pillar or wall.

The following consideration of stresses in walls and foundations assumes axial loads and uniform distribution of stress, under the maximum conditions of loading.

120. Stress on pier and its foundations when axially loaded. The following example refers to a $13\frac{1}{2}''$ square pier supporting the eaves beam of a long, lean-to shed, adjoining the workshop which was illustrated and studied in Vol. I.

A reference to Vol. I will show that the pier supports a large area of the roof, viz. "half the span of the beam on each side of the pier" (the total of which is $10'\ 10\frac{1}{2}''$), multiplied by "half the width of the sloping roof plus the overhang at the eaves", *i.e.* $6'\ 6''$. This area is therefore 70·68 sq. ft.

The load of slating, rafters, battens, eaves beam and gutter may be shown to be about 16 lbs. per sq. ft., hence the load supported by the pier $= 70·68 \times 16 = $ say, **1130** lbs. which we shall assume to be an axial load.

Let the pier be 8 ft. high from top of footings to the template, and $1'\ 1\frac{1}{2}''$ square on plan, and let the material have a density of 112 lbs. per cu. ft.; its

$$\text{total weight} = \text{volume of pier} \times \text{density per cu. ft.}$$
$$= (8 \times 1\frac{1}{8} \times 1\frac{1}{8}) \times 112 = \textbf{1134} \text{ lbs.}$$

It is assumed that the stone template has the same density as the brickwork.

It may be shown that the weight of square footings of standard proportions can be obtained from

$$W_F = ·08T^2\ (T + \tfrac{1}{4}),$$

where $\qquad W_F = \text{weight of footings in tons,}$

$\qquad\qquad T = \text{thickness of square pier in feet.}$

Substituting the value of $T = 1\frac{1}{8}$ ft.

then $W_F = \cdot 08 \times 1\frac{1}{8}^2 (1\frac{1}{8} + \frac{1}{4})$,

$\qquad\qquad W_F = \cdot 139$ ton, or $\cdot 139 \times 2240 = \textbf{311}$ lbs.

It may also be shown that the weight of concrete for the standard foundation to a pier can be obtained from

$\qquad\qquad W_C = \cdot 2T^2 (T + 1)$,

where $W_C =$ weight of concrete in tons,

$\qquad\qquad T =$ thickness of square pier in feet (as before),

$\therefore\; W_C = \cdot 2 \times 1\frac{1}{8}^2 (1\frac{1}{8} + 1)$,

$\qquad W_C = \cdot 538$ ton, or $\cdot 538 \times 2240 =$ say $\textbf{1204}$ lbs.

The total load on the earth is therefore

$$\begin{aligned}
\text{Weight of roofing} &= 1130 \text{ lbs.} \\
\text{Weight of pier} &= 1134 \text{ ,,} \\
\text{Weight of footings} &= 311 \text{ ,,} \\
\text{Weight of concrete} &= 1204 \text{ ,,} \\
\hline
\text{Total weight} &= \textbf{3779 lbs.}
\end{aligned}$$

Consider the pressure per sq. ft. on various parts of the structure represented in detail No. 42. For this purpose we must know the areas of pier, footings and concrete:

$\qquad$ Area of pier $= T^2 = 1\frac{1}{8}$ (ft.)$^2 = 1 \cdot 265$ sq. ft.

Area of footings at base $= (2T)^2 = 2\frac{1}{4}$ (ft.)$^2 = 5 \cdot 062$ sq. ft.

The concrete is $(2T + 1)$ ft. square and its area is therefore

$$[(2 \times 1\frac{1}{8}) + 1]^2 = 3\frac{1}{4}^2 = 10 \cdot 56 \text{ sq. ft.}$$

The pressure per sq. ft. at any horizontal section

$$= \frac{\text{weight above the section}}{\text{area of section}}.$$

Then pressure on top courses of pier at A (*under the template*)

$$= \frac{\text{roof load}}{\text{area of pier}} = \frac{1130}{1 \cdot 265} = \textbf{893} \text{ lbs. per sq. ft.}$$

Pressure on base of pier and top course of footings, at B

$$= \frac{\text{roof load} + \text{weight of pier}}{\text{area of pier}} = \frac{1130 + 1134}{1 \cdot 265} = \frac{2264}{1 \cdot 265}$$

$= \textbf{1789}$ lbs. per sq. ft.

Pressure on base of footings and top of concrete, at C

$$= \frac{\text{roof load} + \text{weight of pier and footings}}{\text{area of base of footings}}$$

$$= \frac{1130 + 1134 + 311}{5 \cdot 062} = \frac{2575}{5 \cdot 062} = \textbf{508} \text{ lbs. per sq. ft.}$$

Note. The results of calculations are mainly approximations, being slide rule results.

Pressure on base of concrete and upon the earth foundation, at D

$$= \frac{\text{roof load} + \text{weight of pier, footings and concrete}}{\text{area of base of concrete}}$$

$$= \frac{1130 + 1134 + 311 + 1204}{10 \cdot 56} = \frac{3779}{10 \cdot 56} = \textbf{358} \text{ lbs. per sq. ft.}$$

The above type of calculation is commonly embodied in and expressed by the simple formula

$$p = \frac{W}{A},$$

where p = pressure per unit of area,
W = total load above the section at which p is required,
A = area of section.

Generally p and W are taken in tons and A in sq. ft., hence p would be the pressure in "tons per sq. ft."

121. Deductions from results. From these results we should note the following:

(*a*) The load per unit of area calls for an equal resistance from the material and the state of active resistance to that load is called "stress".

(*b*) In the example the stress on the pier is smallest at the top and increases at every section downwards towards the base, due to its own increasing weight; at the base it reaches a maximum.

(*c*) The stress on the brickwork at the base of the pier is reduced upon the concrete by the spread of the footings which increases the supporting area.

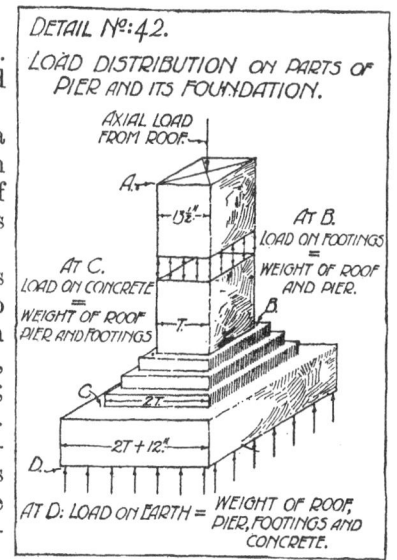

DETAIL Nº:42.

LOAD DISTRIBUTION ON PARTS OF PIER AND ITS FOUNDATION.

AXIAL LOAD FROM ROOF

AT B.
LOAD ON FOOTINGS
=
WEIGHT OF ROOF AND PIER.

AT C.
LOAD ON CONCRETE
=
WEIGHT OF ROOF PIER AND FOOTINGS

AT D: LOAD ON EARTH = WEIGHT OF ROOF, PIER, FOOTINGS AND CONCRETE.

(*d*) The stress on the earth has been further reduced by the extended area of the concrete.

The above results should conduce to a clearer understanding of the purpose of a foundation and it should now be seen that the total load upon the earth must be reduced to a unit pressure—or stress—which will ensure the safety of the structure erected upon it. At the same time the stress upon any intermediate section of the structure must not exceed what has been proved by experience or experiment to be the safe limit.

Every foundation of natural earth and every part of a structure settles or changes form under these stresses, and in practice it is desired to ensure that no undue or irregular settlement will occur, which might develop considerable changes of form.

122. Allowable unit pressure on walls and foundations. The London County Council Code of Practice, formulated under the London Building Act, 1930, allows the following maximum pressures on *natural foundations*:

$\frac{1}{2}$ ton per sq. ft. on alluvial soils, made ground, and very wet sand.

1 ton per sq. ft. on soft clay, wet sand, or loose sand.

2 tons per sq. ft. on ordinary fairly dry clay, fairly dry fine sand, and sandy clay.

3 tons per sq. ft. on firm dry clay.

4 tons per sq. ft. on compact sand or gravel, London blue clay, or similar hard compact clay.

6 tons per sq. ft. on hard solid chalk.

In accomplishing the reduction of pressure to the above unit amounts through walls, piers and footings, the *stress on the brickwork* must not exceed:

On bricks of *first* strength set in		3 to 1 cement mortar		20 tons per sq. ft.
,, ,,	*second* ,,	,,	3 to 1 ,, ,,	15	,,	,,
,, ,,	*third* ,,	,,	4 to 1 ,, ,,	10	,,	,,
,, ,,	*fourth* ,,	,,	4 to 1 ,, ,,	8	,,	,,
,, ,,	,, ,,	,,	2 to 1 lime mortar	4	,,	,,

The above pressures may be exceeded by 20 per cent. where such increased pressure is only of a local nature, as at girder bearings.

Bricks are classified according to their crushing strength under test, thus:

To be classified as "*first strength*" the crushing stress shall be a minimum of   10,000 *lbs. per sq. in.*

To be classified as "*second strength*" the crushing stress shall be a minimum of   5,000 ,, ,,

To be classified as "*third strength*" the crushing stress shall be a minimum of   3,000 ,, ,,

To be classified as "*fourth strength*" the crushing stress shall be a minimum of   1,500 ,, ,,

There are conditions as to testing, and there are limitations of the ratio of height to thickness of isolated piers and walls, but these matters are outside the range of study for this volume.

123. Use of footings. Footings may serve to fulfil the following purposes:

(a) To spread the load over a sufficiently wide surface of concrete, as would be necessary if the concrete were weaker than the brickwork upon it.

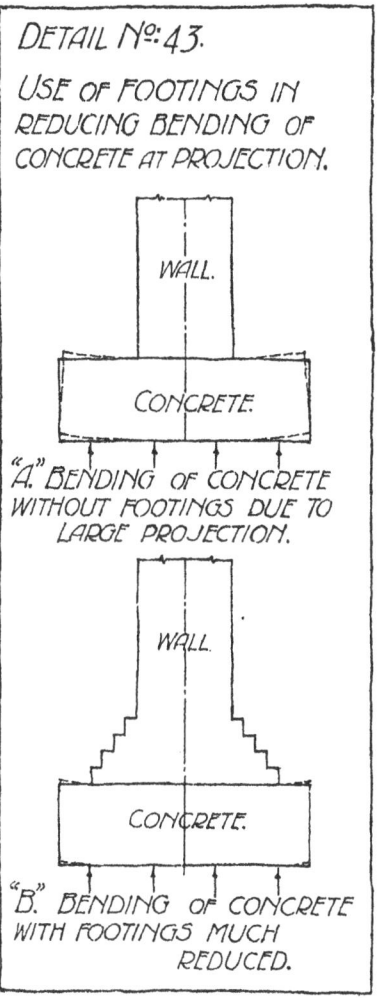

DETAIL Nº:43.

USE OF FOOTINGS IN REDUCING BENDING OF CONCRETE AT PROJECTION.

WALL.

CONCRETE.

"A." BENDING OF CONCRETE WITHOUT FOOTINGS DUE TO LARGE PROJECTION.

WALL.

CONCRETE.

"B." BENDING OF CONCRETE WITH FOOTINGS MUCH REDUCED.

This occurs with lime concretes of the poorer class, but with portland cement concrete it is unnecessary to spread the load because the concrete is of equal strength to the best brickwork.

(b) To transmit the load more uniformly over the foundation concrete by providing a gradual increase of base.

Hence, in foundations to ordinary two-storey buildings (houses and the like) footings can be entirely omitted when good portland cement concrete is used.

(c) To reduce the projection of the foundation concrete beyond that part of the structure which it directly supports. This projection would, in many cases, become excessive if the footings were omitted and would only be capable of resisting the bending tendency illustrated in detail No. 43 at A, if made of considerable thickness.

The insertion of footings reduces the bending tendency on the concrete by curtailing the projection, and also ensures a more uniform distribution of the load over the concrete, and consequently upon the earth.

124. Special foundations. It will now be understood that in tall buildings, especially those of the warehouse class, where considerable

loads have to be transmitted by the walls and pillars to the
foundations, some special form or treatment of the latter may become necessary; hence the designer of the building becomes
responsible for investigating the conditions and ensuring that the
probable maximum stresses are within the limits allowed by the
regulations in force in the district where the building is to be
erected. In matters of detail where no local regulation controls
the designer, reference to some established building regulations,
such as those of the London County Council, will be a safe guide to
successful work, though in some cases excessive caution will be
recognised by the expert.

**125. Brick piers and foundations to ground floor pillars of ware-
house.** An examination of the general drawings will show that the
ground floor pillars in the centre of the building are supported on
brick piers situated in the line of the back wall to the basement;
see also detail No. 2.

Each pier carries load from three floors above, the load being
collected from one full bay of each floor, as shown at C on the floor
plan of the general drawings.

The estimated load from each bay is 26 tons (see paragraph 249)
giving a total floor load of 78 tons. The brick pier is 11 ft. high and
will have a density (or unit weight) of at least 125 lbs. per cu. ft.;
so long as the size of the required pier is unknown, its weight is also
unknown. But as the weight will be considerable it cannot be
neglected, and if kept of constant size throughout its height the
weight may be expressed thus:

Let W_P = weight of pier.

   w = density of brickwork per cu. ft. in tons.

   h = height of pier in ft.

   A = sectional area of pier in sq. ft.

Then $W_P = w \cdot h \cdot A$ (tons).

If $w = \frac{125}{2240}$ ton, and $h = 11$ ft.

   $W_P = \frac{125}{2240} \times 11A = \cdot 614A$ (ton).

Total weight on pier at base (W_T) = weight from three floors (W)
+ weight of pier (W_P).

$$W = 78 \text{ tons.} \quad W_P = \cdot 614A \text{ (ton),}$$

$$\therefore \quad W_T = 78 + \cdot 614A.$$

But safe unit pressure, $p, = \dfrac{\text{total weight}}{\text{sectional area}} = \dfrac{W_T}{A}$.

Let p be assumed as 12 tons per sq. ft. (blue brick in cement).[1]

---

[1] The value of p will depend upon the regulation or code to be complied with,
hence the necessary figure must be inserted in the calculations.

Then
$$p = \frac{78 + \cdot 614A}{A},$$
$$12A = 78 + \cdot 614A,$$
$$A(12 - \cdot 614) = 78,$$
$$\therefore \ A = \frac{78}{11 \cdot 38} = \textbf{6·86} \text{ sq. ft.}$$

For a square pier $A = b^2$, where $b =$ breadth of pier in feet.
$$\therefore \ b^2 = 6 \cdot 86, \text{ and } b = \sqrt{6 \cdot 86} = \textbf{2·62} \text{ ft.}$$
The nearest brick dimension is **2′ 7½″**.

126. Total load on footings, concrete and earth. For a pier
11 ft. × 2′ 7½″ square, with a density of 125 lbs. per cu. ft. the weight
$$(W_P) = 11 \times 2 \cdot 625^2 \times \tfrac{125}{2240} = \textbf{4·22} \text{ tons.}$$
The weight of the footings (W_F), by formula in paragraph 120, is
$$W_F = \cdot 08T^2 \left(T + \tfrac{1}{4}\right)$$
$$= \cdot 08 \times 2 \cdot 625^2 (2 \cdot 625 + \cdot 25)$$
$$= \textbf{1·58} \text{ tons.}$$
Let $W_C =$ weight of concrete, assuming 6″ projection beyond the
footings, then by formula in paragraph 120,
$$W_C = \cdot 2T^2 (T + 1)$$
$$= \cdot 2 \times 2 \cdot 625^2 (2 \cdot 625 + 1)$$
$$= \textbf{4·98} \text{ tons.}$$
The total load on the earth foundation is therefore
$$W_E = W + W_P + W_F + W_C$$
$$= 78 + 4 \cdot 22 + 1 \cdot 58 + 4 \cdot 98$$
$$= \textbf{88·78} \text{ tons.}$$
The width of the concrete base
$$= 2T + 1 = 2 \times 2 \cdot 625 + 1$$
$$= 6 \cdot 25 \text{ ft.}$$
Therefore the area of the concrete
$$= 6 \cdot 25^2 = 39 \text{ sq. ft. (approx.).}$$
Hence, the unit pressure on the earth, if the standard projection
of the concrete beyond the footing is adopted, will be
$$p_E = \frac{W}{A} = \frac{88 \cdot 78}{39} = \textbf{2·27} \text{ tons per sq. ft.}$$
This exceeds the allowance for ordinary clay or confined sand by
about ¼ ton per sq. ft.

127. To decide the necessary size of concrete foundation. Let 2 tons be the safe limit of pressure, p_E, and A_C = area of base of concrete.

Then $\quad p_E = \dfrac{W_E}{A_C}, \qquad \therefore \ A_C = \dfrac{W_E}{p_E} = \dfrac{88 \cdot 78}{2} = 44 \cdot 39$ sq. ft.

Let b_C = breadth of concrete base.

Then $\qquad\qquad b_C{}^2 = A_c = 44 \cdot 39$,

and $\qquad\qquad b_C = \sqrt{44 \cdot 39} = 6 \cdot 66 = $ **6′ 8″** square,

which is the size required to limit the pressure to 2 tons per sq. ft.

128. Brickwork for supporting large loads. Comparison with steel. To carry a useful load of 78 tons, the brick pier and footings produce an additional load of 5·8 tons, which is about $7\frac{1}{2}$ per cent. of the useful load carried or 7 per cent. of the entire load on the concrete. In this case the best class of brickwork was employed, safely supporting 12 tons per sq. ft., hence with brickwork of a lower quality, say common stocks in cement supporting 5 tons per sq. ft., the useless dead weight would be largely increased because a much larger pier would be required.

For supporting ground floor and basement loads, piers may be usefully employed in many instances, but for upper floors they are too bulky to justify their use, taking up valuable internal space.

129. Relative economy of weight—brickwork and steel. In an earlier section of this volume the constructional details of iron and steel pillars are given, as employed for supporting large loads; these should be studied and their sizes noted for comparison.

Stock brickwork, density approximately 1 cwt. per cu. ft. may support 5 tons per sq. ft. upwards if built in cement. Comparing with steel, the latter has a density of $\frac{490}{112} = 4 \cdot 37$ cwts. per cu. ft., and at a very low average may support 4 tons per sq. in. or 576 tons per sq. ft.

The relative economy of weight is expressed by

$$\frac{\text{weight carried per sq. ft.}}{\text{density per cu. ft.}} \ .$$

For brickwork, this value is $\dfrac{5 \ \text{(tons)}}{1/20 \ \text{(ton)}} = 100.$

For steel, the value is $\dfrac{576 \ \text{(tons)}}{4 \cdot 37/20 \ \text{(tons)}} = 2636,$

and on this basis steel is $\frac{2636}{100}$ or 26·36 times as economical as common brickwork for reducing useless dead weight on the foundation. This is a rough comparison taking no account of pillar bases, caps, and connections, hence the economy is reduced below the figure given.

It does not consider the relative costs of the materials and labour of construction, but relates only to weight.

130. Foundations to party wall of warehouse. Now examine the party wall construction and estimate the probable stresses upon the basement wall, footings and foundation; particulars of the wall and foundation may be obtained from detail No. 3 and the general drawings.

Consider a portion of this party wall and assume that a similar property exists on the opposite side to the warehouse. The height of the wall is 52' 6", made up as follows:

(a) Basement wall: 10' 6" high × 2' 3" thick.
(b) Ground storey wall: 14' 0" high × 1' 10½" thick.
(c) First storey wall: 12' 0" high × 1' 6" thick.
(d) Second storey and parapet walls: 16' 0" high × 1' 1½" thick.

Assume the weight of the brickwork to be 1 cwt. per cu. ft.

Then the weight of 1 ft. of length of the party wall in tons would be:

$$\frac{1}{20} \times \text{volume of } (a+b+c+d)$$

$$= \frac{1}{20}\left(\frac{21}{2}\times\frac{9}{4} + 14\times\frac{15}{8} + 12\times\frac{3}{2} + 16\times\frac{9}{8}\right)$$

$$= \frac{1}{20}\left(\frac{189+210+144+144}{8}\right) = \frac{1}{20}\times\frac{687}{8} = 4\cdot29 \text{ tons (approx.).}$$

The floors of the building will be designed to carry 2 cwts. per sq. ft. of area in addition to their own weight, which is averaged for the present purpose at ½ cwt. per sq. ft. Floor girders are spaced along the party wall at varying distances according to the type of floor selected, but, however they are spaced, the portion of load transferred to the party wall from one bay 15 ft. long and 12' 6" wide measured at right angles to the wall, is one-half of the load on the bay and can be considered as uniformly spread over 15 ft. length of base if the brickwork is good and satisfactorily bonded. The other half is transferred to the steel stanchions within the building.

The load transmitted from one bay is therefore:

$$\left(\frac{15\times12\frac{1}{2}}{2}\right)\frac{2\frac{1}{2}}{20} = 11\cdot72 \text{ tons.}$$

The load will be more than this for certain special types of floor given in the Chapter on Floor Construction.

There are three such floors on *each* side of the party wall, hence the total load on 15 ft.

$$= 6\times11\cdot72 = 70\cdot32 \text{ tons,}$$

and the *load per ft. of length*

$$= \frac{70 \cdot 32}{15} = 4 \cdot 68 \text{ tons.}$$

The roofs may be shown to transmit a load of approximately ·45 ton per ft. run.

Footings, as shown in detail No. 2, would weigh at least 1 cwt. per ft. cu. Therefore

$$\text{weight} = \text{volume} \times \frac{1}{20} \text{ (tons)}$$

$$= \text{average breadth} \times \text{depth} \times \frac{1}{20}$$

$$= \frac{4\frac{1}{2} + 2\frac{5}{8}}{2} \times \frac{3}{2} \times \frac{1}{20}$$

$$= \cdot 267 \text{ (say } \cdot 27) \text{ ton.}$$

The concrete bed, $5\frac{1}{2}$ ft. wide $\times 22\frac{1}{2}''$ thick, at 140 lbs. cu. ft., would weigh:

$$(5\frac{1}{2} \times 1\frac{7}{8}) \frac{140}{2240} = \cdot 645 \text{ ton per ft. run.}$$

131. Stress on base of wall. The stress on the base

$$= \frac{\text{total load on base}}{\text{area of base}}.$$

Load = weight of wall, floors and roof per ft. run
$$= 4 \cdot 29 + 4 \cdot 68 + \cdot 45 = 9 \cdot 42 \text{ tons.}$$

Area of base = thickness of wall $\times$ 1 (because 1 ft. of length only has been considered) = 2·25 sq. ft.

Therefore stress on base of wall

$$= \frac{9 \cdot 42}{2 \cdot 25} = 4 \cdot 18 \text{ tons per sq. ft.,}$$

which is safe for ordinary brickwork in cement mortar.

132. Stress on base of footings. Stress on base of footings

$$= \frac{\text{total load on concrete}}{\text{area of footings}}.$$

Load = 9·42 + weight of footings
$$= 9 \cdot 42 + \cdot 27 = 9 \cdot 69 \text{ tons.}$$

Area of footings = breadth of footings $\times$ 1
$$= 4\frac{1}{2} \text{ sq. ft.}$$

Therefore stress on base of footings

$$=\frac{9 \cdot 69}{4 \cdot 5}=2 \cdot 15 \text{ tons per sq. ft.,}$$

which is much less than concrete can bear—see paragraph 596.

133. Stress on earth foundation.

$$\text{Stress on earth}=\frac{\text{total load on earth}}{\text{area of concrete}}.$$

$$\text{Load}=9 \cdot 69 + \text{weight of concrete}$$
$$=9 \cdot 69 + \cdot 645 = 10 \cdot 33 \text{ tons.}$$

$$\text{Area of concrete}=\text{breadth of concrete} \times 1$$
$$=5\tfrac{1}{2} \text{ sq. ft.}$$

Therefore stress on earth foundation

$$=\frac{10 \cdot 33}{5 \cdot 5}=1 \cdot 87 \text{ tons per sq. ft.,}$$

which is suited to ordinary clay or fairly dry sand—see paragraph 122.

If the foundation were of soft clay or loose sand capable of supporting only 1 ton per sq. ft. a concrete bed would be required, 10 ft. wide, to spread the load and reduce the unit pressure to 1 ton. Such a concrete foundation would require reinforcing with steel in order to resist the upthrust on the projecting foot, which would overhang the footings by 2' 9" on each side.[1]

Note. The above calculation assumes that the load is axially placed, viz. over the centre of the base, and that the wall is symmetrical about a centre line. These conditions would be practically met when the maximum allowable load exists on floors and roofs on both sides of the party wall.

[1] The nature of this construction may be gathered from detail No. 23.

CHAPTER SIX

STRUCTURAL DESIGN

Principles of Beam Design

134. Beam. In its simplest form a beam is a horizontal member in a structure, spanning an opening and resting upon separate supports, its function being to receive loads upon its length and transmit them to such supports.

For this purpose the beam needs to possess:

(a) *Strength*, which enables it to carry the load without damage to the material.

(b) *Stiffness*, which enables it to carry the load without undue change of form. Loads cause sagging or *deflection* between the supports, which, if excessive, may damage other parts of the structure. Serious vibration under moving loads is due to lack of stiffness.

Lintols, bressummers, floor joists and girders, purlins and pole plates, are all examples of beams.

Terms previously employed. In Vol. i, and the earlier chapters of this volume, many of the terms have been explained which are used in reference to beams, e.g. load, stress, compression, tension and shear, and it may therefore be assumed that these terms will be generally understood.

Strain. This term is often misused for stress, which is the internal force set up in a material in resistance to load. *Strain* is the *change of form* which always accompanies stress and is proportional thereto, provided that the stress is kept within certain limits which must not be exceeded in practical loading.

The strain of a structural member indicates the nature of the stress to which the material is subjected.

135. Classification of beams. For theoretical consideration of their structural value beams are divided into the following classes:

(a) *Cantilevers*, which are members having one end freely projecting and the other supported and fixed by building into a wall, bolting to a stanchion or some equivalent arrangement.

(b) *Freely supported beams*, which are freely laid upon two supports, generally at the extreme ends of the beam.

(c) *Fixed beams*, which have their ends rigidly held down upon their supports.

(d) *Continuous beams*, which have intermediate supports and therefore continue in one length over two or more spans.

Several conditions of support may also exist which are compounds of those described.

Fixed and continuous beams have certain economical advantages under satisfactory conditions of support. In this volume applications are assumed to include the first two classes only.

Cantilevers. Cantilevers are employed in building for, amongst others, the following purposes: (a) projecting ends of roof timbers, (b) stone corbels for supporting projecting walls, piers, etc., (c) stone hearth slabs built into chimney breasts, (d) hanging stone steps, (e) trimming timbers in stair landings, (f) supports to balcony floors, (g) hoods over doorways.

In all these cases the useful load is carried on the free projection.

Freely supported beams. These are in most common use in lintols, floor joists, floor girders, roof purlins, steps, etc., and may be constructed of timber, steel, reinforced concrete or stone, according to their suitability for the purpose.

SHEAR FORCE

136. Vertical shear force. In a loaded beam or cantilever the principal forces are vertical, and "shear" at any section is a measure of the tendency of these forces to cut the beam in the plane of the section parallel to the forces. Thus, if the cantilever of detail No. 44 is firmly held and supports 2 tons at the free end, at any vertical section such as *ab* the material between the load and *ab* resists being moved vertically downwards by the load. Now the load of 2 tons is transferred through the section *ab* to the support without any movement of the member, hence on the immediate right of the

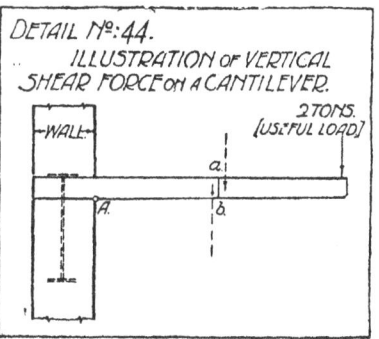

DETAIL Nº:44.
ILLUSTRATION OF VERTICAL SHEAR FORCE ON A CANTILEVER.

section there must be a downward force of 2 tons and on the immediate left an upward force of 2 tons, and similarly for any other section between the load and the support. Such pairs of forces, acting parallel to each other and in opposite directions on any vertical section, constitute a vertical shear force of a magnitude equal to the external forces on one side of the section, in this case 2 tons.

Vertical shear force at any section may therefore be defined as the *nett sum of the vertical forces* on *one side* of the section. For example, if there were 2 tons, 3 tons and 1 ton at different points to the right of *ab* the shear force on *ab* would be 6 tons, no matter how it was distributed.

If the 2 ton load acted upwards, while the 3 and 1 ton loads pressed downwards the nett sum would be $3 + 1 - 2 = 2$ tons downwards, and the shear force on the section $= 2$ tons.

BALANCE AND REACTION

137. Supporting and fixing the cantilever. If the cantilever is strong enough to resist the vertical shear forces, it transfers the load to the wall but in doing so the overhanging part of the cantilever tends to fall, or strictly, to rotate about the edge A of the support. This tendency must be resisted, for which purpose the inserted end of the cantilever is loaded, either by the walling above it or by a bolt anchored down to the mass of walling forming the support.

Obviously, for balance or equilibrium, the anchoring weight must produce a tendency to rotate equal to that caused by the useful load, but must act in the opposite direction.

The tendency to rotate can only be measured by obtaining the product of the acting force and its perpendicular distance—or leverage—from the point about which the tendency is to be measured. Such product is called the *moment of the force* about the point and may be obtained and expressed in any suitable units of weight and distance. If lbs. and feet are employed the moment is expressed in "lbs. feet"; if tons and inches, the moment is in "tons ins." To compare moments they must obviously be expressed in the same units of force and leverage.

A load may be concentrated at a point—which in practice means a very small area—or may be spread over an appreciable surface.

In the first case it is known as a *point load* and in the second as a *distributed load*. If the distribution is uniform per unit of area, the load is said to be *uniform*.

138. Moment of a distributed load. The moment of a distributed load about any point is the product of the whole load and the perpendicular distance of its centre of gravity from the point to the line of load.

If the load be uniform over a given length, the centre of gravity is obviously at the centre of such length.

139. Equilibrium of cantilever. Balance weight. Consider the cantilever of detail No. 45. In this case the 2 ton load (W) is tending to move downwards and inwards about A, as suggested by the

dotted arc, with a *clockwise* moment of 2 (tons) × 6 (ft.) = 12 tons ft.
or, in symbols, $W \times L$.

To resist this clockwise rotary tendency, the balance weight (W_B) must produce an equal but *opposite* tendency, viz. an *anti-clockwise* moment of $W_B \times y$, where y is half the thickness of the wall or $\frac{3}{4}$ ft., because W_B is assumed to be a uniform load on the enclosed end. Hence, to prevent rotation:

$$W_B.y \text{ must} = WL,$$

$$\therefore W_B = \frac{WL}{y},$$

or $$W_B = \frac{2 \times 6}{\frac{3}{4}} = \frac{48}{3} = \mathbf{16} \text{ tons.}$$

Thus, 16 (tons) × $\frac{3}{4}$ (ft.) = 12 tons ft., which is the same moment as before, but in the opposite direction about A.

Thus, if any additional weight is placed on the projecting cantilever it would destroy the balance and cause rotation downwards on the right of A.

For safety in practice, W_B would be two to three times the amount required to produce equilibrium about the edge of the support, but apparently only so much of it would be brought into action as would balance the useful load.

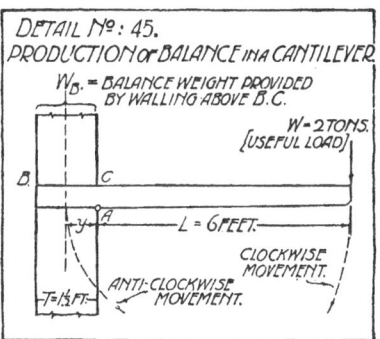

DETAIL N°: 45.
PRODUCTION or BALANCE in a CANTILEVER
W_B = BALANCE WEIGHT PROVIDED BY WALLING ABOVE B.C.
W = 2 TONS. [USEFUL LOAD]
L = 6 FEET.
CLOCKWISE MOVEMENT.
ANTI-CLOCKWISE MOVEMENT.

140. Reaction. It is now evident that the two loads, viz. the useful load and the balance load, are both acting vertically downwards and require support, therefore a supporting force would be required at A equal to $W + W_B$, in order to make the upward and downward forces balance. The upward force is called the reaction because it acts against the loads in an opposite direction. The reaction in this case is 16 + 2 = 18 tons, presumably acting through A.

Observe that only 2 tons of this total is caused by the useful load on the projection, while the rest is due to dead weight of walling which would probably exist in any case. It is therefore usual to state that the reaction of a cantilever equals the load upon its projecting arm.

141. Centre of pressure on the support. If the conditions assumed in the previous paragraphs were practically possible, it would appear that an upthrust of 18 tons would act through the edge of the support at A; this could only happen if the cantilever were just on the point of rotating downwards about A, when all the load would be temporarily thrown on the edge and would crush either the stone template or the bearing surface of the beam.

A little thought will show that the practical effect of providing too much balance weight is to throw the point of real reaction within the wall face until the moment of all the balance weight about that point is equal to the moment of the useful load about the same point. This point is not easily determinable, hence the adoption of the elementary idea of rotation about the edge, security being obtained by employing a load of two or three times the amount required to balance the useful load about the edge of the support. The number 2 or 3 then becomes what is called a "factor of safety".

Bending Moment and Moment of Resistance

142. Disturbing forces causing bending. While at every section of a loaded beam or cantilever in equilibrium there cannot be actual rotary movement, on *each* side of any section there is a *tendency* to move, the two tendencies being necessarily equal but opposite in direction.

The tendency of any piece of the cantilever to rotate about a section is measured by the moments of the forces acting on the piece about such section, and is obvious that in a cantilever this tendency is greatest at the line of support or reaction, hence in detail No. 45 the maximum moment is 2 (tons) × 6 (ft.) = 12 tons ft. tending to cause clockwise rotation on the right.

But balance weight (W_B) produces an equal and opposite moment about the wall edge, therefore the two moments, of equal magnitude, are acting against each other through the section of the material above the edge of the support. The material strains under the action, elongating at the top edge and shortening at the bottom, causes the arm to bend downwards to an amount depending upon its size and elasticity or power of regaining shape after being strained when the load is removed.

Because the moment measuring the tendency to rotate causes the material to strain and bend, it is termed the "bending moment".

In estimating the size of a cantilever or beam for any condition of support and loading, it is necessary to determine the value of the greatest bending moment to which the member is liable and often where such maximum occurs.

143. Distribution of stress. As the cantilever strains, due to the bending moment, it is evident that the material is stressed, for stress and strain accompany each other. Consider the projecting arm of the cantilever in detail No. 46 which is exaggerated in depth and diminished in length to emphasise the conditions.

As bending occurs the lower fibres are evidently shortened and the upper fibres extended. Shortening implies *compression* while extension implies *tension*, hence, we know the kind of stress occurring at the extreme edges of the beam. Somewhere between the two there must be a layer of fibres having no stress upon them, where the change takes place from compression to tension. In elevation this is seen as a curved line at the centre of the depth, bent but not altered in length; it is the edge view of an infinitely thin layer—previously a plane—of fibres, and commonly termed the "neutral

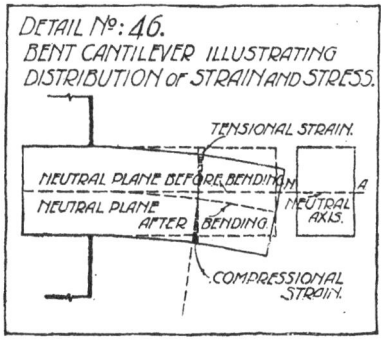

plane". In a vertical section of the beam only the end or edge of the plane can be shown; it is then termed the neutral axis of the section.

The position of the neutral axis depends on two things:

(*a*) the shape of the cross section;

(*b*) whether the material strains or changes length equally under tensional and compressional stresses.

In most structural materials the strains referred to are approximately equal under equal working stresses, within the limits allowed in practice. It may be shown that the position of the neutral axis (N.A.) depends mainly upon the shape of the section. It is at right angles to the direction of the external loads and in such a position that the moments of the sectional areas above and below it are equal.

If a section is symmetrical it is therefore obvious that the N.A. is the centre line which is at right angles to the loads and in general constructional work will be parallel to the bearing surfaces of the supports.

Referring again to detail No. 46, examine the short length of beam near the support. It was originally rectangular but has become wedge-shaped and slightly bent; its change of shape, viz. the strain, is shown by the two shaded triangles and as stress and strain are proportional to each other the stress must also vary in the same way, gradually and directly increasing from zero at the neutral plane to a maximum at the extreme edges, the actual stress per unit area being the same at both edges but of opposite kind. It is also clear that the convex edge is in tension while the concave edge is in compression.

144. Resisting forces. At the wall section in detail No. 47 are represented the forces supplied by the resistance of the material. Conceive the rotary action caused by the load transferred by the material of the projecting arm to the section at the wall plane. On the right of this section are marked forces of varying intensity, pulling above and pushing be-

low. On the left of the section, if the material holds, equal and opposite forces must be supplied, the two sets of forces producing an internal resistance which is called stress and which opposes the bending moment and prevents collapse.

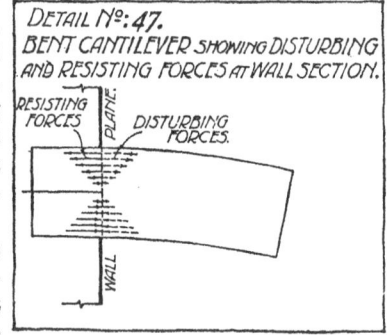

DETAIL Nº: 47.
BENT CANTILEVER SHOWING DISTURBING AND RESISTING FORCES AT WALL SECTION.
RESISTING FORCES   PLANE   DISTURBING FORCES.
WALL

The moment of the forces measuring the tendency to disturb the section on the right will be exactly equal to the bending moment which caused it and the moment of the resisting forces is therefore equal and opposite to the bending moment; the moment produced by the two sets of resisting forces—above and below the neutral axis—is called the "moment of resistance" or "resistance moment" of the section.

It should now be clear that the resistance moment (R) must be equal to the bending moment (B) at every section of the beam, hence, if the beam is the same section throughout, its resistance moment (R) must be made to equal the maximum bending moment (B_M) and at the same time the stress to which the outer fibres are subjected must be within safe limits to prevent deterioration or destruction of the fibres.

A means of expressing the value of R for rectangular beams must now be found in order to equate it to the value of B_M.

145. Graphical representation to obtain resistance moment. Detail No. 48 shows how the forces acting upon the fibres of the beam at the wall plane may be further represented, by building a stress figure upon a vertical plane section of the beam at $BCDE$.

To show the difference between tensional and compressional stresses, the former are shown to the left of the section and the latter to the right, though they both act on one side of the section. On the face of the beam the stress is shown to vary as the ordinates of a triangle, but at any selected level the stress is constant across the breadth. The total stress above or below the N.A. is therefore represented by a solid wedge or triangular prism. *These wedges are not part of the beam* but are adopted as a convenient representation of the forces acting upon the section. If the stress on the outermost fibres is represented by CH, and equals 1000 lbs. per sq. in., it means that the maximum stress is *at the rate of* 1000 lbs. per sq. in., although it only acts at this rate on the extreme layer of fibres.

Now the two wedges of stress represent the total internal forces opposing the rotation of the section $BCDE$ about the neutral axis, which is the axis of rotation. Both groups of forces oppose clockwise rotation, the upper group by pulling and the lower by pushing upon the section plane. Hence the sum of their moments about the N.A. represents the resistance moment (R) of the section.

Consider the tension wedge in the same detail, which is re-drawn and dissociated from the beam. Its volume represents the total tensional force, F_t, and

$$= \frac{b \times \frac{d}{2} \times f}{2} = \frac{b.d.f}{4}.$$

Its moment is obtained by conceiving the whole force acting through the centre of gravity of the wedge, which, being composed of a mass of

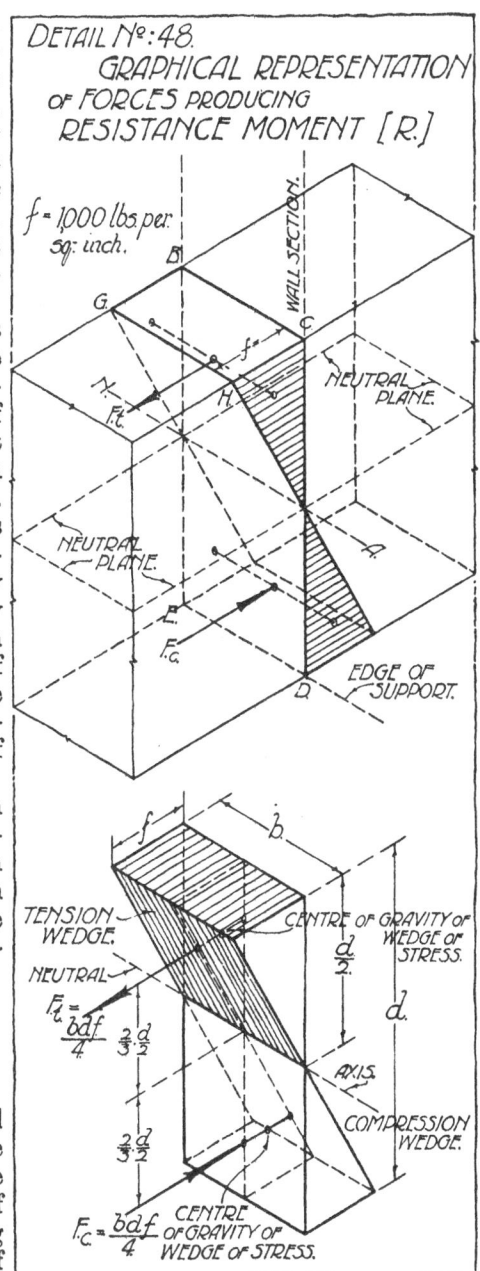

DETAIL N°: 48.
GRAPHICAL REPRESENTATION
OF FORCES PRODUCING
RESISTANCE MOMENT [R.]

$f = 1000$ lbs. per sq. inch.

WALL SECTION.

NEUTRAL PLANE.

NEUTRAL PLANE.

EDGE OF SUPPORT.

TENSION WEDGE.

NEUTRAL

$F_t = \frac{bdf}{4}$ $\frac{2}{3}\frac{d}{2}$

$\frac{2}{3}\frac{d}{2}$

$F_c = \frac{bdf}{4}$ CENTRE OF GRAVITY OF WEDGE OF STRESS.

CENTRE OF GRAVITY OF WEDGE OF STRESS.

$\frac{d}{2}$

d.

AXIS.

COMPRESSION WEDGE.

right-angled triangles grouped in vertical planes, will have its centre of gravity at $\frac{2}{3}$ the height from the vertex and in the centre of the breadth. The height of the triangle being $\frac{d}{2}$, the lever arm of one wedge $= \frac{2}{3} \times \frac{d}{2} = \frac{d}{3}$.

146. Resistance moment. The resistance moment (R) consists of the sum of the moments of the two stress wedges about the N.A.

$$\text{Moment of one wedge} = F \times \text{lever arm}$$

$$= \frac{bdf}{4} \times \frac{d}{3}$$

$$= \frac{bd^2f}{12}.$$

$$\text{Moment of two wedges} = R = 2\left(\frac{bd^2f}{12}\right)$$

$$= \frac{bd^2f}{6}.$$

This is the resistance moment for a rectangular beam and is always expressible in the same terms whatever kind of material may be employed for the member.

The moment is composed of two parts, viz. $\frac{bd^2}{6}$ and f, the former relating only to the dimensions of the section and called the *modulus of section* and the latter to the maximum stress caused by bending. The former would be constant for any given size of beam, but the latter would vary with the allowable stress upon the material employed.

147. Principles of beam design. The elementary design of timber beams usually involves the one operation of equating the maximum bending moment to the moment of resistance.

In the design or selection of steel beams shear stresses must also be considered, but in normal cases of wooden beams shear stresses are too low to become critical, and where sound material is employed may be neglected. The chief danger is the employment of timber which is shaken or defective along the line of the neutral plane, where it may be shown that the shear stress is greatest.

148. Factors affecting the strength of a beam. If the bending moment be equated to the resistance of a beam for the case of a cantilever with an end load,

$$B_M = R,$$

or
$$WL = \frac{bd^2f}{6},$$

$$\therefore\ W = \frac{bd^2}{L} \times \frac{f}{6},$$

where W is the load carried, b and d are dimensions of the section, L is the span, f is the working stress on the outer fibres and is constant for one kind of material, 6 is a constant divisor, and by using one kind of material the load safely carried varies with the quantities b, d and L, $i.e.$:

W increases with an increase of b,

W increases with the square of d,

W decreases with an increase of L.

The strength of a beam therefore varies as $\dfrac{b \cdot d^2}{L}$ and this applies to all forms of loading.

In order to show the method of procedure in design, consider further the simple cantilever of detail No. 46.

Example 1. Suppose such a cantilever, of 7″ × 4″ northern pine, carries an end load, the clear projection being 4 ft. Find the load which may be safely carried, if the working stress may reach 1000 lbs. per sq. in.

In all cases
$$B_M = R.$$

But
$$B_M = W \cdot L,$$

and
$$R = \frac{bd^2f}{6},$$

$$\therefore\ W \times L = \frac{bd^2f}{6},$$

and substituting values
$$W \times 48 = \frac{4 \times 7^2 \times 1000}{6}.$$

Transposing to find the unknown value, and working out,

$$W = \frac{4 \times 49 \times 1000}{48 \times 6} = \mathbf{680 \cdot 5}\ \text{lbs. or}\ \mathbf{6 \cdot 07}\ \text{cwts.}$$

W and f are in the same units (lbs.), both representing forces; L and b and d are in the same units (ins.), all representing linear dimensions.

Example 2. Let the cantilever have a length of 4 ft., carry a load of 8 cwts. at the end, and be constructed of northern pine 4″ broad and

stressed to a maximum of 1000 *lbs. per sq. in. Find the necessary
depth.*

$$B_M = R,$$

$$\therefore \; W \times L = \frac{bd^2f}{6}.$$

In this case d is the unknown factor, therefore collect all the
known factors to one side, then

$$\frac{6 . W . L}{b . f} = d^2.$$

Taking the square root of both sides, placing the unknown quantity first, and substituting

$$d = \sqrt{\frac{6 . W . L}{b . f}} = \sqrt{\frac{6 \times (8 \times 112) \times 48}{4 \times 1000}} = \sqrt{64 \cdot 5},$$

$$\therefore \; d = \text{say } 8''.$$

Presuming that the reader now possesses a general idea of the
principles involved in obtaining the dimensions of timber beams,
it is possible to consider more fully those further conditions which
are likely to occur in practice, and reduce them to an economical
form for regular use.

149. Standard practical cases of shear and bending moment. In
studying the variation of shear and bending moment the changes
are most easily seen by plotting graphs, the vertical ordinates of
which represent the shear or bending moment at any point.

It is convenient to draw a diagram of the beam, showing its
loading and method of support and to arrange the graphs with their
length axis parallel to the beam and to the same scale.

**Notation for reference to shear and bending moment graphs or
diagrams.**

The following notation is employed throughout:

W = the *total load* on beam in any convenient unit.

W_1, W_2, W_3, etc., represent any number of point loads.

w_l = load per unit of length, say per ft. run.

w_s = load per unit of area, say per ft. super.

L = effective span of beam or cantilever. In the former case it is measured
between the centres of the supports.

L_1, L_2, L_3, etc., represent the distances of point loads from a support.

B_S = bending moment at any section S.

B_M = maximum bending moment.

x = distance of any section S from free end of cantilever, or from support
of beam, expressed in any convenient unit.

Cantilever Beams

150. Case I. Cantilever with point load at the free end. Detail No. 49 at A.

Shear. On the right of the support the shear is constant and $= W$. This is shown by the shear graph above the cantilever, any vertical ordinate representing, to scale, the sum of the vertical forces on one side of the corresponding section.

Bending moment. The bending moment at any section S distant x from the free end, is the sum of the moments of all the forces on one side of S, and in this case $B_S = W.x$; x is greatest when S is at the wall face, and the $B_M = W.L$.

It is evident that the bending moment increases directly with the distance from the free end, being zero at that point because $x = 0$. The graph of B is therefore a straight line as shown below the cantilever, and B_S can be scaled off at any section, between the graph and the length axis.

151. Case II. Cantilever with a series of point loads. Detail No. 49 at B.

Shear. When several point loads act on a cantilever the shear at any section is the sum of the shears caused by the individual loads, and is therefore the sum of the loads on one side of the section. The shear graph shows this by the superposed rectangles, producing a stepped outline; the maximum shear is $W_1 + W_2 + W_3$ between W_3 and the support.

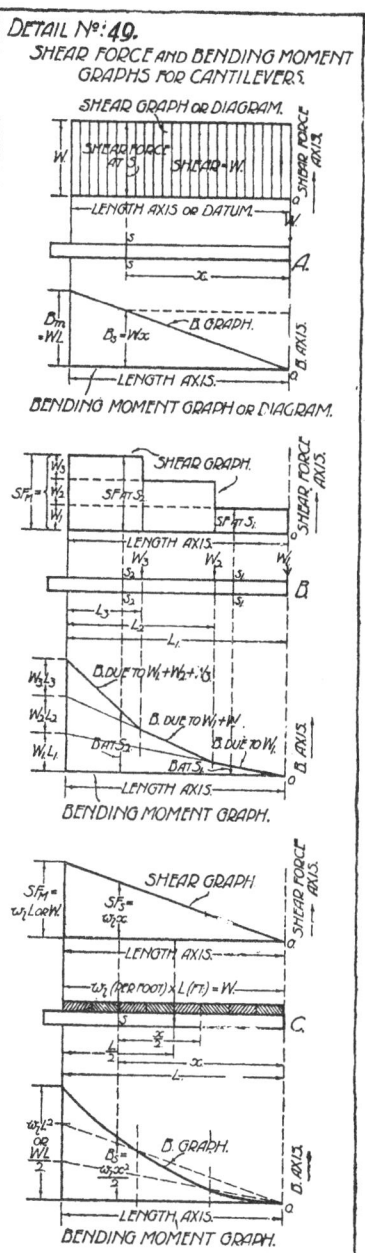

DETAIL Nº: 49.
SHEAR FORCE AND BENDING MOMENT
GRAPHS FOR CANTILEVERS.

Bending moment. As in the case of shear, the bending moment at any section, for a series of point loads, is the sum of the moments due to individual loads on one side of the section.

The effect of each load to cause rotation about the wall section is W_1L_1, etc., and at any intermediate point is proportionate to its distance from the end, hence the

$$B_M = W_1L_1 + W_2L_2 + W_3L_3,$$

and the graph is constructed by setting up these three values at the wall line and placing the zero points on successive parts of the graph immediately under the load points as in the lower part of this detail.

152. Case III. Cantilever with uniform load over the whole length. Detail No. 49 at C.

Shear. The vertical shear force varies, in this instance, from zero at the free end to the total load W (or w_lL) at the wall section. The shear force at any section S distant x from the free end = the load between these points and is therefore w_lx, which evidently increases or decreases directly with x. The shear force graph is a straight line from zero at the end and having its maximum value of w_lL or W at the wall section.

Bending moment. Consider the bending moment at any section S distant x from the end. The load between these points $= w_l x$, and its centre of gravity is $\dfrac{x}{2}$ from S, therefore the moment about S

$$= w_l x \times \frac{x}{2} \quad \text{or} \quad B_S = \frac{w_l x^2}{2} \qquad \text{......(1).}$$

When $x = 0$ there is no bending moment.

When $x = L$ the bending moment $= \dfrac{w_l L^2}{2}$ or $\dfrac{WL}{2}$, and is a maximum.

Hence, the $B_M = \dfrac{WL}{2}$, exactly half the value of Case I, where the load was concentrated at the end; it is therefore clear that twice the end load could be spread uniformly over the length of the cantilever without increasing the bending moment, or, a cantilever beam of half the strength would carry a load equal to the end load if spread uniformly over its length.

From equation (1) it appears that because load and lever arm both increase with x, the bending moment increases as the square of x, (x^2). Thus

$$B_S = \left(\frac{w_l}{2}\right) x^2,$$

which is similar to $y = ax^2$, viz. the equation to a parabola, with its vertex at the origin O. Hence, by drawing a parabola[1] with its vertex at the origin and $\dfrac{WL}{2}$ units on the B scale at the wall line, the bending moment for any section may be measured to the same vertical scale between the length

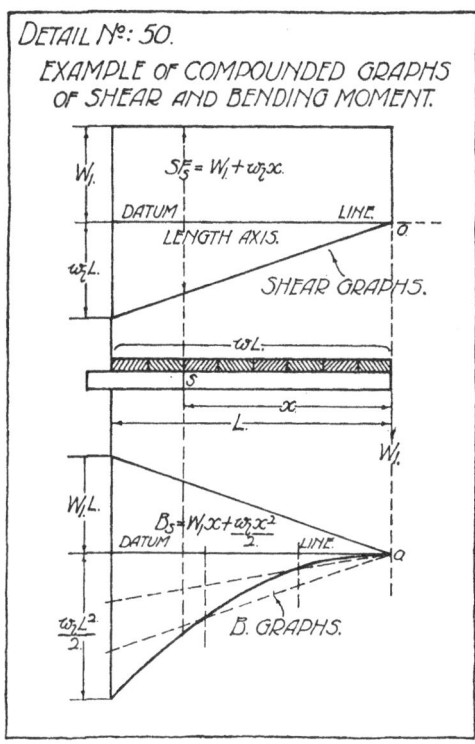

axis and the graph, immediately opposite the section. As an alternative the bending moment may be calculated for a few intermediate sections, and the graph completed by drawing a fair curve through plotted points.

153. Special cases. Other compound cantilever graphs may be constructed on a similar basis, and it will sometimes be found an advantage to arrange graphs for separate loadings on opposite sides of the same length axis.

A good example of this occurs when a point load at the end exists, along with a uniform load. The graphs may then be arranged

---

[1] See details Nos. 49, 50, and 54 for method of setting out.

as in detail No. 50 and the B_S distant x from the end is measured to scale across the whole diagram opposite S, or again, the shear force at S (SF_S) is similarly measured.

BEAMS FREELY SUPPORTED AT BOTH ENDS. (SINGLE SPANS ONLY)

154. When a loaded beam is freely supported at both ends across a single span, the load causes a deflection or sag to occur between the supports due to the latter supplying the necessary upthrust or reaction to resist the load.

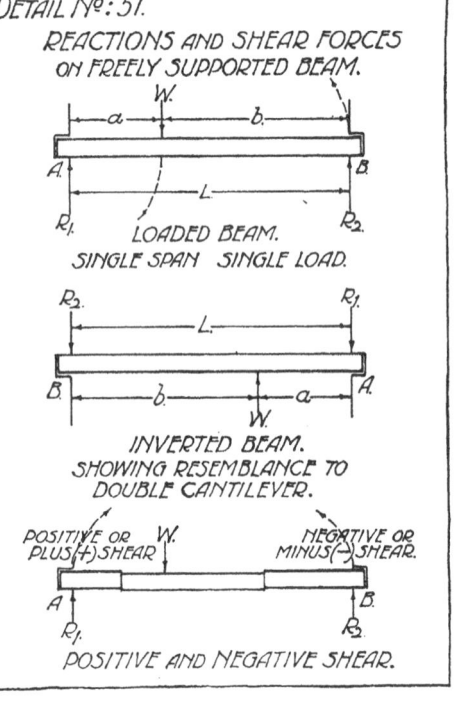

DETAIL Nº: 51.

REACTIONS AND SHEAR FORCES ON FREELY SUPPORTED BEAM.

LOADED BEAM. SINGLE SPAN SINGLE LOAD.

INVERTED BEAM. SHOWING RESEMBLANCE TO DOUBLE CANTILEVER.

POSITIVE AND NEGATIVE SHEAR.

The conditions of equilibrium are exactly the same as if the beam were inverted so that the reactions R_1 and R_2 would appear as loads, and the weight appears as an upthrust in the line about which the reactions produce balance; see detail No. 51.

155. To find the reactions. The first sketch in the detail shows the loaded beam, while the second sketch illustrates the inverted condition.

The beam has become a double cantilever and if equilibrium is to be preserved the moments of R_1 and R_2 about W must be equal, therefore $R_1a = R_2b$. But if W supports R_1 and R_2 it must be the resultant of these forces, and is therefore equal to their sum.

To determine a mode of procedure to find R_1 and R_2, continue thus:

$$R_1a = R_2b.$$

But

$$R_1 + R_2 = W,$$

$$\therefore R_1 = W - R_2.$$

Substituting for R_1

$$(W - R_2)\, a = R_2 b,$$
$$Wa - R_2 a = R_2 b,$$
$$\therefore\ Wa = R_2 b + R_2 a = R_2\,(b + a).$$

But
$$b + a = L,$$
$$\therefore\ Wa = R_2 L.$$

Now consider the original loaded beam. The result obtained is equivalent to taking moments of all the external forces about the support A at R_1, for $W \times a$ (clockwise) must equal $R_2 \times L$ (anticlockwise).

$$\therefore\ R_2 = \frac{W \cdot a}{L}.$$

This leads to the expression of the result as a formula or rule, as follows: To find the reaction of a beam at one of the supports, due to a single load, multiply the load by the distance from the other support and divide by the span.

Thus:
$$R_1 = \frac{W \cdot b}{L}.$$

If there are several loads, apply the rule for each load and sum the results; proceed by taking the products of load and distance first, then sum the products and divide the total by the span.

In obtaining the shear and bending moment for any section of a freely supported beam the reactions must always be calculated first, except where the case is a standard one and the results have been memorised as a formula.

156. Shear force on beams. Consider again the inverted beam of detail No. 51. If this be compared with a simple cantilever it is seen that the shear on either side of the support W is equal to the reaction on that side. Now revert to the original position of the beam; R_1 is a force on the left of W, tending to cut the beam at any section between A and the load. If the tendency developed, any piece which was severed would move upwards and slightly clockwise because of the accompanying bending, while on the right of the load W, severed portions would tend to move upwards and anticlockwise. To distinguish between these shear force tendencies, call them positive, or $+$, and negative, or $-$. Note that "$+$shear" tends to cut off and turn clockwise while "$-$shear" is opposite in effect, as shown in the detail.

In representing shear forces on beams supported at both ends, these $+$ and $-$ effects are distinguished in this volume by plotting the + shear graphs above the length axis, and the − shear graphs below the axis.

Detail No. 52 shows the shear force graph above the beam and it will be evident that the previous definition holds if extended to meet the conditions of the beam; thus, the vertical shear force at any section is the nett sum of the forces acting either on the right or left of that section. In this example the shear force on the left of W is positive and on the right of W is negative.

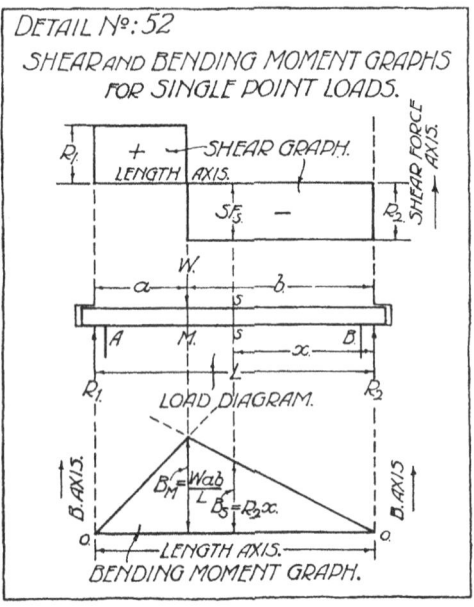

157. Bending moment. The bending moment at any section is measured in the same way as for a cantilever. Just as the wall section of the introductory example, detail No. 45, was balanced by equal moments of the forces on each side of the section and in opposite directions, so every section of a balanced beam has the same nett value of the moments of all the forces on *either side*. Hence, bending moment may be defined as follows: *the bending moment at any section of a beam is the algebraic (or nett) sum of the moments of all the forces on either one side or the other of that section.* If the result is a clockwise tendency it will be called positive, and if anti-clockwise, negative, as in the case of shear.

Consider detail No. 52. Using the symbols a and b for the segments of the beam as divided by the load and taking moments on the left of the load about M,

$$B_M = +(R_1 \times a).$$

But, from paragraph 155, $R_1 = \dfrac{Wb}{L},$

$$\therefore B_M = + \frac{W.b.a}{L} \qquad\qquad \ldots\ldots(1).$$

Taking moments on the right of W

$$B_M = -(R_2 \times b).$$

But
$$R_2 = \frac{Wa}{L},$$

$$\therefore \ B_M = -\frac{W.a.b}{L} \qquad \ldots\ldots(2).$$

Thus the bending moment under the single load is shown in (1) and (2) to equal $+$ or $-\dfrac{W.a.b}{L}$ according to the side of W selected for calculation. (This is also the maximum value, for it can be shown that the maximum always occurs under the load.) As the positive or negative value of the moment has no bearing on the result in practical application further reference to it may be omitted in this volume, which deals only with freely supported beams.

To plot the bending moment graph, it is clear that the bending moment at a support must be zero, because the reaction at that support has no leverage, while W and the opposite reaction have equal moments about the support. Hence, as the moment at any section between A and M increases directly with the length, and similarly from B to M, two straight line graphs meet under the load, where the value is at a maximum of $\dfrac{W.a.b}{L}$. The graphs may be plotted either above or below the datum line as may be convenient.

158. Case IV. Freely supported ends and central point load.

If $a = b = \dfrac{L}{2}$, viz. when W is in the centre of the span,

$$B_M = \frac{W \times \dfrac{L}{2} \times \dfrac{L}{2}}{L} = \frac{WL}{4},$$

which is the formula for finding the maximum bending moment for a central point load.

The shear would be the same on each side of the centre, viz. $\dfrac{W}{2}$, because the reactions are each $\dfrac{W}{2}$.

The diagrams would be similar to those of detail No. 52.

159. Case V. Freely supported ends and point load not at the centre.
This is the case used for study in paragraph 157; the maximum

shear force (SF_M) = the greater reaction, and the maximum bending moment

$$(B_M) = \frac{W \cdot a \cdot b}{L},$$

a and b being the distances of the load from the two supports.

160. Case VI. Freely supported ends and two equal point loads dividing span into three equal parts. Detail No. 53. This is a case of common occurrence in floors of the framed type.

Let the total load be W, then each load is $\dfrac{W}{2}$.

Reactions. Applying the rule from paragraph 155, or taking moments about R_2,

$$R_1 = \frac{\left(\dfrac{W}{2} \times \dfrac{L}{3}\right) + \left(\dfrac{W}{2} \times \dfrac{2L}{3}\right)}{L},$$

$$\therefore R_1 = \frac{3}{6}\frac{WL}{L} = \frac{W}{2}.$$

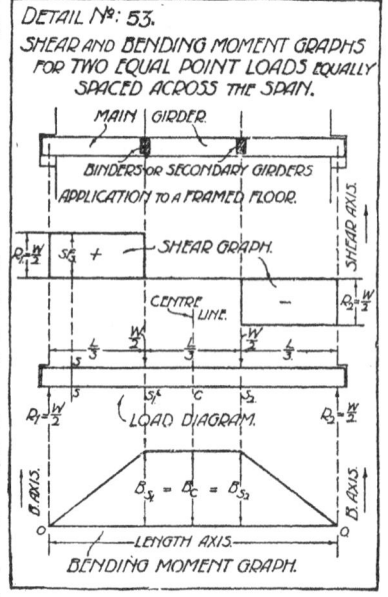

DETAIL Nº: 53.
SHEAR AND BENDING MOMENT GRAPHS
FOR TWO EQUAL POINT LOADS EQUALLY
SPACED ACROSS THE SPAN.

It is evident that the reactions each $= \dfrac{W}{2}$, and this will be found to occur for all cases of loading which are symmetrically disposed about the centre line of the span.

Shear. The shear force in the end bays = the reaction, and as the reaction is equal to one of the loads $\left(\dfrac{W}{2}\right)$ no shear exists across the centre bay because, on either side of any vertical section taken through it, the nett sum of the forces is zero. This is clearly shown in the shear force graph above the load diagram of the detail.

Bending moment. Calculating the bending moment at S_1 under the load. By moments on the left

$$B_{S_1} = R_1 \times \frac{L}{3}$$

$$= \frac{W}{2} \times \frac{L}{3},$$

$$\therefore B_{S_1} = \frac{WL}{6}.$$

Calculating the bending moment at the centre section C.
By moments on the left

$$B_C = \left(R_1 + \frac{L}{2}\right) - \frac{W}{2}\left(\frac{1}{2} \times \frac{L}{3}\right)$$

$$= \left(\frac{W}{2} \times \frac{L}{2}\right) - \frac{WL}{12}$$

$$= WL\left(\frac{1}{4} - \frac{1}{12}\right),$$

and $$B_C = \frac{WL}{6}.$$

It is evident that B_{S_2} would be equal to B_{S_1}, and that the bending moment is constant across the centre bay, being everywhere $= \dfrac{WL}{6}$.

The graph of bending moment is therefore composed of three straight lines intersecting under the loads and passing through O at the ends, as shown in the detail.

161. Case VII. Freely supported ends and three equal point loads dividing the span into four equal parts. By working on the same principles as in the last paragraph, it is found that the reactions $= \dfrac{W}{2}$; the maximum vertical shear force is $\dfrac{W}{2}$, occurring between supports and end loads; the bending moment at the first load $= \dfrac{WL}{8}$, and at the centre $\dfrac{WL}{6}$. Hence, while the B_M is of equal value for the *same total load and span* as in the last case, it occurs only at the centre.

As the number of loads increases—while the total remains the same—the maximum gradually decreases, until, when the loads are very numerous, the value of B_M may, without serious error, be taken as $\dfrac{WL}{8}$, which is the B_M for a uniform load as shown in the next paragraph.

162. Case VIII. Freely supported ends and a uniform load. **Detail No. 54.** This is the commonest practical use.

Reactions. Being a symmetrical load about the centre line, the

reactions are each
$$=\frac{w_l L}{2} \text{ or } \frac{W}{2},$$

where w_l is the load per ft. run and L is the span in ft.

This may be proved by taking moments about either reaction line to find the opposite reaction as in previous cases.

Shear. The shear force graph for a uniformly loaded cantilever was shown to be a straight line of constant slope, representing the constantly increasing load between the free end and the wall section. In the case of a beam with a uniform load, we have seen that the reactions $=\dfrac{W}{2}$, therefore at the supports the shear $=\dfrac{W}{2}$; but, as the supporting forces are met by the distributed load which is gradually increasing in total between the support and the centre, the sum of the upward and downward forces is gradually and constantly diminishing towards the centre, where it becomes zero.

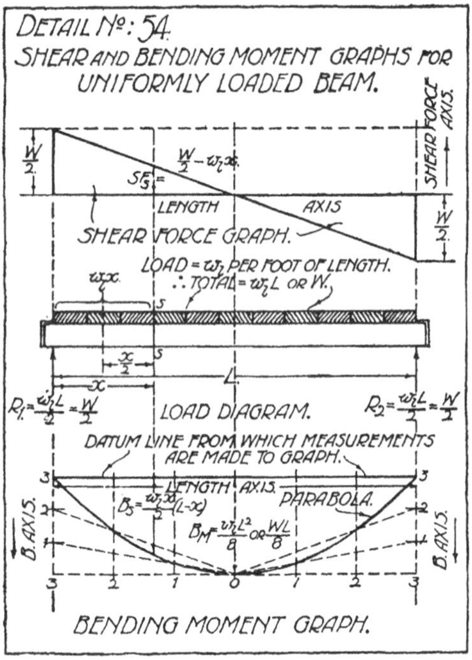

At the centre the sum of the downward forces

DETAIL Nº: 54.
SHEAR AND BENDING MOMENT GRAPHS FOR
UNIFORMLY LOADED BEAM.

on the left has become equal to R_1, hence zero shear. *At any intermediate section the nett force to produce shear is the difference between the reaction, and the load between that section and the support.*

The shear graph is shown above the load diagram of detail No. 54, and is a straight line passing through zero at the centre, through $+\dfrac{W}{2}$ at the left hand support line and through $-\dfrac{W}{2}$ at the right hand support.

Bending moment. Consider the section S. Take moments of all forces on the left. R_1 is acting clockwise with a lever arm x, and a portion of a load $w_l x$ is acting anti-clockwise with a lever arm $\dfrac{x}{2}$.

Let R_1 be expressed as $\dfrac{w_l L}{2}$, then

$$B_S = R_1 x - w_l x \times \frac{x}{2}$$

$$= \frac{w_l L x}{2} - \frac{w_l x^2}{2},$$

$$\therefore B_S = \frac{w_l x}{2}(L-x) \qquad \ldots\ldots(1).$$

This expresses the bending moment at any section S distant x from the support, and, being the equation of a parabola having its vertex and maximum height at the centre of the span, the parabola can be drawn geometrically if its height be known. This height will be the bending moment at the centre. Take S at the centre; then x will $= \dfrac{L}{2}$. Insert this in equation (1) above.

Then
$$B_M = \frac{w_l \dfrac{L}{2}}{2}\left(L - \frac{L}{2}\right) = \frac{w_l L}{4}\left(\frac{L}{2}\right) = \frac{w_l L^2}{8}.$$

But as $w_l L = W$, then $B_M = \dfrac{WL}{8}$.

The parabola is then constructed geometrically as indicated on the detail, No. 54 (in which case the vertex is at O), and the B_S may be scaled off for any section.

163. Case IX. Freely supported ends; point loads of varying amounts and irregularly spaced. Let a beam be loaded as shown in detail No. 55 with three unequal point loads at non-uniform distances. Determine the reactions, the maximum shear force, and maximum bending moment, and draw graphs showing how the shear force and bending moment vary.

Reactions. Take moments about the line of R_2 to obtain R_1.

Then $R_1 \times L = (2 \text{ (tons)} \times 2 \text{ (ft.)}) + (1\frac{1}{2} \text{ (tons)} \times 10 \text{ (ft.)})$
$$+ (1\frac{1}{2} \text{ (tons)} \times 14 \text{ (ft.)}),$$

but $L = 20$ ft.,

$$\therefore R_1 = \frac{4 + 15 + 21}{20} = \frac{40}{20},$$

and $R_1 = \mathbf{2}$ tons.

Now $R_1 + R_2 = W$, and $W = 5$ tons (total load),
$$\therefore R_2 = W - R_1 = 5 - 2 = \mathbf{3} \text{ tons.}$$

But to check, take moments about the line of R_1.

Then $\qquad R_2 \times L = (1\frac{1}{2} \times 6) + (1\frac{1}{2} \times 10) + (2 \times 18),$

$$\therefore R_2 = \frac{9 + 15 + 36}{20} = \frac{60}{20},$$

and $\qquad\qquad R_2 = \mathbf{3}$ tons.

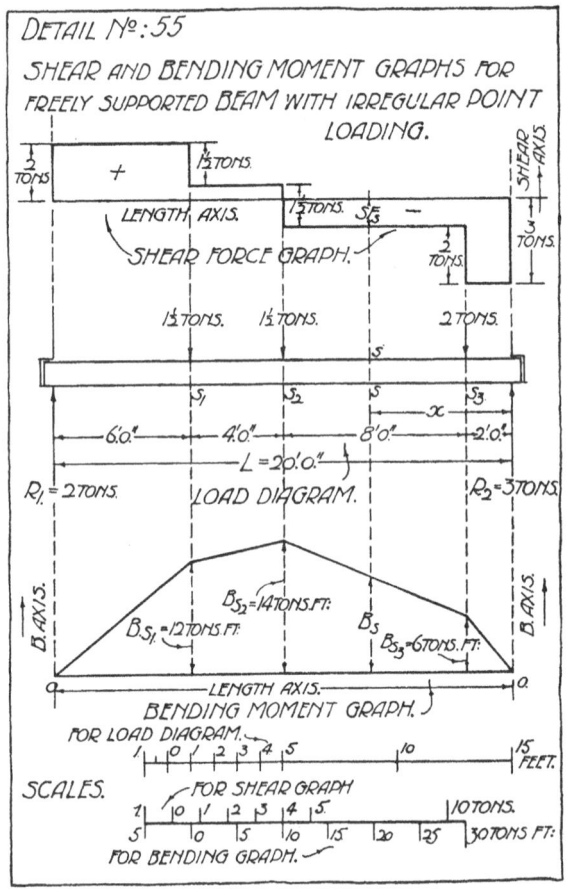

Shear force. The maximum vertical shear force $= R_2 = 3$ tons. The diagram or graph of shear may be plotted above and below the length axis thus: Set up 2 tons at R_1 and keep constant across the first bay, deduct $1\frac{1}{2}$ tons load acting opposite to S_1 and keep constant across the second bay, deduct $1\frac{1}{2}$ tons at S_2 and keep constant across third the bay, leaving a minus shear (below axis) of 1 ton; deduct 2 tons at S_3 producing a minus shear of 3 tons which agrees with the value of R_2. To maintain the conception of $+$ and $-$

shear, on passing the axis from the left, work might have been commenced from the right hand support and the reaction (R_2) set downwards, working the deductions upwards as positive quantities. The process is therefore to add and subtract forces graphically to obtain the nett or algebraic sum at every section.

DETAIL Nº: 56
STANDARD CASES OF LOADING.
SHEAR AND BENDING MOMENTS.
MAXIMUM VALUES.

LOAD DIAGRAM.	S.F.MAX	B_M.
	$W.$	$W.L.$
	$W.$	$\dfrac{WL.}{2.}$
$R_1 = \dfrac{Wb}{L}$ $R_2 = \dfrac{Wa}{L}$	EQUAL TO LARGER REACTION.	$\dfrac{Wab}{L.}$
	$\dfrac{W.}{2.}$	$\dfrac{WL}{4.}$
	$\dfrac{W.}{2.}$	$\dfrac{WL.}{8.}$
	$\dfrac{W}{2.}$	$\dfrac{WL.}{6}$ FROM A TO B.

Bending moment. From the known loads and reactions (viz. all the external forces on the beam) calculate the bending moment under each load.

Taking moments on the left

$$B_{S_1} = R_1 \times 6 \text{ (ft.)} = 2 \times 6 = \mathbf{12} \text{ tons ft.}$$
$$B_{S_2} = R_1 \times 10 \text{ (ft.)} - 1\tfrac{1}{2} \text{ (tons)} \times 4 \text{ (ft.)}$$
$$= (2 \times 10) - (1\tfrac{1}{2} \times 4) = 20 - 6 = \mathbf{14} \text{ tons ft.}$$

Taking moments on the right

$$B_{S_3} = - (R_2 \times 2 \text{ (ft.)}) = - (3 \times 2) = -\textbf{6} \text{ tons ft.}$$

B_{S_3} would be $+6$ tons ft. if moments were taken on the left, hence we shall ignore the sign. The reason for the change from left to right is to reduce the calculation, which becomes longer as the forces between the support and the section increase in number.

These values may now be plotted from the length axis under the loads and the diagram completed by the four straight line graphs terminating at zero at the supports. The graph then shows the bending moment at any section.

164. Use of shear diagrams for freely supported beams. The shear diagram affords a speedy means of deciding where the bending moment becomes a maximum, for B_M *always occurs at that point where the shear diagram shows a zero value of shear force.* In other words, the B_M occurs at that point in the loaded beam where the reaction is equalled by the total load between the point and the support.

It may therefore be shown graphically or may be calculated. The value of this method of determining B_M will be more evident in later work.

Detail No. 56 illustrates in tabulated form the six standard cases of loading commonly occurring in beams and cantilevers, with the max. shear force and bending moment for each case.

GRAPHICAL METHOD OF OBTAINING REACTIONS AND BENDING MOMENT

165. A simple graphical process is available for determining the reactions and the bending moment at any section of a freely supported beam or structure. It is based upon the following principles.

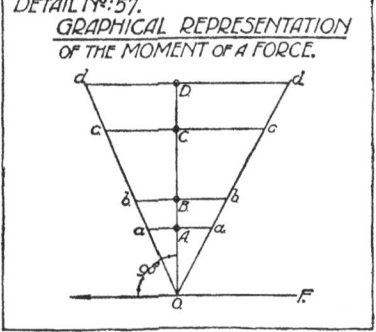

DETAIL Nº:57.
GRAPHICAL REPRESENTATION
OF THE MOMENT OF A FORCE.

1. The moment of a constant force, acting in any given line about a point in space, varies directly with the perpendicular distance of the point from the line of action of the force.

In detail No. 57 let F be a given force acting in the line FO, and A, B, C, D be points in a perpendicular line OD about each of which it is desired to measure the moment of the force F. Let $M_A, M_B \dots M_D$, etc., represent moments about $A, B \dots D$, etc.

Then $M_A = F \times OA,$

                             $M_B = F \times OB,$

                             ....

                             $M_D = F \times OD$, etc.

Now let dd on the line through D be the value of M_D to any scale and in any position, so long as the line is parallel to FO. Join d and d to O. Then the intercepted lengths cc, bb and aa will represent the values of M_C, M_B, M_A to the same scale as M_D, because $dd \dots aa$, etc., are parallel and dOd, cOc, $\dots aOa$ are similar triangles, therefore

$$\frac{dd}{cc} = \frac{DO}{CO}.$$

Hence, dd, cc, etc., increase or decrease in the same ratio as the lever arms DO, CO, etc.

2. Having seen that the ordinates of a triangle having its vertex in the line of force—such ordinates considered parallel to the line of action—may represent the relative values of the moments of the force about any selected point corresponding in position to the ordinate, it is clear that triangles may be utilised for representation, addition and subtraction of moments whether their scale be known or unknown so long as one constant scale is employed throughout the process.

To be useful, some easy means of deciding a scale for measuring results must be employable. This is shown to be the case in the following example.

Example. A beam is shown freely supported at L and R in detail No. 58 and loaded with two point loads EG and GJ, these being 2 cwts. and 4 cwts. respectively. Find a method of determining the reactions, the bending moment at any section and a convenient means of measuring these values.[1]

166. Principle of representing and comparing moments. Let the forces EG and GJ in the space diagram be represented by the scaled values eg and gj on the vertical load line of the force diagram. In the latter figure select the pole O in any position distant h force units measured at right angles from ej; complete the triangles eOg and gOj. From any point in the continued force line EG, draw links parallel to eO and gO to meet the reaction line R_L at e_1 and g_1. Then e_1g_1 to some scale—at present unknown—represents the moment of the force EG about L.

---

[1] The method of lettering the spaces on each side of a force, and referring to the force by the space letters read in a clockwise direction about any point at which a group of forces act, is known as Bow's notation.

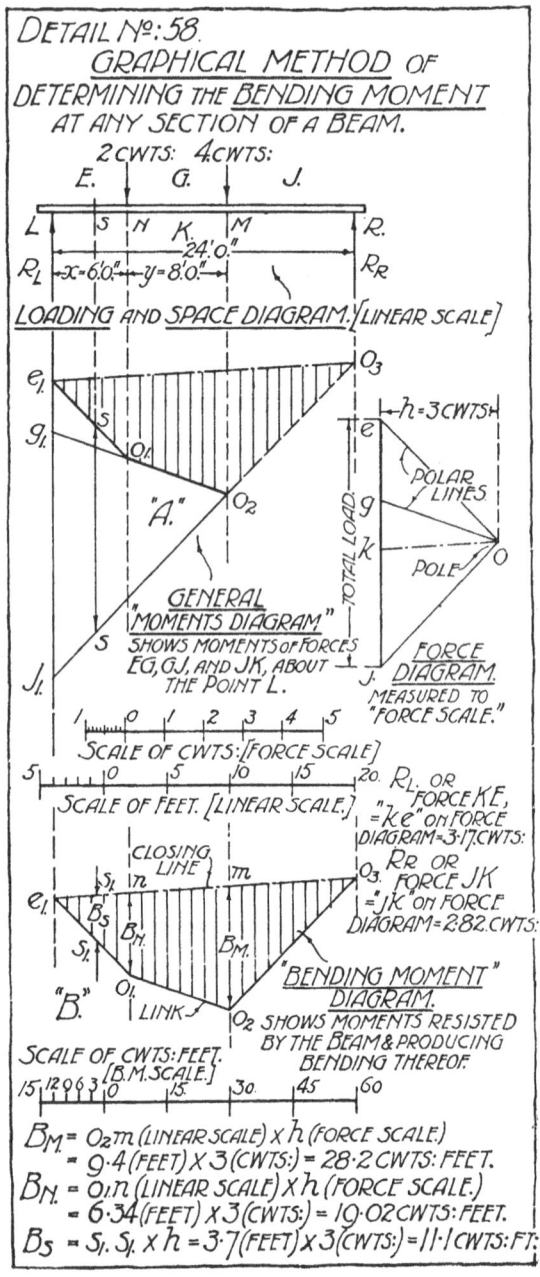

DETAIL N°:58.

GRAPHICAL METHOD OF DETERMINING THE BENDING MOMENT AT ANY SECTION OF A BEAM.

2 CWTS: 4 CWTS:

E.   G.   J.

L   S   N   K   M   R.
          24'.0."
R_L ⊢x=6.0"⊣y=8.0"⊣ R_R

LOADING AND SPACE DIAGRAM. [LINEAR SCALE]

e_1 O_3
g_1 h = 3 CWTS⊢
    S O_1 e   POLAR LINES
    "A." O_2 g   POLE
                    k
        GENERAL   O
    "MOMENTS DIAGRAM"
    S SHOWS MOMENTS OF FORCES   FORCE
J_1 EG, GJ, AND JK, ABOUT   J. DIAGRAM.
    THE POINT L.   MEASURED TO
                "FORCE SCALE."

SCALE OF CWTS: [FORCE SCALE]
0 1 2 3 4 5

SCALE OF FEET. [LINEAR SCALE.]
0 5 10 15 20 R_L OR FORCE KE,
                "ke" on FORCE
        CLOSING   DIAGRAM=3.17 CWTS:
        LINE   m O_3 R_R OR
    S_1 n   FORCE JK
e_1 B_S B_N "jk" on FORCE
    S_1 B_M DIAGRAM=2.82.CWTS:
    "B." O_1 "BENDING MOMENT"
        LINK O_2 DIAGRAM.
                SHOWS MOMENTS RESISTED
SCALE OF CWTS:FEET.   BY THE BEAM & PRODUCING
    [B.M.SCALE.]   BENDING THEREOF.
15 12 9 6 3 0 15 30 45 60

$B_M = O_2 m$ (LINEAR SCALE) x h (FORCE SCALE.)
    = 9.4 (FEET) X 3 (CWTS:) = 28.2 CWTS: FEET.
$B_N = O_1 n$ (LINEAR SCALE) X h (FORCE SCALE.)
    = 6.34 (FEET) X 3 (CWTS:) = 19.02 CWTS: FEET.
$B_S = S_1 . S_1 . x h$ = 3.7 (FEET) X 3 (CWTS:) = 11.1 CWTS: FT:

To this moment diagram add a second diagram to the same scale showing the moment of force GJ about L; this is done by producing g_1O_1 to meet GJ continued at O_2, and from O_2 drawing a line parallel to Oj in the force diagram to intersect R_L at j_1.

Then e_1j_1 represents the sum of the moments of EG and GJ about the support L.

167. Scale of moments for diagrams. The scale of moments depends upon the respective scales of the space and force diagrams. By similar triangles

$$\frac{e_1g_1}{eg} = \frac{x}{h}, \quad \therefore \ e_1g_1 \times h = eg \times x.$$

But eg = the force EG or, say, F_{EG},

$$\therefore \ e_1g_1 \times h = F_{EG} \times x,$$

which is the moment of the force EG about L.

In a similar way

$$e_1j_1 \times h = (F_{EG} \times x) + (F_{GJ} \times \overline{x+y}).$$

But e_1j_1 is drawn to the scale of the space diagram (linear scale) while h is drawn to the scale of the force diagram (force scale).

Suppose $h = 3$ cwts., and the space diagram is drawn to a scale of 8 ft. to 1 inch, then the moment scale for measuring e_1j_1, or lines parallel thereto, would be 3 (cwts.) × 8 (ft.) = 24 cwts. ft. *per inch of depth*. As the triangles $e_1O_1g_1$ and $g_1O_2j_1$ are "moment diagrams", the vertical depth of the whole diagram at any selected section of the beam represents the total moment of the forces involved, about that section. Thus, the length $ss \times h$ represents the sum of the moments of F_{EG} and F_{GJ} about the section S of the beam.

Note. The scale is constant for all ordinates because the polar distance h is constant for all the force triangles in the polar force diagram.

168. To determine the reactions. Prolong j_1O_2 as shown by broken line to intersect the R_R line in O_3; join O_3 to e_1 as shown by stroke and dot line. Then $e_1O_3j_1$ is a triangle, any vertical ordinate of which would represent the moment of R_R about the corresponding point of the beam.

For equilibrium, it is shown in paragraph 155 that the reaction at R must have a moment about L equal and opposite to the sum of the moments of EG and GJ about L. The triangle $e_1O_3j_1$ has the same ordinate as the two first triangles, viz. j_1e_1 is equal and opposite to $e_1g_1 + g_1j_1$. It is required to know what force acting in the line of R_R will produce the moment represented by j_1e_1. This will be obtained on the load line by reproducing the triangle similar to $e_1O_3j_1$ having the same pole O, in order to maintain the same scale.

The triangle required is seen to be completed at kOj on the force diagram if a line be drawn from O parallel to e_1O_3 intersecting ej at k. Then, by Bow's notation on the space diagram and the corresponding letters on the force diagram, it is seen that the magnitude of jk refers to the upward force at R, viz. F_{JK} or R_R, while the magnitude of ke refers to the upward force at L, viz. F_{KE} or R_L.

169. Graphical procedure for obtaining reactions of parallel force. To obtain the reactions of a beam or structure, freely supported and carrying a series of parallel forces, note that the reactions are parallel to the forces and that their sum is equal and opposite in direction to the sum of the series. Then the graphical procedure to determine the magnitude of the reactions, and carried out in detail No. 58 at B, is:

(*a*) Draw a space and loading diagram to a convenient scale, showing the position and direction of the loads and use Bow's notation with capital letters for reference to these forces.

(*b*) To a convenient scale set down a line of loads, as the latter occur read clockwise round the structure, and letter these to correspond with the space diagram but using small letters.

(*c*) Select any point outside the load line and join each division of the latter to it.

(*d*) Prolong the lines of action of the forces on the space diagram; then draw links across each space parallel to the polar line having the corresponding letter to that in the space between any two forces, letting the links join end to end until they bridge the space between the two extreme forces.

(*e*) Join the extreme ends of the links (between their terminations on the lines of the reactions) by a "closing line". From the polar point O draw a line parallel to the closing line to intersect the load line; this point of intersection divides the total load into two parts which are the reactions at the ends having corresponding letters. The letter for this point of division on the load line is always similar to the one placed between the reactions on the space diagram.

Note. The above mode of procedure should be carried out for several examples, checked by calculation, and memorised for general use, because it can be applied with slight variations for any system of loading in one plane. See application to roof trusses with inclined forces, Chapter Ten.

170. Bending moment diagrams. Referring back to diagram A of the same detail, it should now be observed that if the triangle $e_1O_3j_1$ and the figure $e_1O_1O_2j_1$ represent moments of certain forces acting in opposite directions and therefore partially neutralising each other, that portion of one diagram which does not coincide with the other but overlaps it, must represent the nett active

moments which have to be resisted by the beam or structure between the supports.

In this case the unbalanced moments are represented by the
shaded part of the figure at A or by the separately drawn figure at B,
which is, in fact, a bending moment diagram of the same form as
the graphs previously plotted from calculated values; the only
difference is, that the graphically obtained diagrams are not usually
on a horizontal base.

To find the bending moment at any section it is only necessary
to measure the depth of the ordinate intercepted across the
diagram *in the direction of action of the loads*, at the corresponding
position, and to read this value to the bending moment scale; or,
as an alternative, measure the depth of the ordinate to the linear
scale of the space diagram and multiply it by the polar distance h
measured to the force scale. Using a slide rule this is a speedy
operation and gives good practical results for checking calculations.

It is possible by making h an easy factor such as 10, 20, 50 or 100
units of force, to avoid awkward multiplications which might arise
by placing the pole O indiscriminately.

171. Graphical process of constructing bending moment diagram.
On examination of the process of finding reactions graphically, as
described in paragraph 168, it will be seen that the bending
moment diagram is produced at the same time as the reactions are
found, being merely the "difference diagram" between the moments
of the forces and of the reactions, about the opposite support.

CHAPTER SEVEN

STRUCTURAL DESIGN

DESIGN AND SELECTION OF BEAMS

Practical distribution of loads on beams

Having shown how to obtain the shear force and bending moment at any section of a loaded beam, it is now necessary to consider how the incidence of practical loading agrees with the cases studied and how to recognise them.

172. Concentrated or point loads. Where a load has been represented upon a beam by a single arrow, it has been assumed that the load is brought to bear upon the beam over a small area, or length of it. The load may then be conceived, without serious error, to be concentrated in the line of the arrow. Hence, (*a*) a stanchion standing upon a beam; (*b*) the end of a roof truss or partition supported by a girder; (*c*) a joist or a secondary girder resting upon a main girder, are all practical examples of point loads.

It is not possible to have a point load in practice, but only reasonable approximations, because every load must be transmitted by sufficient area to prevent the bearing surface being crushed.

173. Uniform loads. Practical loading can seldom be truly uniform but it is convenient to treat many structures on the assumption that they *may* be uniformly loaded to a defined maximum load per ft. of length, or per ft. super.

Lintols and bressummers carrying portions of walls across window and door openings and across shop fronts appear to be subject to uniform loading if the wall is uniform and solid above. Actually, if the member is very stiff, resisting deflection so that it is scarcely appreciable, the load carried is approximately uniform, but as will be seen in later work, if the member sags freely much less load is supported because the mass of walling tends to arch itself over the opening, leaving a roughly triangular mass immediately on the member to be supported.

To ensure safety in calculations it is wise (in the early stages of study) to assume a uniform load over the whole span.

174. Loads on floors. In designing floors, it is usual to assume that the flooring may receive a certain uniform load over the whole surface, the amount per foot super varying with the intended use of the floor. The intensity of the load adopted—which is thought of as a dead load—is high enough to allow for the effect of probable moving loads upon the given class of floor. In addition to this *useful* or superimposed load, the dead weight of the floor framing and covering must also be allowed for; in concrete floors this weight is considerable.

175. Table of superimposed loads for which floors of different classes should be designed.[1] The following are considered as equivalent dead loads for the purposes of design.

Rooms for domestic purposes, hotel bedrooms, hospital rooms and wards ...	50 lbs. per ft. sup.
Offices, churches, schools, reading rooms, art galleries, retail shops and garages for cars of not more than 2 tons dead weight	80 ,, ,,
Assembly halls, workshops, dance halls, cinemas, staircases and landings, etc. ...	100 ,, ,,
Warehouses, book and stationery stores and garages for heavy motor vehicles ...	200 lbs. per ft. sup. at least but actual load to be calculated.

Also

Flat roofs (inclination not more than 20°)	50 lbs. per sq. ft.

Note. Where partitions are placed anywhere on floors without special provision for support, the floor must be designed to carry an additional 20 lbs. per sq. ft.

176. Loads transmitted to joists and binders. Detail No. 59 shows a portion of a double floor, and indicates flooring, joists and binder.

Suppose the entire flooring covered by a load $= w_s$ (per ft. super). Then *each board* spanning across two joists would transmit half its total load to each joist, and *each joist* would receive a similar half load from the board on each side of it, as shown by the rectangles A and B.

If the joist is completely covered by the continuous boards it eventually receives a load equivalent to that supported by the rectangle C. Each joist resting upon a girder transmits half its load to the girder, hence, if the joist is continuous across the girder or a pair of joists rest thereon, the rectangle D represents the area of equivalent load transmitted to the girder at the load point P.

---

[1] These are the values contained in the British Standard Specification No. 449—1935, and apply particularly to steel-framed buildings. In some items, the above figures are in excess of the requirements of the Code of Practice of the London County Council, in which 40 lbs. per sq. ft. is given for domestic buildings, 50 lbs. for upper floors of offices and 70 lbs. for churches, schools, etc.

The above statement is obviously true for full uniform loading only.
If point loads or partial uniform loads are transmitted and the
loading is unsymmetrical, the reactions at the end of each member
must be calculated and summed up to obtain the load at any
bearing.

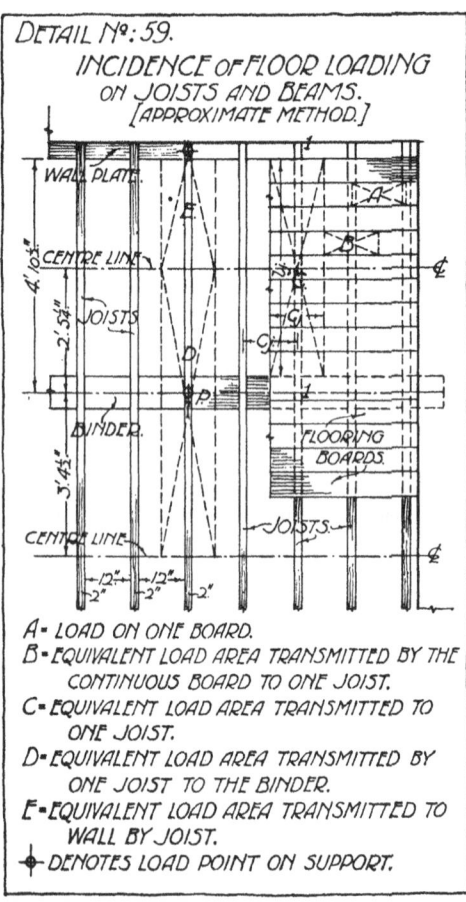

DETAIL Nº: 59.

INCIDENCE of FLOOR LOADING
ON JOISTS AND BEAMS.
[APPROXIMATE METHOD.]

A = LOAD ON ONE BOARD.
B = EQUIVALENT LOAD AREA TRANSMITTED BY THE
    CONTINUOUS BOARD TO ONE JOIST.
C = EQUIVALENT LOAD AREA TRANSMITTED TO
    ONE JOIST.
D = EQUIVALENT LOAD AREA TRANSMITTED BY
    ONE JOIST TO THE BINDER.
E = EQUIVALENT LOAD AREA TRANSMITTED TO
    WALL BY JOIST.
✛ DENOTES LOAD POINT ON SUPPORT.

177. Load on joist of detail No. 59.

Let w_s = superimposed or useful load per sq. ft.
    w_d = dead weight of joists, flooring and ceiling, per sq. ft.
    l_j = clear span of joist.
    c_j = width spacing of joists, centre to centre.

Then the total uniform load carried by the joist $1.1 = W_j$, and
$W_j = l_j \times c_j (w_s + w_d)$, which for timber, is conveniently in lbs. or
cwts.

178. Load on binder supporting joists at regular intervals. The joists rest upon the binder, each bearing surface transmitting half the joist load to the beam, or to the wall (as at panel E). Hence, the entire load on the binder is composed of a series of point loads like P of the panel D, and with uniform spacing of joists and binders the total load is

$$W_b = \text{no. of joists} \times W_j \qquad \ldots\ldots(1).$$

Or, again,

if l_b = span of binder,
and c_b = width spacing of binders, centre to centre,
and w_d = dead weight of all floor timbers, etc. averaged at per ft. sup.,

then $$W_b = l_b \times c_b \, (w_s + w_d) \qquad \ldots\ldots(2).$$

Such a binder, with joists closely spaced and continued right up to the wall at each end, as in detail No. 59, may be considered to carry a uniform load as assumed in equation (2).

179. To obtain the weight of a floor. The dead weight of a floor or of its parts will be unknown until their dimensions have been settled, *but sufficiently near approximations on the side of safety can be predetermined.*

Flooring and joists. Let northern pine be employed for these, and take the density as 35 lbs. per cu. ft. Find the weight per ft. sup. of $\frac{7}{8}''$ flooring boards and the equivalent load per ft. sup. of $9'' \times 2''$ joists at $14''$ centres.

$\frac{7}{8}$ in. boarding per ft. sup. contains $\frac{7}{8} \times \frac{1}{12} = \frac{7}{96}$ cu. ft.

Hence, 1 ft. sup. weighs $\frac{7}{96} \times 35$ (lbs.) = **2·55** lbs. per sq. ft. (say $2\frac{1}{2}$).

If boards were $1''$ thick the weight would be nearly **3** lbs. sq. ft.
If boards were $1\frac{1}{8}''$ thick the weight would be about **3·28** lbs. sq. ft. (say $3\frac{1}{4}$).

One ft. run of joist supports $\frac{14}{12}$ (ft.) $\times 1$ (ft.) $= \frac{7}{6}$ ft. sup. of flooring.

The joist weighs

$$\frac{9 \text{ (ins.)} \times 2 \text{ (ins.)} \times 35 \text{ (lbs.)}}{144 \text{ (ins. per sq. ft.)}} = \mathbf{4\cdot375} \text{ lbs. per ft. run.}$$

Hence the weight of joist covering 1 ft. sup. $= 4\cdot375 \times \frac{6}{7} = \mathbf{3\cdot75}$ lbs.

A plain unceiled floor of $\frac{7}{8}''$ boards and $9'' \times 2''$ joists would thus average, say $2\cdot5 + 3\cdot75 = \mathbf{6\cdot25}$ lbs. per sq. ft. over the entire floor, or, with $1\frac{1}{8}''$ boards $3\cdot25 + 3\cdot75 = \mathbf{7}$ lbs. per sq. ft.

Ceilings. Lime plaster as used for ceilings varies in density, but may be taken at 140 lbs. per cu. ft., which would be reduced considerably by the laths. Taking its thickness as $\frac{3}{4}''$ and ignoring the key plaster as against the laths, the *weight per ft. sup.*

$$= \left(\frac{3}{4} \times \frac{1}{12}\right) \text{ (cu. ft.)} \times 140 \text{ (lbs. per cu. ft.)} = \textbf{8·75 lbs.}$$

Total load. For an ordinary dwelling house floor, single joisted and ceiled, the weight would be, say, $6\cdot25 + 8\cdot75 = \textbf{15}$ lbs. per ft. sup. *For dwelling houses and domestic buildings generally* the minimum superimposed load allowed for design is 50 *lbs. per sq. ft.* as stated in paragraph 175.

With timber joisted floors, however, if the usual thickness of $2''$ to $3''$ be adopted for joists, a floor so designed may have the necessary strength while lacking in rigidity (see next paragraph). Until deflection of beams has been studied (after which all the conditions for any given case can be taken fully into account) it is wise to base the calculations on a higher figure of loading. The value taken for all the succeeding load calculations for the examples relating to the house is 70 lbs. per sq. ft.

Taking into account the average weights of floors and ceilings the following figures have been used:

$\qquad$ (1) $\qquad\qquad w_s = 70$ lbs. (superimposed load)
$\qquad\qquad\qquad\quad w_d = 7$ lbs. (dead load)

$\qquad \therefore$ Total $w = 77$ lbs. per sq. ft. for unceiled floors.

$\qquad$ (2) $\qquad\qquad w_s = 70$ lbs. (superimposed load)
$\qquad\qquad\qquad\quad w_d = 15$ lbs. (dead load)

$\qquad \therefore$ Total $w = 85$ lbs. per sq. ft. for ceiled floors.

If joists are heavier than the average, 80 lbs. may be adopted for unceiled and 90 lbs. for ceiled floors.

In designing main beams to carry the joists of a floor—particularly for bedrooms—it is unnecessary to design them for the purpose of carrying the above named unit loads over the entire floor. For *this* purpose 50 to 60 lbs. per sq. ft. is sufficient to allow. For further explanation and an application of this principle see paragraph 185.

TIMBER BEAMS

180. Sizes of joists. Many Bye-laws specify the sizes of joists to be employed for given spans in different classes of buildings but some of these are very faulty, being uneconomical in material and not rigid enough to prevent vibration. The important points to note in selecting joists are:
(*a*) strength,
(*b*) rigidity,

(c) breadth enough for secure nailing and for receiving heading joints in the flooring,

(d) sufficient depth to allow of X bracing or herring-bone strutting in spans over 6 ft.

Usually (d) is satisfactory if (b) is provided for.

Strength. By the strength of a beam is meant its capacity to carry load without overstressing the fibres. It has already been shown that strength increases directly with the breadth and with the square of the depth, and decreases directly with an increase of span; see paragraph 148.

Rigidity. By the rigidity of a beam is meant its capacity to resist deflection or sag under load. Rigidity increases directly with the breadth and with the cube of the depth and decreases with the cube of the span.

It is therefore important to pay attention to rigidity because floors should not noticeably vibrate at every movement, as is so often the case.

The subject of deflection cannot rationally be dealt with in this volume, but any serious errors can be guarded against by selecting a high factor of safety for the material.

With average sizes of well proportioned joists of northern pine an economical result may be expected and a rigid floor obtained by adopting a uniform breadth of 2″ and limiting the maximum stress on the material to 1000 lbs. per sq. inch.

181. Working stresses for common timbers.

From the authors' experiments on a large number of good average specimens of northern pine building timber, the mean stress on the outer fibres at which rupture occurred was about 8000 lbs. sq. in., and on pitch pine about 10,600 lbs. sq. in. Hence, if working stresses of 1000 lbs. sq. in. be adopted for the northern pine and 1350 lbs. sq. in. for pitch pine, the factor of safety is about 8. The student should note that this is an elementary way of guarding against undue deflection, *not merely to ensure strength*, and also to make allowance for hidden defects in timber. In most cases, for strength, a factor of 5 would be sufficient.

If spruce fir is to be used its stress should be limited to 900 lbs. sq. in., and oak on an average to 1300 lbs. sq. in.

182. Rule for depth of 2″ joists, of northern pine, 12″ apart.

Let $w_s + w_d = w_t$ (total weight per ft. sup.). In this case 85 lbs. per sq. ft. will be adopted as the inclusive designing load. Then for a uniform load, see paragraph 162,

$$B_M = \frac{WL}{8} \qquad \ldots\ldots(1).$$

Also $$W = l_j \times c_j \times w_t.$$

But l_j and L are practically the same.

$$\therefore \ W = L \text{ (ft.)} \times \frac{14}{12} \text{ (ft.)} \times 85 \text{ (lbs.)}$$

$$= \frac{595}{6} L \text{ (lbs.)}.$$

Inserting this value in (1),

$$B_M = \frac{595L}{6} \times \frac{12L}{8} = \frac{595L^2}{4} \text{ (lbs. ins.)} \qquad \ldots\ldots(2).$$

Note. L is multiplied by 12 to reduce lever arm to inches.

But $\qquad B_M = R,$

and $\qquad R = \dfrac{bd^2f}{6}$ (see paragraph 146).

Let $\qquad f = 1000$ lbs. per sq. in.

and $\qquad b = 2$ ins.

Then $\qquad R = \dfrac{2 \times 1000d^2}{6}$

$$= \frac{1000}{3} d^2 \qquad \ldots\ldots(3).$$

Equating (2) and (3),

$$B_M = R.$$

$$\therefore \ \frac{595}{4} L^2 = \frac{1000}{3} d^2,$$

$$\frac{3 \times 595 L^2}{4 \times 1000} = d^2,$$

$$\therefore \ d^2 = \cdot 446 L^2,$$

and $\qquad d = \sqrt{\cdot 446} . L = \cdot 667 L$, say $\cdot 7L$.

Hence the rule: *depth of joist in inches equals* ·7 *of the span in feet.* The rule strictly applies for 2″ joists at 12″ apart and 14″ centres, but is safely applicable up to 15″ centres.

Detail No. 60 shows the graph of the above rule, from which the depth of a 2″ joist for any given span up to 16 ft. can be directly read, if the joists are space at 14″ centres or thereabouts. Two additional graphs on the same detail show the relation of depth to span for 2½″ and 3″ joists under similar conditions.

183. Effective span. For obtaining the leverage in calculating bending moment the effective span L should be employed, viz. the distance between centres of bearings as indicated in detail

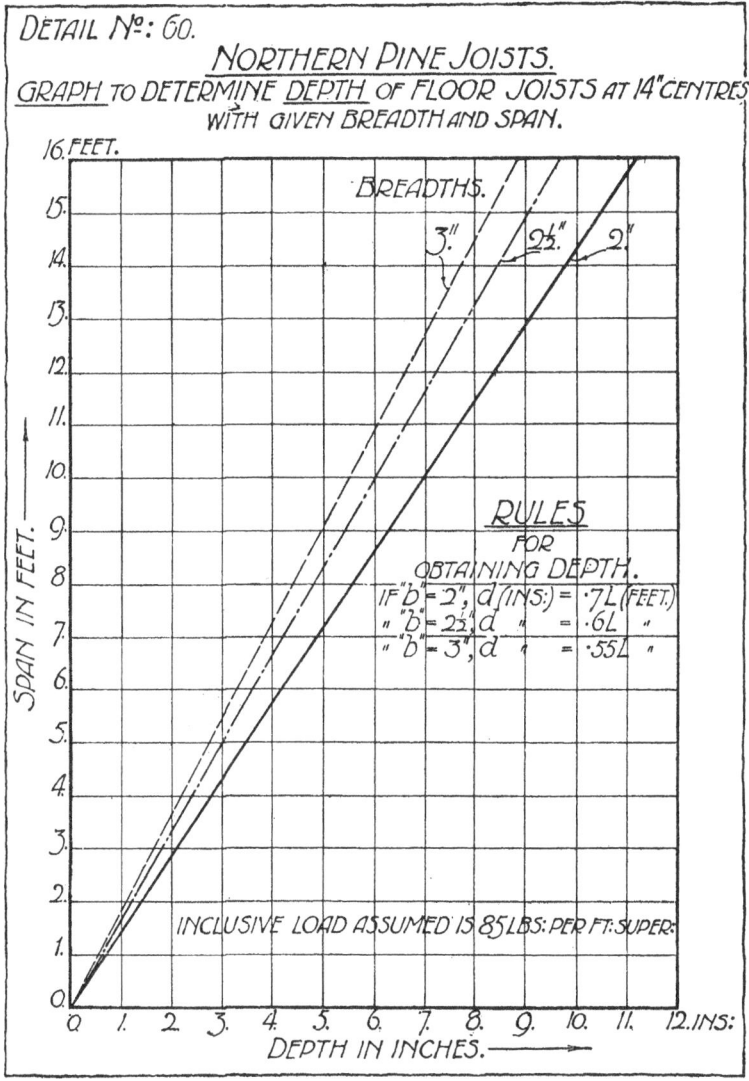

DETAIL No: 60.

NORTHERN PINE JOISTS.

GRAPH TO DETERMINE DEPTH OF FLOOR JOISTS AT 14"CENTRES
WITH GIVEN BREADTH AND SPAN.

RULES
FOR
OBTAINING DEPTH.
IF "b" = 2", d (INS:) = ·7L (FEET.)
 " "b" = 2½", d   "   = ·6L "
 " "b" = 3", d   "   = ·55L "

INCLUSIVE LOAD ASSUMED IS 85 LBS: PER FT: SUPER:

Note. Students are recommended to redraw the graphs given in Details
Nos. 60, 82, 83 and 86 upon suitable squared sectional paper, in order that
intermediate dimensions to those given may be obtained.

No. 61. For calculating the loading the length l_b is required. Usually the difference is not great, but it is wise to differentiate between them because odd cases occur where the difference is considerable. Having obtained the correct load, treat it as if uniform over the whole span, which gives a close approximation to the actual conditions.

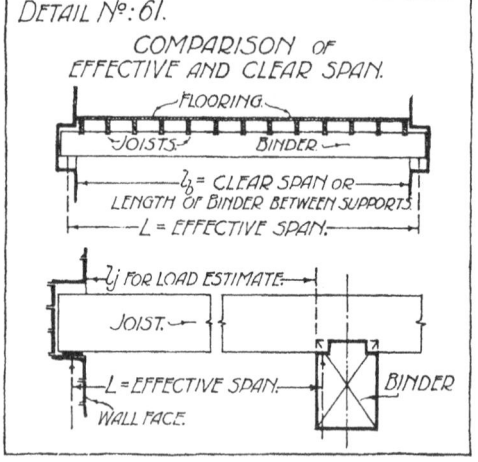

DETAIL Nº: 61.

COMPARISON OF EFFECTIVE AND CLEAR SPAN.

FLOORING.
JOISTS.   BINDER.
l_b = CLEAR SPAN OR LENGTH OF BINDER BETWEEN SUPPORTS
L = EFFECTIVE SPAN.

l_j FOR LOAD ESTIMATE.
JOIST.
L = EFFECTIVE SPAN.   BINDER
WALL FACE.

184. Selection of joists for the house. By using the graph of detail No. 60 the depth of the joists required for the floors of the house may be selected; these joists are all 2″ thick and spaced at approximately 14″ centres. The sizes selected are as follows and represent the nearest standard scantlings usually obtainable.

Floor levels	Room	Approx. span	Graph size	Practical size
Ground floor	Dining and drawing rooms	5′ 0″	3·5″	4″
,,	Kitchen	6′ 0″	4·2″	4″ or 4½″
First floor	Bedrooms Nos. 1 and 2	6′ 3″	4·3″	4½″
,,	Bedroom No. 3	12′ 0″	8·4″	9″
,,	,, 4 and bathroom	8′ 0″	5·6″	6″
,,	Landing	14′ 10″	10·4″	11″
Attic floor	Bedroom No. 5	14′ 0″	9·8″	9″*
,,	,, 6	6′ 3″	4·3″	4½″
,,	,, 7	12′ 0″	8·4″	9″
,,	Cistern room	8′ 0″	5·6″	6″
,,	Landing	9′ 6″	6·7″	7″
,,	Box room	11′ 0″	7·7″	8″

* 9″ joists are used here because part of the floor area is not boarded, and one end is fixed by the roof timbers.

Note. It is usual, where spans vary on one floor level, to select the size that will be safe for the larger span and employ this throughout if the change of size would cause practical inconvenience.

185. Design of binders for double floor over drawing room. (Level ceiling.) The binders of the double floor do not carry a symmetrical load because the middle or "case" bay is larger than the end or "tail" bays; detail No. 62. Also, the tail bay extends across the full width of the room while the case bay is shortened by the hearth and chimney breast. The hearth is, however, partially carried on the front joist which transmits load to the binder and this load is greater over the hearth than at other parts of the floor. Considering these points, it is justifiable to assume that a uniform load over the portion crossed diagonally in the detail will be fully equivalent to the actual load.

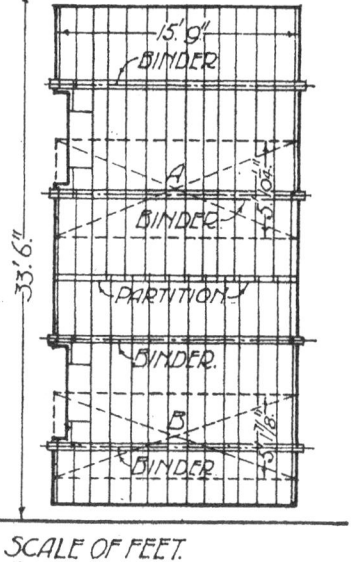

DETAIL N°:62.
THE HOUSE.
LINE DIAGRAM OF DOUBLE FLOOR OVER DINING AND DRAWING ROOMS SHOWING EQUIVALENT LOAD AREAS FOR BINDERS.

SCALE OF FEET.

A = EQUIVALENT LOAD AREA ON BINDER OVER DRAWING ROOM
B = EQUIVALENT LOAD AREA ON BINDER OVER DINING ROOM.

The joists for this floor were designed for a load of 85 lbs. per sq. ft. inclusive, because any one joist might be well loaded, but for the whole area a load of 65 lbs. per sq. ft. has been assumed, as it cannot be conceived that any great percentage of the floor of an ordinary bedroom will be subjected to the greater load.

The design is for a flush ceiling; northern pine beams, built up of $11'' \times 3''$ or $11'' \times 4''$ planks bolted together, are proposed, as shown in Vol. II; therefore $d = 11''$, and b is required.

Assuming the load as 65 lbs. per ft. sup. over the area shown,

$$W = l_b \times c_b \times w_t.$$

l_b = length of binder between walls = $15\frac{3}{4}$ ft.
c_b = distance between centres of *bays* = $5'$ 10″ or 5·83 ft.
w_t = 65 lbs. per sq. ft.

$$\therefore \quad W = 15\cdot75 \times 5\cdot83 \times 65$$
$$= 5968, \text{ say } \mathbf{6000 \text{ lbs.}}$$

Then, taking effective span, L, as $15'\ 9'' + 6\frac{3}{4}''$ = say $16\frac{1}{4}$ ft., approximately,

$$B_M = \frac{WL}{8} = \frac{6000 \times (16\cdot25 \times 12)}{8} = 9000 \times 16\cdot25 \text{ lbs. ins.}$$

But $\qquad B_M = R$, and $R = \dfrac{bd^2 f}{6}$.

$$f = 1000 \text{ lbs. per sq. in.}$$
$$d = 11''.$$

Substituting values,

$$\therefore\ 9000 \times 16 \cdot 25 = \frac{b \times (11 \times 11) \times 1000}{6}.$$

Hence $\qquad\qquad b = \textbf{7·25}$ inches,

and the nearest practical size, larger, is $8''$.

As the weight of the beam is unknown, though small, adopt, say, two $11'' \times 4''$ planks to form the beam.

Note. Weight of beam of length L is approximately

$$\frac{16\frac{1}{4} \times 8 \times 11 \times 35}{144} = 347 \cdot 5 \text{ lbs., say } \textbf{350 lbs.,}$$

which is 5·8 per cent. of the load carried. It is possible to find the exact breadth to allow for the weight of the binder, but the calculation is more difficult and quite unnecessary when all the possible variations of conditions are considered.

186. Design of binder for double floor over dining room. (Panelled ceiling.) In this case a panelled ceiling is required, the plaster being attached to the soffit of the joists and the binders wrought and moulded to stand below the ceiling as in Vol. ii. Pitch pine or oak beams may suitably be used. The former is available in large square balks, is an excellent rigid timber for beams, and is adopted in this example, with a working stress of 1350 lbs. per sq. in.

This is much below the working strength of the material, but if much exceeded in beams of ordinary proportions excessive deflection is produced.

The equivalent load area carried by the binder is shown in detail No. 62 and is a rectangle $l_b \times c_b$.

$$l_b = \text{length of binder between walls} = 15\tfrac{3}{4} \text{ ft.}$$
$$c_b = \text{distance between centres of bays} = 5\tfrac{1}{4} \text{ ft.}$$
$$w_t = 65 \text{ lbs. per sq. ft.}$$

Hence
$$\begin{aligned} W &= l_b \times c_b \times w_t \\ &= 15 \cdot 75 \times 5 \cdot 25 \times 65 \\ &= \textbf{5375} \text{ lbs. (approximately).} \end{aligned}$$

$$B_M = \frac{WL}{8}$$

$$L = 16\tfrac{1}{4} \text{ ft.} = 195''.$$

$$= \frac{5375 \times 195}{8} \text{ lbs. ins.}$$

But $\qquad B_M = R$, and $R = \dfrac{bd^2 f}{6}$.

$$f = 1350 \text{ lbs.}$$
$$d = 12'' \text{ (as a trial).}$$
$$b \text{ is required.}$$

Substituting values,

$$\therefore \ \frac{5375 \times 195}{8} = \frac{b \times 12^2 \times 1350}{6},$$

$$\therefore \ b = 4\cdot04''.$$

As before, the weight of the beam has been neglected, hence the thickness is increased to $4\frac{1}{2}''$ or $5''$. Such a beam would have to be cut from, say, a $12''$ square balk and to get clean wrought surfaces should be sawn or "slabbed", as shown in Vol. II, which would produce two $11\frac{1}{2}'' \times 5\frac{1}{2}''$ beams, clean on all visible surfaces. On testing these they are found to be a little larger than necessary, but have been adopted in this case as the nearest practicable sizes.

187. Example of design when ratio of breadth to depth is given, applied to irregular point loading. As an example of procedure, let an oak beam be required to carry the loads of detail No. 55, for the case illustrated. Let the working stress be 1300 lbs. per sq. in., and the beams be cut out of a square balk so that the breadth is half the depth, viz. $b = \dfrac{d}{2}$.

The B_M is known from paragraph 163, viz. 14 tons ft.

As the working will be in units of lbs. and ins., the value given in tons ft. must be converted to lbs. ins.

Then $B_M = 14 \times 12 \times 2240$ (lbs. ins.),

$$R = \frac{bd^2f}{6},$$

and, substituting $\dfrac{d}{2}$ as the value of b, an expression is obtained, containing only one unknown, viz. d.

Then $R = \dfrac{d}{2} \cdot \dfrac{d^2f}{6} = \dfrac{d^3f}{12}.$

$$f = 1300 \text{ lbs. per sq. in.}$$

Substituting and equating: $B_M = R$,

$$\therefore \ 14 \times 12 \times 2240 = \frac{d^3 \times 1300}{12}.$$

Evaluating and transposing: $d^3 = 3474$ (approximately),

and $d = \sqrt[3]{3474} = 15\cdot15''.$

$$\therefore \ b = \frac{d}{2} = 7\cdot57''.$$

The nearest practicable beams are $15'' \times 7\frac{1}{2}''$, and $16'' \times 8''$.

With practice on variations of the preceding examples, the student should now be able to follow the routine design of a wooden beam, and to obtain its dimensions when given the amount and manner of loading, working stress and sufficient information to determine the breadth or the depth, or both.

STEEL BEAMS

188. Resistance to bending. A steel joist of **I** section resists bending moment in a similar manner to a wooden beam, that is to say, it deflects under the load, straining uniformly at any section from zero at the neutral axis to a maximum at the extreme edges; then, as stress and strain are proportional to each other the stress distribution above and below the neutral axis may be represented by a figure such as detail No. 63. In a freely supported beam the upper portion represents compressional stress and the lower portion tensional stress, while the ordinates f and f_1 represent the intensity of stress per square inch at their respective positions.

189. Approximate "resistance moment". It should be clear from this diagram when compared with detail No. 48 that the principal resistance is offered by the flanges and it is commonly assumed —

DETAIL Nº:63.

STRESS DISTRIBUTION and INERTIA DIAGRAMS FOR I AND J BEAMS.

F_c = AREA OF FLANGE MULTIPLIED BY f.

LEVER ARM = $\frac{d}{2} - \frac{t}{3}$ OR ROUGHLY $\frac{d}{2}$ BECAUSE $\frac{t}{3}$ IS SMALL.

F_t = AREA OF FLANGE MULTIPLIED BY f_1

BEAM SECTION.   CHANNEL SECTION.

for purposes of rough estimation—that the moment of resistance (R) of the beam

$$= 2 \times \left(\text{area of the flange} \times \text{stress per sq. inch} \times \frac{d}{2} \right).$$

This means that R = the sum of the moments of the total stresses in the flanges about the N.A. of the beam.

Let A = area of *one* flange,
                              f = maximum fibre stress,
                              d = full depth of beam.

Then $R = 2\left(A \times f \times \dfrac{d}{2}\right)$,

or $R = A \times f \times d$.

Actually the stress f does not act over all the flange area A, neither does all the stress in the flange act with the lever arm $\dfrac{d}{2}$, but the web, which has been neglected, compensates somewhat for the over-estimation of the flange resistance.

190. Accurate "resistance moment". Steel beams are made to many standard sizes, the full properties of which have been officially determined and tabulated, hence the accurate value of R can be quickly obtained by the use of these tables. Lists of the properties of British Standard Beams are given in the Appendices, being defined by each beam in the standard list, its outside dimensions, weight per ft. run, area of section in sq. in., thickness of flanges and of web, and the three special properties known as the "moment of inertia", the "radius of gyration" and the "modulus of section".

The modulus of section (SM, or Z) is a measure of the capacity of the *form* and *dimensions* of the section to resist bending and on reference to paragraph 146 it will be recalled that the Z of a rectangular beam = $\dfrac{bd^2}{6}$. In a general way, this is obtained from the mathematical value of the physical property called the moment of inertia (I) of the rectangular section about its neutral axis, the value being expressed by $I = \dfrac{bd^3}{12}$,* where d is always the dimension at 90° to the axis. It can be shown that $Z = \dfrac{I}{y}$, where y is the distance from the axis to the extreme fibres, in this case $\dfrac{d}{2}$,

$$\therefore\ Z = \frac{bd^3}{12} \div \frac{d}{2} = \frac{bd^2}{6},$$

as before.

Note. The moment of inertia is obtained by multiplying $b \times d \times d \times d$ and dividing by the constant, 12. The result is therefore in dimension units to four factors, and if the dimensions are in inches would be stated as, say, 1200 inch⁴ units.

The modulus of section contains three dimension factors and would therefore be stated as, say, 120 inch³ units.

* See Chapter Eight for proof.

191. Moment of inertia of I and ⊐ sections. The moment of inertia for an **I** section about the neutral axis parallel to the flanges is the sum of the quantities $\frac{bd^3}{12}$ for all the rectangles composing the section, *provided that each rectangle has its neutral axis on the* N.A. *for the whole section* (see paragraphs 201 to 205 for explanation of Moment of Inertia).

In detail No. 63 the section can be conceived as being formed of the enclosing rectangle less the two rectangles between flanges and web, each rectangle complying with the above condition.

Hence $I_{NA} = I$ for whole rectangle $- I$ for the rectangles omitted, and using the notation shown:

$$I_{NA} = \frac{BD^3}{12} - \frac{2\frac{b}{2}d^3}{12}$$

$$= \frac{1}{12}(BD^3 - bd^3).$$

The moment of inertia of a channel about the N.A. parallel to the flanges is expressed in the same way, viz.

$$I_{NA} = \frac{1}{12}(BD^3 - bd^3).$$

192. Section modulus of I and ⊐ sections. The modulus of section (Z) of the standard beam or channel about its neutral axis parallel to the flanges is the value $I_{NA} \div \frac{D}{2}$, and can therefore be expressed by

$$Z = \frac{I_{NA}}{\frac{D}{2}} = \frac{\frac{1}{12}(BD^3 - bd^3)}{\frac{D}{2}} = \frac{1}{6}\frac{(BD^3 - bd^3)}{D}.$$

The numerical value of Z in inch³ units is given for each standard section in the Appendices.

193. Moment of resistance of I and ⊐ sections. The moment of resistance of any section is expressed by $R = Zf$; but B_M must equal R, therefore in any beam design $B_M = Zf$ or $\frac{B_M}{f} = Z$, where f is the allowable extreme fibre stress and may be taken as 8 tons per sq. inch for structural steel.

To select a steel section the process is therefore:

(a) Determine the maximum bending moment (B_M).

(b) Divide B_M by the allowable working stress to obtain the required value of Z.

(c) Select a section from the list having the least weight per ft. run, for the nearest value of Z.

Examples of design from the house and warehouse will now be taken in the order of difficulty.

SELECTION OF SUITABLE STEEL SECTIONS FOR GIVEN CONDITIONS OF LOADING

The whole of the steelwork in this section has been designed for extreme fibre stresses of 8 tons per sq. in. (See list of working stresses for steel on page 424.)

194. Lavatory floors to warehouse. Details of the steel and concrete floors to the landings and lavatories of the warehouse were given in Vol. II.

The space to be bridged is about 14 ft. long × 8 ft. 9 ins. between walls, and steel joists are placed across the shorter direction as shown in detail No. 64. Assuming them to enter the walls $4\frac{1}{2}''$ at each end and the effective span to be measured between the centres of the bearings, then the effective span $= 8'\ 9'' + 4\frac{1}{2}'' = 9'\ 1\frac{1}{2}''$ (say 110″). The joists are spaced at $2'\ 2\frac{1}{2}''$ centres, say 2·2 ft., and the load supported by each joist is the equivalent of the total load upon the crossed bay in the detail.

The floor is designed to carry 100 lbs. per sq. ft.[1] in addition to its own weight. The joists are widely spaced and form a small proportion

DETAIL Nº: 64 THE WAREHOUSE.

KEY PLAN.—LAVᵞ FLOOR.

of the cross section of the floor; therefore assume the weight of steel and concrete to be 144 lbs. per cu. ft., then the weight per ft. sup. at 7″ thick would be $\frac{7}{12} \times 144 = \mathbf{84}$ lbs.

[1] See paragraph 175.

The total designing load is therefore $100+84=184$ lbs. per ft. sup. and each joist may be assumed to transmit the load on one bay $2 \cdot 2$ ft. wide and $8 \cdot 75$ ft. long; hence

$$\text{load} = \frac{2 \cdot 2 \times 8 \cdot 75 \times 184}{2240} = 1 \cdot 58 \text{ tons (uniform).}$$

Then
$$B_M = R,$$
$$\therefore \frac{WL}{8} = Zf.$$

But
$$W = 1 \cdot 58 \text{ tons.}$$
$$L = 110 \text{ ins.}$$
$$f = 8 \text{ tons per sq. inch.}$$
(Code of Practice)

$$\therefore \frac{1 \cdot 58 \times 110}{8 \times 8} = Z,$$
and
$$Z = 2 \cdot 72 \text{ inch}^3 \text{ units.}$$

Referring to the list of B.S.B.'s in Appendix I, the nearest size is $4\frac{3}{4}'' \times 1\frac{3}{4}'' \times 6 \cdot 5$ lbs. per ft., having $Z = 2 \cdot 83$ inch units.[1]

Observe that no account has been taken of the strength of the concrete in the above computation, but it is assumed that the concrete will be strong enough to transmit the whole panel load to the joists. But if the concrete slab is supported on the walls at one pair of edges and on joists at the other pair it will transmit part of its load to the walls. Further, the concrete adheres to the steel and compels the two materials to act together. Provided that the spacing of joists is reasonable another procedure is allowed by the Code of Practice. This Code provides that for filler beams in concrete floors, the strength of the floor slab (including the beams) may be calculated as a reinforced concrete floor, if the beams are fully encased in the concrete. Or, the concrete may be neglected, and, to allow for its assistance in strengthening the steel joists, the extreme fibre stress in the steel may be calculated at 9 tons per sq. inch instead of 8. In this particular case, the same size of joist would be necessary if selected from the British Standard sizes.

195. Alternative method of selecting and spacing steel joists. An alternative and very useful method of practical procedure is to select the size of joist which is thought suitable for the desired thickness of floor and to calculate the centre spacing of the steel joists.

Thus
$$B_M = R,$$
$$\therefore \frac{WL}{8} = Zf,$$
or
$$W = \frac{8Zf}{L}.$$

[1] All future references to values of Z will be given in *inch units* (or $''$ units), the degree of three dimension factors being understood.

Suppose that a B.S.B., $4'' \times 1\frac{3}{4}'' \times 5$ lbs. per ft. run is selected, having a section modulus $(Z) = 1 \cdot 834$ inch units. To enclose this joist the floor would be suitably about $6''$ thick and the designing load per ft. sup. (w_t)

$$\doteqdot 100 + \frac{6 \times 145}{12} = 172 \cdot 5 \text{ lbs., say } \mathbf{172} \text{ lbs.}$$

Then
$$W = \frac{8Zf}{L}.$$

But
$Z = 1 \cdot 834$ inch units.
$f = 8$ tons per sq. inch.
$L = 110$ ins.

$$\therefore \ W = \frac{8 \times 1 \cdot 834 \times 8}{110},$$

or
$$W = \mathbf{1 \cdot 07} \text{ tons.}$$

But
$$W = l_j \times c_j \times w_t,$$
$$\therefore \ c_j = \frac{W}{l_j \times w_t}.$$

Also
$l_j = 8 \cdot 75$ ft.,
$w_t = \dfrac{172}{2240} = \cdot 077$ ton.

$$\therefore \ c_j = \frac{1 \cdot 07}{8 \cdot 75 \times \cdot 077} = \mathbf{1 \cdot 59} \text{ ft.}$$

Such joists may therefore be spaced at $1'\ 6''$ centres.

196. Lavatory floors—alternative form of joist. It is sometimes preferable to employ two channels in lieu of the solid $\mathbf{I}$ section. These would be placed web to web, packed apart by steel disc packings and riveted or bolted together at about 3 ft. centres, as illustrated and described in Vol. ii.

To select the size of channel proceed as for the $\mathbf{I}$ beam.

Thus, the required Z from paragraph 194 is $2 \cdot 72$ inch units and this must be supplied by two channels, therefore each requires Z to equal $1 \cdot 36$ inch units. Referring to a list of British Standard Channels, the nearest is B.S.C., $3'' \times 1\frac{1}{2}'' \times 4 \cdot 6$ lbs. per ft. run, $Z = 1 \cdot 21$ inch units.

If the horizontal thickness of floor slab be retained this channel has apparently too low a value of Z, while the next one, $4'' \times 2''$ $\times 7 \cdot 09$ lbs., is much too large having $Z = 2 \cdot 53$ inch units. But, considering the depth of the smaller channel is only $3''$, the thickness of the floor slab can be reduced to $4\frac{1}{2}''$ or $5''$; for any case the cover below the bottom flange need only be $1''$ while little or no cover is *necessary* above the top flange, though it may be a convenience for levelling up to receive the wearing-surface coat of cement, tiles, etc. In addition the floor concrete adds appreciably to the strength of the channel joists as will be found in later studies.

Assume a finished floor thickness of 5″ (which will give a reduced weight) and check the spacing in the manner adopted in the last paragraph:

$$\frac{WL}{8} = Zf,$$

or
$$W = \frac{8Zf}{L}$$

But
$$Z = 2 \times 1\cdot33 = 2\cdot66 \text{ ″ units.}$$
$$f = 8 \text{ tons per sq. in.}$$
$$L = 110 \text{ inches.}$$

$$= \frac{8 \times 2\cdot66 \times 8}{110}.$$

$$\therefore W = \mathbf{1\cdot54} \text{ tons.}$$

Now W also $= l_j \times c_j \times w_t,$

But
$$w_t = 100 + \frac{5 \times 145}{12} = \text{say } \mathbf{160} \text{ lbs. (as against 184 lbs. in the calculation for the thicker floor).}$$

$$\therefore w_t = \frac{160}{2240} = \mathbf{\cdot071} \text{ ton,}$$
$$l_j = 8\cdot75 \text{ ft.}$$

and
$$c_j = \frac{W}{l_j \times w_t},$$

$$\therefore c_j = \frac{1\cdot54}{8\cdot75 \times 0\cdot07} = \mathbf{2\cdot48} \text{ ft.}$$

Hence the double channels may be spaced at 2′ 6″ centres, which is more than the spacing of 2′ 2½″ conveniently adopted in the details.

Note. The fault of these shallow channels is their excessive deflection if fully loaded.

197. Use of Structural Handbooks giving tabular loads for uniform loading. In practice a large number of problems involve uniform loading and the safe load for any beam can be speedily obtained from tables contained in the various handbooks of structural sections issued by steel firms.

The tables available are generally given as:

(a) Safe uniform load for a 1 ft. span,

or (b) Safe uniform load for a tabulated span,

and in many cases both tables are given.

The safe loads in the second form of table are read off directly, the load being given in tons for given spans in feet, while the values in the first named table require conversion.

Example. The load carried by the $4\frac{1}{2}'' \times 2'' \times 7$ lbs. B.S.B. used in the lavatory floors of the warehouse.

Referring to Redpath, Brown and Co.'s handbook,[1] page 4, the uniform load is approximately 1·7 tons on a 9 ft. span or from page 5 (column 10) the safe uniform load for a 1 ft. span is 15·7 tons for a working maximum stress of 8 tons per sq. inch.

But the B_M increases directly with the span and the safe load decreases similarly, hence $\dfrac{\text{tabular load}}{L \text{ (feet)}} = $ the load W required.

Hence $W = \dfrac{15 \cdot 7}{9 \cdot 1} = 1 \cdot 72$ tons,

being the load which may be carried by each joist, and from which the spacing can be calculated. (Students should test the actual joists used by this means.)

Again, testing the channels employed as an alternative, from Redpath, Brown and Co.'s handbook, page 59 (column 10), it is found that the safe uniform load on a B.S.C., $3'' \times 1\frac{1}{2}'' \times 4\cdot6$ lbs., is 6·4 tons and therefore 12·8 tons on two such channels, fixed together.

Hence, as before,

$$W = \frac{\text{tabular load}}{L \text{ (feet)}} = \frac{12 \cdot 8}{9 \cdot 1} = 1 \cdot 406 \text{ tons,}$$

which is a little less than the previous value given in paragraph 196, as would be expected because the joists alone are a little under strength.

Note. For uniform loads and a working stress of 8 tons per sq. inch (or other tabular basis for f) this method of procedure is the most practical. We must, however, develop and apply the initial method of procedure, which is *general* and has been used to obtain the tabular loads. For irregular loading it is preferable to a second conversion of the tabular load, and quite as speedy.

HALL FLOOR OF HOUSE—SELECTION OF STEEL MEMBERS

Refer now to the detail No. 65, which shows the arrangement of the hall floor to the house, the construction of which has already been described in Vol. II.

The plan shows the positions of the beams and joists forming a double floor trimmed for a staircase.

198. Steel joists to hall floor. These members run lengthwise of the hall across the beams, and have their maximum span across the $7' \ 7\frac{1}{2}''$ wide bays, giving the load area *abcd*. Assume, for calculation, that the joists are in single spans of $7' \ 6''$ effective length which is approximately the centre to centre spacing of girder and wall bearing in the end bay.

In Vol. II the finished floor is shown to be about $6''$ thick and its

[1] 1928 Edition.

weight may be taken at 145 lbs. per cu. ft. or, say 72 lbs. per ft. sup. Designing for a useful load of 70 lbs. per sq. ft. the total designing load is $72 + 70 = 142$ lbs. per ft. sup.

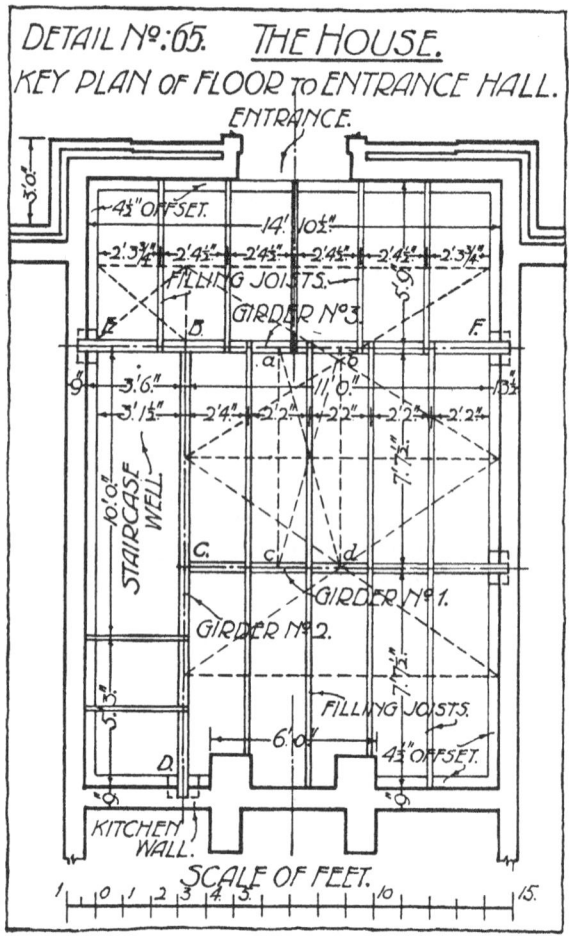

For trial, adopt a B.S.B. $4'' \times 1\frac{3}{4}'' \times 5$ lbs. per ft., having $Z = 1 \cdot 83$ inch units, and find the maximum allowable spacing of the joists.

$$B_M = R.$$

$$\therefore \ \frac{WL}{8} = Zf,$$

and

$$W = \frac{8Zf}{L}.$$

But

$Z = 1 \cdot 83$ inch units,

$f = 8$ tons per sq. inch,

and

$L = 90$ inches (approximately).

$$\therefore \ W = \frac{8 \times 1 \cdot 83 \times 8}{90}$$

$$= 1 \cdot 3 \text{ tons.}$$

But

W also $= l_j \times c_j \times w_t.$

Now

$l_j = 7' \ 6'' = 7 \cdot 5$ ft.,

and

$w_t = \dfrac{142}{2240} = \cdot 063$ ton.

Transposing and inserting values

$$c_j = \frac{W}{l_j \times w_t} = \frac{1 \cdot 3}{7 \cdot 5 \times \cdot 063}$$

$$= 2 \cdot 74 \text{ ft.}$$

This would be met by using three joists giving a spacing of $2 \cdot 75$ ft., but it is wise not to exceed a spacing of $2 \cdot 6''$, since the floor is 11 ft. clear width, therefore four joists are used at a uniform spacing of $\dfrac{11}{5} = 2 \cdot 2$ ft. centres or, say, $2' \ 2\frac{1}{2}''$. Joists of the same size are employed across the shorter span adjoining the front of the building, with a slightly wider spacing.

Student's exercise. Lighter joists might have been employed for the smaller span and the student should determine whether a less size is available for these joists, allowing for the same total depth of concrete as shown in the details.

199. Girder No. 1 to hall floor. The girder No. 1 in detail No. 65 may be considered to carry the amount of load which would cover the half bay on each side of it, as shown by dotted lines. This neglects the reduction of load due to the projecting fireplace and would therefore give a greater margin of safety if the selected beam has a modulus of section equal to the calculated value.

Omitting the weight of the beam, the load supported is approximately

$W = l_b \times c_b \times w_t,$

where

$l_b = 11$ ft.

$c_b = 7' \ 7\frac{1}{2}'' = 7 \cdot 625$ ft.

and

$w_t = 142$ lbs.

$$\therefore \ W = 11 \times 7 \cdot 625 \times \frac{142}{2240}$$

$$= 5 \cdot 32 \text{ tons.}$$

Then

$B_M = R,$

$$\therefore \ \frac{WL}{8} = Zf,$$

and

$$Z = \frac{WL}{8f} \ .$$

But $\qquad\qquad\qquad\qquad\qquad\qquad$ $L = 11' \ 4\frac{1}{2}''$, say $136''$.

$\qquad\qquad\qquad\qquad\qquad\qquad\qquad\quad$ $f = 8$ tons per sq. inch.

$$\therefore \ Z = \frac{5\cdot32 \times 136}{8 \times 8} = 11\cdot3 \text{ inch units.}$$

Referring to the list of B.S.B.'s (Appendix I), B.S.B. $8'' \times 4'' \times 18$ lbs., has $Z = 13\cdot9$ inch units, is the nearest beam available of adequate strength, and makes an allowance for the dead weight of the beam which has been neglected.

200. Girder No. 2 to hall floor. This member supports the end of girder No. 1, which transmits a point load near the centre, of

$$\frac{W \text{ (on No. 1)}}{2} = \frac{5\cdot32}{2} = 2\cdot66 \text{ tons,}$$

and also a uniform load from the rectangular bay on the left of the fireplace amounting to half the total on the bay and spread over a length of $4' \ 10\frac{1}{2}''$. This load

$$= \frac{\text{area of bay}}{2} \times \frac{142}{2240} = \frac{4\cdot875 \times 3\cdot125}{2} \times \frac{142}{2240} = \cdot48 \text{ ton.}$$

The diagram representing the loading is therefore as shown in detail No. 66, and, not being a standard case, there is no formula applicable. It is necessary therefore to proceed as described in paragraph 163, viz.:

(a) Determine the reactions.

(b) Calculate the B_M.

Taking moments about the line of the support at D in detail No. 66, and applying the rule obtained in paragraph 155:

$$R_B = \frac{2\cdot66 \text{ (tons)} \times 7' \ 7\frac{1}{2}''}{15\cdot25 \text{ (ft.)}} + \frac{\cdot48 \text{ (ton)} \times 2' \ 10''}{15\cdot25 \text{ (ft.)}}$$

$$= \frac{2\cdot66 \times 7\cdot625 + \cdot48 \times 2\cdot81}{15\cdot25}$$

$$= \frac{20\cdot3 + 1\cdot35}{15\cdot25} = 1\cdot42 \text{ tons (approximately),}$$

and $\qquad\quad R_D = \text{Total load} - R_B$

$\qquad\qquad\qquad = 3\cdot14 - 1\cdot42 = 1\cdot72 \text{ tons.}$

It is advisable to check this latter result by taking moments about B and equating as for R_B.

The B_M will occur under the point load because the shear diagram passes through zero at this point, as shown in detail No. 66 and explained in paragraph 164.

Taking moments of the forces on the left of the point load and about its line of action,

$$B_M = R_B \times 7' \ 7\tfrac{1}{2}'' = 1\cdot42 \times 91\cdot5$$
$$= 130 \text{ tons ins.}$$

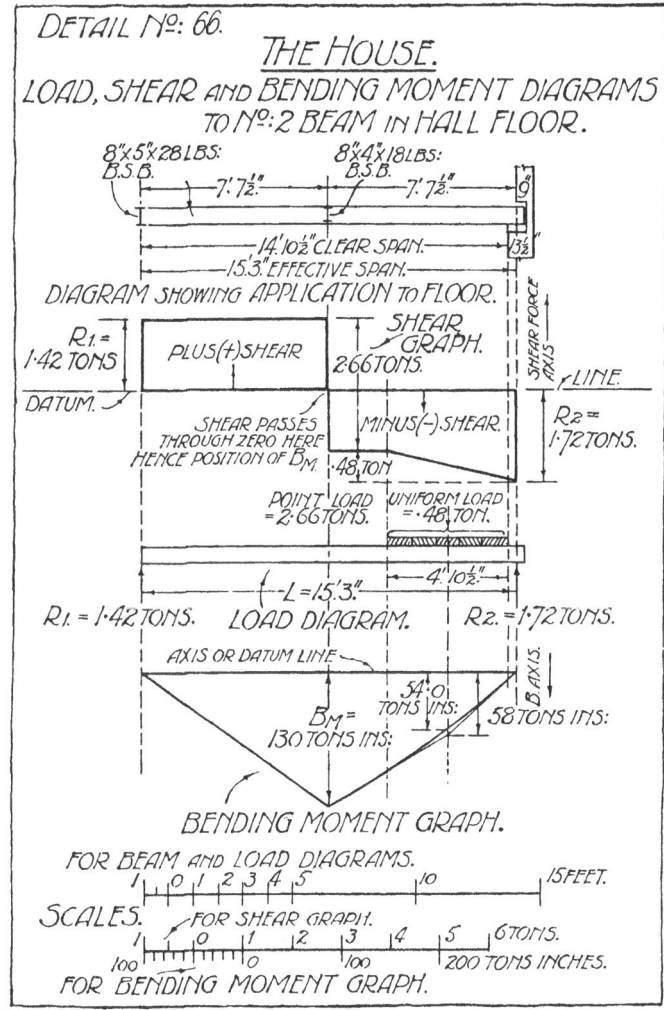

DETAIL Nº: 66.

THE HOUSE.

LOAD, SHEAR AND BENDING MOMENT DIAGRAMS TO Nº:2 BEAM IN HALL FLOOR.

Also $Z = \dfrac{B_M}{f} = \dfrac{130}{8} = \mathbf{16\cdot25}$ inch units.

The nearest B.S.B. for this purpose is a $9'' \times 4'' \times 21$ lbs. per ft., having $Z = 18$ inch units; but it is considered better to employ a B.S.B. $8'' \times 5'' \times 28$ lbs., having $Z = 22 \cdot 3$ inch units, in order to connect satisfactorily to girder No. 3 which is $8''$ deep.

It is sometimes possible to find a beam, not of British standard section, which will better serve the purpose. If such is available, without sacrificing economy in other ways, it should be employed.

The shear and bending moment graphs are plotted in the detail and may be checked by the student from previous examples.

201. Girder No. 3 to hall floor. This girder supports the end of girder No. 2 supplying the reaction of $1 \cdot 42$ tons, and also supports uniform loads gathered from the bays on each side of EF. The load on the portion BF is gathered from the bays on each side and is equivalent to the full load on the two half bays shown dotted in detail No. 65. The area is 11 ft. $\times$ 6 ft. 6 ins. and at 142 lbs. per sq. ft. gives

$$W_{BF} = 11 \times 6\tfrac{1}{2} \times \frac{142}{2240} = 4 \cdot 53 \text{ tons.}$$

The load on BE is equivalent to that carried on the half bay adjacent to it, which is $3' \; 1\tfrac{1}{2}''$ long $\times 2' \; 8\tfrac{1}{4}''$ wide, and is therefore approximately

$$W_{BE} = 3 \cdot 125 \times 2 \cdot 67 \times \frac{142}{2240} = \cdot 53 \text{ ton.}$$

Assembling these loads on the span, the load diagram becomes as shown in detail No. 67. Proceed as in the last case.

To find R_F take moments about the line of support at E; then

$$R_F = \frac{4 \cdot 53 \times 9}{14 \cdot 87} + \frac{1 \cdot 42 \times 3 \cdot 5}{14 \cdot 87} + \frac{\cdot 53 \times 2}{14 \cdot 87}$$

$$= \frac{40 \cdot 77 + 4 \cdot 97 + 1 \cdot 06}{14 \cdot 87},$$

$$\therefore \; R_F = 3 \cdot 15 \text{ tons.}$$

Student's exercise. As an exercise check the result by obtaining the reaction at the support E, and comparing the sum of the reactions with the total load.

From paragraph 164 shear is zero when the bending moment reaches its maximum, hence find a point P, x ft. from the end of the load near F, such that the load upon it $= R_F$. Then, as load per ft. on this part $= \dfrac{4 \cdot 53}{11} = \cdot 412$ ton,

$$x \times \cdot 412 = 3 \cdot 15,$$

and

$$x = \frac{3 \cdot 15}{\cdot 412} = 7 \cdot 66 \text{ ft.}$$

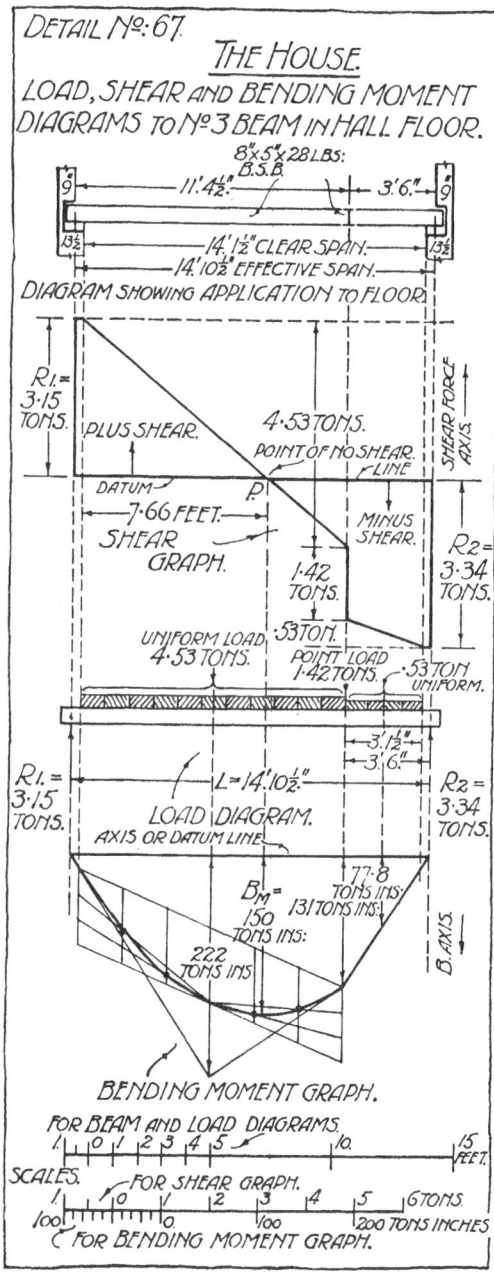

DETAIL Nº: 67.

THE HOUSE.

LOAD, SHEAR AND BENDING MOMENT DIAGRAMS TO Nº 3 BEAM IN HALL FLOOR.

8"×5"×28 LBS:
B.S.B.

9"   11'.4½"   3'.6".   9"
13½"   14' 1½" CLEAR SPAN.   13½"
14'.10½" EFFECTIVE SPAN.

DIAGRAM SHOWING APPLICATION TO FLOOR.

R1.=
3·15
TONS.   PLUS SHEAR.   4·53 TONS.   SHEAR FORCE AXIS.
                            POINT OF NO SHEAR.
                            LINE
DATUM   P.
7·66 FEET.   MINUS SHEAR.
SHEAR   R2=
GRAPH.   1·42   3·34
            TONS.   TONS.
            ·53 TON.
UNIFORM LOAD 4·53 TONS.   POINT LOAD   ·53 TON UNIFORM.
                          1·42 TONS.

3·1½"
3'.6."

R1.=   L= 14'.10½"   R2=
3·15   3·34
TONS.   LOAD DIAGRAM.   TONS.
         AXIS OR DATUM LINE.

                              77·8
                            TONS INS:
B.M=   131 TONS INS:
150   B. AXIS.
TONS INS:
       222
       TONS INS.

BENDING MOMENT GRAPH.

FOR BEAM AND LOAD DIAGRAMS.
1.   0 1 2 3 4 5   10.   15
                                       FEET.
SCALES.   FOR SHEAR GRAPH.
1.   0   1   2   3   4   5   6 TONS.
100   0   100   200 TONS INCHES
         FOR BENDING MOMENT GRAPH.

Hence, obtain B_M by taking moments on one side of P, say the left, then

$$B_M = 3 \cdot 15 \times 8 \cdot 03 - (\cdot 412 \times 7 \cdot 66) \times \frac{7 \cdot 66}{2}$$

$$= 25 \cdot 3 - 12 \cdot 05,$$

and $B_M = 13 \cdot 25$ tons ft. or **159** tons ins.

Now $Z = \dfrac{B_M}{f} = \dfrac{159}{8},$

$$\therefore Z = 19 \cdot 87 \text{ inch units.}$$

The same size of beam will therefore serve as was used for girder No. 2, viz. B.S.B. $8'' \times 5'' \times 28$ lbs.

CONNECTIONS OF GIRDERS

The connections between steel girders or joists usually consist of standard angle cleats fixed by bolts or rivets through the webs of the girders, as already described in Vol. II.

202. Reasons for web connections. When a girder or joist supports a secondary member its own strength must not be diminished unnecessarily, hence the flanges must not be cut nor damaged and the connections are made to the web which is usually strong enough for the purpose.

The cylindrical pin fixings—rivets or bolts—tend chiefly to fail by shear or by bearing; see Vol. II.

The girder No. 1 of the hall floor transmits half its total load, viz. $\dfrac{5 \cdot 32}{2} = 2 \cdot 66$ tons, to girder No. 2. The rivets necessary for the joint between the webs and angles depend upon safely supplying shear resistance of this value.

Shear resistance of rivets.[1] The shear resistance (R_S) of a rivet is its sectional area × the unit resistance to shear

$$= \frac{\pi}{4} d^2 \times f_S.$$

where $d =$ diameter of rivet in inches,
$f_S =$ safe shear per square inch
$= 6$ tons for steel.

Therefore one $\frac{3}{4}''$ rivet has $R_S = \dfrac{11}{14} \left(\dfrac{3}{4} \right)^2 \times 6$

$$= 2 \cdot 65 \text{ tons,}$$

---

[1] See also Vol. II.

and if shearing acts in two planes of a rivet simultaneously, called double shear,

$$R_{S(D)} = 2 \times 2 \cdot 65 = 5 \cdot 3 \text{ tons.}$$

203. Bearing resistance of rivets. The bearing resistance of a rivet (R_b) is the diameter of rivet × thickness of thinnest plate (or plates) against which bearing takes place in one direction × safe bearing stress per sq. in.

$$\therefore R_b = d \times t \times f_b,$$

where
      d = diameter of rivet in inches,
      t = thickness of thinnest bearing plate (or plates),
      f_b = safe bearing stress per sq. inch
         = 12 tons for steel.

Applying the bearing resistance with $\frac{3}{4}''$ rivets, it is found that the web of the $8'' \times 5'' \times 28$ lbs. B.S.B. = $\cdot 35''$ thick.

$\therefore R_b$ (one rivet) = $\frac{3}{4} \times \cdot 35 \times 12 = 3 \cdot 15$ tons (for the larger girder).

But the web of the $8'' \times 4'' \times 18$ lbs. B.S.B. is only $\cdot 28''$ thick.

$\therefore R_b$ (one rivet) = $\frac{3}{4} \times \cdot 28 \times 12 = 2 \cdot 52$ tons (for the smaller girder).

204. Number of rivets required.[1] Analysing the joint, in connecting the angles to the smaller girder ($8'' \times 4''$) the rivets are in double shear, capable of resisting $5 \cdot 3$ tons each, but in bearing will resist only $2 \cdot 52$ tons, hence bearing determines the number.

The load transmitted is $2 \cdot 53$ tons, and, while one rivet is strong enough, two rivets must be used for fixing purposes. At the connection of the angles to the larger girder ($8'' \times 5''$) the rivets are in single shear, capable of resisting $2 \cdot 65$ tons each, while in bearing they will resist $3 \cdot 15$ tons, hence shear determines the number. The reaction is only $2 \cdot 53$ tons which would be met by one rivet, but again, for practical reasons two must be employed.

This is a case where the reactions are small and the number of rivets required is less than that employed in the standard angle cleats; the standard connections are, however, used. It is always desirable to have at least two rivets to every wing of a girder connection in order to ensure the rigidity of the joint.

Reference to structural handbooks should be made for sizes of manufacturers' standard connections and rivets.

If ordinary black bolts are used they should be calculated to resist 4 tons in single shear and 8 tons in bearing.

---

[1] There are other stresses besides the *direct* bearing and shear to take into account. These additional stresses must be considered, but are necessarily left for later study.

LOADING SHED TO WAREHOUSE. DESIGN OF FLOOR

Before leaving the simpler forms of floor construction it is convenient here to consider the loading shed floor.

It is constructed of steel girders, timber joists and plank flooring 2″ thick and is intended to carry a useful load of 2 cwts. per sq. ft. in addition to its own weight.

The flooring need not be calculated, being much in excess of the thickness required to support the load, in order to allow for hard wear.

205. Joists of northern pine to loading shed floor. The joists have a span of 8′ 2″ between the centres of the steel angles which provide their support. Allowing 10 lbs. per ft. sup. over all, to cover weight of flooring and joists, the total designing load per ft. sup. (w_t)

$$= 2 \times 112 + 10 = 234 \text{ lbs.}$$

As the floor may be loaded over the whole area, crossing the supports of the joists, the load on each may be

$$W = l_j \times c_j \times w_t.$$

But l_j = span between centres of support = 8′ 2″ or 8·17 ft.
$c_j = 1\cdot25$ ft.

$$\therefore \ W = 8\cdot17 \times 1\cdot25 \times 234$$
$$= \text{say } \mathbf{2390} \text{ lbs.}$$

Try a northern pine joist 2″ broad and find the required depth, taking f as 1000 lbs. per sq. in.

Then $$B_M = R.$$

$$\therefore \ \frac{WL}{8} = \frac{bd^2 f}{6},$$

and $$d = \sqrt{\frac{WL}{8} \times \frac{6}{bf}}.$$

$$L = 98''.$$

Substituting values, $$d = \sqrt{\frac{2390 \times 98 \times 6}{8 \times 2 \times 1000}}$$
$$= \mathbf{9\cdot36''}.$$

Joists 9″ × 2″ have been employed as being the nearest practicable size; the working stress on these when fully loaded would not exceed 1100 lbs. per sq. in.

206. Girders to loading shed floor. If strictly considered these girders would present difficulty, because they are really continuous over two spans. This condition alters the position of the maximum bending moment, causing it to occur at the intermediate support. With *two equal spans*, as in the case of the longer girder, the maximum bending moment agrees in *value* with that for a single freely supported span similarly loaded, where W is calculated for a single bay. Hence, proceed as if the girders were in short lengths from pier to pier.

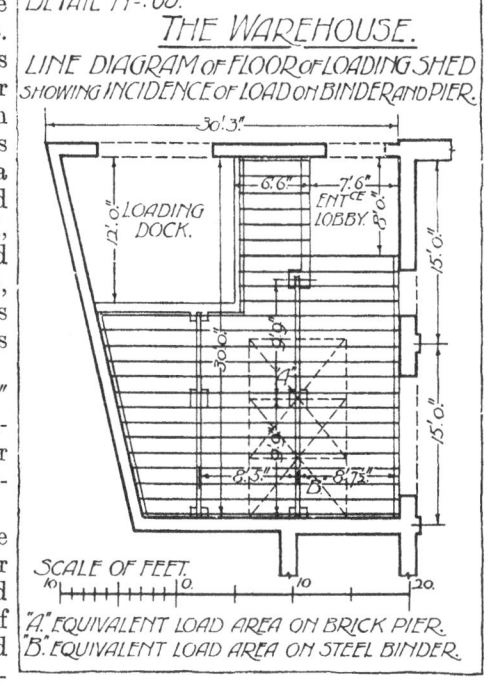

DETAIL Nº: 68.

THE WAREHOUSE.

LINE DIAGRAM of FLOOR of LOADING SHED
SHOWING INCIDENCE of LOAD on BINDER and PIER.

30'.3"

LOADING DOCK.

6'.6"

7'.6"
ENT-CE
LOBBY.

SCALE OF FEET.

A." EQUIVALENT LOAD AREA ON BRICK PIER.
B." EQUIVALENT LOAD AREA ON STEEL BINDER.

The clear span is 8' 6" allowing for the chamfered edges of the pier templates, and the effective span about 9 ft.

In this case, because loading is assumed over the entire floor, the load carried by one span of the girder is determined by the area of the rectangle "B" shown in detail No. 68, viz. effective span × centre to centre spacing of the girders. Hence, the approximate load on one length of girder is

$$W = 9 \text{ (ft.)} \times 8 \cdot 41 \text{ (ft.)} \times 234 \text{ (lbs.)},$$

or

$$W = \frac{9 \times 8 \cdot 41 \times 234}{2240} = \textbf{7·9} \text{ tons.}$$

Then the required

$$Z = \frac{WL}{8f}.$$

$$L = 108'',$$
$$f = 8 \text{ tons.}$$

$$\therefore Z = \frac{7 \cdot 9 \times 108}{8 \times 8}$$

$$= \textbf{13·3} \text{ inch units.}$$

The nearest available beam is a B.S.B., 8" × 4" × 18 lbs. per ft., having a section modulus of 13·9 inch units.

207. Piers to support girders. The pier dividing the longer girder into two spans will carry a rectangular panel of flooring measuring 9′ 9″ long between centres of girder spans and 8′ 5″ wide between centres of bays, hence

$$W \text{ (on pier)} = 9{\cdot}75 \times \frac{8{\cdot}41 \times 234}{2240}$$

$$= 8{\cdot}57 \text{ tons (approximately)}.$$

If built in common stocks and 4 to 1 cement mortar, the safe stress (f) is 4 tons per sq. ft. Therefore area required (A)

$$= \frac{W}{f} = \frac{8{\cdot}57}{4} = 2{\cdot}14 \text{ sq. ft.}$$

But $A = d^2$, where d is the diameter of square pier.

$$\therefore \ d = \sqrt{A} = \sqrt{2{\cdot}14} = 1{\cdot}46 \text{ ft.}$$

The nearest brick dimension is 1′ 6″, which has been adopted.

WAREHOUSE FLOORS

The various floor constructions available as alternatives in the warehouse have already been described in Chapter Three.

208. Conditions for design. All the floors are designed to carry a superimposed load of 2 cwts. per ft. sup. in addition to their own weight, a little above the minimum required by the London County Council; the maximum working stress on the steel is to be 8 tons per sq. inch, as previously adopted.

While many of the steel beams appear to be securely fixed in position at their supports, none of these is sufficiently rigid to produce ends which can be accepted as "fixed in position and direction". This is important, because, if so fixed, the maximum bending moment would be of less value for the same loading on a freely supported beam. *All beams are therefore designed for freely supported ends* but it is justifiable to employ a beam which falls slightly short in strength because of the small reduction in bending moment due to the connections.

GROUND FLOOR OF WAREHOUSE

209. Reinforced concrete floor slab supported by steel girders. Referring to the key plan and sections of detail No. 30 the floor is shown to consist of large rectangular bays, *abcd*, divided lengthwise by main girders and crosswise by secondary girders. At their junction these pairs of girders are supported by steel stanchions or by cast iron columns. Each main bay is subdivided into three smaller bays by intermediate secondary girders supported by the main girders and walls.

The main bays vary somewhat in size, from $12' \times 15'$ at the centre to $12'\ 6\frac{3}{4}'' \times 15'\ 0''$ at the party wall and end bays. The clear span of the main girders in the end bays does not reach 15 ft. however, because of the loss due to the diameter of the column or stanchion to which they are attached. Similarly, the maximum clear span of the secondary girders is about $12'\ 2''$, while that of the intermediate girders is $12'\ 6\frac{3}{4}''$.

To simplify the calculations an average of $15' \times 12'\ 6''$ for the dimensions of the main bays is taken and equal divisions are assumed for the subsidiary bays.

Reinforced concrete floor slab. The floor slab is divided into bays approximately $12'\ 6'' \times 5'\ 0''$, and though in one continuous piece may be assumed, for an elementary example, as being freely supported along the $12'\ 6''$ edges; their span is therefore 5 ft. centre to centre of intermediate girders.

From the Expanded Metal Company's pamphlet, we find that a concrete slab of $1+2+4$ concrete is required, $4''$ thick, reinforced near the base with the Company's No. 6 rib-mesh expanded metal, weighing $5\frac{1}{4}$ lbs. per yard super.

Detail No. 30 shows the arrangement and the mode of finishing the floor.

Dead weight of floor. The density of reinforced concrete is usually taken at 144 to 150 lbs. per cu. ft., though it varies considerably. Allowing for $\frac{1}{2}''$ cement floating, the floor slab becomes $4\frac{1}{2}''$ thick and weighs

$$\frac{4\frac{1}{2}}{12} \times 150 = 56 \cdot 25 \text{ lbs. per ft. sup.}$$

The floor is finished with $2''$ wood block flooring, which, at 35 lbs. per cu. ft., weighs $\frac{2}{12} \times 35 = 5 \cdot 83$ lbs. per ft. sup.

The total dead load is therefore $56 \cdot 25 + 5 \cdot 83 = $ say **62** lbs. per ft. sup., and the designing load (w_t) is $224 + 62 = $ **286** lbs. per ft. sup.

210. Secondary girders. Ground floor of warehouse. The secondary and intermediate girders support a uniform load equivalent to the load on one floor slab or on the area of the crossed rectangle which is

$$W = l_g \times c_g \times w_t$$

$l_g = $ clear span of girder,
$c_g = $ distance between centres of girders.

or

$$W = 12 \cdot 5 \times 5 \times \frac{286}{2240}.$$

$$\therefore\ W = 7 \cdot 98, \text{ say } \mathbf{8} \text{ tons.}$$

Then

$$B_M = R,$$

or

$$\frac{WL}{8} = Zf.$$

But

$L = $ effective span
$= 12'\ 10\frac{1}{2}''$, or, say **155''**.

$$\therefore\ Z = \frac{WL}{8f} = \frac{8 \times 155}{8 \times 8},$$

$$\therefore\ Z = \mathbf{19 \cdot 37} \text{ inch units.}$$

The nearest steel beam is a B.S.B., $10'' \times 4\frac{1}{2}'' \times 25$ lbs. per ft., having $Z = 24.46$ inch units.

211. Main girders. Ground floor of warehouse. The main or longitudinal girder supports the ends of two pairs of intermediate girders, each pair transmitting a load equivalent to that on one bay as indicated by diagonals, over the load point P, and found in the last paragraph to be 8 tons.

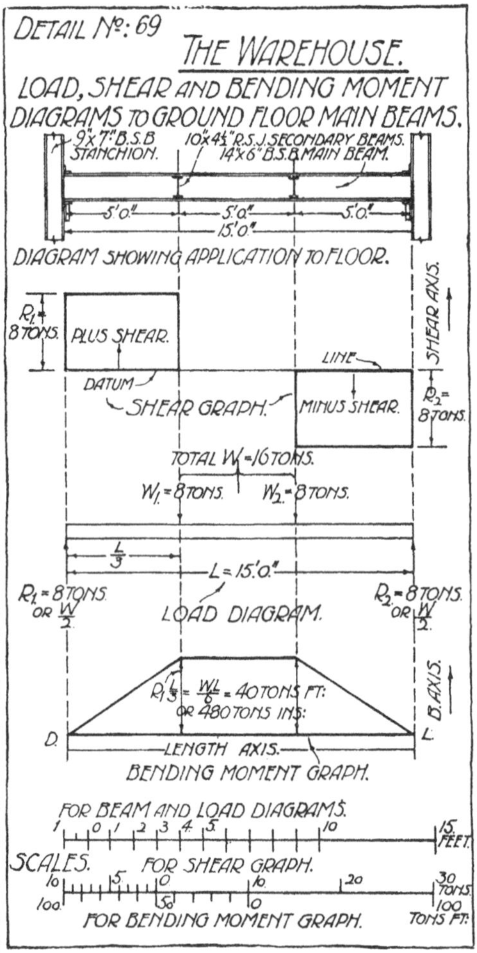

While the spacing of the secondary and intermediate girders is uniform over the whole floor and at 5 ft. centres, these members do not divide the main girder into equal sections, hence this girder is not quite symmetrically loaded. For practical design symmetrical loading will be assumed, thus obtaining a load diagram and graphs, as shown in the adjoining detail.

The maximum bending moment for this case has been shown at

DETAIL No: 69

THE WAREHOUSE.

LOAD, SHEAR AND BENDING MOMENT DIAGRAMS TO GROUND FLOOR MAIN BEAMS.

paragraph 160 to be $\dfrac{WL}{6}$, where W is the total load, viz. 16 tons.

(The shear force and bending moment diagrams are shown in detail No. 69.)

Then
$$B_M = R.$$
$$\therefore \quad \frac{WL}{6} = Zf,$$

and
$$Z = \frac{WL}{6f}$$

But $L =$ effective span and may be taken as 15 ft. (180″).

or $$Z = \frac{16 \times 180}{6 \times 8}.$$

$$\therefore \ Z = 60 \text{ inch units.}$$

B.S.B. $15'' \times 6'' \times 45$ lbs. per ft., having $Z = 65 \cdot 58$ inch units, is the nearest size available.

Note. The weights of the beam and the concrete haunches have been neglected in the above calculation. The weights are small as compared with the total load; *e.g.* the approximate weight of the main girder and concrete haunches is $\frac{1}{4}$ ton or 2 per cent. of the total load. This has also been compensated for by taking the floor slabs of maximum size to determine dead weight of floor. The beam selected is strong enough to allow for dead weight without exceeding the allowable stress.

FIRST FLOOR OF WAREHOUSE

For this floor are shown four methods of construction, any one of which might be suitably employed for all the floors of the warehouse.

212. Method I. Homan and Rodgers' floor. This construction is shown applied to a wall bay $15' \times 12'$ $6\frac{3}{4}''$, and is spanned in the shorter direction by steel joists spaced at $1'$ $6\frac{1}{2}''$ centres, running parallel to the secondary beams. Fireclay lintols span across the joists at right angles as shown in detail No. 33.

Steel joists for Homan and Rodgers' floor. Assume an average density of 130 lbs. per cu. ft. for the fireclay lintols and concrete taken together, and making allowance for the spaces in the former; for a floor $7''$ thick the weight is about 69 lbs. per sq. ft. Adding 5 lbs. for bearers and flooring and 7 lbs. for steel joists—as an approximation—gives a total dead load, for an unplastered ceiling, of 81 lbs. per sq. ft.

By employing special concrete made with breeze, pumice stone or light clinker, the weight could be further reduced to balance the additional weight of the plaster.

Proceeding on the basis of 81 lbs. per sq. ft., the total designing load (w_t) is $224 + 81 = 305$ *lbs. per ft. sup.*

The load carried by one joist is uniform and equals "the area of one joist panel $\times w_t$."

$$\therefore \ W = l_j \times c_j \times w_t$$

$$= 12' \ 6\frac{3}{4}'' \times 1' \ 6\frac{1}{2}'' \times 305 \text{ lbs.}$$

$$= 12 \cdot 56 \times 1 \cdot 54 \times \frac{305}{2240} = 2 \cdot 634 \text{ tons.}$$

Then $$Z = \frac{WL}{8f}.$$

But $L = $ say $\mathbf{155''}$.

$$\therefore \; Z = \frac{2 \cdot 634 \times 155}{8 \times 8} = \mathbf{6 \cdot 37} \text{ inch units.}$$

From Appendix I a B.S.B., $6'' \times 3'' \times 12$ lbs., having $Z = 6 \cdot 99$ inch units, is found suitable, though its deflection might be slightly excessive. The next beam in order of suitability is $7'' \times 3\frac{1}{2}'' \times 15$ lbs., $Z = 10 \cdot 25$ inch units, which is too large for economy.

Checking weight of joists. Each joist weighs, say, $12\frac{1}{2}$ (ft.) $\times 12$ (lbs.) $= 150$ lbs. Its load is distributed over $12\frac{1}{2}$ (ft.) $\times 1 \cdot 54$ (ft.) $= 19 \cdot 25$ ft. sup.; therefore the weight per sq. ft. averages $\frac{150}{19 \cdot 25} = 7 \cdot 8$ lbs., as against 7 lbs. assumed.

Secondary beams for Homan and Rodgers' floor. Because the secondary beams run parallel to the joists and support the ends of two sets of lintols, they only receive the same weight as the joists plus the weight of the concrete haunches. They are necessary to secure rigidity in the stanchions or columns at the floor levels and also add to the transverse stiffness of the floor framing; $10'' \times 4\frac{1}{2}''$ beams are suitable.

Main beams of Homan and Rodgers' floor. As the main beam receives and supports the ends of the joists which are uniformly spaced throughout its length, it will practically carry a uniform load equal to that on a full bay, the area of which is $15' \times 12'\ 6''$, and the load per ft. sup. 305 lbs.

$$\therefore \; W = 15 \times 12 \cdot 5 \times \frac{305}{2240} = 25 \cdot 5 \text{ tons.}$$

Then $$Z = \frac{WL}{8f}.$$

                                        $L = $ say $180''$.

$$\therefore \; Z = \frac{25 \cdot 5 \times 180}{8 \times 8}$$

$$= \mathbf{71 \cdot 7} \text{ inch units.}$$

From Appendix I the most suitable beam is a B.S.B., $16'' \times 6'' \times 50$ lbs., $Z = 77 \cdot 26$ inch units.

213. Method II. Fram floor. In this example the load is all transmitted to the main girders, because the panel of hollow bricks is reinforced in the lines of joints parallel to the secondary girder, thus converting each line of blocks into a beam spanning the shorter way of the panel; see detail No. 34 for the construction.

Averaging the density of brick, cement jointing, etc., as 130 lbs. per cu. ft. and making allowances for the spaces, for a floor $7''$ thick the weight per ft. sup. is about 65 lbs.

Main girder of " Fram" floor. The total designing load (w_t) is therefore $224 + 65 = 289$ lbs. per ft. sup.

The uniform load on the main girder equals "area of panel $\times w_t$."

$$\therefore \quad W = 15 \times 12\tfrac{1}{2} \times \frac{289}{2240} = \mathbf{24 \cdot 2} \text{ tons,}$$

say 24·7 tons to include weight of girder and haunches.

Then $\qquad Z = \dfrac{WL}{8f} = \dfrac{24 \cdot 7 \times 180}{8 \times 8} = \mathbf{69 \cdot 5}$ inch units.

From Appendix I the nearest steel beam is a B.S.B., $16'' \times 6''$ $\times 50$ lbs. as in the previous example.

214. Method III. Armoured Tubular floor. This consists of light reinforced concrete beams or "webs" of $\perp$ section on which rest tubular lintols; the construction is shown in detail No. 36.

The beams span the whole width of the panel and, to carry 2 cwts. per sq. ft. in addition to their own weight, require to be $8''$ deep $\times 3''$ wide at the base, as given in the Company's pamphlets.

The weight of the floor, based on a density of 100 lbs. per cu. ft. for light concrete, would be about 52·5 lbs. at $9''$ thick. Adding 3·25 lbs. for $1\tfrac{1}{4}''$ flooring a total dead weight of say 56 lbs. or $\tfrac{1}{2}$ cwt. per ft. sup. is obtained.

The total designing load is therefore $2\tfrac{1}{2}$ cwts. per sq. ft.

Secondary girders for Armoured Tubular floor. Because the tee beams run parallel to the secondary girders they get little or no load, but are still necessary to give rigidity to the floor framing in a transverse direction. They serve to produce fixity of the stanchions or columns at the floor levels and $8'' \times 5''$ to $10'' \times 5''$ beams would be employed.

Main girders of Armoured Tubular floor. The main girders carry a uniform load from half the panel on each side—measuring approximately $15' \times 12' \ 6''$—at $2\tfrac{1}{2}$ cwts. per sq. ft.

Hence $\qquad W = 15 \times 12\tfrac{1}{2} \times \dfrac{2\tfrac{1}{2}}{20} = \mathbf{23 \cdot 44} \text{ tons,}$

say **24** tons, to include weight of girder and concrete haunches.

Then $\qquad\qquad Z = \dfrac{WL}{8f},$

$L = 180''$ as before.

$$\therefore \quad \frac{Z = 24 \times 180}{8 \times 8} = \mathbf{67 \cdot 5} \text{ inch units.}$$

From Appendix I the nearest steel beam is a B.S.B., $16'' \times 6''$ $\times 50$ lbs. per ft., having $Z = 77 \cdot 26$ inch units.

215. Method IV. Siegwart floor. The steel framing for this floor (detail No. 36) is arranged in a manner similar to that adopted for the ground floor, except that only one intermediate secondary girder is employed dividing the bay into two panels.

The Siegwart beams (described in paragraph 97) run parallel to the main girder and are supported by the secondary girders, except in the end bays where they rest upon the wall. The Siegwart beams are selected from the Company's catalogue and require to be 5″ deep to carry 2 cwts. per sq. ft. in addition to their own weight, over a span of 7′ 6″. The dead weight of the floor per ft. super is given as 34 lbs. To this must be added the weight of flooring, 3·25 lbs., weight of fixing blocks 3 lbs., and weight of cement screeding 4·5 lbs., making a total of, say, 45 lbs. per ft. sup.

The designing load (w_t) is therefore

$$224 + 45 = \textbf{269} \text{ lbs. per sq. ft., say 270 lbs.}$$

Secondary girders for Siegwart floor. Each secondary and intermediate girder carries approximately the uniform load from one panel.

$$\therefore\ W = l_g \times c_g \times w_t$$
$$= 12\cdot5 \times 7\cdot5 \times \frac{270}{2240} = \textbf{11·3} \text{ tons.}$$

But $L = 155''.$

Then $Z = \dfrac{WL}{8f} = \dfrac{11\cdot3 \times 155}{8 \times 8}.$

$$\therefore\ Z = \textbf{27·37} \text{ inch units.}$$

From Appendix I the nearest steel beam is a B.S.B., 10″ × 5″ × 30 lbs. per ft., having $Z = 29\cdot25$ inch units.

Main girders for Siegwart floor. The load transmitted to the centre of these girders from the ends of two intermediate secondary girders = a full panel load as obtained in the last paragraph and to allow for weight of girders, say 11·6 tons.

For the approximately central load $B_M = \dfrac{WL}{4}.$

$$\therefore\ \frac{WL}{4} = Zf,$$

and $Z = \dfrac{WL}{4f}.$

But $L = 180''.$

$$\therefore\ Z = \frac{11\cdot6 \times 180}{4 \times 8} = \textbf{65·25} \text{ inch units.}$$

From Appendix I the nearest steel beam is a B.S.B., 15″ × 6″ × 45 lbs. per ft., having $Z = 65\cdot58$ inch units.

CHAPTER EIGHT

STRUCTURAL DESIGN

MOMENT OF INERTIA AND MOMENT OF RESISTANCE OF UNSYMMETRICAL AND COMPOUND SECTIONS

216. Moment of inertia. Introduction. The following explanation is intended to make clear the meaning and use of the property known as the "Moment of Inertia" or "Second Moment" of a plane section of a beam, and to explain its use in determining the Moment of Resistance.

In the detail No. 48 the distribution of stress in a bent beam was represented by solid wedges built upon the section of the beam. The section considered was symmetrical and the wedges were equal in size and volume because the stress on the extreme fibres of the beam was the same at each side owing to the N.A. being centrally placed between the upper and lower surfaces.

Each wedge represented the *total stress* acting in the fibres either in tension or compression, hence the total tensile force was equal to the compression force, and as the forces were acting in parallel but opposite directions they formed a *couple*, the moment of which represented the moment of resistance of the beam.

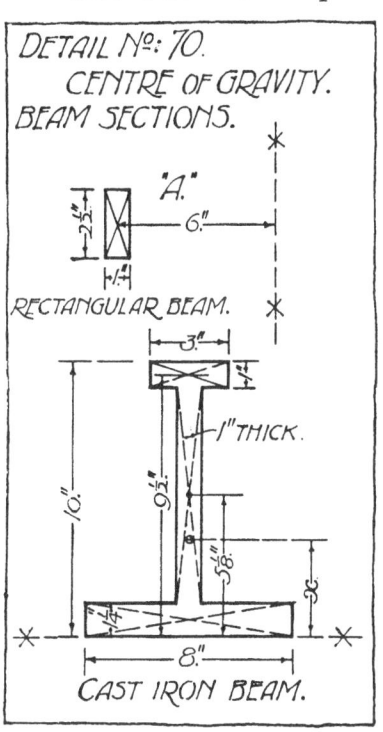

DETAIL Nº: 70.
CENTRE OF GRAVITY.
BEAM SECTIONS.

"A."
6."
RECTANGULAR BEAM.
3."
1" THICK.
10."
9½."
5⅝."
8."
CAST IRON BEAM.

217. Centre of gravity and neutral axis. When a section is unsymmetrical the resisting forces must form a similar couple acting in opposition to the bending couple (see paragraph 219). Consider a section such as that used for cast iron beams, detail No. 70; because the lower flange is larger than the upper flange it is obvious that with the neutral axis assumed at the centre of the depth, the volumes of the stress wedges would not be equal if the strain and stress are still assumed to vary uniformly above and below the axis as in previous considerations (see detail No. 72). The neutral axis must therefore be lower and in such a position that the volumes of the wedges above and below the N.A. are equal, viz. the total force on each side is of the same magnitude.

It can be shown that the volumes are equal when the N.A. passes through the centre of gravity of the section. Now the centre of gravity is a point in the plane containing the figure, about which the areas on each side of any line passing through it are balanced, or, the point at which the whole area could be massed so that it would have the same moment about any point in its plane as the "sum of the moments of the elementary areas composing it." Its position may be found by balancing the figure across a knife edge in two directions and noting the point of intersection of these directions when balanced, or by taking moments about one edge of the figure and equating *total area* × *x* to *the sum of the moments of the rectangles*; *x* = the distance of the C.G. from the edge about which moments are taken.

Thus, the moment of the rectangle at A in the detail, about the line XX, is the area ($2\frac{1}{2}$ sq. ins.) × the perpendicular distance of its centre from XX (6″).

Therefore moment = $2\frac{1}{2} \times 6 = 15$ inch units.

218. Centre of gravity of cast iron beam. Refer to the cast iron beam section of detail No. 70. The beam is symmetrical about the vertical centre line of the web and therefore has its C.G. on this line. The distance from one edge—say the base—is to be determined.

The areas of the parts are:

$$\begin{aligned}
\text{Area of top flange} &= 3 \text{ sq. ins.}\\
\text{Area of web} &= 7\tfrac{3}{4} \text{ ,,}\\
\text{Area of bottom flange} &= 10 \text{ ,,}\\
\hline
\text{Total area } (A) &= 20\tfrac{3}{4} \text{ ,,} \quad (20\cdot75).
\end{aligned}$$

Then

Moment of top flange about XX $= 3 \times 9\frac{1}{2} = 28\cdot5$ inch units
Moment of web about XX $= 7\frac{3}{4} \times 5\frac{1}{8} = 39\cdot72$,, ,,
Moment of bottom flange about $XX = 10 \times \frac{5}{8} = 6\cdot25$,, ,,

Total moment $= 74\cdot47$,, ,,

Equating $A \times x$ to the total moment of the parts,

$$20\cdot75x = 74\cdot47,$$

$$\therefore \ x = \frac{74\cdot47}{20\cdot75} = 3\cdot58 \text{ inches.}$$

If the flanges are set horizontally and the load is vertical then the N.A. would be horizontal, passing through a point 3·58″ from the base—see detail No. 71.

A check may be obtained upon the above calculation by finding the moments of the areas on each side of the axis; their values should be equal.

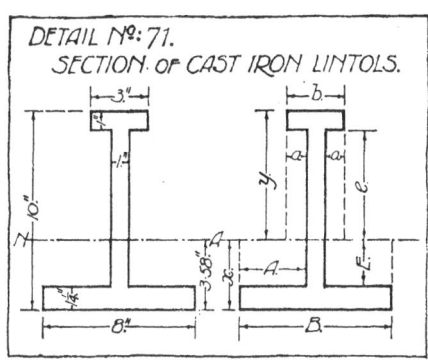

DETAIL Nº: 71.
SECTION of CAST IRON LINTOLS.

219. Linear or inertia wedges for cast iron beam.

In previous work on stresses in bent beams it has been shown that the stresses at different points in the section vary directly with their distances from the neutral axis (N.A.).

Detail No. 72 at A gives the side elevation of a pair of wedges of stress, where f_y and f_x show the magnitudes of the stresses per unit of area on the top and bottom fibres of the section of a cast iron beam.

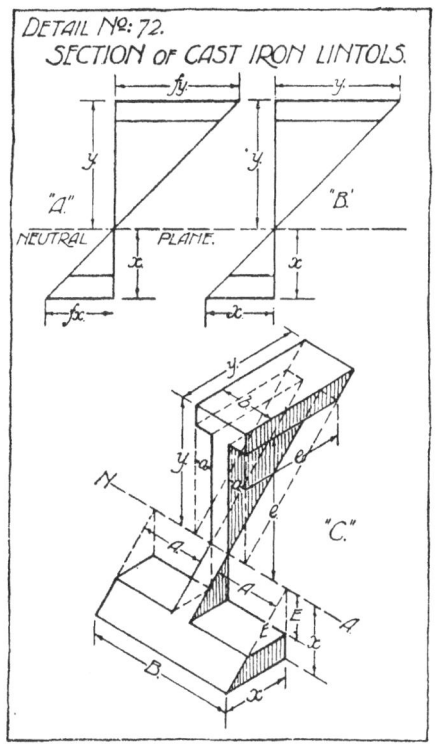

DETAIL Nº: 72.
SECTION of CAST IRON LINTOLS.

At B in the same detail the stresses f_y and f_x are replaced by the linear distances y and x, which still indicate the distribution of stress. These linear wedges are reproduced in isometric detail at C, and it should be clear that if the wedges which are built on the beam section above and below N.A. can be proved to be of equal volume, they will be equal for all such wedges where the conditions are similar and will guarantee that the total force is the same on each side of the axis.

Using the notation of details Nos. 71 and 72 C, compare the moments of the areas above and below N.A. with the volumes of the linear wedges.

By the method of differences:

From detail No. 71:

Moment of areas above N.A. $= (b \times y)\, \dfrac{y}{2} - 2\,(a \times e)\, \dfrac{e}{2}$

$$= \frac{by^2}{2} - ae^2 \qquad \qquad \text{......(1).}$$

Moment of areas below N.A. $= (B \times x)\, \dfrac{x}{2} - 2\,(A \times E)\, \dfrac{E}{2}$

$$= \frac{Bx^2}{2} - AE^2 \qquad \qquad \text{......(2).}$$

From detail No. 72, considering volumes:

Volume of wedge above N.A. $= \dfrac{1}{2}\,(b \times y \times y) - \dfrac{1}{2}\,(2a \times e \times e)$

$$= \frac{by^2}{2} - ae^2 \qquad \qquad \text{......(3).}$$

Volume of wedge below N.A. $= \dfrac{1}{2}\,(B \times x \times x) - \dfrac{1}{2}\,(2A \times E \times E)$

$$= \frac{Bx^2}{2} - AE^2 \qquad \qquad \text{......(4).}$$

Comparing results we see that (1) and (3) are equal, and also (2) and (4) are equal, or, the moment of the area equals the volume of the corresponding wedge. Further, the method of obtaining the centre of gravity ensures that expressions (1) and (2) are equal.

Hence, when the axis of division (or N.A.) between the tensile and compressive areas passes through the C.G. of the section, the total tensional stress equals the total compressional stress.

220. Moment of inertia of cast iron beam. In order to arrive at an expression of the moment of resistance of a section which is *not* a single rectangle as was described at paragraph 146 it is found convenient to use the wedges illustrated at No. 72 C, retaining temporarily the linear dimension replacing stress. The sum of the moments may then be obtained in the same manner as previously employed; thus, again using the method of differences:

Moment of upper wedges about N.A. = volume × leverage.
Leverage = distance of C.G. of wedge from axis.

$$\therefore \ M_{(\text{upper})} = \frac{by^2}{2} \times \frac{2y}{3} - 2\left(\frac{ae^2}{2} \times \frac{2}{3}\,e\right)$$

$$= \frac{1}{3}\,(by^3 - 2ae^3) \qquad \ldots\ldots(5).$$

Moment of lower wedges about N.A. $= \dfrac{Bx^2}{2} \times \dfrac{2x}{3} - 2\left(\dfrac{AE^2}{2} \times \dfrac{2}{3}\,E\right)$

$$\therefore \ M_{(\text{lower})} = \frac{1}{3}\,(Bx^3 - 2AE^3) \ \ldots\ldots(6).$$

As both wedges assist in resisting bending, adding these expressions we get the sum of the moments about N.A.

$$= \frac{1}{3}\,[(by^3 + Bx^3) - (2ae^3 + 2AE^3)],$$

and this value is called the "moment of inertia" or "second moment" of the section, and generally represented by I. If I be referred to for any given axis suitable letters are attached; thus, I_{NA} means the moment of inertia about the neutral axis.

221. Re-conversion of linear wedges to stress wedges. To estimate the moment of resistance (R) of the section just considered, it is only necessary to convert the linear wedges employed to represent distribution of stress into their previous form, inserting the value of f to be employed as working stress on the extreme fibres, and dividing out y (or x), the linear length of wedge temporarily employed.

Then resistance moment $(R) = 1 \times \dfrac{f}{y}$.

When a material has an equal working stress in tension and compression (which is approximately the case with steel) f does not vary and y is always the distance to the extreme fibres—those furthest away from the axis on either side; but if the material be cast iron having a working stress in compression $(f_c) = 7\frac{1}{2}$ tons per sq. in., and in tension $(f_t) = 1\frac{1}{2}$ tons per sq. in., it is necessary to estimate the value of R for both extremes.

Let the distance to the compressed edge be y. Then

$$R = I\frac{f_c}{y} \ \text{for compression.}$$

If the distance to the tension edge be x, then

$$R = I\frac{f_t}{x} \ \text{for tension.}$$

In assessing the value of R for any given section the lower of these values must be employed.

222. Economical sections in cast iron. From the above consideration it should be clear that, theoretically, an economically designed cast iron beam—with freely supported ends, which ensures no change of stress from tension to compression along the same flange—should have its N.A. in such a position that

$$I\frac{f_c}{y} = I\frac{f_t}{x},$$

or

$$\frac{f_c}{y} = \frac{f_t}{x},$$

or

$$\frac{f_c}{f_t} = \frac{y}{x}.$$

Let $\qquad f_c = 7\frac{1}{2}$ tons per sq. in.

and $\qquad f_t = 1\frac{1}{2}$,, ,,

Then

$$\frac{7\frac{1}{2}}{1\frac{1}{2}} = \frac{5}{1} = \frac{y}{x},$$

and $\qquad y = 5x.$

Hence the N.A. should be about one-sixth of the depth from the tensile edge for a freely supported beam or cantilever.

In practice it is seldom possible to design cast iron beams of theoretically economical section, owing to difficulties of moulding, effects of cooling, necessary practicable proportions, etc., hence most such beams are found to have their N.A.'s in position at from one-fourth to one-third their depth and *would therefore fail by tension*. In all cases the design should approximate as near as possible to that of the economical section.

223. Mathematical expression of the moment of inertia of a section. Let the wedge illustrated at detail No. 73 be a wedge of stress built on a rectangular section above the axis XX. The stress per unit area at any layer of fibres is proportional to its distance from XX and ultimately reaches its maximum f on the outermost fibres.

Let f_1 be the stress at a distance y_1 from the axis, then, by similar triangles,

$$\frac{f_1}{f} = \frac{y_1}{y}, \text{ or } f_1 = \frac{fy_1}{y} \qquad \ldots\ldots(1).$$

The moment of the strip of width b, thickness δy, and stress f_1 about the axis XX

$$= (b\,\delta y f_1)\,y_1 \qquad \ldots\ldots(2).$$

Substituting in (2) the value of f_1 from expression (1):

$$\text{Moment of strip} = \left(b\,\delta y \frac{fy_1}{y}\right) y_1 = \frac{bf}{y}\,\delta y\,y_1{}^2 \qquad \ldots\ldots(3).$$

By summing the moments of all similar strips composing the wedge, the moment of resistance (R) of the wedge is obtained. Thus

$$R = \int_{y_1=y}^{y_1=0} \frac{bf}{y}\,(dy\,y_1{}^2)$$

$$= \frac{bf}{y} \times \frac{y^3}{3}$$

$$= \frac{by^3}{3} \times \frac{f}{y}.$$

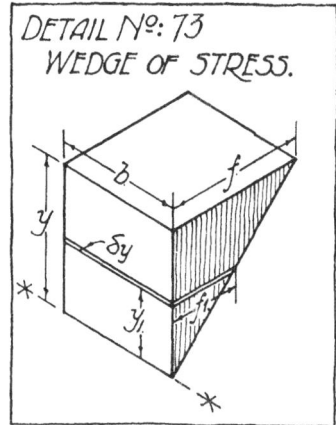

DETAIL Nº: 73
WEDGE OF STRESS.

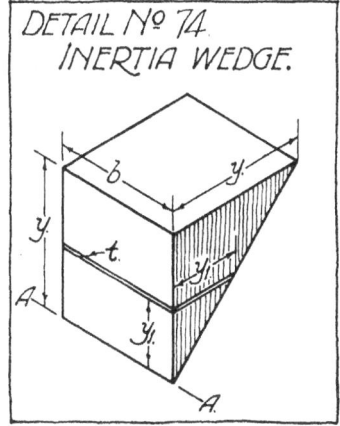

DETAIL Nº 74
INERTIA WEDGE.

The value $\frac{by^3}{3}$ is called the moment of inertia (I) of the rectangular section and is a measure of the resistance to bending moment due to the *form* of the section.

The moment of resistance of any section can be shown to be $I\frac{f}{y}$, where I is the sum of "the products of all the elements of area, into the squares of their distances from an assigned axis". In each case the axis giving the least value of I must be employed, viz. the *neutral* axis.

Let I be obtained in *inch units*, y in linear inches, and f in tons per sq. inch, then R would be expressed in *inch tons*. If foot units be employed and tons per sq. ft., R would be in *foot tons*.

For students possessing no knowledge of calculus methods the following method of reasoning will probably be sufficient to meet elementary needs.

224. Alternative method of expressing moment of inertia for compound rectangular sections. Let the sketch at detail No. 74

represent an inertia wedge of the type described in paragraph 219; its length and depth $=y$.

Consider a very thin layer of thickness t at a distance y_1 from the edge AA.

The moment of this layer about AA = volume of layer × lever arm

$$= (bty_1) \times y_1$$
$$= bty_1^2.$$

Now $b \times t$ is an element of area of the section and y_1 is its distance from AA, and the moment is seen to equal "the element of area multiplied by the *square* of its distance from AA". By the definition in the preceding paragraph the total of the moments of all such thin layers = the moment of inertia. It is easy to see that this *sum* is the moment of the whole wedge, viz. volume of wedge × the distance of its c.g. from AA.

$$\therefore I_{AA} = \frac{by^2}{2} \times \frac{2}{3} y = \frac{by^3}{3},$$

which agrees with the previous result.

In words, the value of I of a rectangle *about one edge* is

$$\frac{\text{breadth} \times \text{depth cubed}}{3},$$

and this value for any section composed of rectangles can be obtained by calculating the value of the sum of the I's for all the rectangles. *In each case the rectangle must have one edge coinciding with the axis.*

It is unnecessary to learn formulae for these cases; the statement can be written down at sight from a dimensioned section.

225. Moment of inertia, etc. of a rectangle about its neutral axis parallel to a side. This is the common case applicable to all rectangular sections of beams in which the material is equally elastic in tension and compression, a condition applying to all ordinary structural materials within working limits.

From the last two paragraphs the I of a rectangle about one edge is $\frac{by^3}{3}$.

Hence the I of the *two* rectangles A and B in detail No. 75 $= 2 \frac{by^3}{3}$.

Let
$$y = \frac{d}{2}.$$

Then
$$I_{NA} = \frac{2b \left(\frac{d}{2}\right)^3}{3} = \frac{bd^3}{12},$$

an expression only applicable when the N.A. is at the centre of depth of the rectangle considered.

From paragraph 221 $R = I \dfrac{f}{y}$ or $\dfrac{I}{y} f.$

Now y in this case is $\dfrac{d}{2}$,

$$\therefore \ R = \frac{2I}{d} f,$$

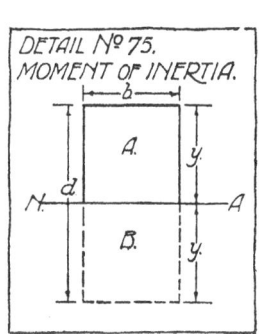

DETAIL Nº 75.
MOMENT OF INERTIA.

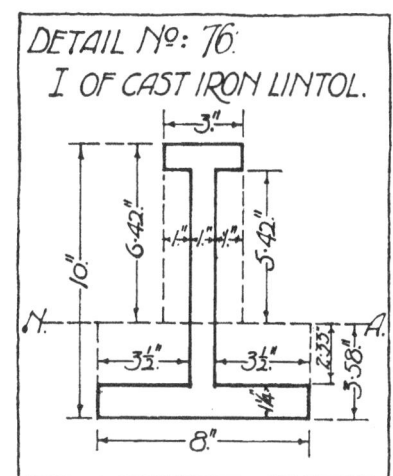

DETAIL Nº: 76.
I OF CAST IRON LINTOL.

and inserting the value of I_{NA} for the rectangle

$$R = \frac{2}{d} \frac{bd^3}{12} f$$
$$= \frac{bd^2}{6} f,$$

the same expression as deduced at paragraph 146 in dealing with the resistance moment of rectangular beams.

Note that $\dfrac{I}{y}$ in any case is called the "modulus of section" and is a measure of the capacity of the form and dimensions of the section to resist bending, whatever the form and conditions may be. It is often referred to by Z (or in some notations by SM).

226. "Moment of inertia" of cast iron beam section. (Example.)
For the cast iron beam of detail No. 76 the moment of inertia (I_{NA}) can be obtained as follows:

Dimension the section as shown, after determining the position of the N.A.

The I_{NA} = the sum of the moments of inertia of all the rectangles composing the section and having their edges in the N.A.

Using the method of differences

$$I_{NA} = \frac{3 \times 6 \cdot 42^3 - 2 \times 1 \times 5 \cdot 42^3}{3} \text{ (for the rectangles above)}$$

$$+ \frac{8 \times 3 \cdot 58^3 - 2 \times 3\frac{1}{2} \times 2 \cdot 33^3}{3} \text{ (for the rectangles below)}$$

$$= \frac{1}{3} \left[(3 \times 6 \cdot 42^3 + 8 \times 3 \cdot 58^3) - (2 \times 5 \cdot 42^3 + 7 \times 2 \cdot 33^3) \right]$$

$$= \frac{1}{3} \left[(793 \cdot 7 + 367 \cdot 1) - (318 \cdot 4 + 88 \cdot 7) \right] *$$

$$= \mathbf{251 \cdot 2} \text{ inch}^4 \text{ units.}$$

227. "Moment of resistance" of cast iron beam section. Let the working stress in tension on cast iron $(f_t) = 1\frac{1}{2}$ tons per sq. inch, and the distance to the outermost tension fibre (x) be $3 \cdot 58''$, then the moment of resistance of the section (R)

$$= \frac{I_{NA} f_t}{x} = \frac{251 \cdot 2 \times 1 \cdot 5}{3 \cdot 58} = \mathbf{105 \cdot 5} \text{ inch tons.}$$

As the opposite edge of the section is $6 \cdot 42''$ from the N.A., the stress thereat would be in the same ratio to the stress at the tension edge as the ratio of the distances of the extreme fibres from the N.A. Since the maximum allowable stress at the bottom edge is $1 \cdot 5$ tons per sq. inch, the stress at the top edge would be $1 \cdot 5 \times \dfrac{6 \cdot 42}{3 \cdot 58} = 2 \cdot 69$ tons per sq. inch in compression.

Every structural section subject to bending may have the stresses on the extreme fibres determined by the above process. The moment of inertia must first be ascertained; then, generally, $R = I \dfrac{f}{y}$, where y is the distance of the extreme fibres from the N.A. of the section. (In symmetrical sections the N.A. passes through the centre and is an axis of symmetry.)

To obtain the moment of inertia of sections not consisting of rectangles the formula expressing the value must be obtained by purely mathematical methods, but in all cases it is the "*sum of the products of all the elements of area, into the squares of their respective distances from the assigned axis*".

228. Moments of inertia of common structural sections.

The moments of inertia of several common forms of cross-section are given in detail No. 77; see also paragraphs 190 and 191.

* Approximate results.

229. Cast iron bressummers to loading shed doorway of warehouse.

The opening has a clear span of 9′ 9″ and the bearings of the bressummers are 9″ long. Hence, the effective span will be 10′ 6″.

It would be possible in this case to fix the ends of the bressummers horizontally by careful construction, but the intention here is to assume the ends free, which could occur under certain conditions and would give the maximum bending moment.

It has previously been shown that the bonding of brickwork enables it to distribute pressure along the joint lines due to the overlap of bricks in consecutive courses and it is found that, provided a bressummer is not very rigid in cross section, the amount of brickwork above an opening which it may be called upon to support does not exceed that enclosed within an equilateral triangle constructed upon the effective span, as shown in detail No. 78 at A.

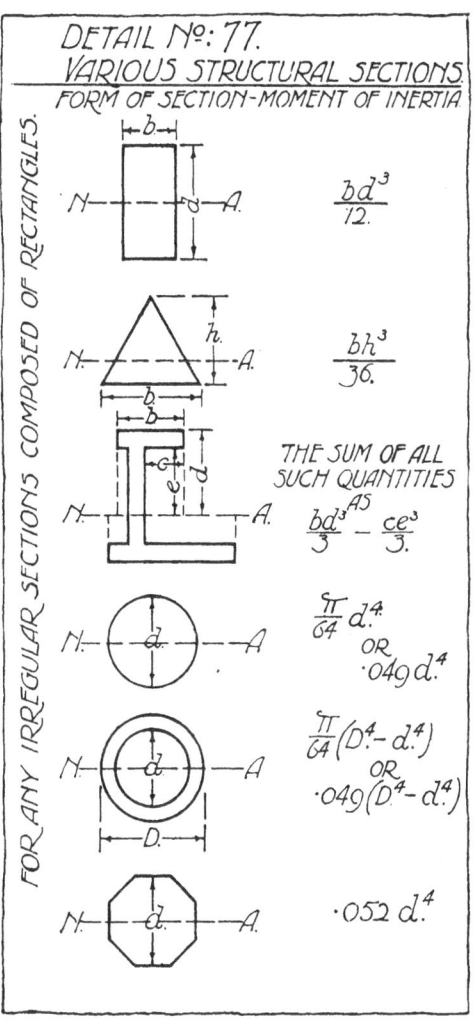

230. Load supported. Let the brickwork have a density of 120 lbs. per cu. ft. and the wall be 18″ thick. An equilateral triangle with base 10′ 6″ long has a height of

$$\cdot866 \times 10\cdot5 = 9\cdot09 \text{ ft.}$$

The volume of the triangular prism of brickwork supported is therefore

$$\frac{\text{base} \times \text{height}}{2} \times \text{thickness} = \frac{10 \cdot 5 \times 9 \cdot 09}{2} \times 1 \cdot 5 = 71 \cdot 5 \text{ cu. ft.}$$

$$\text{Its weight} = \frac{71 \cdot 5 \times 120}{2240} = \mathbf{3 \cdot 83} \text{ tons.}$$

This is carried upon two girders, hence, load on each

$$= \mathbf{1 \cdot 91} \text{ tons.}$$

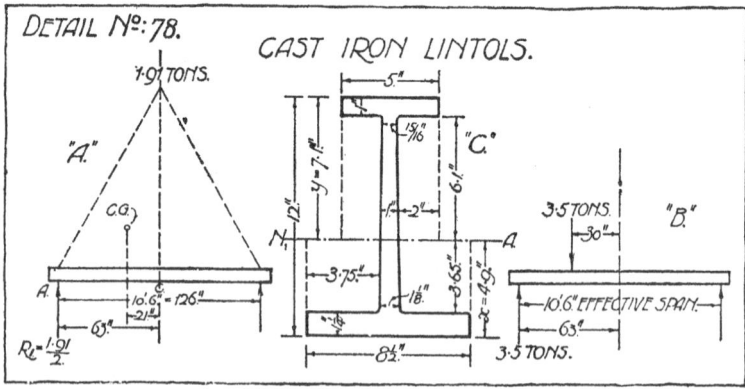

The bending moment at the centre of the span (taking moments on left of C) = (Reaction at A × half span) − (half load × distance of c.g. of △ from C),

$$\therefore B_C = \frac{1 \cdot 91}{2} \times 63 - \frac{1 \cdot 91}{2} \times 21.$$

Taking out common factors

$$B_C = \frac{21 \times 1 \cdot 91}{2} (3 - 1) = \mathbf{40 \cdot 1} \text{ tons inches.}$$

In addition to the brickwork, the bressummer may be called upon to carry one or more floor girders upon the inner half of the wall. Assume two such girders occurring at 5 ft. centres and placed centrally along the bressummer. The loads are gathered from bays approximately 12 ft × 5 ft. and, according to the type of floor selected, the latter may have to support an inclusive load of 2·35 to 2·75 cwts. per ft. sup. An equivalent to the load on one-half of the bay is transmitted to each support, hence the loads above the bressummer may be from 3·5 to 4·1 tons.

Assuming the lower value, the bressummer is additionally loaded as illustrated in the detail at B. Applying the principles explained in Chapter Six, and taking moments about the centre of the span,

$$B_M = 3 \cdot 5 \times 63 - 3 \cdot 5 \times 30$$
$$= 115 \cdot 5 \text{ tons inches.}$$

The designing value of the bending moment is the sum of the two values obtained above, viz.

$$115 \cdot 5 + 40 \cdot 1 = 155 \cdot 6 \text{ tons inches.}$$

Now $B_M = \dfrac{I}{x} f_t$, therefore $\dfrac{I}{x} = \dfrac{B_M}{f_t}$ and if f_t be $1\frac{1}{2}$ tons per sq. inch

$$\frac{I}{x} = \frac{155 \cdot 6}{1 \cdot 5} = 103 \cdot 8 \text{ inch units.}$$

Note. The quantity $\dfrac{I}{x}$ is the tensile modulus of section of the beam and is often referred to by the symbol Z_t.

Selecting the section illustrated in detail No. 78 at C and averaging the thickness of flanges and web, the areas are:

Top flange	5	sq. ins.
Web	9·75	,,
Bottom flange	10·62	,,
Total area	25·37, say **25·4** sq. ins.	

Taking moments about the base:

then $\qquad x = \dfrac{5 \times 11 \cdot 5 + 9 \cdot 75 \times 6 \cdot 12 + 10 \cdot 62 \times \cdot 625}{25 \cdot 4}$,

or $\qquad x = 4 \cdot 87$, say **4·9″**,

and $\qquad y = 12 - 4 \cdot 9 = \textbf{7·1″}.$

Summing the moments of inertia of the several rectangles composing the section:

$$I_{NA} = \frac{1}{3} \left[(5 \times 7 \cdot 1^3 + 8 \cdot 5 \times 4 \cdot 9^3) - (2 \times 2 \times 6 \cdot 1^3 + 2 \times 3 \cdot 75 \times 3 \cdot 65^3) \right]$$

$$= \textbf{506 inch}^4 \text{ units,}$$

$$Z_t = \frac{I_{NA}}{x} = \frac{506}{4 \cdot 9} = \textbf{103·2 inch units,}$$

and therefore just satisfies the condition that the tensile stress does not exceed the value of $1\frac{1}{2}$ tons per sq. inch.

If the cast iron bressummer were really fixed at the ends[1]—an unlikely condition with so short a bearing—the maximum bending moment would occur at the ends. In this case the top flange would be in tension. To determine the conditions and the stress requires a knowledge of fixed beams—a subject for later study.

[1] If the ends were deliberately fixed the top flange would be most highly stressed in tension over the supports.

231. Cast iron girders or bressummers, of the type illustrated in Vol. II, are designed or checked by exactly the same method as explained above. The position of the web, as regards the width of the beam, does not affect the value of I, but such beams must be well stiffened because the load tends to twist them if used singly owing to the transmission from the upper flange not being axial. In this example the pair of bressummers employed to span the loading shed opening are filled with concrete which assists in correcting the distribution of the load; the addition of concrete in the mass as applied here should not be considered to add strength to the bressummer.

STRENGTH OF FLITCHED BEAMS

232. When a beam or lintol is built up of several slabs placed in vertical planes and adequately bolted side by side, with the bolts not nearer to the top and bottom edges of the beam than one-sixth of the depth, the strength of the built-up member may be safely taken as the sum of the strengths of the component parts.

The bolts employed should not be less than $\frac{5}{8}''$ diameter, especially if the load is not uniformly distributed over the whole member. The function of the bolts is to prevent displacement and to transmit the stresses as uniformly as practicable over the several parts.

If the component slabs are not of the same depth, the shallower portions of the beam are not so effective because their outer fibres cannot be stressed to the full working limit; the latter condition only exists in the extreme fibres of the deeper parts.

233. Loading shed opening. (Example.) Let a bressummer of northern pine be composed of two $11'' \times 3''$ planks and one $9'' \times 3''$ deal, securely bolted together and carrying a uniform load of 4·5 tons over an effective span of 10 ft. Detail No. 79.

Find the working stress per sq. inch on the extreme fibres.

From first principles:
$$B_M = R,$$
$$\therefore \frac{WL}{8} = \frac{I}{y} f.$$

$$I_{NA} = \frac{2 \times 3 \times 11^3 + 3 \times 9^3}{12} = 848 \text{ inch}^4 \text{ units (approx.)},$$

$y = 5\frac{1}{2}$ ins.,
$W = 4·5$ tons,
$L = 120$ ins.

$$\therefore \frac{4·5 \times 2240 \times 120}{8} = \frac{848}{5·5} \times f,$$

or
$$15100 = 154f \qquad \qquad \qquad \dots\dots(a).$$

$$\therefore f = \frac{15100}{154} = 980 \text{ lbs. per sq. inch (approx.).}$$

This result may also be obtained by using the modulus of section of each piece and noting that the maximum fibre stress on the centre portion varies with the half depth as compared with that of the outer portions.

Thus: let f = maximum stress on outer fibres of $11''$ portion,

$$f_1 = \quad \text{,,} \quad \text{,,} \quad \text{,,} \quad 9'' \quad \text{,,}$$

then $\quad f_1 = \dfrac{4 \cdot 5}{5 \cdot 5} f.$

Now the R of a rectangular section $= \dfrac{bd^2 f}{6}$, and for all three rectangles

$$= \frac{2 \times bd^2 f}{6} + \frac{bd_1^2 f_1}{6},$$

where d_1 is depth of lesser piece $= 9''$.

Hence $\quad R = \dfrac{2 \times 3 \times 11 \times 11 \times f}{6} + \dfrac{3 \times 9 \times 9 \times 4 \cdot 5 f}{6 \times 5 \cdot 5}.$

$\therefore \; R = \dfrac{f}{6} (726 + 199) = \mathbf{154} f$, which compares with (a) above.

This is the same value as obtained from the more general formula, viz.

$$\frac{I}{y} f = R.$$

234. Steel flitched beams. Steel flitched beams are not in common use; they have been replaced by rolled steel beams of **I** or other sections in the erection of new structures. It is useful to be able to assess their strength, because flitches may need to be added in improving the stability of existing structures during repairs and alterations.

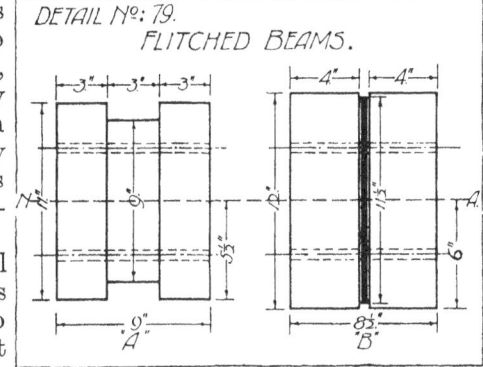

DETAIL Nº: 79.
FLITCHED BEAMS.

The mechanics of steel flitched timber beams cannot be entered into fully here, but a short summary of the main principles is given, with an application to a fully worked example.

It can be shown that when two materials are so securely fixed together as to strain at the same rate at sections equidistant from the neutral axis, the stress at such sections varies as the modulus of elasticity of the materials. On an average, steel has an elastic modulus twenty times that of timber. Hence, for beams in which the steel flitches are approximately the same depth as the timber, the stress on the extreme fibres of the steel is approximately twenty times the stress on the extreme fibres of the timber.

Using a similar method to that in the previous paragraph, let f_s = the maximum stress on the steel, and f = the maximum stress on the timber, then

$$R = \frac{bd^2f}{6} + \frac{b_s d^2 f_s}{6}.$$

But $f_s = 20f.$

$$\therefore \ R = \frac{bd^2f}{6} + \frac{b_s d^2 \times 20f}{6}$$

$$= \frac{d^2f}{6}(b + 20b_s).$$

235. Example. Let a steel flitched beam of the section shown in detail No. 79, support a uniform load of 8 tons over a span of 12 ft. (effective). Find the maximum stress on the timber and on the steel.

$$B_M = R,$$

$$\frac{WL}{8} = \frac{d^2f}{6}(b + 20b_s),$$

$$\frac{8 \times 2240 \times 144}{8} = \frac{144 \times f}{6}\left(8 + 20 \times \frac{1}{2}\right),$$

$$2240 = 3f,$$

$$f = \textbf{746·7} \text{ lbs. per sq. inch on the timber.}$$

Then $f_s = 20f.$

Therefore the stress on the steel $= \dfrac{746 \cdot 7 \times 20}{2240}$

$$= \textbf{6·67} \text{ tons per sq. in.}$$

Suppose the stress on the steel had reached 7 tons per sq. in., which is quite safe, then the stress on the timber would be one-twentieth of 7 tons, viz.

$$\frac{7 \times 2240}{20} = 784 \text{ lbs. per sq. in.,}$$

or 78 per cent. of the 1000 lbs. usually employed for northern pine. This result shows that, while a steel flitch may add considerable strength to a beam it prevents the full working stress on the timber being developed except by over-stressing the steel. Thus, if the timber be stressed to 1000 lbs. per sq. inch, the stress on the steel would reach 8·93 tons, which is 11·6 per cent. over the 8 tons per sq. in. allowed by the Code of Practice. It is common, though, to work steel structures to a stress of 9 tons per sq. inch when subjected to constant dead loads.

Steel flitches should, where possible, be firmly clasped between slabs of timber and not placed on the outside of the timber. In the latter case their liability to buckle, if comparatively thin, causes a great loss in effective resistance to bending.

Note. An example of a flitched beam is given in Vol. II. The student may usefully calculate the moment of resistance of this beam as applied to the loading shed opening above referred to.

CHAPTER NINE

STRUCTURAL DESIGN

DESIGN OF STANCHIONS, COLUMNS AND STRUTS

236. A pillar is a structural member initially under compressional stress and occupying a vertical position.

If loaded axially[1] the stress induced appears to be pure compression, and such would be the case except for practical errors of loading, and defects in shape and uniformity of quality of the material, which make a truly axial distribution of load impossible.

When considering piers of brickwork, masonry or plain concrete the effects of slight eccentricity are small and generally negligible owing to the stiffness which the considerable thickness in relation to length confers upon them under normal conditions,[2] but, when stronger materials are employed, or those of a fibrous nature suitable for use in comparatively long, thin pieces, any eccentricity of loading whether accidental or intentional becomes of vital importance, because it at once induces bending or buckling of the members.

This condition applies, not only to pillars, but also to compressional members of all kinds, e.g. struts and rafters in partitions and roof trusses.

237. Buckling of compression members. A strut or pillar with an apparent axial load may begin to buckle as shown in detail No. 80 from one of several causes. The following are the most important:

(*a*) Accidental eccentricity of loading; probably slight.

(*b*) Lack of initial straightness which is emphasised and increased by the load if centrally placed on the end of the member.

(*c*) The member, if a pillar, not being in a vertical position; or, if an inclined strut, tending to bend by its own weight.

(*d*) Non-uniformity of cross section, especially liable to occur in cast iron columns and stanchions.

(*e*) Inequality in elasticity of the material. All materials are

---

[1] See paragraph 118.

[2] No isolated brick or masonry pillar shall be less than $13\frac{1}{2}''$ wide; if loaded to produce the tabulated stresses (Code of Practice) the height shall not exceed six times the least width. If the height exceeds six times the least width the working stress shall be reduced proportionately so that when the height reaches twelve times the least width the stress shall be only 40 per cent. of the tabulated stress. No pillars shall exceed in height twelve times the least width.

defective in this respect, steel being the most reliable and timber very unreliable owing to the change of condition during growth, etc.

It is clear that if buckling commences under any of these possible conditions, not only will direct compression continue to act, but bending stresses will be developed of the same character as previously described for beams, compression due to bending acting on the concave side of the buckled member and tension on the opposite side.

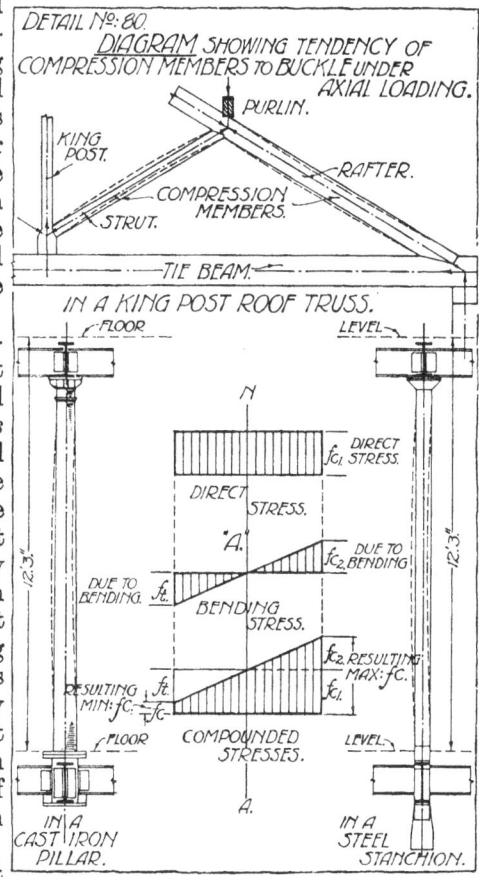

DETAIL Nº: 80.
DIAGRAM SHOWING TENDENCY OF COMPRESSION MEMBERS TO BUCKLE UNDER AXIAL LOADING.
PURLIN.
KING POST.
RAFTER.
COMPRESSION MEMBERS.
STRUT.
TIE BEAM.
IN A KING POST ROOF TRUSS.

The tendency to induce tension *may* not actually cause tensional stress, and generally *does* not, because the original compression over the whole area due to the vertical load must first be reduced to zero by the bending stress on the tension side; but in some cases of long slender pillars or struts the bending stress may be larger than the direct compression, in which case the convex side of the member would be in tension.

The detail No. 80 at A shows the effect of compounding direct compression and bending stress. The intention of the designer of a pillar is to ensure that the resultant of the compounded stresses (max. f_c) shall not exceed the safe compression on the material. If tension occurs on the opposite edge—due to the direct stress being smaller than the bending stress caused by buckling—then this tension must also be kept within safe limits.

238. Steel, cast iron and timber pillars. Steel and wrought iron will resist tension better than compression, and timber is of the same character, though less dependable, but cast iron is weak in tension and it is therefore undesirable to allow tension to develop in any pillar or strut of cast iron. The criterion of safety for *any* compression member is that the total compression stress due to "direct pressure" and "bending" shall not exceed a safe value, and also, in the case of cast iron members, no tension should be permitted.

To deal with the computation of probable stresses at the edges of pillar sections, is beyond the scope of this volume, but herewith is given the method adopted in the L.C.C. Code of Practice for deciding upon the suitability of sections proposed to be employed as pillars or struts subjected to compressional load.

239. Flexibility of pillars. The resistance of a pillar to bending depends chiefly upon its length and the disposition of the material in its cross section. Thus, two pillars of the same length and having the same amount of material in their cross sections may vary much in flexibility, the stiffness increasing as the material is placed further from the geometrical centres of the section. Let a solid rectangle of 4″ side be used as a pillar section and compare it with a hollow square of 8½″ side and ¼″ thick; both have an area of 16 sq. ins. but the hollow square would be much less flexible than the solid one. A solid circle may be compared with a hollow one in the same way.

Previous study has shown that resistance to bending is dependent on the moment of inertia of the section, yet in this case inertia cannot be used as the criterion of stiffness because two sections with equal inertia may have different areas.

It is therefore necessary to employ both inertia and area to determine the factor which governs flexibility.

240. Radius of gyration. The least moment of inertia or second moment of a section has been previously defined as the sum of the products of all the elements of area into the squares of their respective distances from the neutral axis of that section, and it has been shown that the stress on any particle caused by bending varies directly with the distance from the axis. If we suppose that a section is trying to rotate about the neutral axis when strained by bending, there is an average distance from the axis at which all the material in the section could be massed to produce the same resistance.

Hence, if I represents the measure of moment of inertia or resistance to rotation of a section having an area A, and g represents the distance from the axis at which the whole area could be massed to produce the same effect; then $A \times g^2$ would equal I, because the latter is the sum of the second moments of similar but small quantities of area disposed over the whole section and composing it.

Then $$Ag^2 = I, \quad g^2 = \frac{I}{A}, \quad \text{and} \quad g = \sqrt{\frac{I}{A}},$$

which provides a means of finding the average radius at which the whole area may be conceived to act, to produce a given moment of inertia; the value of g is called the *radius of gyration*, which is recognised as the dimension of the cross section governing the resistance to buckling.

241. Slenderness of a pillar. The slenderness of a pillar is gauged by the ratio of its effective or virtual length to its radius of gyration, viz. by the value of $\frac{l}{g}$, both dimensions being usually taken in inches, and necessarily in the same units.

Within certain limits the stress, *averaged over the whole cross section*, is reduced proportionately as $\frac{l}{g}$ increases.

This reduction of stress can be calculated for an assumed set of conditions, and formulae for calculation can be deduced according to the conditions assumed—chiefly depending upon $\frac{l}{g}$.

At this stage the knowledge of the average student does not permit a study of the formulae relating to pillars, hence for the time being, the consideration will be confined to the interpretation of information provided as a guide to practical designers.

The succeeding paragraphs are compiled with this end in view.

242. Effect of end conditions in pillars. A brief consideration will show that if a pillar or strut be held rigidly so that the axis remains straight and in the original unloaded position at the ends, as in detail No. 81 at A, it can only buckle freely over about one-half its length at the centre (marked l_b), the end segments bending in the opposite direction.

If one end is rigidly held while the other is so lightly held that it can tilt or rotate as shown at B, the freely buckled portion is about two-thirds the whole length.

Again, if both ends can tilt, the whole member can buckle freely, as shown at C.

It is necessary to make some allowances for these variable end conditions which greatly affect the stresses induced, though it is difficult in practice to determine under which heading any case shall be classed.

In steel framed buildings the very variable conditions of loading on the adjoining bays of flooring on each side of a pillar may cause more deflection of the connected beam on one side than on the other, especially when overloading of the floor occurs—an accidental condition which must be considered. The deflected beam is restrained

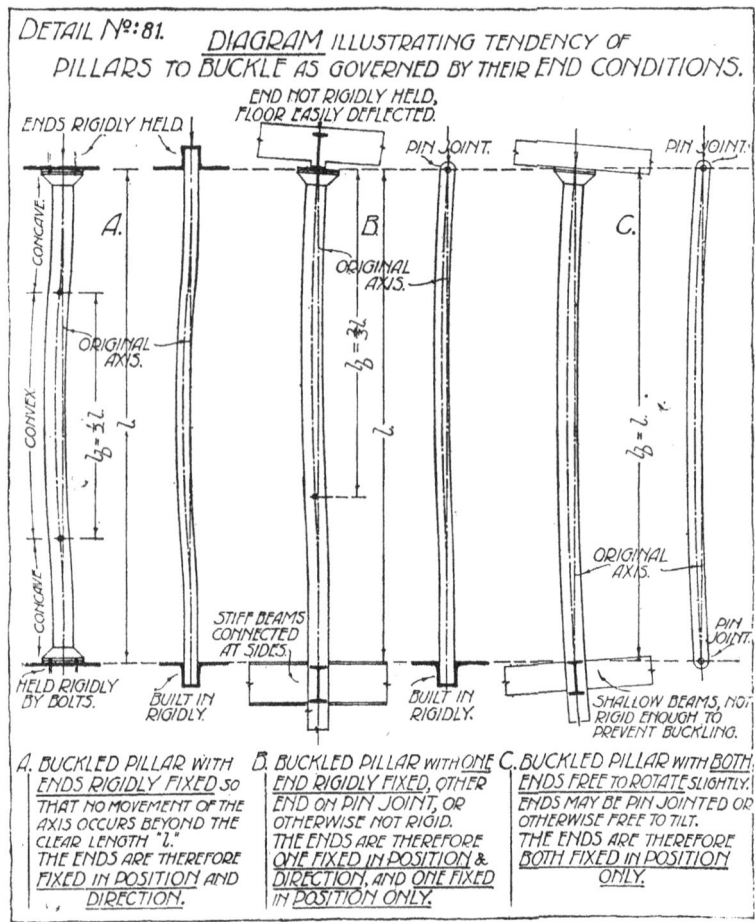

This diagram represents theoretical conditions only.

by the pillar, but as the beam and its connections are often stiffer than the pillar the latter is deflected sidewise and follows the beam. It is therefore unlikely that any ordinary pillar—especially in the upper storeys of a building—can be depended upon to retain the original vertical position of the axis. Hence the buckled segment of the pillar which constitutes the effective length may, under certain conditions, be considerably more than the theoretical value for the several conditions illustrated in detail No. 81.

The L.C.C. Code of Practice for steel framed buildings refers to the freely buckled length of a pillar as the "effective length" and the effective length depends upon the end conditions and the restraint applied (a) to maintain the original position of the end of the pillar, (b) to maintain the original direction of the axis. According to the extent of the restraint under these headings the effective length may vary from ·75 of the actual pillar length between supports to the full length between supports.

The following table shows the conditions laid down in the L.C.C. Code of Practice for determining the effective length of a pillar:

Type of pillar		Effective pillar length
Pillars of one storey	Adequately restrained at both ends in *position* and *direction*	0·75 of the actual pillar length
	Adequately restrained at both ends in position but *not* in direction	Actual pillar length
	Adequately restrained at one end in position and direction and imperfectly restrained in both position and direction at the other end	A value intermediate between the actual pillar length and twice that length, depending upon the efficiency of the imperfect restraint
Pillars continuing through two or more storeys	Adequately restrained at both ends in position and direction	0·75 of the distance from floor level to floor level
	Adequately restrained at both ends in position and imperfectly restrained in direction at one or both ends	A value intermediate between 0·75 and 1·00 of the distance from floor level to floor (or roof) level, depending upon the efficiency of the directional restraint

243. Tabulated values of average stresses on mild steel pillars. The following figures and a graph relating to average stresses in pillars are based on the L.C.C. Code of Practice and the British Standard Specification for the use of Structural Steel in Building—No. 449—1935.

The attached table shows the working loads in tons per sq. in. of gross section of a pillar for a given ratio of effective pillar length to the least radius of gyration of the section.

In the table l = effective pillar length

g = radius of gyration

F_1 = average stress *or* working loads in tons per sq. in.

$\dfrac{l}{g}$	F_1	$\dfrac{l}{g}$	F_1
The limit of $\dfrac{l}{g}$ is 150 for pillars and struts forming part of the main structure of a building		These values allowed only for subsidiary members in compression	
		L.C.C. Limit	
20	7·2	160	1·8
30	6·9	170	1·6
40	6·6	180	1·5
50	6·3	190	1·3
60	5·9	200	1·2
70	5·4	B.S.S. Limit	
80	4·9	210	1·10
90	4·3	220	1·01
100	3·8	230	0·93
110	3·3	240	0·86
120	2·9		
130	2·6		
140	2·3		
150	2·0		

244. Graph of pillar stresses—L.C.C. Code of Practice. Detail No. 82 shows a graph of the values tabulated above.

Intermediate values of F_1 for any given value of $\dfrac{l}{g}$ can be quickly read from this graph or such values may be obtained from the table by interpolation.

In order to determine the working (average) stress for any given case, two factors have to be determined:

(*a*) the radius of gyration (*g*) of the section of the pillar;

(*b*) the effective length of the pillar for the conditions of restraint which are imposed.

Radius of gyration (referred to under (*a*) above), has already been discussed in paragraph 240, but some general values of *g* for pillar sections are determined in paragraph 246.

The conditions governing (*b*) are laid down in paragraph 243, although they may need special interpretation where there is a doubt as to the efficiency of directional restraint. This matter

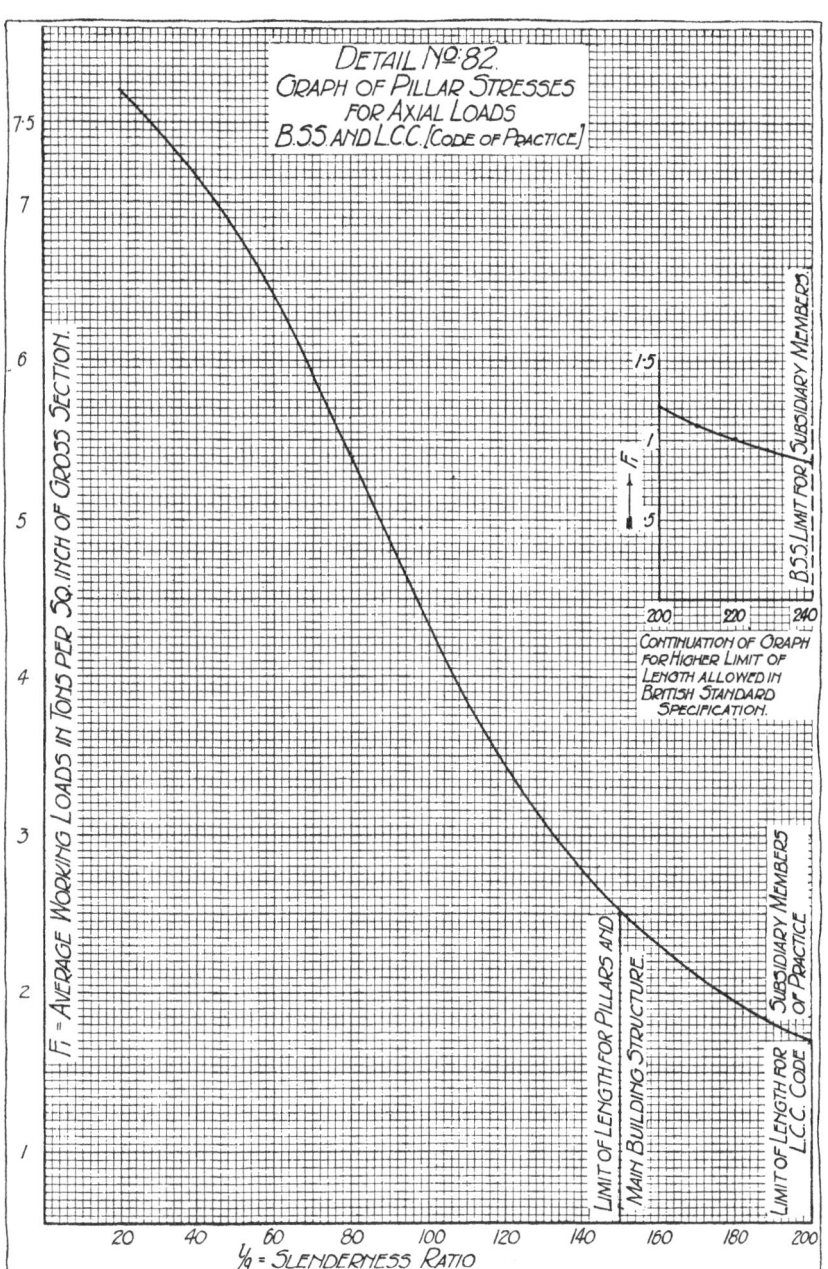

DETAIL Nº 82.
GRAPH OF PILLAR STRESSES
FOR AXIAL LOADS
B.SS AND L.C.C. [CODE OF PRACTICE]

CONTINUATION OF GRAPH
FOR HIGHER LIMIT OF
LENGTH ALLOWED IN
BRITISH STANDARD
SPECIFICATION.

B.S.S. LIMIT FOR SUBSIDIARY MEMBERS

LIMIT OF LENGTH FOR PILLARS AND
MAIN BUILDING STRUCTURE.

SUBSIDIARY MEMBERS
OF PRACTICE.

LIMIT OF LENGTH FOR
L.C.C. CODE

F_1 = AVERAGE WORKING LOADS IN TONS PER SQ. INCH OF GROSS SECTION.

l/g = SLENDERNESS RATIO

will be discussed as occasion arises in connection with examples which follow.

245. Cast Iron Pillars—L.C.C. Regulations. Under the existing regulations of the London County Council the average working stresses on axially loaded cast iron pillars are represented in graphs Nos. 1, 2 and 3 of detail No. 83. The graphs are straight lines showing a maximum average stress of $5\frac{1}{2}$ tons per sq. in. in graph No. 1, 5 tons

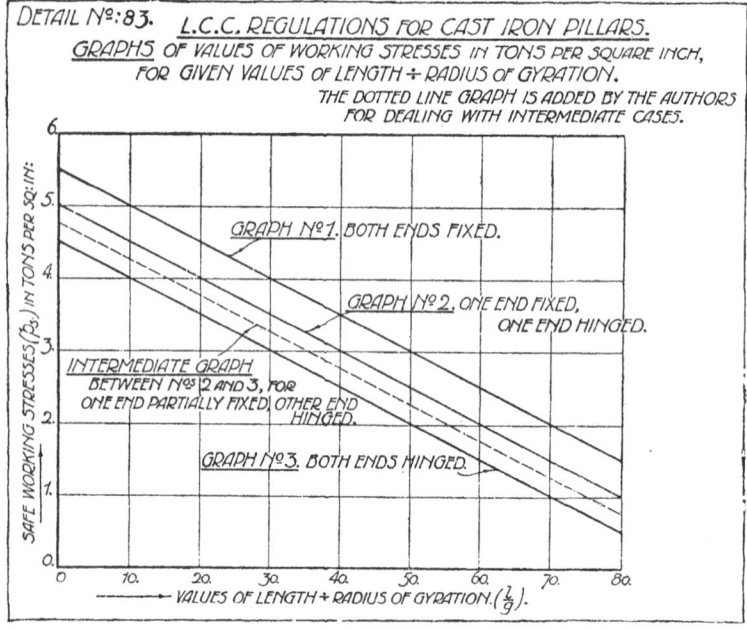

DETAIL Nº:83. L.C.C. REGULATIONS FOR CAST IRON PILLARS.
GRAPHS OF VALUES OF WORKING STRESSES IN TONS PER SQUARE INCH, FOR GIVEN VALUES OF LENGTH ÷ RADIUS OF GYRATION.
THE DOTTED LINE GRAPH IS ADDED BY THE AUTHORS FOR DEALING WITH INTERMEDIATE CASES.

GRAPH Nº1. BOTH ENDS FIXED.
GRAPH Nº2. ONE END FIXED, ONE END HINGED.
INTERMEDIATE GRAPH BETWEEN Nºs 2 AND 3, FOR ONE END PARTIALLY FIXED, OTHER END HINGED.
GRAPH Nº3. BOTH ENDS HINGED.

SAFE WORKING STRESSES (f_s) IN TONS PER SQ. IN.

VALUES OF LENGTH ÷ RADIUS OF GYRATION. $\left(\frac{l}{g}\right)$.

sq. in. in graph No. 2, and $4\frac{1}{2}$ tons per sq. in. in graph No. 3. The limit of length for any pillar is defined by $\dfrac{l}{g} = 80$ (l is the actual length of the pillar). This limit is seldom approached in pillars of *practical proportions*.

Graph No. 1 applies for *both* ends fixed in position and direction.

Graph No. 2 applies for *one* end fixed in position and direction and *one* end fixed in position only.

Graph No. 3 applies for *both ends fixed in position only.*

The dotted line graph is added by the authors for dealing with intermediate cases where the L.C.C. regulations may (or do) not apply.

If the usual mathematical analysis be applied to the series of graphs, it will be found that the slope is $-\dfrac{1}{20}\dfrac{l}{g}$ and that the formulae for the three main graphs are:

(1) Average working stress $= 5 \cdot 5 - \dfrac{1}{20}\dfrac{l}{g}$;

(2)   „   „   „ $= 5 \cdot 0 - \dfrac{1}{20}\dfrac{l}{g}$;

(3)   „   „   „ $= 4 \cdot 5 - \dfrac{1}{20}\dfrac{l}{g}$;

and for the intermediate graph

Average working stress $= 4 \cdot 75 - \dfrac{1}{20}\dfrac{l}{g}$.

For the determination of the average working stress for any given case, the slenderness is obtained as for steel pillars—bearing in mind the maximum of 80—and the allowable stress for that slenderness read from the graph or calculated by formula.

246. Some particular values of the radius of gyration. For some common sections the value of the radius of gyration (g) may be definitely stated as a constant ratio of the diameter. Thus, for

Rectangular sections $g = \sqrt{\dfrac{I}{A}} = \sqrt{\dfrac{bd^3}{12bd}} = \sqrt{\dfrac{d^2}{12}} = \cdot 289d$,

Circular sections $g = \sqrt{\dfrac{I}{A}} = \sqrt{\dfrac{\dfrac{\pi}{64}d^4}{\dfrac{\pi}{4}d^2}} = \sqrt{\dfrac{d^2}{16}} = \dfrac{d}{4}$ or $\cdot 25d$.

Hollow rectangular section (thin walls)
$$g = \cdot 4d_m \text{ (approximately)},$$
Hollow circular section (thin walls)
$$g = \cdot 35d_m \text{ (approximately)},$$

where d_m is the mean diameter of the hollow section in each case.

The exact value of g for any case is easily obtained, given the shape and dimensions of the section. The procedure is:

(a) Calculate the area.

(b) Obtain the value of the moment of inertia about the neutral axis.

Then $g = \sqrt{\dfrac{I}{A}}$.

247. Radius of gyration of standard steel sections. In the case of standard steel sections employed as pillars or struts the value of g

for either principal axis can usually be read off directly from manufacturers' section books, and
much labour of calculation is
saved thereby. In other cases
the area and moment of inertia in
each principal direction are given
and g is quickly determined.

Example. The $4'' \times 4'' \times \frac{3}{8}''$ Tee
section of mild steel, shown in
detail No. 84, is to be employed
as a vertical strut. Its length is
6 ft. and the ends are pin jointed.
The area is 2·872 sq. ins.,

DETAIL N°: 84.
ROLLED STEEL TEE,
TO BE EMPLOYED AS A STRUT.

$$I_{N_1A_1} = 4·189 \text{ inch units}, \quad I_{N_2A_2} = 1·901 \text{ inch units}.$$

Find the maximum ratio of $\dfrac{l}{g}$.

The greatest tendency to buckle is in the direction at right angles
to the axis N_2A_2;[1] find g about this axis.

$$g_{(N_2A_2)} = \sqrt{\frac{1·90}{2·87}} = \sqrt{·662} = ·814''.$$

Hence maximum $\dfrac{l}{g} = \dfrac{72}{·814} = 88·5$ (say 88).

STANCHIONS AND COLUMNS FOR WAREHOUSE

248. Stanchions to warehouse. The construction of the stanchions, etc., for providing internal support to the warehouse floors,
has been described in Chapter Two.

It is now possible to show how the sizes of these stanchions were
determined.

For this purpose let it be assumed that the ends of all beams
connected to the stanchions are not rigidly fixed, and that the
stanchions receive their maximum stresses when the floors are
completely loaded to 2 cwts. per ft. sup. in addition to their own
dead weight. To simplify the calculation, $\frac{3}{4}$ cwt. per ft. sup. will
be taken as the average inclusive weight of the floor, which has
been shown to be somewhat in excess of the actual average load.
This excess will ensure some allowance for the dead weight of the
stanchions and their protective coverings without the tedium of
detailed calculation, which must be made as a check on all practical
designs. The object here is to show the mode of procedure.

[1] This assumes that the end joints restrain the buckling tendency to the
principal axes. Otherwise, in long struts, bending may occur about an inclined
axis across the section.

249. Load on stanchions. Each floor will transmit load to a stanchion equal to that supported by a rectangle enclosing half the bay in each direction around it. Taking the bays as approximately 15 ft. × 12′ 6″, the load

$$= \frac{2 \cdot 75^{(c)}}{20} \times 15 \times 12 \cdot 5 = \mathbf{25 \cdot 78} \text{ tons, } say \text{ 26 } tons.$$

The basement stanchions carry the weight from three floors, viz. ground, first and second floors, the total load being 78 tons. Ground floor stanchions carry first and second floors, total 52 tons. First floor stanchions carry the second floor only and the load is 26 tons.

The stanchions are actually continuous through three floors as shown in detail No. 28. This detail gives the floor to floor heights necessary for calculation.

Note. In the example adopted for the second floor, in which timber joists are employed, the dead weight is much less than would exist for a fire-resisting floor. The resulting stanchions will therefore allow for the employment of fire-resisting floors throughout, if desired.

250. Basement stanchion. This length of the continuous stanchion, has considerable restraint at both ends, but as a first calculation let the effective length be taken as 10 ft.—the height from floor to floor.

Selecting a 9″ × 7″ × 50 lbs. B.S.B. from Appendix I, its sectional area is found to be 14·71 sq. ins., and its least radius of gyration (about axis YY) is 1·65 ins., hence the slenderness ratio

$$\frac{l}{g} = \frac{120}{1 \cdot 65} = 72 \cdot 7.$$

Referring to the graph—detail No. 82—or to the table in paragraph 243, the value of F_1 for a slenderness ratio of 72·7 is approximately 5·26 tons sq. ins.

The allowable axial load on the pillar is

$$P = A \times F_1$$

where A = sectional area in sq. ins.

and F_1 = average working stress in tons per sq. in.

$$\therefore \ P = 14 \cdot 71 \times 5 \cdot 26 = \mathbf{77 \cdot 37} \text{ tons} \qquad \dots\dots(1)$$

for ends unrestrained in direction, but adequately restrained in position.

Now assume adequate restraint in position and direction. The effective length is then ·75 of the actual length, viz. 7·5 ft. or 90 inches.

Then $$\frac{l}{g} = \frac{90}{1 \cdot 65} = 54 \cdot 5.$$

For this ratio $F_1 = 6\cdot1$ tons sq. in. (see graph or table).

$$\therefore\ P = 14\cdot71 \times 6\cdot1 = \mathbf{89\cdot73}\ \text{tons} \qquad \ldots\ldots(2).$$

The results (1) and (2) are for the limiting conditions, from that of adequate fixing in position and direction to that of no directional restraint. The actual conditions are probably nearer the perfect than the imperfect and the effective length of the pillar should be taken at $\cdot8$ to $\cdot85$. Using $\cdot85$, the effective length is $\cdot85 \times 120^{(\prime\prime)} = 102$.

Then
$$\frac{l}{g} = \frac{102}{1\cdot65} = 61\cdot8.$$

For this ratio $\qquad\qquad F_1 = 5\cdot5$ tons sq. in.

and $\qquad\qquad P = 14\cdot71 \times 5\cdot8 = \mathbf{85\cdot32}$ tons.

The $9'' \times 7'' \times 50$ lbs. B.S.B. is therefore satisfactory.

For practical reasons the same section is desirably carried through at least two storeys of the height in order to avoid undue length jointing, which is often more costly than allowing a section which is larger than necessary to be continued without a joint.

In this case, however, a calculation must be made to check the necessary size of the ground floor pillar because the height of the ground floor is 14 ft. (much higher than the basement) and the slenderness ratio will be greater, although the load to be carried is less.

251. Ground floor stanchion. The ground floor storey is 14 ft. high from floor to floor, and the stanchions are held sidewise by two pairs of beam connections at each floor level. The main floor beams are connected to the flanges and the $10''$ secondary beams to the webs, so that these latter beams supply the restraining effect on the direction of the axis of the pillar. They are not sufficiently stiff to be considered as giving adequate restraint and the effective length will therefore be taken as $\cdot85$ of the actual length.

Then
$$\frac{l}{g} = \frac{\cdot85 \times 168^{(\prime\prime)}}{1\cdot65} = 86\cdot5.$$

For the ratio $\qquad\qquad F_1 = 4\cdot5$ tons sq. in.

$$\therefore\ P = 14\cdot71 \times 4\cdot5 = \mathbf{62\cdot52}\ \text{tons}.$$

The load to be carried—see paragraph 249—is 52 tons so that, apart from practical convenience, the $9'' \times 7''$ section must be continued through at least two storeys because the next size smaller would be inadequate.

252. First floor stanchion. The height of the first floor storey is about 12 ft. from floor to floor. The stanchions terminate at the second floor level and cannot be said to approach adequate restraint at the top. The base is reasonably restrained because of the plated and riveted joint connecting it to the head of the ground floor pillar and the two pairs of beam connections thus ensuring continuity, and a measure of restraint. The effective length of this pillar will therefore be taken at ·9 of the actual length.

Try an $8'' \times 6'' \times 35$ lbs. B.S.B. having a sectional area of 10·29 sq. ins.

The least radius of gyration is 1·37 ins.

Then
$$\frac{l}{g} = \frac{\cdot 9 \times 144}{1 \cdot 37} = 94 \cdot 6.$$

For this ratio $F_1 = 4 \cdot 1$ tons sq. in.

$$\therefore \ P = 10 \cdot 29 \times 4 \cdot 1 = \mathbf{42 \cdot 2} \text{ tons.}$$

From paragraph 249, the load on this top length of stanchion is 26 tons. Hence try the next smaller size of B.S.B.

B.S.B. is $6'' \times 5'' \times 25$ lbs. and has a sectional area of 7·35 sq. in. Its least radius of gyration is 1·15″.

Then
$$\frac{l}{g} = \frac{\cdot 9 \times 144}{1 \cdot 15} = 112 \cdot 7.$$

For this ratio $F_1 = 3 \cdot 15$ tons sq. in.

$$\therefore \ P = 7 \cdot 35 \times 3 \cdot 15 = \mathbf{23 \cdot 15} \text{ tons,}$$

which is less than the load to be carried. In addition the effective length taken for the above two calculations might be considered by the local authority to be on the generous side. The $8'' \times 6'' \times 35$ lbs. stanchion is therefore used.

Note. While there may be no practical saving in employing the $8'' \times 6'' \times 35$ lbs. B.S.B. for the first floor as against carrying the $9'' \times 7''$ section to the top, the former has been adopted in order to show the connection between the lengths of stanchion, as illustrated in detail No. 26.

It should also be noted that a good practical procedure, often adopted where the total height of the stanchion (up to 40 ft.) is obtainable in one length, is to select a section which is economical for the upper storeys and to increase the rigidity and bearing power of the section for the lower storeys by adding riveted flange plates in proportion to the load and stiffness required. Such a section is shown in detail No. 21 at A.

In this case the value of g would be found from $g = \sqrt{\dfrac{I}{A}}$, I being calculated as explained in paragraph 191.

253. Cast iron columns. Cast iron columns have been previously considered and detailed as an alternative construction for supporting the principal floor beams of the warehouse.

The dimensions of these columns have been selected by employing the values of working stresses for cast iron pillars laid down in paragraph 245 and shown in the graphs of detail No. 83.

The method of checking the suitability of these columns will now be dealt with in detail. The process is to select what appears to be a suitable column and to test it by regulation requirements.

254. Basement column. This column, shown in detail No. 16, has a diameter of $11''$ at the base and $10''$ at the top or a mean external diameter of $10\frac{1}{2}''$, which is adopted for these calculations. The thickness of the metal is $\frac{7}{8}''$, giving a mean internal diameter of $8\frac{3}{4}''$, and the length of the column is 10 ft.

Let D = external diameter = $10\frac{1}{2}''$,

$\quad d$ = internal diameter = $8\frac{3}{4}''$,

$\quad d_m$ = mean diameter of section = $\dfrac{D+d}{2} = \dfrac{10\frac{1}{2}+8\frac{3}{4}}{2} = 9\cdot625''$.

From paragraph 240 $\qquad\qquad g = \sqrt{\dfrac{I}{A}}.$

But $\qquad\qquad I$ for a hollow circle $= \dfrac{\pi}{64}\,(D^4 - d^4)$,

$\qquad\qquad\qquad\qquad\qquad\qquad$ (see detail No. 77)

and $\qquad\qquad A$ for a hollow circle $= \dfrac{\pi}{4}\,(D^2 - d^2)$,

$$\therefore\; g = \sqrt{\dfrac{\dfrac{\pi}{64}\,(D^4 - d^4)}{\dfrac{\pi}{4}\,(D^2 - d^2)}} = \sqrt{\dfrac{1}{16}\,(D^2 + d^2)} = \dfrac{1}{4}\sqrt{D^2 + d^2}.$$

Inserting values for the above case

$$g = \dfrac{1}{4}\sqrt{10\cdot5^2 + 8\cdot75^2} = \dfrac{\sqrt{186\cdot8}}{4} = 3\cdot4''.$$

Note. This result agrees fairly well with the approximation given in paragraph 246, where g is stated for a hollow circle as $\cdot35d_m$; d_m (see above) = $9\cdot625''$, therefore g (approximately) = $\cdot35 \times 9\cdot625 = 3\cdot37''$. This is slightly below the accurate value, which will always be the case and therefore on the safe side in designing.

Taking the value of g as $3\cdot4''$, then $\dfrac{l}{g} = \dfrac{120}{3\cdot4} = 35\cdot3$.

Let p_s be the safe pressure per sq. inch (average stress) on the metal. (This symbol is used to avoid confusion with F_1 for steel, since the basis of calculation is different.)

Using graph No. 2 (assuming one end fixed p. and d., and the other fixed p. only), then,

$$p_s = 3\cdot23 \text{ tons per sq. in.}$$

Checking by formula,

$$p_s = 5 - \frac{35 \cdot 3}{20} = 5 - 1 \cdot 765 = \mathbf{3 \cdot 23} \text{ tons per sq. in.}$$

Let $d_m = $ mean diameter of section $= 9 \cdot 62''$,

$t = $ thickness of section $= \frac{7}{8}'' = \cdot 875''$.

Then sectional area $= \pi \cdot d_m \cdot t = 3 \cdot 14 \times 9 \cdot 62 \times \cdot 875 = 26 \cdot 4$ sq. ins.

The total safe load

$$= p_s \times A = 3 \cdot 23 \times 26 \cdot 4 = \mathbf{85 \cdot 2} \text{ tons,}$$

as against 78 tons required.

By reducing the diameter a little or by reducing the thickness of metal in the column the total safe load is brought nearer to the mark, but it is unwise to attempt too close a value in cast iron work. The following considerations have led to the adoption of the design in its present form:

(a) to guard against inaccuracies in the thickness of the casting;

(b) to retain good architectural proportions allowing of a satisfactory detail for an unprotected column;

(c) to comply with the L.C.C. requirement relating to thickness, viz. the least thickness of a cast iron pillar shall be $\frac{3}{4}''$ and in no case shall it be less than one-twelfth of the external diameter of the pillar; in this case $10\frac{1}{2}'' \div 12 = \frac{7}{8}''$.

It is convenient to note here that no pillar may have a less external diameter than $5\frac{1}{2}''$.

255. Ground floor column. The ground floor column shown in details Nos. 16, 17, 18 and 20 has a length of 14 ft. and is $10''$ external diameter at the base and $8\frac{3}{4}''$ at the head, with an average thickness of $1''$.

The mean external diameter is $\dfrac{10'' + 8\frac{3}{4}''}{2} = 9 \cdot 37''$; the internal diameter is $9 \cdot 37'' - 2'' = 7 \cdot 37''$; and the mean sectional diameter

$$= \frac{9 \cdot 37 + 7 \cdot 37}{2} = 8 \cdot 37''.$$

The approximate value of

$$g = \cdot 35 d_m = \cdot 35 \times 8 \cdot 37 = 2 \cdot 93'', \quad \therefore \; \frac{l}{g} = \frac{168}{2 \cdot 93} = 57 \cdot 4.$$

Assuming both ends *partially* fixed and roughly equivalent to the conditions applying to graph No. 2, then

$$p_s = 2 \cdot 13 \text{ tons per sq. in.}$$

Checking by formula,

$$p_s = 5 - \frac{57 \cdot 4}{20} = 5 - 2 \cdot 87 = 2 \cdot 13 \text{ tons per sq. in.}$$

The sectional area of the column

$$= \pi \cdot d_m \cdot t = 3 \cdot 14 \times 8 \cdot 37 \times 1 = 26 \cdot 3 \text{ sq. ins.,}$$

and the total safe load

$$= p_s A = 2\cdot13 \times 26\cdot3 = \mathbf{56} \text{ tons,}$$

as against 52 tons required, which is satisfactory.

256. First floor column. The first floor column, also illustrated in details Nos. 16 to 20 is 11 ft. high, $8\frac{1}{2}''$ external diameter at the base, and $7''$ at the top, and has an average thickness of $\frac{3}{4}''$. The mean external diameter is $\dfrac{8\frac{1}{2}'' + 7''}{2} = 7\frac{3}{4}''$; the internal diameter is $7\frac{3}{4}'' - 1\frac{1}{2}'' = 6\frac{1}{4}''$; and the mean sectional diameter is $\dfrac{7\frac{3}{4}'' + 6\frac{1}{4}''}{2} = 7''$. Then g (approximately) $= \cdot35 \times 7 = 2\cdot45''$, and

$$\frac{l}{g} = \frac{132}{2\cdot45} = \mathbf{54}.$$

The top end of this pillar terminates at the floor level and is only held in position; it has no fixity of direction because the beams rest upon its cap and are only lightly bolted down. Its base may be considered as *partially* fixed in direction. Hence, using a formula intermediate between those representing the equations to graphs Nos. 2 and 3, see dotted line:

$$p_s = 4\cdot75 - \frac{1}{20}\frac{l}{g},$$

and for this example

$$p_s = 4\cdot75 - \frac{54}{20} = 4\cdot75 - 2\cdot7 = 2\cdot05 \text{ tons per sq. in.}$$

The sectional area of the column

$$= \pi . d_m . t = 3\cdot14 \times 7 \times \cdot75 = 16\cdot48 \text{ sq. ins.}$$

The total safe load $= p_s A = 2\cdot05 \times 16\cdot48 = 33\cdot7$ tons, as against 26 tons required.

257. Note regarding fixed ends to pillars and struts. It should be noted by students that very few ends of pillars in actual practice can be really "fixed" in direction.

When two members such as a beam and a pillar are well connected, any movement of one member is communicated in some degree to the other, depending upon their relative stiffness and the rigidity of the connection.

Thus, if the beam on one side of the pillar is loaded it deflects and communicates its bending moment to the pillar in proportion to the capacity of the latter to resist it. The point to be borne in mind is: that no security or rigidity of connection of the *beams* will absolutely fix the *pillar*.

CHAPTER TEN

STRUCTURAL DESIGN

STRESSES IN FRAMED STRUCTURES

In Vol. II an introduction was given to the study of forces in the members of loaded triangulated frames, such as roof trusses. The loads in the introductory examples were assumed.

258. Loads on roof frames. The nature and distribution of the loads which roof frames, purlins and rafters have to support may now be investigated.

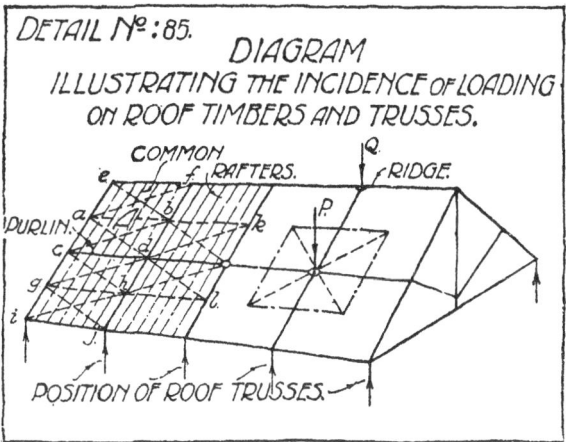

DETAIL Nº:85.
DIAGRAM ILLUSTRATING THE INCIDENCE OF LOADING ON ROOF TIMBERS AND TRUSSES.

The common rafters are inclined beams transmitting the weight of the roof covering which rests upon them to the purlins, ridge and wall plate. In detail No. 85 the common rafters in bay A transfer their weight to the purlin and ridge, hence the latter members become horizontal beams almost uniformly loaded, and the equivalent amount of load transferred to each from this bay alone is approximately represented by the rectangles *abcd* and *abef* respectively. But the purlin and ridge also receive loads from the half bays on the opposite sides of the ridge and purlin, hence the load on a purlin becomes that supported upon an area equal to its own length and a breadth made up of *half the length of the bay on each side* measured along the slope of the roof, as indicated by the area lettered *b, h, k, l*.

The ridge e to f carries the load on its own length and across half the first bay down each slope, while the half bay at the foot of the slope has its equivalent load transmitted directly to the wall plate and does not affect the truss.

The loads transmitted to the roof truss are evidently concentrated or "point" loads transferred from the ends of the ridge and purlins, each pair of these transferring half of their loads to the truss, hence—approximately—the load on each purlin point of the truss is the weight supported upon a rectangle extending half-way across each surrounding bay, as shown at P in the detail: in the same way the load on the ridge point is gathered from a similar area extending down each slope, as at Q.

259. Practical loading of roofs. The amount of the load on any load point of a roof truss will depend upon the spacing of the trusses, on the distance apart of the horizontal timbers and on the nature of the covering.

It must be borne in mind that the whole fabric of the structure is dead weight, and is sufficiently well distributed to be thought of as uniform.

In order to follow the usual mode of procedure the average covering weights of parts of the structure will be expressed in terms of the equivalent uniform load spread over the whole sloping area of the roof, except in the case of roof trusses, which are more conveniently given as total weights of trusses, to be divided uniformly over the number of load points as shown in the examples which follow.

260. Weights of roofing materials and framework.

Material	Average weight per ft. sup. distributed over the slope lbs.
Plain tiles, $10\frac{1}{2}'' \times 6\frac{1}{2}'' \times \frac{3}{8}''$ thick (laid to $2\frac{1}{2}''$ lap) ...	10
,, ,, ,, $\frac{1}{2}''$,, ,, ,, ...	13
,, ,, ,, $\frac{5}{8}''$,, ,, ,, ...	16
Countess slates, $20'' \times 10'' \times \frac{3}{16}''$,, (laid to $4''$ lap) ...	$6\frac{3}{4}$
,, ,, ,, $\frac{1}{4}''$,, ,, ,, ...	9
,, ,, ,, $1\frac{5}{16}''$,, ,, ,, ...	$11\frac{1}{4}$
,, ,, ,, $\frac{3}{8}''$,, ,, ,, ...	$13\frac{1}{2}$
Pantiles, $13\frac{1}{2}'' \times 9\frac{1}{2}'' \times$ about $\frac{1}{2}''$ thick, and torched at head (laid to $3\frac{1}{2}''$ lap)	10
Pantiles, bedded solid in mortar (laid to $3\frac{1}{2}''$ lap)	14–16
Roof boarding $1''$	3
,, $\frac{7}{8}''$	$2\frac{1}{2}$
,, $\frac{3}{4}''$	$2\frac{1}{8}$
,, $\frac{5}{8}''$	$1\frac{3}{4}$
Tiling battens ($4''$ gauge) $2'' \times 1''$	$1\frac{1}{2}$
,, ,, $2'' \times \frac{3}{4}''$ or $1\frac{1}{2}'' \times 1''$	$1\frac{1}{8}$
,, ,, $1'' \times 1''$	$\frac{3}{4}$
(Other gauges proportionate)	

Material	Average weight per ft. sup. distributed over the slope lbs.
Felt (bituminous) $\frac{1}{16}''$ thick	$\frac{1}{2}$
,, ,, $\frac{1}{8}''$,, ...	1
,, ,, $\frac{1}{4}''$,, ...	2
Asbestos slates, $\frac{1}{4}''$ thick (laid to a $2\frac{1}{2}''$ diagonal lap)	4
Asbestos corrugated sheets, $\frac{1}{4}''$ thick (6″ lap and side lap of one corrugation)	5
Asbestos Trafford Tiles	3
Common rafters of northern pine:	
$3'' \times 2''$ at 14″ centres	$1\frac{1}{4}$
$4'' \times 2''$,, ,, ...	$1\frac{5}{8}$
$5'' \times 2''$,, ,, ...	2
$6'' \times 2''$,, ,, ...	$2\frac{1}{2}$
Northern pine purlins, based on 7 ft. centres:	
$7'' \times 4''$ or $9'' \times 3''$	1
$9'' \times 4''$ or $11'' \times 3''$	$1\frac{1}{4}$
$11'' \times 4''$	$1\frac{1}{2}$
Steel L purlins (approximately)...	$1\frac{1}{2}$–2

Note. Snow is often allowed for on flat pitched roofs at 3 lbs. per sq. ft.

261. Weights of roof trusses. The weights of roof trusses vary considerably because the spacing, form and detailed construction is so varied.

For average trusses in steel and timber spaced at centres not exceeding 10 ft., the following formulae[1] give sufficiently approximate weights for practical purposes. Let W_T = weight of truss in lbs., S = effective span in feet, then for steel trusses $W_T = \cdot 64S^2$, and for timber trusses $W_T = \cdot 9S^2$, the timber trusses averaging 45 per cent. more weight than the steel but assumed to be capable of carrying ceiling loads on the tie beam. The graphs of detail No. 86 represent the above equations and allow the weight to be read off directly for any span up to 50 ft. for timber trusses and 60 ft. for steel trusses.

Composite roofs have a weight which is roughly the mean between timber and steel, their weight may therefore be approximated by using the dotted line graph, or by calculating from $W_T = \cdot 77S^2$.

262. Allotment of weight of truss to load points of purlins. It is sufficiently accurate in solving the forces in the bars of framed structures, to assume that the weight of the frame is uniformly divided between the number of purlins and ridge points + 1 or,

$$w_p = \frac{W_T}{n+1},$$

[1] These formulae have been obtained by averaging and plotting weights of trusses on squared paper and deducing the approximate law of the graph obtained.

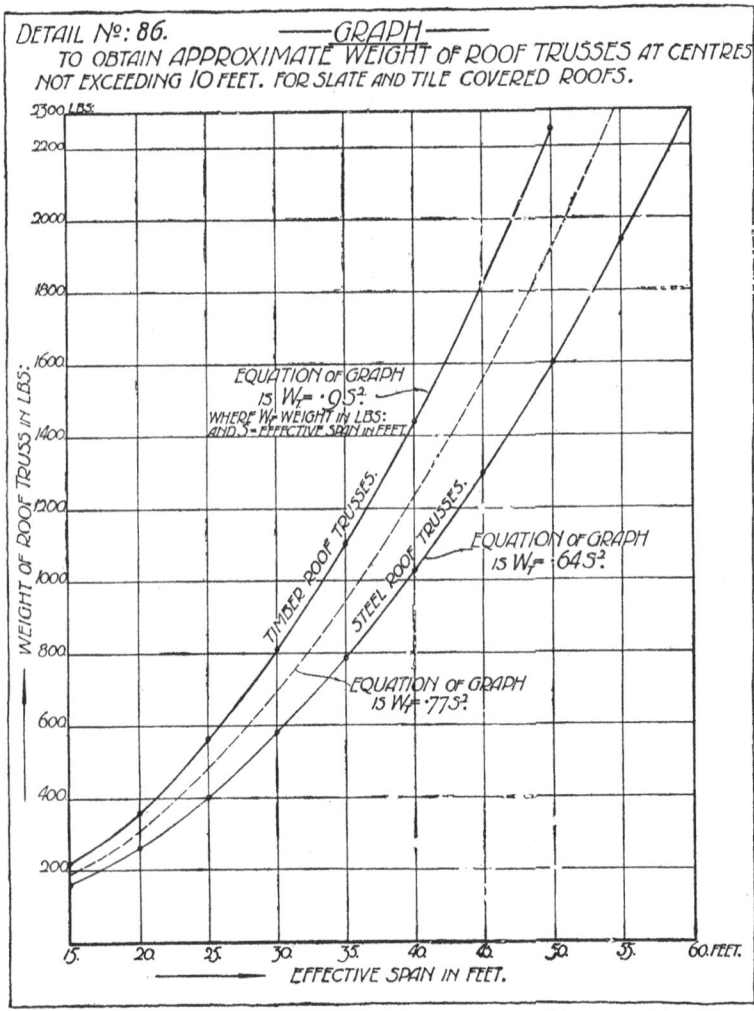

DETAIL Nº: 86.   ——GRAPH——
TO OBTAIN APPROXIMATE WEIGHT OF ROOF TRUSSES AT CENTRES
NOT EXCEEDING 10 FEET. FOR SLATE AND TILE COVERED ROOFS.

EQUATION OF GRAPH
IS $W_T = \cdot 9 S^2$.
WHERE W_T WEIGHT IN LBS:
AND $S =$ EFFECTIVE SPAN IN FEET.

TIMBER ROOF TRUSSES.

STEEL ROOF TRUSSES.

EQUATION OF GRAPH
IS $W_T = \cdot 64 S^2$.

EQUATION OF GRAPH
IS $W_T = \cdot 775 S^2$.

WEIGHT OF ROOF TRUSS IN LBS:

EFFECTIVE SPAN IN FEET.

where $w_p =$ weight per load point; $n =$ number of intermediate load
points. (Part of the load is transferred directly to the wall without
stressing the frame, the proportion being roughly equivalent to one
half bay on each side, hence the addition of 1 point to the number n.)

In the line diagram, detail No. 87 (in which a steel bolt has been
inserted instead of the usual timber post), there are three intermediate load points, therefore the value of

$$w_p = \frac{W_T}{3+1}.$$

Let the truss weigh 520 lbs., as obtained by the mean point between the graphs, for a composite roof truss and a span of 26 ft.; then

$$w_p = \frac{520}{4} = 130 \text{ lbs.}$$

263. Stresses in king-bolt truss. As an example of the method of solution, consider this truss since it is perfectly triangulated. The span is 25′ 8″ and the centre line diagram is reproduced in the detail. The trusses are at 8′ 3″ centres, purlins are 7″ × 4″, common rafters 4″ × 2″, roof boarding 1″ thick, bituminous felt $\frac{1}{8}$″ thick, slating laths 2″ × $\frac{3}{4}$″, and countess slates laid to a 4″ lap.

Apart from the trusses (see last paragraph) the roof load per ft. sup. is made up of the following items obtained from the table in paragraph 260:

		lbs.
Purlins and ridge	... ...	1
Common rafters ...	... ...	$1\frac{5}{8}$
Roof boarding ...	... ...	3
Felt ($\frac{1}{8}$″)	... ...	1
Slating laths (2″ × $\frac{3}{4}$″ at 8″ gauge)		$\frac{5}{8}$
Slates ($\frac{1}{4}$″ thick) ...	... ...	9
And, possibly, snow	... ...	3
	Total	$19\frac{1}{4}$

The area supported by the ridge is $8\frac{1}{4}$ ft. × $6\frac{1}{2}$ ft. = 53·6 sq. ft., and the load at head of truss = 53·6 × $19\frac{1}{4}$ = **1030** lbs.

The area supported by a purlin is $8\frac{1}{4}$ × ($4\frac{1}{4}$ + $3\frac{1}{4}$) = $8\frac{1}{4}$ × $7\frac{1}{2}$ = 61·9 sq. ft., and the load at the purlin point = 61·9 × $19\frac{1}{4}$ = **1190** lbs.

Adding to each of these values 130 lbs. as the proportionate weight of the truss allotted thereto, and obtained in paragraph 262, the total loads are: ridge **1160** lbs. and purlins **1320** lbs.

(*Note.* The loads on the lower half bays are spread uniformly over the *full length of wall* and do not enter into the direct reaction at the bearing of the truss. They are therefore neglected in the load diagram, both eaves being assumed of the ordinary type.)

264. To solve the forces acting at the joints of the frame. To solve the forces acting at the joints proceed as follows, in reference to detail No. 87.

(*a*) Draw the loaded frame diagram to scale, mark the direction and magnitudes of the loads, calculate the reactions and letter the spaces.

(*b*) Seek for a point where there are only two unknown forces; this occurs at the supports in connection with rafter and tie beam. Solve the forces by drawing the vector triangle for the point, as in diagram No. 1.

(*c*) Find the next point where there are only two unknowns; this occurs at the purlin point *AB*, the forces in the upper part of rafter and in the strut being required, because the force in the

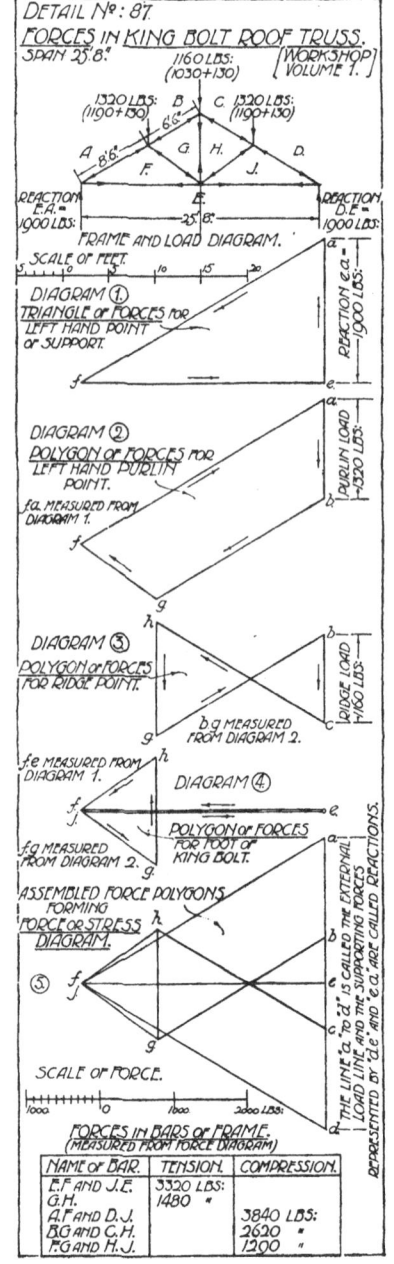

DETAIL N°: 87.

FORCES IN KING BOLT ROOF TRUSS.
SPAN 25'8."   1160 LBS: [WORKSHOP]
            (1030+130)   [VOLUME 1.]

1320 LBS: B   C 1320 LBS:
(1190+130)   (1190+130)
        6'6"
A   8'6"   G.   H.   J.   D.
        F.   E.
REACTION   REACTION
E.A.=   D.E.=
1900 LBS: —25'8."—   1900 LBS:
    FRAME AND LOAD DIAGRAM.
SCALE OF FEET.
5   0   10   15   20

DIAGRAM ①.
TRIANGLE OF FORCES FOR
LEFT HAND POINT
OR SUPPORT.

REACTION e.a.=
1900 LBS:

DIAGRAM ②.
POLYGON OF FORCES FOR
LEFT HAND PURLIN
POINT.
f.a. MEASURED FROM
DIAGRAM 1.

PURLIN LOAD
1320 LBS:

DIAGRAM ③.
POLYGON OF FORCES
FOR RIDGE POINT.
b.g MEASURED
FROM DIAGRAM 2.

RIDGE LOAD
1160 LBS:

f.e MEASURED FROM
DIAGRAM 1.
f.g MEASURED
FROM DIAGRAM 2.
DIAGRAM ④.
POLYGON OF FORCES
FOR FOOT OF
KING BOLT.

ASSEMBLED FORCE POLYGONS
FORMING
FORCE OR STRESS
DIAGRAM.
⑤

SCALE OF FORCE.
1000   0   1000   2000 LBS:

THE LINE "a." TO "d." IS CALLED THE EXTERNAL
LOAD LINE AND THE SUPPORTING FORCES
REPRESENTED BY "d.e" AND "e.a." ARE CALLED REACTIONS.

FORCES IN BARS OF FRAME.
(MEASURED FROM FORCE DIAGRAM)

NAME OF BAR.	TENSION.	COMPRESSION.
E.F AND J.E.	3320 LBS:	
G.H.	1480 "	
A.F AND D.J.		3840 LBS:
B.G AND C.H.		2620 "
F.G AND H.J.		1200 "

DETAIL N°:88.

FORCES IN KING POST ROOF TRUSS.
SPAN 25'8."   [WORKSHOP]
                   [VOLUME 1.]
1030+130
=1160 LBS:
        1320 LBS:
    1090+130   Wa =99 LBS:
              a+b
        N 38°   W=674 LBS:
                      Wb =575 LBS:
                      a+b
                    —65 LBS:
                    TOTAL =
                M. 640 LBS:
LOAD W. ON POLE PLATE IS DISTRI-
BUTED BY RAFTER – ACTING AS A BEAM –
TO M AND N.

            1160 LBS:
    1320 LBS: 1320 LBS:
         C.   D.
640 LBS: B   J. K.   E. 640 LBS:
A.   30° H.   L.   F.
                G.
REACTION   REACTION
G.A.=   F.G =
2540 LBS: —25'8."—   2540 LBS:
    FRAME AND LOAD DIAGRAM.
SCALE OF FEET.
5   0   10   15   20   25

FORCE OR STRESS DIAGRAM.

SCALE OF FORCE.
1000   0   1000   2000 LBS:

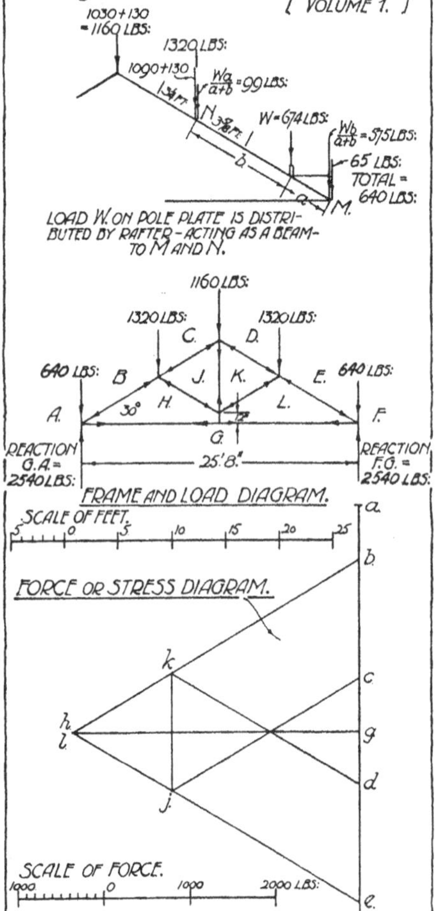

NOTE: THE LOADS OF 640 LBS: ARE TRANS-
MITTED DIRECTLY TO THE SUPPORTS AND
DO NOT AFFECT THE FORCES IN THE BARS
OF THE FRAME. THEY MERELY INCREASE
THE REACTIONS "f.g" AND "g.a."

FORCES IN BARS OF FRAME.
(MEASURED FROM FORCE DIAGRAM.)

NAME OF BAR	TENSION	COMPRESSION
B.H AND L.E.		3840 LBS:
C.J AND K.D.		2520 "
H.J AND K.L.		1320 "
J.K.	1320 LBS:	
G.H AND L.G.	3320 "	

lower part of rafter was solved in the last operation. This solution is shown in the vector polygon of diagram No. 2.

(*d*) Proceed in the same way for all points, in every case *seeking for a point with only two unknown magnitudes to solve.* These will be found to be "ridge point", and "foot of king-rod", in the order given.

Now examine the separate vector polygons from Nos. 1 to 4, and note that there would be similar ones to Nos. 1 and 2, reversed for the points on the opposite half.

265. To draw the complete force diagram for the frame. To *assemble* these separate diagrams into one force diagram amend the procedure thus, as referred to diagram No. 5:

(*a*) Set down the load line or sum of all the external forces to scale from *a* to *d*; measure *de* to represent the right hand reaction, leaving *ea* to represent the left hand reaction.

(*b*) Commencing with any convenient point where there are only two unknown forces and using the forces set out in the load line, construct the successive polygons by adding them on the same diagram.

(This method of procedure will be followed throughout all succeeding examples.)

(*c*) On the solution of any load point, proceed *immediately* to determine the directions of the forces by clockwise reading round the point of the frame and comparison of the direction on the force diagram.

(*d*) To find whether a bar is in tension or compression, remember that the frame has been investigated for the *forces at its points of assemblage*, hence, arrows pointing towards the ends of a member denote thrust or compression while arrows pointing away from the ends denote pull or tension.

(*e*) Finally, measure off the magnitudes of the forces in all the bars and write them on the frame using a + sign for compression and a − sign for tension; or, make a table of the forces for the complete frame as shown at the foot of the detail No. 87.

It should here be noted that the total reactions at the supports will be due to the loads transmitted through the purlin and ridge points of the frame plus the portion of weight of truss transferred directly to the support; hence, the total reaction

$$= \text{say, } 1900 + \frac{130}{2} = \mathbf{1965} \text{ lbs.}$$

266. King-post truss. The same vector method may be used for approximating the forces in a timber frame, though additional forces will occur owing to the nature and disposition of the joints.

Let diagram No. 88 represent a king-post truss with parapet eaves as shown in the workshop roof detail of Vol. I, having the

same span and covering conditions as the last example. In this case
a pole plate is used at the foot of the roof slope and this bears on
the trusses only, thus transferring the load upon the back of the
rafter near the foot.

The pole plate rests upon the rafter at a point 1′ 3″ up the slope
of the centre line, and thus delivers its load on a sloping beam which
transfers it to the purlin point N and the support M in proportion
to the distances from the load to the opposite ends as explained in
paragraph 155.

The loads on the rafter will be made up as follows:

Ridge load (as before) 1030 + 130 (for truss) = **1160** lbs.

Direct purlin load is gathered from an area

$8\frac{1}{4}$ (centres of trusses) × $(3\frac{1}{4} + 3\frac{5}{8}) = 8\frac{1}{4} \times 6\frac{7}{8} = 56.75$ sq. ft.;

at $19\frac{1}{4}$ lbs. per sq. ft. the load = **1090** lbs.

Load on lower half of slope is gathered from an area

$= 8\frac{1}{4} \times 3\frac{5}{8} = 29.9$ sq. ft.; $29.9 \times 19\frac{1}{4} = 575$ lbs.,

which is transmitted direct to the pole plate.

Now the gutter with its lead covering and bearers transfers about
12 lbs. per ft. run to the pole plate or a total of $12 \times 8\frac{1}{4} = 99$ lbs.

The direct load on the pole plate is therefore 575 + 99 = **674** lbs.
This load is transmitted to M in the ratio of

$$\frac{Wb}{a+b} = \frac{674 \times 7.25}{8.5} = \textbf{575} \text{ lbs.}$$

and to N in the ratio of

$$\frac{Wa}{a+b} = \frac{674 \times 1.25}{8.5} = \textbf{99} \text{ lbs.}$$

The total purlin load at N is therefore

1090 + 99 + 130 (for truss) = 1319, say **1320** lbs.,

and the load on the support

$$= 575 + \frac{130}{2} \text{ (for truss)} = \textbf{640} \text{ lbs.}$$

Placing these vertical loads on the points of the frame and
solving in detail as before, the diagrams of detail No. 88 are obtained.

In addition to the thrust or compressional stress in the rafter LE,
there is a "bending stress" due to the pole plate being supported
upon it. The effect of such combined stresses is a problem for
advanced study.

MOMENT PROBLEMS WORKED GRAPHICALLY. RESULTANT OF A SERIES
OF FORCES IN THE SAME PLANE BUT NOT PARALLEL

267. Introduction. Because forces are vectors the direction of
the resultant of a series of non-parallel forces may be obtained by
adding the scalar representations of the forces graphically, as shown
in detail No. 89.

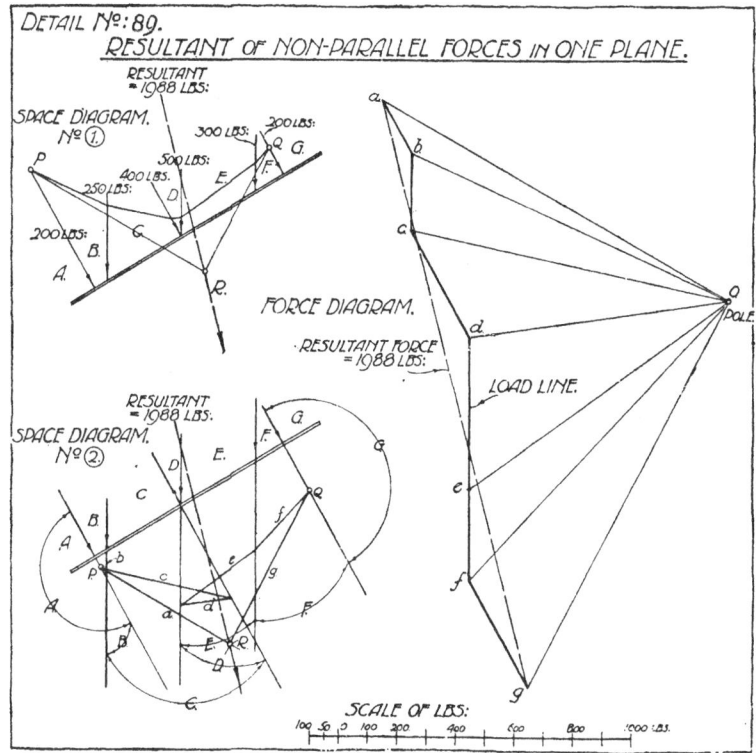

DETAIL Nº: 89.
RESULTANT OF NON-PARALLEL FORCES IN ONE PLANE.

The forces AB, BC, ... FG act upon an inclined beam, as shown in
the space diagram; they are added in the force diagram, to a suitable
scale, from a to g, and the resultant force in magnitude and direction equals ag.

Now the resultant is defined as that force which equals the sum
of the series and has the same moment about any point in space.
To fulfil the latter condition the position must be obtained on the
space diagram, which is accomplished by an application of the same
principle as applied in detail No. 58.

Take any pole O outside the load line and join each point in the latter to this pole.

Select any point P in the force line AB of diagram No. 1, and draw connected links across the successive spaces, BC, etc., to F, terminating at Q on the force line FG, through P draw inwards parallel to aO and through Q also inwards, parallel to gO; the intersection of these links in R gives a point through which the resultant passes.

Through R draw the strong dotted line parallel to ag. This line represents the line of action of the resultant force which could replace the series of forces and have the same moment about any point. The proof is too lengthy to include here, but a simple verification can be made by calculating the sum of the moments of the series about P (adding together the products of force by perpendicular distance from P) and comparing with the moment of the resultant about P.

A fair approximation is obtainable by the above graphical process.

As the starting-point may be selected anywhere on the force lines the space diagram is repeated at No. 2 in order to show the process of drawing the links when the point P is selected so low on the force line AB, that the links pass below the intersections of the force lines, which cross each other. The spaces below and at the ends are shown by connecting arcs and lettered to correspond with the spaces above. It thus follows that the links c and e cross the other lines DE and CD, while d is drawn backwards across the overlapping space. The initial and terminal points are P and Q as before, and the links a and g intersect at R which, on testing, will be found to lie in the same line as before.

268. Reactions of a beam supporting non-parallel forces. Let the beam of detail No. 90 support a series of non-parallel forces, which are contained in a vertical plane. The reactions cannot be vertical in this case, but if the ends are freely supported and prevented from slipping, the supporting forces will be parallel to the resultant of the series. Hence, find the resultant hl as previously explained and further illustrated in this detail. Now the reactions will be parallel to and have a sum equal to hl if other conditions allow, because in this direction the total force supplied by the two reactions will be the least possible which can balance the series, being equal and opposite to the resultant.

Place the directions of the reactions therefore through the supporting points as shown, and to decide how much force acts in each line use the series of links shown from P to Q and complete the polygon by producing the outer links h and l inwards until they intersect at R. We know that the resultant passes through R and is equal to hl.

Let this resultant replace the series, then by first principles, the reaction LM (to the right) must produce the same moment about the point P as that of the resultant. Produce QR to S. Then PS represents the moment of R about P. Join PQ and draw parallel to PQ from O in the force diagram to meet hl in m. Then lm is the reaction at LM, for its moment diagram is PQS and PS is its moment about P and is equal to the moment of the resultant R. Hence, the remainder of the resultant mh is the reaction at MH, because the reactions or equilibrating forces lm and mh equal the resultant force hl and are opposite thereto.

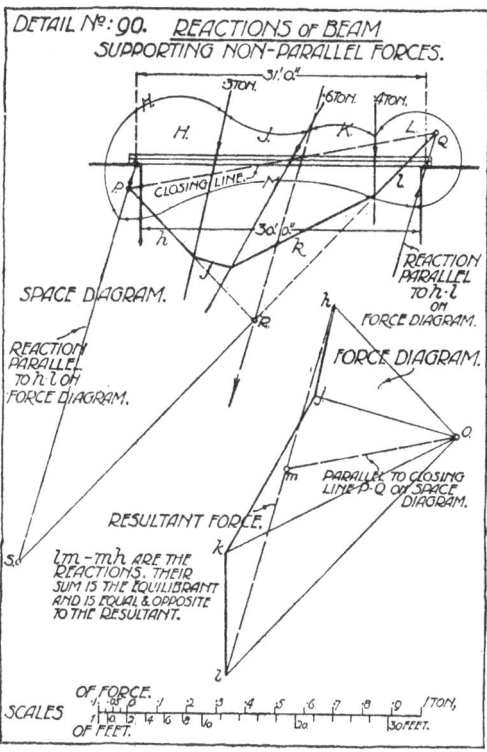

It should now be noticed that if the reactions were assumed in any other directions than those parallel to hl they would, in total, be of greater amount, because any two lines passing through h and l and not lying in the line hl must have a greater total length than $lm+mh$.

269. Wind pressure on roofs. When wind acts upon a roof its effective pressure is felt normally or at right angles to the roof surface, because air is a fluid and wind is merely air in motion and the principles of hydrostatics show that the pressure of fluids acts at right angles to the containing vessel or supporting surface. The theory of wind pressure cannot be investigated here, nor can the various formulae be discussed which are in use for computing its value under varying conditions of velocity and frictional effect upon sloping surfaces.

It must suffice at present to apply the normal L.C.C. regulations for steel framed structures, in which it is laid down that if a roof or other structure has a slope exceeding 20° from the horizontal, the structure must be designed to support a normal pressure of 28 lbs. per ft. sup. over the entire sloping surface. For this purpose the pressure on any joint of support of a roof truss may therefore be obtained by determining the area of the panel in sq. ft. supported by the framing and multiplying by 28, giving normal point pressure in lbs. *But see next paragraph.*

(The area employed for the wind pressure in any given case is that estimated for the dead load, see paragraph 258.)

270. Incidence of wind load on roofs. The wind load on a roof is usually conceived to act on one side only, producing a direct pressure at the panel points of the truss on the windward side. It has been known for some time, however, that the effect of the wind passing across the ridge is to relieve the atmospheric pressure on the leeward side, producing a suctional effect and therefore reducing the primary effect of the dead load on the framing; this effect is largely varied by some external conditions of the building. The L.C.C. Code of Practice and the British Standard Specification No. 449 now require that roof frames shall be designed to resist a pressure of 15 lbs. per sq. ft. on the windward side and an outward lift of 10 lbs. per sq. ft. on the leeward side.

As an introduction to the use of inclined loads on roof frames several cases are treated according to the original regulations of 1909 and one example of the method required by the Code of Practice, 1933, is included in detail No. 95 and the accompanying description.

Students should become familiar with the procedure of solving forces due to pressure on one side of a roof, before attempting the case with downward and upward pressures on the two slopes respectively.

As in the case of dead loads, the equivalent wind pressure on the lower half of the eaves panel is transmitted directly to the wall plate and does not affect the stresses in members of the frame, it has therefore not been included; it must, however, be taken into account when determining the effect of wind pressure to overturn the walls.

Detail No. 91 shows the incidence of wind and dead loading for the king-rod roof truss applied to the loading shed of the warehouse.

STRESSES IN "LOADING SHED ROOF TRUSSES"

271. Loading of truss. Alternative forms of trusses are shown in this and the following detail and both are placed at 8' 6" centres; the span is 29' 7½" clear, or 30' 4½" effective between centres of bearings. The length of rafter may be measured off the drawing, or may be calculated from $l_r = \dfrac{\text{span}}{2 \cos \theta}$, where $\theta =$ the angle of slope to the horizontal, in this case 30°.

Assuming equal divisions of the rafter by the strut, the approximate length of panel supported at the centre of the slope is

$$\frac{1}{2} \cdot \frac{\text{span}}{2 \cos \theta} = \frac{30 \cdot 375}{4 \times \cdot 866} = 8 \cdot 77 \text{ ft.,}$$

which would be somewhat excessive in a heavily covered roof.

Using the table given in paragraph 260 the dead load would be made up as follows:

3″ × 3″ × ⅜″ steel purlins and ridge with timber fillings (say)	1½	lbs. per ft. sup.
5″ × 2½″ northern pine rafters	2½	,, ,,
2″ × 1″ tiling battens at 10½″ gauge	¾	,, ,,
Pantiles at 10½″ gauge	10	,, ,,
Snow	3	,, ,,
	17¾	,, ,,
Total (say)	18	,, ,,

The weight of the steel truss may be roughly estimated by using the graph of detail No. 86. For 30 ft. span the weight is **580** lbs., hence, as per paragraph 262, the weight per intermediate load point is

$$\frac{580}{3+1} = 145 \text{ lbs.}$$

Each roof panel measures 8·77 ft. × 8·5 ft., therefore the uniform dead load per panel will be $8 \cdot 77 \times 8 \cdot 5 \times 18 = 1342$ lbs.

The *total* dead load per purlin and ridge point is

$$145 + 1342 = 1487 \text{ lbs.,}$$

or, say ·67 ton. The wind pressure per load point to be allowed for is at the rate of 28 lbs. per sq. ft. over the same panel area as included for the dead load; hence, the amount per load point will be $8 \cdot 77 \times 8 \cdot 5 \times 28 = 2087$ lbs., or, say ·93 tons. As the wind is active on one side only the pressure communicated to the ridge point will be equivalent to that upon a half panel, viz. ·465 ton.

272. Solution of stresses in king-rod truss. In detail No. 91 at A the force diagram for the truss is shown, when supporting the dead load of covering and snow only. This should require no explanation if the previous work is understood; it is only necessary to omit the wind loads on the right hand side of the "space and loading" diagram and to assemble the force polygons for the several points of connection.

At B is shown the force diagram for the dead load and wind forces acting together. The wind is assumed to blow from the right

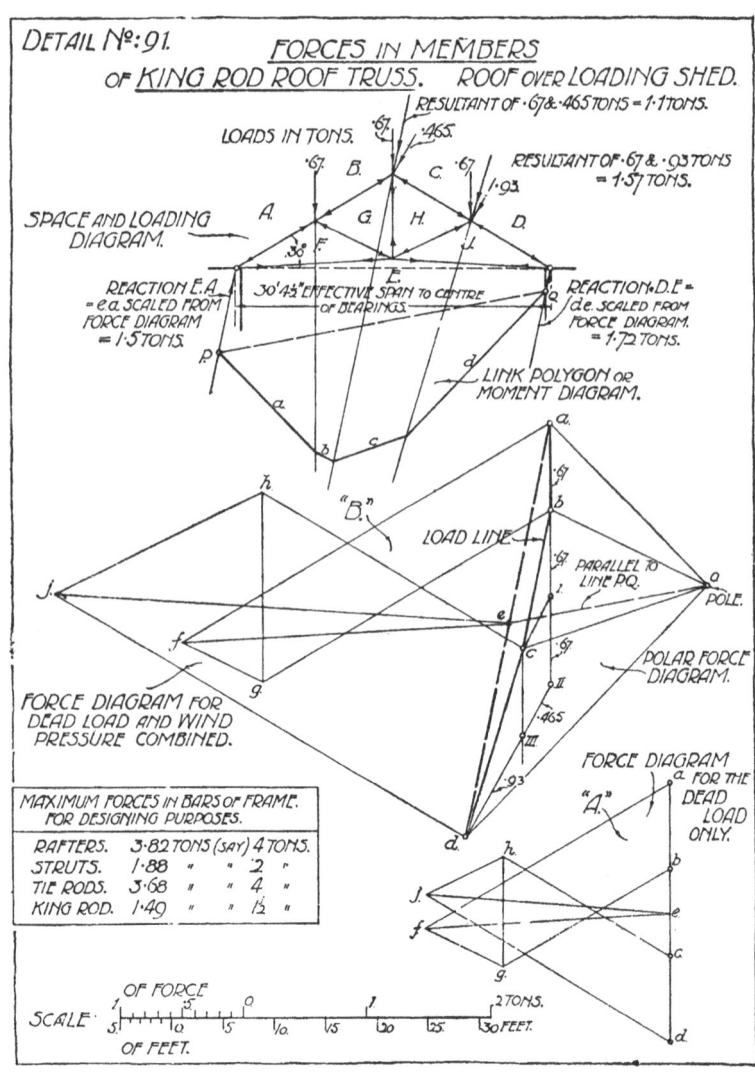

DETAIL Nº:91.
FORCES IN MEMBERS
OF KING ROD ROOF TRUSS. ROOF OVER LOADING SHED.

and as two forces then act at the ridge and purlin points they have been compounded into one resultant force at each point; this simplifies the working. Such forces may be resolved by a parallelogram drawn on the space diagram, or better, by setting out the load line of the force diagram as shown. The whole of the vertical loads are first set down in clockwise order and followed by the normal wind pressures in the same order. By projecting the pairs of corresponding forces, as b I and II, III in directions respectively vertical and normal to the slope, the resultant force bc is obtained, and similarly for I, II and IIId, the resultant force cd. These directions may then be transferred from the load line to the point of the frame through which they act, as shown by the thick arrows in the space diagram.

Reactions of truss. The reactions of forces on a frame are exactly similar to those for a beam, provided that the end conditions are similar in each case. Most roof trusses are borne upon flat bearings, bedded upon stone—or other—templates and bolted to prevent side or lateral movement; both ends are therefore capable of offering resistance to the horizontal tendency to shift under wind pressure.

Determine, therefore, the inclination and magnitudes of the reactions by the application of the principles laid down in paragraph 268. Select a pole O, and connect a, b, c and d thereto. Take any point P on one of the reactions, say EA, and construct a link polygon, the links being parallel to the polar force lines having the same letters as the spaces crossed thereby, and terminating at Q. Join PQ and draw parallel from the pole O to intersect the resultant load line ad at e. Then de, to scale, represents the reaction DE, and ea represents EA.

Using this complete series of external forces, viz. loads and reactions, solve the several force polygons for the points of assemblage of the frame, commencing at, say, the left hand support where only two unknown magnitudes of force exist in the directions of the rafter and tie.

Having solved this point and the forces in AF and FE, proceed from point to point so that only two unknown magnitudes are sought at each step of the solution. The order will be found to be: (1) left hand support; (2) left hand purlin point; (3) head of truss; (4) foot of king-rod. It is also possible to work from the right hand support in a similar way, or to work from both supports to the centre of the frame.

The directions of the forces should be marked on the frame, as shown, to determine whether the bar is in tension or compression.

273. Forces in bars for purposes of design. To design a truss it is necessary to know the maximum force to which any bar is subjected. An examination of the "dead-load diagram" at A, and the "combined-load diagram" at B, will show that the effect of wind pressure is to increase the forces in most segments of the frame and in no case to diminish them. (The latter point does not hold in all frames; cases arise where certain bars have less stress under wind pressure than without it.)

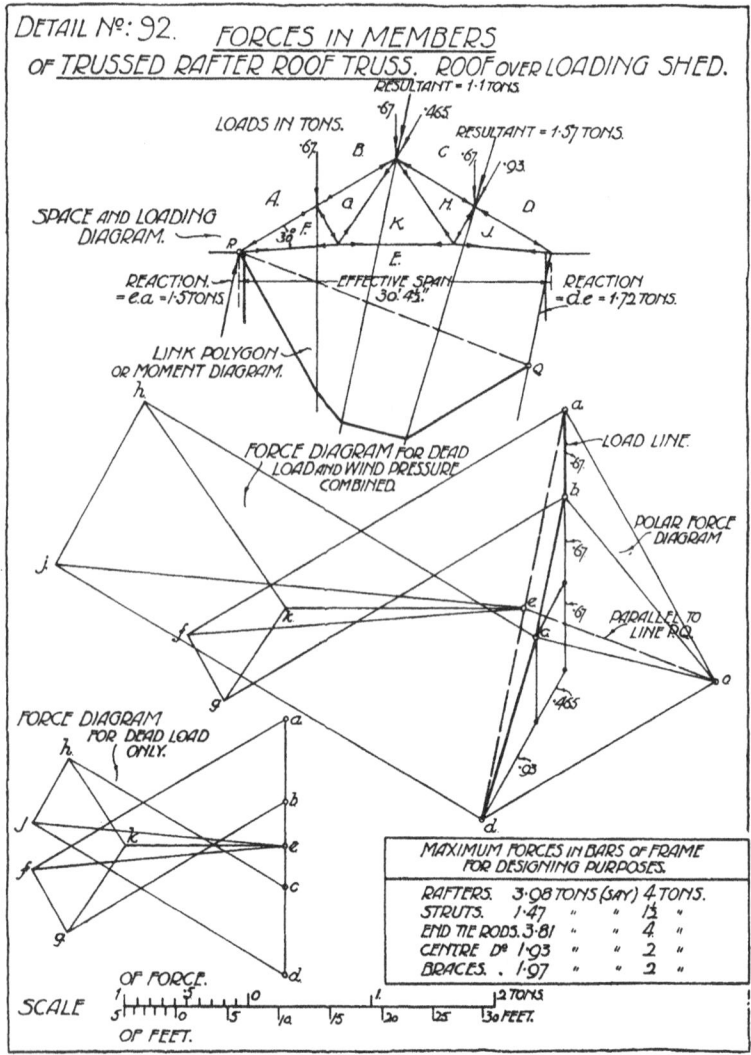

If the wind changed to the left, and acted with the same force, it is obvious that the conditions would be reversed, those bars which at present receive the greater forces being relieved and the corresponding bars on the other side proportionately increased.

Further, as the rafters are continuous, their strength must provide for the greater force. It is necessary, therefore, to tabulate only the *maximum* forces to which each kind of bar is subjected, as shown in the detail.

274. Solution of stresses in trussed rafter roof. Detail No. 92 shows the solution of stresses in the alternative form of trussed rafter roof adopted for the loading shed. The procedure is the same as described for the last example, and should be worked to a larger scale by the student. Slight variations have been introduced, *e.g.* the pole is taken in a different and lower position and the link polygon commenced at the left hand point of support P. There is no difference in the resulting reactions when compared with detail No. 91. The maximum forces are tabulated on the detail.

On comparing these forces certain differences in value are found, as compared with those of the king-rod roof for the same span. There is a slightly increased value of force in the main rafter, about 4·2 per cent.; a decrease in the force on the strut of about 21·8 per cent.; an increase of 3·5 per cent. in the end ties, but a decrease of 47·6 per cent. in the centre length; the braces *each* have an increase of 32·2 per cent. over the force in the king-rod.

In small spans the king-rod form of truss is the more economically stressed.

The most important thing in larger spans is to keep the stresses in struts and rafters as low as possible and to reduce the lengths of these members, thereby decreasing their tendency to buckle. At the same time round bar tension members should not be unduly long between points of support or attachment, or sagging will occur to an extent causing bending stresses of an appreciable amount.

Flat bar ties on edge are efficient to resist bending for long lengths or cases where small intermediate loads must be attached to the tie. If ceilings or other considerable loads are to be carried on the ties L or ⊥ sections become necessary.

275. Queen-post trusses. Queen-post trusses cannot have their stresses solved directly without special artifices, owing to their tendency to collapse across the central rectangle when subjected to wind pressure. The tendency to collapse is resisted by metal fastenings at the rectangular joints or by continuous rafters and tie beam. It is only possible to deal with symmetrical dead loading in this volume.

Detail No. 93 shows the force diagram for uniform dead load only, which is easily solved on first principles. This is only a first step, however, in the consideration of the strength of the structure.

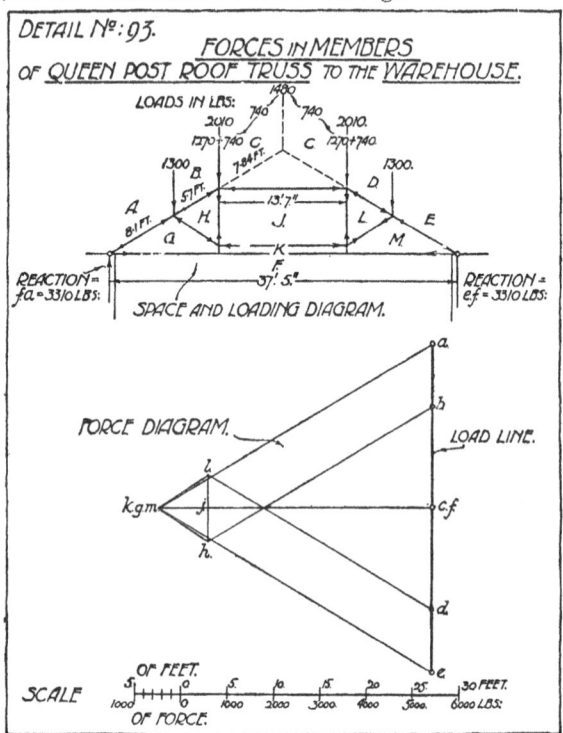

276. **Loads on queen-post roof truss for warehouse.** The effective span is 37′ 6″ approximately, and the dead loads are assessed as follows:

9″ × 4″ purlins of northern pine ...	$1\frac{1}{4}$ lbs. per ft. sup.	
4″ × 2″ rafters of northern pine ...	$1\frac{5}{8}$,,	,,
1″ boarding of northern pine ...	3 ,,	,,
Countess slates to 4″ lap ($\frac{5}{16}$″ thick)	$11\frac{1}{4}$,,	,,
Snow	3 ,,	,,
Total	$20\frac{1}{8}$,,	,,

say a total of 20 lbs. per ft. sup.

Taking one of the 9′ $4\frac{1}{2}$″ bays (centre to centre of trusses) and measuring the lengths of the panels centre to centre up the slope the areas for obtaining the point loads read from the right are as follows:

No. 1. 6·9 × 9·37 = 64·6 sq. ft., and load = 20 × 64·6 = 1292, say **1300 lbs.**

No. 2. 6·77 × 9·37 = 63·4 sq. ft., and load = 20 × 63·4 = 1268, say **1270 lbs.**

No. 3. 7·84 × 9·37 = 73·4 sq. ft., and load = 20 × 73·4 = 1468, say **1480 lbs.**

to allow of easy division (see later).

On the left hand side the pole plate causes some variation in the distribution of the load, but as the change would cause unsymmetrical loading and consequently render this truss incapable of direct solution it has not been made and an ordinary case has been assumed where both have overhanging eaves and the loads from the lower ends of the rafters are transmitted to a wall plate and thus spread over the whole length of bay.

It should now be observed that while the roof as a whole is a span roof with the common rafters passing to the central ridge and supporting the load on the upper slope, there is no part of the truss supporting the ridge directly. Hence, the load of 1480 lbs. which reaches the ridge is re-transmitted as thrust down the upper segment of the rafters and is ultimately transferred to the first purlin points in equal quantities as shown by the arrows. These purlin points, therefore, have loads of $1270 \text{ lbs.} + \dfrac{1480}{2} = \mathbf{2010}$ lbs.

The solution can now proceed as in previous cases, each reaction being equal to half the total load. Commence the solution of the force polygon at the support and seek for two unknowns at each step.

The straining cill has its centre line above the centre of the tie and its function is evidently to receive the thrust from the struts GH and LM. The force in the tie beam is thus constant throughout, which is not the case if the tenons at the feet of the queen-posts have to resist the thrust of the struts; this occurs if the straining cill is omitted.

277. Queen-rod roof truss. Merely as an example of the extended principle of the king-rod roof truss, detail No. 94 is included as an alternative to the queen-post truss for the warehouse roof. It is similar to the king-rod roof truss, but with additional vertical ties and struts on each side. Such trusses have often been used both in steel and in timber.

In most cases the rafter would be divided into equal portions for spacing the purlins but the spacing used for the queen-post truss has been adhered to in order to utilise the same point loads.

To assess the reactions correctly the load to be transmitted from the pole plate to the truss has been taken into consideration.

This load is gathered from half the bay on each side of point X, and for an approximation it will be considered that the uniform load of 20 lbs. per ft. sup. allows for the gutter, etc., if measured on the slope. Then, $b = 6.5$ ft., $\dfrac{b}{2} = 3.25$; $a = 1.6$ ft. and $\dfrac{a}{2} = .8$. Hence, the load area for point X becomes $(3.25 + .8) \ 9.37 = \text{say } 38$ ft. sup. At 20 lbs. the load is $20 \times 38 = 760$ lbs.

By the principle of moments—see paragraph 155—the load transferred to point Y from the lower bay

$$= \frac{760 \times a}{a+b} = \frac{760 \times 1.6}{8.1} = \mathbf{150} \text{ lbs.,}$$

and to point Z from the same bay

$$= \frac{760 \times b}{a+b} = \frac{760 \times 6 \cdot 5}{8 \cdot 1} = \mathbf{610} \text{ lbs.}$$

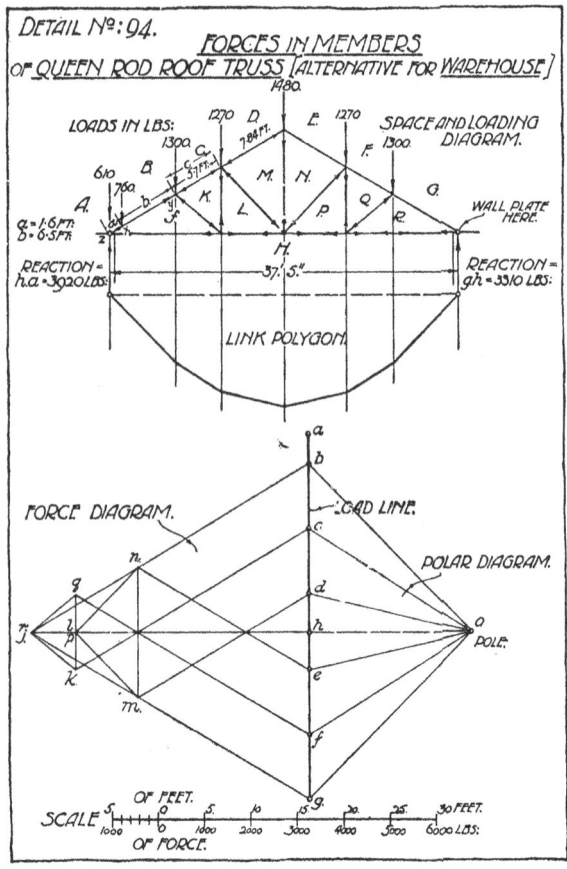

The total load at Y is 150 lbs. + the load gathered from half the bay on each side of it, viz. $\dfrac{b+c}{2} = \dfrac{6 \cdot 5 + 5 \cdot 7}{2} = 6 \cdot 1$ ft.; then $6 \cdot 1 \times 9 \cdot 37 \times 20 = 1143$ lbs.; hence, load at $Y = 150 + 1143 = 1293$, say **1300** lbs., which agrees exactly with the last case.—*This will always occur when the load is considered uniform over the slope.* But note that 610 lbs. at the foot is transmitted through the bearing of the truss instead of being spread along the wall, because the pole plate rests directly upon the rafter.

Further, the pole plate causes extra stress in the rafter due to bending, which has not been accounted for in the present treatment. To obtain the reactions, either the polar diagram and link polygon may be employed as before, see detail, *or* as follows: Apart from the 610 lbs. on the left hand support, the load is symmetrical; ignoring the 610 lbs., the reactions would therefore be equal and half of the total load, viz. 3310 lbs. Add 610 lbs. for the left hand reaction, thus making 3920 lbs.

The point Z is obtained by setting out 3310 lbs. from g on the load line. The solution is then direct, as in previous cases.

Student's exercise. As an exercise the student should solve this last example when wind pressure to the value of 28 lbs. per sq. ft. acts on the left hand slope.

278. Outward forces acting on roof structures. The L.C.C. Code of Practice requires that, in the design of a roof structure having a slope greater than 20°, superimposed loads of at least 15 lbs. per sq. ft. on the windward side and at least 10 lbs. per sq. ft. on the leeward side shall be assumed to act normal to the roof surface.

The latter value is intended to allow for the suctional—or lifting—effect of the wind on the leeward side of the sloping roof, and the former provides for direct wind pressure.

Detail No. 95 shows the procedure adopted for solving the forces in the members of the trussed rafter roof, span 30 ft. $4\frac{1}{2}$ in., already considered with wind load on one side only in para. 274.

Ignoring loads at the "foot joints" as before—since these do not affect the forces in members of the frame—and calculating on the same area of roof panels, viz. $8\cdot77 \times 8\cdot5 = 74\cdot5$ sq. ft., the direct wind load per panel point is $\dfrac{74\cdot5 \times 15}{2240} = \cdot5$ ton (approx.), and the outward wind force per panel point is $\dfrac{74\cdot5 \times 15}{2240} = \cdot33$ ton (approx.). For the half panels at the head of the truss these point loads become $\cdot25$ and $\cdot165$ ton respectively.

These loads, plus the dead loads for truss and covering previously used, are shown in position on the outline of the truss. A load line is set out as before. Taking all the loads in order and noting their magnitudes and directions, they are set out continuously from a to d in the force diagram. The resultant of each group of forces at each load point is obtained from the load line and treated as single loads. These are shown on the truss and are used to obtain the reactions by drawing a polar force diagram and link polygon as in previous cases. Observe that the latter figure must start and terminate on the lines of reaction (taken parallel) and that the reactions are themselves parallel to the resultant of all the loads, marked *a–d* in the force diagram. The closing line (*PQ*) of the link polygon determines at *e* the division of the resultant into the magnitudes of the two reactions, by drawing *OC* parallel to *PQ*.

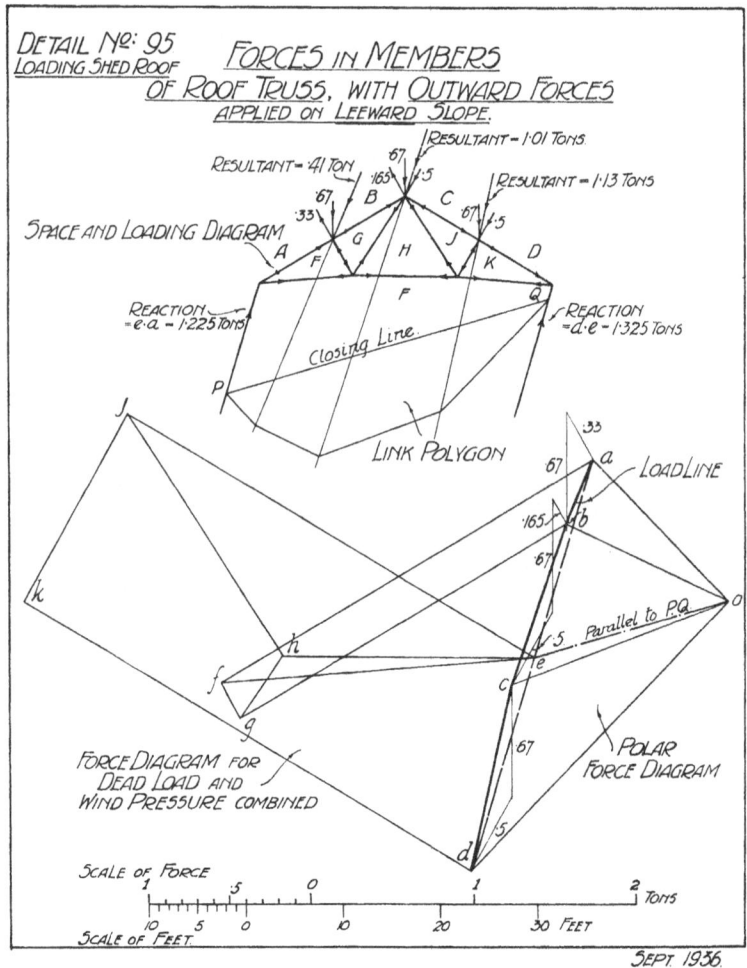

The force diagram can then be drawn for the whole truss by following the same procedure as in previous examples.

The reader should tabulate the results and compare the forces in the bars with those previously obtained. In making this comparison, however, it must be remembered that whereas in the earlier example of para. 274 a wind load of 28 lbs. per sq. ft. was used, it is reduced to 15 lbs. per sq. ft. on the windward side and that a "lift" of 10 lbs. per sq. ft. has also been added.

CHAPTER ELEVEN

STRUCTURAL DESIGN

PLATE GIRDERS

279. Plate girders are built-up girders of **I** or **II** section, made by riveting together plates and angles of steel or wrought iron. At the present time mild structural steel plate is almost exclusively employed of the kind already referred to, and used for rolled steel sections.

It is uneconomical to employ plate girders for positions where simple compound sections of rolled steel can be made to serve the purpose, nor where such sections strengthened by riveted flange plates can be employed.

In this chapter the object is to show how a plate girder of **I** section is designed and constructed and for this purpose it is assumed that the first floor of the warehouse is to be free from pillars. This condition requires that some form of girder shall be adopted which will carry the floor load safely across the full span of the warehouse.

280. Type of floor supported by plate girders. Let the same type of floor be adopted, viz. with timber joists and flooring, secondary beams of rolled **I** section running parallel to the side walls, and main plate girders at 15 ft. centres across the building, their bearings being extended by the use of brick piers as in the previous arrangement where pillars were employed; see key plan, detail 96 at A.

The clear span of the main girders will be 35′ 3″ between piers, and if the bearings are 1′ 6″ long the span between centres of bearings becomes 36′ 9″ or, say, 37 ft. effective span.

281. Secondary beams and joists. The secondary beams must be spaced so that timber joists will be available of an economical and market size.

From paragraph 205 the inclusive designing load for the floor is 234 lbs. Let the joists selected be 9″ × 2″, of northern pine; find a suitable span, which may be adopted as the maximum spacing of the beams.

Then $$B = R.$$

$$\therefore \ \frac{WL}{8} = \frac{bd^2f}{6}.$$

For this case L must be in feet, and as B is in lbs. inches $12L'$ may be inserted to convert to the required unit.

$$\therefore \frac{12WL}{8} = \frac{bd^2f}{6}.$$

With joists 12″ apart, viz. 14″ centres,

$$W = \frac{14}{12} \times L \times 234$$

$$= 273L \text{ (lbs.)}$$

Let $f = 1000$ lbs. sq. ins.

$$\therefore \frac{12 \times 273L^2}{8} = \frac{2 \times 81 \times 1000}{6}.$$

$$\therefore L^2 = 66,$$

and $L = 8.12$ ft.

A suitable division of the span of 37 ft. would therefore be five bays, the three centre spans being 7′ 6″ each and the end spans 7′ 3″ each.

To select a steel secondary beam, determine the total load to be supported and proceed as described in paragraph 193.

Total load = load per ft. sup. × area of one bay.

Load per ft. sup. = 234 lbs. or ·1045 ton.
Length of bay = 15 ft.
Span of bay = 7′ 6″.

$$\therefore W = ·1045 \times 15 \times 7\tfrac{1}{2} = 11·75 \text{ tons.}$$

Then $B_M = \dfrac{WL}{8} = \dfrac{11·75 \times (15 \times 12)}{8} =$ approximately, 264 tons ins.

But $B_M = Zf.$

Let $f = 7·5$ tons per sq. in.

Then $Z = \dfrac{B_M}{f} = \dfrac{264}{8} = 33$ inch units.

From the table in Appendix I the most suitable beam is B.S.B. 12″ × 5″ × 30 lbs. having $Z = 34·48$ inch units or, for stiffness of floor, a 12″ × 5″ × 32 lbs. B.S.B., $Z = 36·84$ inch units.

282. Bending moment on plate girder. The load carried by the plate girder is transferred to it at four points, as shown by the load diagram of detail No. 96 at B, each pair of secondary beams transmitting a load equal to that on one complete bay, plus half of their own weight.

One secondary beam weighs $15 \times 30 = 450$ lbs., say 500 lbs. or ·223 ton, with connections.

Hence the load at each load point

$$= 11·75 + ·223 = 11·973, \text{ say } \mathbf{12} \text{ tons,}$$

or a total load W on the girder of **48** tons.

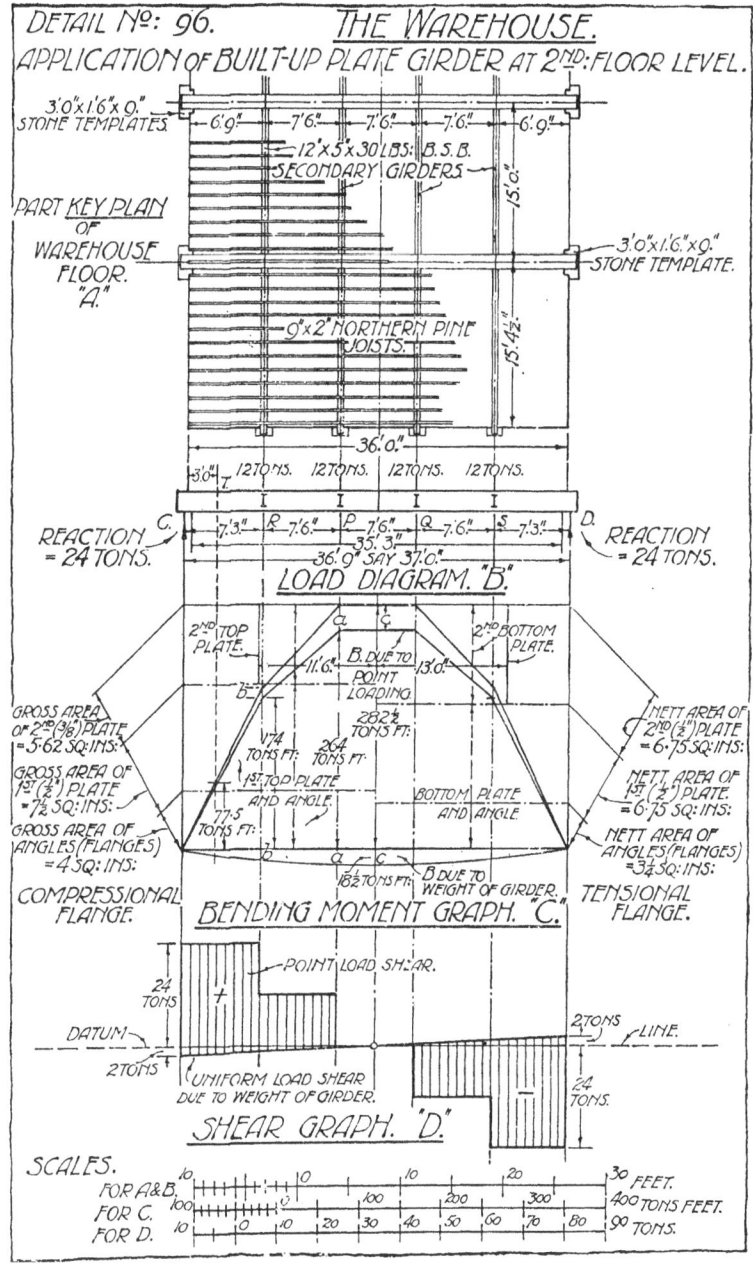

DETAIL Nº: 96.   *THE WAREHOUSE.*
APPLICATION of BUILT-UP PLATE GIRDER AT 2ND. FLOOR LEVEL.

3'.0"x1'.6"x 9."
STONE TEMPLATES.

6'.9"   7'.6"   7'.6"   7'.6"   6'.9"

12'x5'x30 LBS. B.S.B.
SECONDARY GIRDERS

15'.0."

PART KEY PLAN
OF
WAREHOUSE
FLOOR.
"A."

3'.0"x1'.6"x 9."
STONE TEMPLATE.

9'x2' NORTHERN PINE
JOISTS.

15'.4½."

36'.0."

3'.0"   12TONS.   12TONS.   12TONS.   12TONS.

REACTION C.   7'.3" R 7'.6" P 7'.6" Q 7'.6" S 7'.3"   D. REACTION
= 24 TONS.   35'.3"   = 24 TONS.
                 36'.9" SAY 37'.0."
              LOAD DIAGRAM. "B."

2ND TOP
PLATE.   2ND BOTTOM
                                          PLATE.

GROSS AREA   a   c
OF 2ND (⅜) PLATE   B. DUE TO   13'.0."
= 5·62 SQ: INS: 11'.6"   POINT
                  B.   LOADING
                  282½
GROSS AREA OF   V74   264
1ST (½) PLATE   TONS FT. TONS FT.
= 7½ SQ: INS: 1ST TOP PLATE
                  AND ANGLE   BOTTOM PLATE
                  77·5   AND ANGLE
GROSS AREA OF   TONS FT.
ANGLES (FLANGES)
= 4 SQ: INS: b   a   c
                         182TONS FT. B DUE TO
COMPRESSIONAL   WEIGHT OF GIRDER.
FLANGE.   *BENDING MOMENT GRAPH. "C."*

NETT AREA OF
2ND (½) PLATE
= 6·75 SQ: INS:

NETT AREA OF
1ST (½) PLATE
= 6·75 SQ: INS:

NETT AREA OF
ANGLES (FLANGES)
= 3½ SQ: INS:

TENSIONAL
FLANGE.

POINT LOAD SHEAR.

24
TONS

DATUM   2TONS LINE.

2TONS
          UNIFORM LOAD SHEAR   24
          DUE TO WEIGHT OF GIRDER. TONS.
          SHEAR GRAPH. "D."

SCALES.
FOR A&B.   10   0   10   20   30 FEET.
FOR C.   100   0   100   200   300   400 TONS FEET.
FOR D.   10   0 10 20 30 40 50 60 70 80 90 TONS.

From previous considerations it is known that the bending moment is constant across the centre bay of the main girder, hence we may obtain the maximum by calculating the nett moments of all the forces on one side of the boundary to the middle bay. Thus, taking moments about P on the left, we have

$$B_P = Re_1 \times 14\tfrac{3}{4} \text{ ft.} - \frac{W}{4} \times 7\tfrac{1}{2} \text{ (ft.)}$$

$$= 24 \times 14\cdot75 - 12 \times 7\cdot5 = 12\,(29\cdot5 - 7\cdot5) = 12 \times 22,$$

$$\therefore\ B_P = \mathbf{264}\text{ tons ft.}$$

This value is also B_M and occurs from P to Q.

In order to plot a diagram or graph of the varying bending moment, the values must be calculated for points R and S. The beam being symmetrically loaded these values are equal,

$$\therefore\ B_R \text{ or } B_S = \text{reaction} \times 7\tfrac{1}{4} \text{ (ft.)}$$
$$= 24 \times 7\cdot25 = \mathbf{174}\text{ tons ft.}$$

283. Weight of girder. In all such calculations the bending moment due to the weight of the girder must be allowed for, therefor some method must be adopted to determine approximately the dead weight.

Let W_g = weight of girder,
W_u = weight on girder which, if uniform, would produce the same B_M,
L = effective span.

Then $W_g = \dfrac{W_u L}{540}$ (approximately).

But the value of $B_M = 264$ tons ft.,

and a uniform load produces a B_M of $\dfrac{W_u L}{8}$ (tons ft.).

$$\therefore\ \frac{W_u L}{8} = 264,$$

and $\quad W_u = \dfrac{264 \times 8}{L} = \dfrac{264 \times 8}{37} = \mathbf{57}$ tons (say).

Then $\quad W_g = \dfrac{W_u L}{540} = \dfrac{57 \times 37}{540} = 3\cdot91$, say $\mathbf{4}$ tons.

Note. The above work might be shortened, if care be taken in substituting, thus:
$$W_g = \frac{W_u L}{540}.$$

But $\quad B_M = \dfrac{W_u L}{8}.$

$$\therefore\ W_u = \frac{8 \cdot B_M}{L}.$$

Inserting value of W_u

$$*W_g = \frac{8 \times B_M \times L}{540 \times L} = \frac{8B_M}{540} = \frac{8 \times 264}{540} = 3 \cdot 91 \text{ tons (say 4 tons).}$$

The additional bending moment due to the dead weight of the girder, assumed uniformly distributed over the length, will therefore be

$$\frac{W_g L}{8} = \frac{4 \times 37}{8} = 18\tfrac{1}{2} \text{ tons ft.,}$$

and the maximum (central) bending moment becomes

$$264 + 18\tfrac{1}{2} = 282\tfrac{1}{2} \text{ tons ft.}$$

284. Graph of bending moment for plate girder. By plotting two graphs, one for the point loads and the other for uniform weight of girder, above each other, as shown in detail No. 96 at C, and to the same scale, the bending moment can be measured at any section of the girder. For a further purpose, viz. curtailing the flange plates, it is convenient to have this diagram on a level base which was obtained in the detail by adding the ordinates of the parabola below the datum line to those of the point load diagram above.

The final outline of the corrected graph is really a series of curves, but they approximate so closely to straight lines that they may be drawn as such.

285. Principle of determining moment of resistance for plate girders. To express the approximate moment of resistance of any thin webbed girder, make use of the fact that the resistance of the web to bending is small compared with that of the flanges and also that the flanges have their sectional areas concentrated near the extreme edges of the depth.

If d represents the depth measured over the first flange plate it will approximate to the position of the centre of pressure in the flange. Then, if an average working stress f, of 8 tons per sq. inch, be taken for mild steel, the total resistance offered by the flange is fA or $8A$, where $A = $ the *gross* area for compressional stress and the *nett* area where the flange is subjected to tensional stress.

But the moment of resistance, R, is the sum of the moments of the total compression and total tension about the neutral axis, the lever arm of each set of forces being $\frac{d}{2}$ (see paragraph 189), therefore

$$R = f_c A_c \times \frac{d}{2} + f_t A_t \times \frac{d}{2}$$

$$= \frac{d}{2} (f_c A_c + f_t A_t).$$

* Approximately $W_g = \dfrac{B_M}{70}$ (tons).

But f_c and f_t are the working stresses in compression and tension respectively and are usually considered equal, therefore let each be called f. Also A_c and A_t are equal effective areas.

$$\therefore \; R = \frac{d}{2}(2fA),$$

or
$$R = fAd.$$

Note. A is the area of *one flange* and d is the depth measured outside the first flange plate. The angles used to connect the flanges to the web provide an important portion of the tensile and compressive resistance; for shallow girders the effective addition to the flanges provided by the web is taken as the area of the horizontal arms of the angles working at full stress, while the vertical arms are ignored.

It is usual to express d in feet, f in tons per sq. inch, and A in sq. inches; then, as fA = total force in one flange in tons, fAd = resistance moment (R) in tons ft.

286. Area of flange required for plate girder. Before determining the flange area, a convenient and economical depth for the girder must be decided; otherwise there would be two unknown factors in the equation.

It is found that d should range from $\frac{L}{12}$ to $\frac{L}{16}$ for average work, unless the girder is to be designed for a maximum deflection.

With uniform loading, $d = \frac{L}{12}$ ensures a deflection not exceeding $\frac{L}{700}$; if $d = \frac{L}{16}$, deflection $= \frac{L}{500}$; if $d = \frac{L}{20}$, deflection $= \frac{L}{400}$.

All the above are roughly approximate values with a maximum working stress of 8 tons per sq. inch.

In this case it is desired to keep the girder shallow, thus avoiding undue encroachment on the height of the first floor storey of the warehouse; trying the lower value, viz. $\frac{L}{16}$, gives $\frac{37}{16}$ or $2\frac{5}{16}$ ft., which is approximately $2'\,3''$ outside the first flange plate.

Now $\qquad B_M = R$, and $R = fAd$.

$$\therefore \; 282\tfrac{1}{2} \text{ (tons ft.)} = 8 \text{ (tons)} \times A \text{ (sq. ins.)} \times 2\tfrac{1}{4} \text{ (ft.)}.$$

Hence, $\qquad A = \dfrac{282.5}{8 \times 2.25} = \mathbf{15.7}$ sq. ins.

287. Compressional flange. Let $4'' \times 4'' \times \frac{1}{2}'' \times 12\frac{3}{4}$ lbs. B.S.E.A.'s be used to connect the web to the flanges. In this consideration the rivets employed for the connection are to be assumed to fill the holes in the plates and angles perfectly, and to be capable of transmitting compressional stress equally with these parts; hence, no deduction of sectional area is required for this flange.

Towards the total area of the flange the angles provide two $4 \times \frac{1}{2} = 4$ sq. ins. $A - 4 = 15 \cdot 7 - 4 = 11 \cdot 7$ sq. ins. to be provided by plates (A_p).

Let the breadth of the flange be $\dfrac{L}{30} = \dfrac{37}{30} = 1 \cdot 23$ ft. or, say, $15''$; then

$$A_p = t_f \times b_f$$

where
$$t_f = \text{thickness of flange},$$
$$b_f = \text{breadth of flange},$$
$$\therefore \ 11 \cdot 7 = t_f \times 15,$$

and
$$t_f = \frac{11 \cdot 7}{15} = \mathbf{\cdot 78''}.$$

Use one $\frac{1}{2}''$ and one $\frac{3}{8}''$ plate making a total thickness of $\cdot 875''$, placing the $\frac{1}{2}''$ plate against the angles and continuing it to the ends.

It may be more economical to place the $\frac{3}{8}''$ plate against the angles and carry to the ends, but such treatment results in a lack of rigidity of the compressional flange.

288. Tension flange. This flange must provide the same net area as the *gross* area of the compression flange. Allowance must be made for rivet holes, the areas of these being deducted because tensional strength is lost by drilling or punching the holes through plate and angles. The rivets connecting the angles to the flange plates and those securing the edges of the overhanging flanges should be chequered to avoid more than two rivets occurring in any cross section of the flange, where the metal is working at *full stress*.

Using $\frac{3}{4}''$ rivets the effective area of the angles to resist tension

$$= 2 \left(\left[4 - \frac{3}{4} \right] \cdot \frac{1}{2} \right) = 3 \tfrac{1}{4} \text{ sq. ins.}$$

Nett $A = 15 \cdot 7$ sq. ins., therefore nett
$$A_p = 15 \cdot 7 - 3 \cdot 25 = \mathbf{12 \cdot 45} \text{ sq. ins.}$$

Deducting the two holes in the plates, their nett effective breadth becomes $15 - (2 \times \frac{3}{4}) = 13\frac{1}{2}''$.

Then, as
$$\text{nett } A_p = t_f \times \text{nett } b_f,$$
$$12 \cdot 45 = t_f \times 13 \cdot 5,$$

and
$$t_f = \frac{12 \cdot 45}{13 \cdot 5} = \mathbf{\cdot 912''}, \text{ viz. } \mathbf{1''} \text{ thick.}$$

Hence, two $\frac{1}{2}''$ plates for the tension flange may be used.

289. Curtailment of flange plates. Because the bending moment diminishes towards the supports, as indicated by the graph of detail No. 96 at C, while the depth of the girder remains practically constant, the total force to be provided by the flanges varies directly as the bending moment for any section, hence the second or outer plate may be curtailed at the section where the stress in the remaining flange area will not exceed 8 tons per sq. inch. To determine this section, use the bending moment graph and divide its height into parts proportional to the effective areas of the plates and angles.

Thus, by setting up an inclined line, marking off the effective areas to any suitable scale and projecting the divisions on to the perpendicular height as shown in the detail, the points of curtailment are determined by the intersection of the horizontal lines with the graph. This is done for the tensile areas on the right and compressional areas on the left of the bending moment graph.

As a margin of safety the curtailed plates are carried at least two rivet pitches beyond the theoretical point of termination.

If more than two flange plates are employed all except the inner plates are curtailed.

290. Thickness of web of plate girder. It has been previously shown, that the chief function of the web of a thin webbed girder is to transform the vertical shear stress into direct stress, viz. into the pull and thrust in the flanges. The web is first called upon to resist the vertical shear force and, if it does this successfully, and is satisfactorily connected to the flanges, it transfers the horizontal forces (due to its own tendency to distort into rhomboidal form) to the angles and thus to the plates.

Whilst resisting shear it tends to buckle, because of its height and comparative slenderness, hence the working value of shear allowed for short rigid pieces (viz. 6 tons per sq. in.) cannot be adopted. The limit is found to be from 2 to $3\frac{1}{2}$ tons per sq. in. according to the depth and proportions of the girder; $2\frac{1}{2}$ tons per sq. inch is adopted averaged over the section.

The maximum vertical shear force occurs at the support and is the sum of the useful and dead load shears, summed up in detail No. 96 at D; the total reaction is $24 + 2 = 26$ tons, which is the shear $(+ \text{ or } -)$ at each support.

As the shear stress does not vary largely over the section of the web, a close approximation to the correct shear may be obtained from

$$S_{F(\text{max.})} = A_w \times f_{sw} = t_w \times d_w \times f_{sw},$$

where

$S_{F(\text{max.})} = $ maximum shear force $= 26$ tons.

$\quad A_w = $ area of web (sq. ins.).

$\quad d_w = $ depth of web $= 27'' - (2 \times \frac{1}{2}'') = 26''$.

$\quad t_w = $ thickness of web (inches).

$\quad f_{sw} = $ allowable average shear stress on web $= 2\frac{1}{2}$ tons per sq. in.

For the example in question:

$$S_{F(\max.)} = t_w \times d_w \times f_{sw}.$$

$$\therefore \ t_w = \frac{S_{F(\max.)}}{d_w \times f_{sw}} = \frac{26}{26 \times 2\frac{1}{2}} = \cdot 4''.$$

Use a $\frac{7}{16}''$ plate if available, giving $\cdot 4375''$ as against $\cdot 4''$ required. As this girder is employed within the building where it can be well protected a $\frac{3}{8}''$ plate could be employed, which would produce a shear stress of about $2\cdot67$ tons per sq. in. instead of $2\cdot5$.

291. Pitch of rivets in web to angle connection. There are several methods of determining the pitch of rivets. Two well-known methods are, briefly, as follows:

(a) Determine maximum horizontal force tending to cause failure of the rivets near the end, by shear or by bearing.

(b) Determine the total horizontal force to be transmitted by a series of rivets extending over a selected length near the supports.

In this example the selected length may be any portion or the whole of the end bay, because the rate of increase of bending moment is practically constant, as shown by the shear diagram.

Where bending moment is caused by a uniform load it is wise to use method (a), or as an alternative, method (b) with a selected length not exceeding $\frac{1}{10}$th of the span.

In all cases the greatest rate of increase of horizontal stress occurs near to or at the support if the weight of the beam is taken into account, hence the shorter the length considered, the more accurate will be the provision against the over-stressing of the rivets and plates.

292. Factors determining pitch of rivets. If the web transmits all the horizontal stress to the flanges, the addition of such stress in a given length can be determined, and hence the bearing and shearing forces to be taken by the rivets and plates in the selected length. The increase of stress is proportional to the increase of bending moment between any two sections.

Selecting 3 ft. of length from support C to the section T; $B_T - B_C$ represents the increase. By calculation, or by measurement from the bending moment graph, or again by obtaining the area of the shear graph between C and T, it will be found that $B_T = 77\cdot5$ tons ft., while $B_C = 0$. Hence, the increase from C and $T = 77\cdot5$ tons ft.

From previous work $B = Afd$ generally and $\dfrac{B}{d} = Af = $ total stress in flange, or horizontal force (F_H) in the length of 3 ft.

If B_i represents the increase and d_r the depth between centres of lines of rivets connecting web to angles, then $\dfrac{B_i}{d_r} = F_H$.

Now $d_r = 27'' - (2 \times \tfrac{1}{2}'' + 2 \times 2\tfrac{1}{4}'') = 21\tfrac{1}{2}''$ or **1·8** ft. approximately.

Hence $\qquad F_H = \dfrac{77\cdot5 \text{ (tons ft.)}}{1\cdot8 \text{ (ft.)}} = \textbf{43 tons.}$

293. Shear and bearing strength of rivets. Each rivet passing through the web and flange angles is subjected to double shear in the planes of junction of angles to web. A $\tfrac{3}{4}''$ rivet has a sectional area

$$= \frac{\pi}{4}\,d^2 = \cdot785 \times \left(\frac{3}{4}\right)^2 = \cdot442 \text{ sq. ins.}$$

Allowing double the value of the single shear resistance, see Vol. II, and 6 tons per sq. inch for f_s, each rivet resists $2 \times \cdot442 \times 6 = 5\cdot3$ tons.

Then the number of rivets required to resist shear

$$= \frac{\text{total force transmitted}}{\text{resistance of one rivet}} = \frac{F_H}{r_s} = \frac{43}{5\cdot3} = 8\cdot1, \text{ say } \textbf{9 rivets.}$$

The same series of rivets may fail by bearing. Each rivet has a bearing area equivalent to its own diameter × thickness of plate (or plates) against which it bears.

In this case the smaller area is against the web and

$$a_b = \frac{3}{4} \times \frac{7}{16} = \frac{21}{64} = \cdot328 \text{ sq. ins.}$$

Let $\qquad a_b =$ effective bearing area of one rivet,
$\qquad\qquad f_b =$ safe bearing stress
$\qquad\qquad\quad = 12$ tons for mild steel.

Then each rivet resists $a_b \times f_b = \cdot328 \times 12 = \textbf{3·93 tons.}$

No. of rivets required $= \dfrac{\text{total force transmitted}}{\text{resistance to one rivet}} = \dfrac{43}{3\cdot93}$
$$= \textbf{11 rivets (say).}$$

The greater number (11) is required to resist bearing and the rivets are to be placed in a length of 36″.

Hence $\qquad$ pitch of rivets $= \dfrac{36}{11} = \textbf{3·27″.}$

Employing a $\tfrac{7}{16}''$ web adopt a practical pitch of 3″ over the length from the support to the nearest secondary girder, after which the pitch may be increased as calculated (see next paragraph). The rivets would then be safe *on the basis of no allowance for friction,* which is the ordinary designing practice.

294. Rivet pitch over second bay of plate girder. To determine this pitch, having noticed that the *rate of increase of the bending moment* in this example is practically constant over the whole bay, measure the total value of B at each end of the bay.

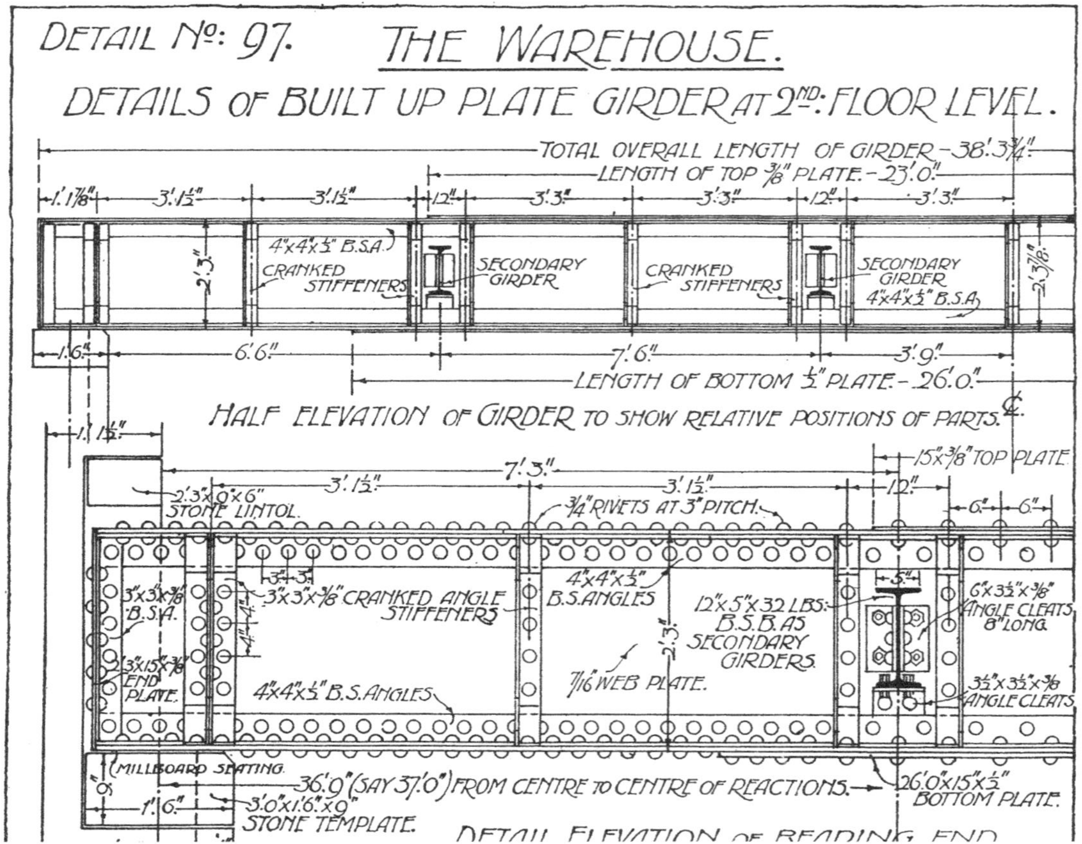
DETAIL No: 97.
THE WAREHOUSE.
DETAILS OF BUILT UP PLATE GIRDER AT 2ND: FLOOR LEVEL.
TOTAL OVERALL LENGTH OF GIRDER - 38'.3¾".
LENGTH OF TOP ⅜" PLATE. - 23'.0".
1'.1⅞"
3'.1½"
3'.1½"
12"
3'.3"
3'.3"
12"
3'.3"
4"x4"x½" B.S.A.
CRANKED STIFFENERS
SECONDARY GIRDER
CRANKED STIFFENERS
SECONDARY GIRDER
4"x4"x½" B.S.A.
2'.3"
2'.3⅞"
1'.6"
6'.6"
7'.6"
3'.9"
LENGTH OF BOTTOM ½" PLATE. - 26'.0".
HALF ELEVATION OF GIRDER TO SHOW RELATIVE POSITIONS OF PARTS.
1'.1½"
15"x⅜" TOP PLATE.
7'.3"
3'.1½"
3'.1½"
12"
6"
6"
2'.3"x0'x6" STONE LINTOL.
¾" RIVETS AT 3" PITCH.
3"x3"x⅜" B.S.A.
3"x3"x⅜" CRANKED ANGLE STIFFENERS
4"x4"x½" B.S.ANGLES
12"x5"x32 LBS: B.S.B. AS SECONDARY GIRDERS.
6"x3½"x⅜" ANGLE CLEATS 8" LONG.
2'.3"
13"x15"x⅜" END PLATE.
7/16" WEB PLATE.
4"x4"x½" B.S.ANGLES
3½"x3½"x⅜" ANGLE CLEATS
MILLBOARD SEATING.
36'.9" (SAY 37'.0") FROM CENTRE TO CENTRE OF REACTIONS.
26'.0"x15"x½" BOTTOM PLATE.
1'.6"
3'.0"x1'.6"x9" STONE TEMPLATE.
DETAIL ELEVATION OF READING END

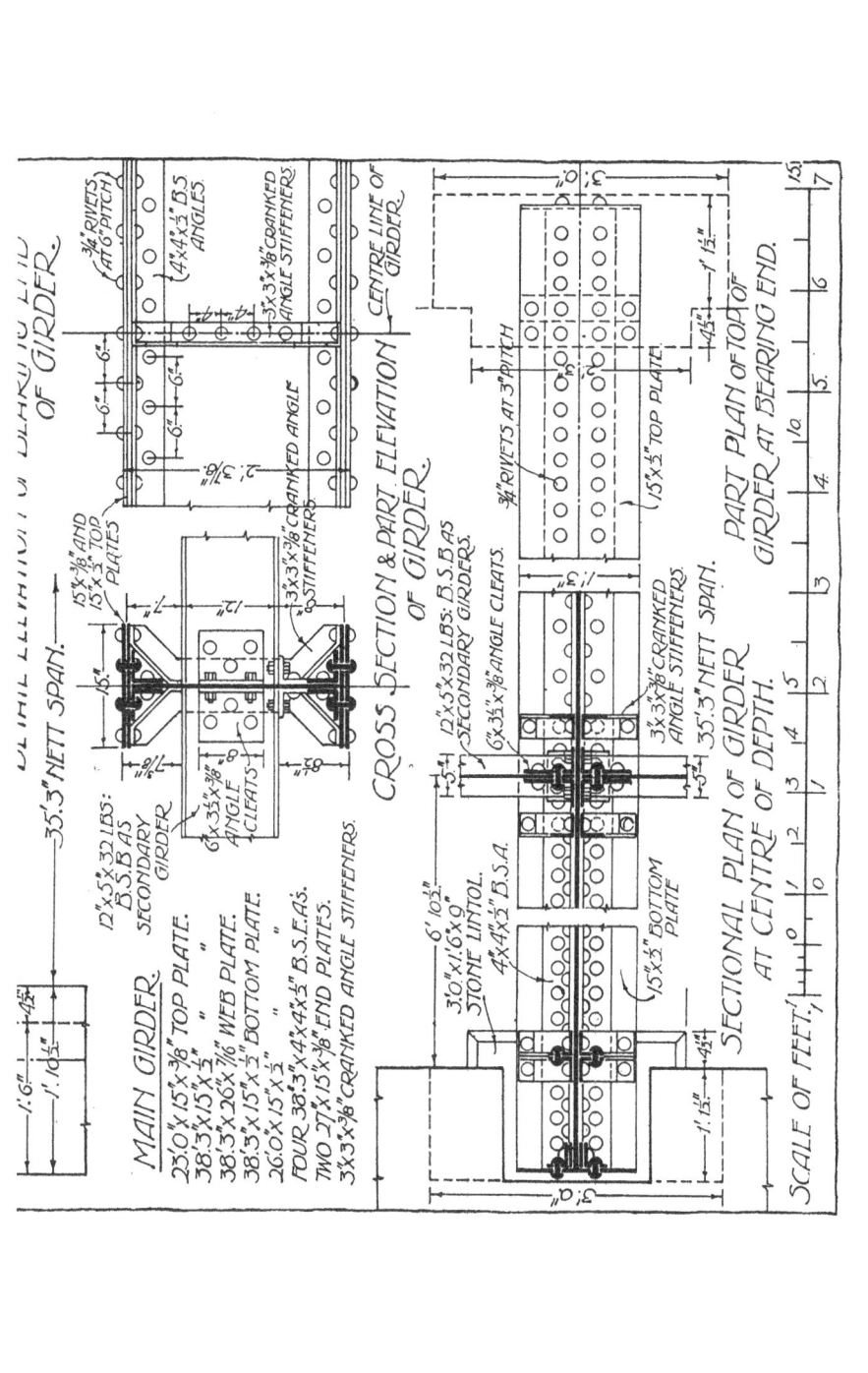

PART ELEVATION OF BEARING END OF GIRDER.

35'-3" NETT SPAN.

3/4" RIVETS (AT 6" PITCH)

(4"×4"×5/8" B.S ANGLES

3"×3"×3/8" CRANKED ANGLE STIFFENERS

15"×3/8" AND 15"×1/2" TOP PLATES

CENTRE LINE OF GIRDER.

2'-3 9/16".

3"×3"×3/8" CRANKED ANGLE STIFFENERS

CROSS SECTION & PART ELEVATION OF GIRDER.

12"×5"×32 LBS: B.S.B AS SECONDARY GIRDER.

6"×3½"×3/8" ANGLE CLEATS

15".

MAIN GIRDER.

25'·0"×15"×3/8" TOP PLATE.
38'·3"×15"×½" " " "
38'·3"×26"×7/16 WEB PLATE.
38'·3"×15"×5" BOTTOM PLATE.
26'·0"×15"×½" " " "
FOUR 38'·3"×4"×4"×5" B.S.E.A's.
TWO 27"×15"×3/8" END PLATES.
3"×3"×3/8 CRANKED ANGLE STIFFENERS.

3'·0".

¾" RIVETTS AT 3" PITCH

15"×½" TOP PLATE

1'·1½".

1'·3".

12"×5"×32 LBS: B.S.B AS SECONDARY GIRDERS.

3"×3"×3/8" CRANKED ANGLE STIFFENERS

6"×3½"×3/8" ANGLE CLEATS.

35'·3" NETT SPAN.

3'·10"×16×9" STONE LINTOL.
4"×4"×5" B.S.A.

15"×5" BOTTOM PLATE

SECTIONAL PLAN OF GIRDER AT CENTRE OF DEPTH.

6'·10½".

1'·1½". 4'·4½".

PART PLAN OF TOP OF GIRDER AT BEARING END.

3'·0".

SCALE OF FEET.

1'·6".
1'·10½". 4'½".

At P the value is **282** tons ft. and at R the value is **186** tons ft., therefore the increase is **96** tons ft. in $7\frac{1}{2}$ ft.

Then $\dfrac{B_i}{d_r} = \dfrac{96}{1\cdot8} = 53\cdot4$ tons total force transmitted by the rivets.

With a $\frac{7}{16}''$ web the bearing resistance of a $\frac{3}{4}''$ rivet has already been found to $= 3\cdot93$ tons, hence the number of rivets required $= \dfrac{53\cdot4}{3\cdot93}$ $= \mathbf{13\cdot6}$, and the pitch $= \dfrac{90}{13\cdot6} = 6\cdot6$, say **6″**, as a practical pitch.

The above result shows that the pitch may be 6″ for the two bays nearest the centre and the same pitch would be retained for the centre bay; a larger pitch would not ensure sufficiently close contact between the assembled angles and plates.

With a view to avoiding the buckling of compression plates it is usual to restrict the pitch to a maximum = 16 × thickness of thinnest plate assembled in the group. The web is $\frac{7}{16}''$ and the angles $\frac{1}{2}''$ thick. Hence $16 \times \frac{7}{16} = 7''$, which is greater than the calculated pitch for shear; 6″ may therefore be adopted.

295. Connection between angles and flange plates. The lever arm for this connection is greater than that for the web rivets, the bearing area being $\frac{3}{4}'' \times \frac{1}{2}''$ for the angle and plate as against $\frac{3}{4}'' \times \frac{7}{16}''$ for the web, while the two lines of rivets are *each* in single shear; hence, it is evident that the same pitches may be adopted for these lines as for the web. The connection will be stronger than necessary, but it is impracticable to alter the pitch.

296. Stiffeners to web and flanges. In order to assist the web against the tendency to fail by buckling or crumpling due to its slenderness angle iron stiffeners are employed at intervals throughout the length. These members are bent to fit the recessed faces of the girder and are riveted to the web and flanges, as shown in detail No. 97.

Stiffeners should be placed at or near all load points such as the junctions of secondary girders and over the supports, and also at intervals not exceeding twice the depth of the girder. In deep girders the spacing of stiffeners near the supports should not exceed the depth of the girder, so that the web becomes divided into approximately square panels, but they may be more widely spaced towards the centre as the shear diminishes, where their chief function is to offer assistance to the flanges against side bending at the overhanging edges and to prevent the web bending over sidewise should the floor be loaded on one side of the girder only.

297. Stopped end of girder. The ends of plate girders are usually terminated by a rectangular cover plate, $\frac{1}{4}''$ to $\frac{3}{8}''$ thick, secured to the web by riveted angles; this adds to the rigidity over the support, and provides for a more uniform transmission of the load thereto. If the flanges are broader than 15″, the connecting angle for the end plate should be shaped like the stiffeners, as shown in the detail, or horizontal angles may be employed between flanges and plate.

Support stiffeners are often made of T or double L section instead of the single L employed on the span. In any case the regular pitch should not be interfered with by the insertion of stiffeners, if it is possible to avoid it.

298. Size of bearing. Let the stone template under the girder be of hard York stone; this may safely support 20 tons per sq. ft. if a felt pad be inserted between the template and girder. The proposed length of bearing is $1'\ 6''=1\frac{1}{2}$ ft., and the breadth of the girder is $15''=1\frac{1}{4}$ ft. The load on each support is $24+2=26$ tons, hence the pressure per sq. ft. is

$$\frac{W}{A}=\frac{26}{1\frac{1}{2}\times 1\frac{1}{4}}=\frac{26\times 2\times 4}{3\times 5}=\text{say }\mathbf{13\cdot 87}\text{ tons.}$$

If desired to reduce this unit pressure for softer stones without lengthening the bearing a stout bearing plate of greater width would be riveted on the under flange of the girder. Suppose this plate to be 18″ long, then

$$p_{(\text{sq. ft.})}=\frac{26}{1\frac{1}{2}\times 1\frac{1}{2}}=\mathbf{11\cdot 5}\text{ tons.}$$

The rivets on the underside of the flange—or bearing plate if employed—must be countersunk and chipped off flush to avoid undue pressure at isolated points.

299. Dimensions of template. The function of the stone template is to distribute the load uniformly over the brickwork at the allowable stress for the class of walling provided.

Let blue bricks in cement be employed, then the pressure should not usually exceed 12 tons per sq. ft. and the area of the base of the template is found thus:

$$p=\frac{W}{A},\quad \therefore\ A=\frac{W}{p},\quad \text{and}\quad A=\frac{26}{12}=2\cdot 16\text{ sq. ft.}$$

This would require a double notched template as shown in the detail No. 97. The area of the middle rectangle is $1'\ 6''\times 1'\ 6''=2\cdot 25$ sq. ft., and of the two projections $=2\times 4\frac{1}{2}''\times 9''=\cdot 562$ sq. ft., a total of (say) 2·81 sq. ft. The outside dimensions of the template are $2'\ 3''\times 1'\ 6''$ with notches $4\frac{1}{2}''$ square cut out of the front angles. The thickness of the template should be at least equal to the projection beyond the side of the girder, which in this case is 6″.

300. Connections of secondary girders to plate girder. These are made by vertical angle cleats, with sufficient rivets to resist the shear stresses due to the load transmitted through the joint, viz. half the load carried by the girder. A horizontal "support cleat" is shown to facilitate the erection.

Stiffeners are placed one on each side of the secondary girder connection and as near to the latter as practicable.

CHAPTER TWELVE

PERMANENT CARPENTRY—ROOFS

In the previous volumes, examples of the three main divisions of roof construction have been considered.

The roof planes employed were all of a simple character and although some intersections occurred forming hips and valleys the constructional details of these parts were omitted. In this chapter are provided somewhat similar, though more difficult, cases which will be dealt with in full detail.

301. Double or purlin roofs. While a roof may be a difficult one to construct and to erect, because of the irregular outlines of its surfaces and the numerous changes in direction of the roof planes, it may very often be constructed without the aid of roof trusses by utilising the party and external walls and partitions for support of the purlins, hip rafters and valley rafters.

The roof to the house is one of this class, and the general form of this roof may be gathered from the perspective sketch detail given in Vol. II and reproduced here as detail No. 98.

302. Formation of the house roof. The plan of the house is a rectangle, 42′ 3″ long by 33′ 6″ wide, with a rectangular projection at the entrance portion, 17′ 6″ wide by 3 ft. deep, on the front of the building. The roof follows this general plan outline but is extended to project 1′ 3″ beyond the face of the brickwork at the eaves.

As the roof is to be covered with tiles and a pitch of 45° is desired,[1] it is found, on setting up the cross section and projecting the roof outline, that owing to the considerable depth of the building from front to back, the height of the ridge is excessive and produces a badly proportioned elevation.

To avoid this defect and to economise by reducing the area of the sloping surface, the following disposition is adopted. A uniform slope of 45° is given to the lower surface, this slope terminating at the head against a lead flat about 31′ 4″ long × 13 ft. broad, and as the inclination is carried across the end of the building hips are formed at the back and front angles.

Over the front projection is a gable with its roof surfaces also at 45° and its eaves at the same level, but because the span is less than

---

[1] See Vol. I for pitches of roofs.

twice the plan width of the main slopes, it does not reach the same
height as the flat and the ridge occurs about 1 ft. below the curb
of the flat.

There are three bedrooms, a box room, a cistern room and stair
landing on the attic floor, and these are lighted respectively by
dormer windows, a gable window, and skylights.

DETAIL Nº:98.   THE HOUSE.
PERSPECTIVE SKETCH OF ROOF SHOWING
PLAIN TILING AND LEAD
COVERED FLAT.

303. General construction of the house roof. The main part of the
roof is illustrated in detail No. 98 A, and is constructed as follows:

To form the long edges of the flat portion, two $11'' \times 5\frac{1}{2}''$ pitch
pine curbs run parallel to the front and back of the building and
are supported on the $13\frac{1}{2}''$ party wall and the $9''$ staircase walls.
A cross curb, of $7'' \times 4''$ deal, is framed to the main curbs at the
angles.

From the angles of these curbs to the quoins of the building two
$11'' \times 2''$ deal rafters are framed to form the hips and to receive the
common rafters and purlins, their feet resting upon special bearings
which are described later in paragraph No. 308.

Where the front gable intersects the main roof two valleys are
formed with $11'' \times 2''$ deal rafters to receive the common rafters
and purlin ends in the same way as the hip rafters.

To form the flat, $7'' \times 2''$ deal joists span across the shorter direction and derive intermediate support from the head of a timber partition which divides the bedrooms and also from a $9''$ wall at the back of the stair landing.

To support the common rafters (or spars) at intermediate points on the roof slopes, $9'' \times 3''$ deal purlins are employed, though their

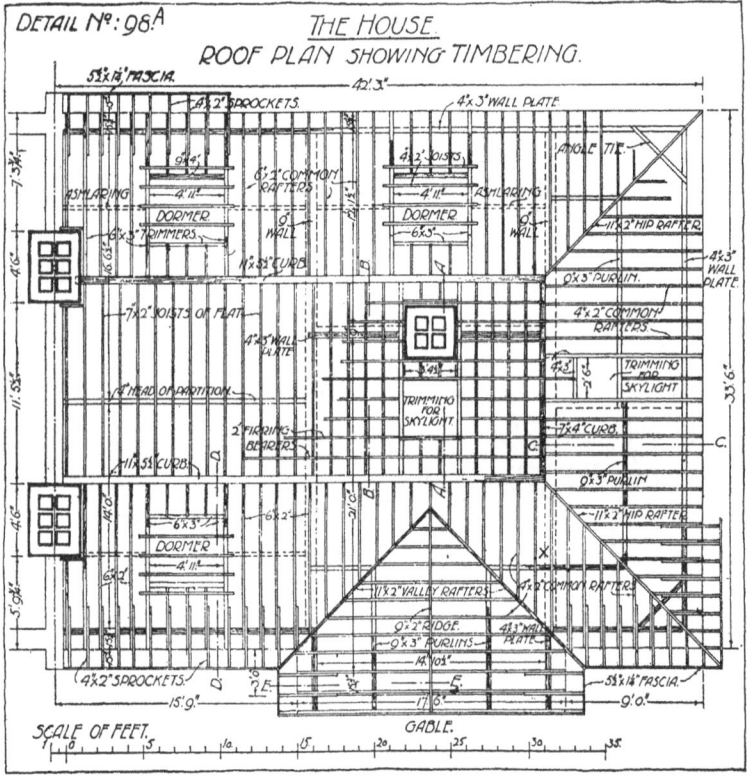

size may be varied with the span and the loads they are required to support. All are placed in vertical planes for the convenience of bearing solidly on the walls and for cutting to fit against the vertical sides of the hip and valley rafters. The feet of the rafters are supported, as usual, upon $4'' \times 3''$ wall plates.

The foot of the roof slope is reduced in pitch by fixing short sprocket rafters to the sides of the common rafters and continuing them over the wall to form the projecting eaves.

Before entering upon the detailed study of isolated portions of the roof the student should carefully peruse the general plans of the house, together with the roof plan here included, and so follow the disposition of all the main timbers; the following paragraphs may then profitably be studied and the illustrations of detailed construction prepared to scale.

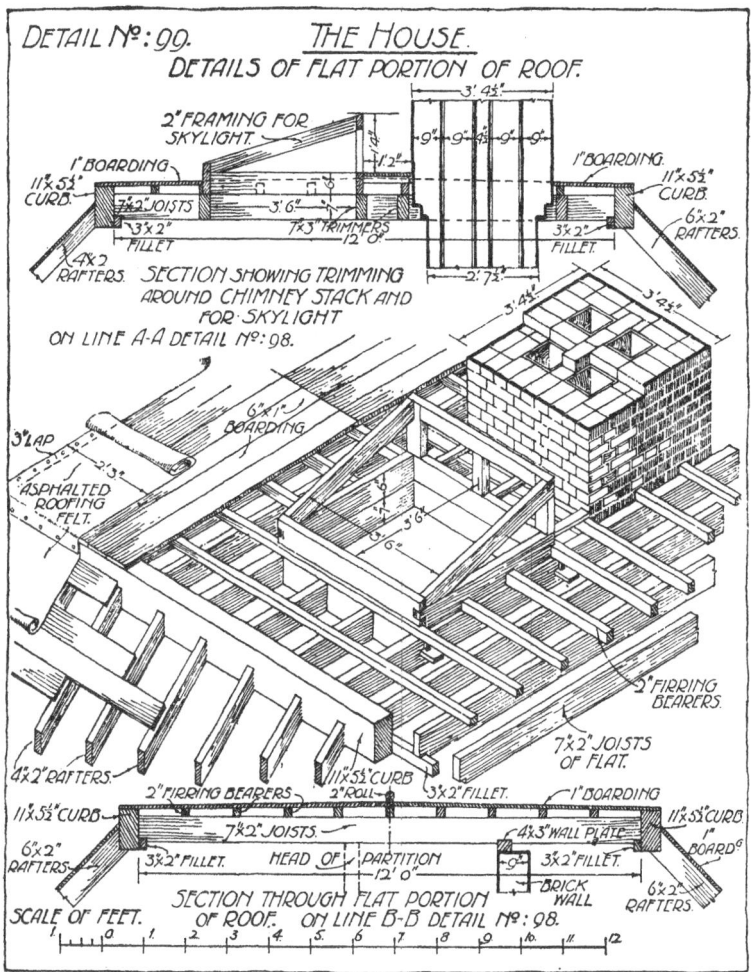

304. Construction of flat roof. Detail No. 99 shows the method of framing the flat roof over the stair landing and bedrooms.

A chimney stack containing the kitchen and bedroom flues penetrates this portion of the roof and a skylight is provided in front of the stack to light the stair landing.

The detail deals only with the timber framing and boarding in preparation for the joiners' and plumbers' work.

Two sections across the flat are given, the upper one across the chimney stack and skylight opening and the lower one in the same direction but clear of the trimming round the openings.

The flat is supported by $7'' \times 2''$ deal joists spanning from curb to curb, these being notched upon $3'' \times 2''$ deal fillets spiked on the inside lower edge. The necessary fall to the flat roof is given from the centre of the span towards the curbs and the boarding to receive the lead must have its length in this direction.

The method adopted is to "fir" the joists by timbers $2''$ thick, at right angles to the joists—the firring varying in depth from $3''$ at the centre to $2''$ at the edges—and to lay the boards upon these in the direction of the fall. An alternative method is to place wedge-shaped packings upon the joists—or otherwise taper these from larger stuff to the falls required—and lay firrings of uniform depth across these packings to receive the boarding.

The boarding should be of uniform thickness, with tongued and grooved joints. Upon the boarding is placed a layer of asphalted felt with $3''$ lapped joints in the direction of the fall; at the crown of the flat a $2''$ wood roll provides for dividing the falls and jointing the lead sheets.

305. Trimming to skylight opening and chimney stack. The opening for the skylight is made the same width as the chimney stack, which enables a pair of trimming joists to be placed near to the sides of the chimney, and so form the width of the skylight opening; cross trimmers are tusk tenoned into the trimming joists or dovetail notched thereto to complete the framing.

306. Skylight frame. The skylight consists of a glazed sash fixed upon an inclined timber frame which surrounds the opening and rests upon the trimming timbers referred to in the last paragraph. This frame consists of a $7\frac{1}{2}'' \times 2''$ front curb, two spandril sides and back framing, the sides and back being framed up with $2''$ material and dovetailed together at the external angles as illustrated in the detail. By this method the unit may be completed in the workshop and fixed in position around the opening by skew nails, before the firring is laid.

307. Jointing of rafters to curb. Two sizes of rafters are employed in the roof because some lengths are supported at an intermediate point by purlins, while others span from eaves to curb without such support; $4'' \times 2''$ material is used for the former condition and $6'' \times 2''$ for the latter.

Wherever possible the heads of the rafters should be birdsmouthed to the curb, as shown in several of the details; direct support is thus provided for the curb by transmitting some of its load as thrust through the rafters.

308. Eaves construction and finish. The eaves finish consists of clothing the ends of the rafters by boarding and wood mouldings to form a cornice, the detail of which is given in No. 100; this also shows the attic floor joists and how advantage is taken of the position of the latter effectively to tie the rafter feet and to resist spreading and avoid thrust on the wall.

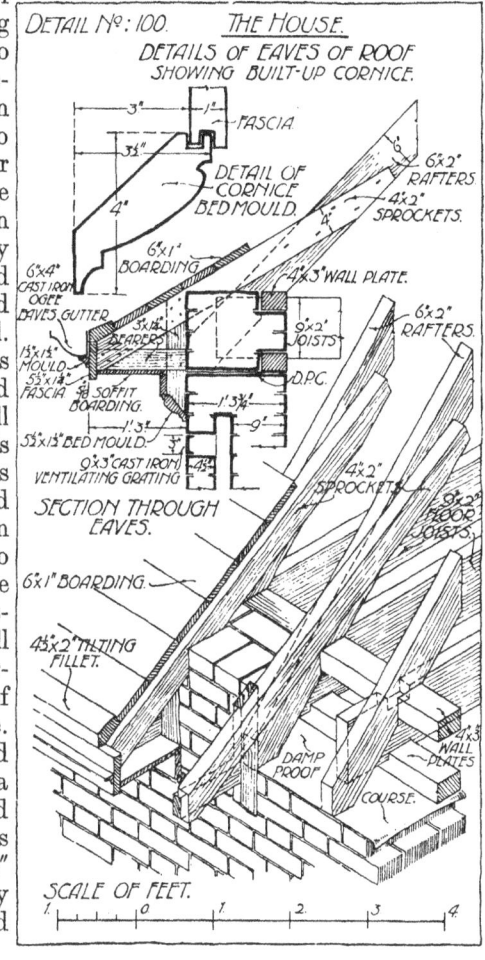

The 9″ × 2″ floor joists are notched over and spiked to a 4″ × 3″ wall plate; a second plate is notched over the joists to a depth of 2″ and spiked down to them and is then ready to receive the feet of the rafters which are birdsmouthed to the wall plate and continued forward to the level of the damp proof course. Each rafter is placed against the side of a joist and firmly nailed thereto, while to the sides of the rafters 4″ × 2″ sprockets are similarly secured at the required slope.

309. Cornice and cradling. The eaves cornice is of an outline with which the student should be already familiar.

The corona is formed by a 5½″ × 1¼″ fascia board and a 6″ × 4″ cast iron eaves gutter bearing upon a 1½″ × 1½″ cavetto mould, and with ⅝″ perforated soffit boards on the bed of this projection. The bed mould is constructed from a $3\frac{1}{4}'' \times 1''$ fascia and a $4'' \times 3\frac{1}{2}''$ solid mould which are tongued together.

To support the cornice, timber "cradling" is required and is shown in detail No. 100; it consists of light angle brackets framed up from $3'' \times 1\frac{1}{4}''$ deal, halved together to form an irregular cross and secured by notching and nailing to the sprockets.

The ends of the vertical pieces are left long enough to secure the bed mould at the base and to bear against the wall behind.

If such cradling enters the wall it should be treated with wood preservative before building in or otherwise securing, and where possible the inserted ends of cradling and joists should have enough clearance to ventilate the encased parts.

310. Dormer window. A dormer window differs from a skylight in that the latter is a light formed in a roof surface or in a plane approximately parallel thereto, while the dormer is a means of admitting light to a room wholly or partially within a roof space, through the vertical sides of a structure raised above the plane of the roof.

The sides and front may be glazed into fixed sashes, but the front is usually arranged with casements or sliding sashes according to preference, and in a larger dormer portions of the sides may also be made to open.

Side glazing is objectionable in terrace houses where dormers are closely spaced along the roofs, and in addition reduces the architectural value of the feature, but where it is necessary to admit the maximum of light, both the sides and the roof are often glazed.

Every dormer has a separate roof, which may be either a pitched roof covered with slates or tiles, or a lead covered flat, or glazed sash framing.

311. Arrangement of dormer framing for the house. Detail No. 101 is an isometric projection, showing the disposition of the whole of the dormer framing in association with roof timbers, attic floor, and the vertical studding at the sides of the attic bedroom.

To follow the outlines clearly, observe the shape of the room in plan, where the space is enclosed by a row of $3'' \times 2''$ studs resting upon a $3'' \times 3''$ cill and notched over the sides of the common rafters.

At the dormer opening this row of studs is turned at right angles and brought forward a distance of $1' 8''$ into the opening, where a $4'' \times 4''$ angle post resting upon a $4'' \times 3''$ cill supports the trimming rafters.

To form the projecting part of the structure two spandril side frames are seated upon the trimming rafters, and the roof is formed as a lead flat supported upon joists and boarding.

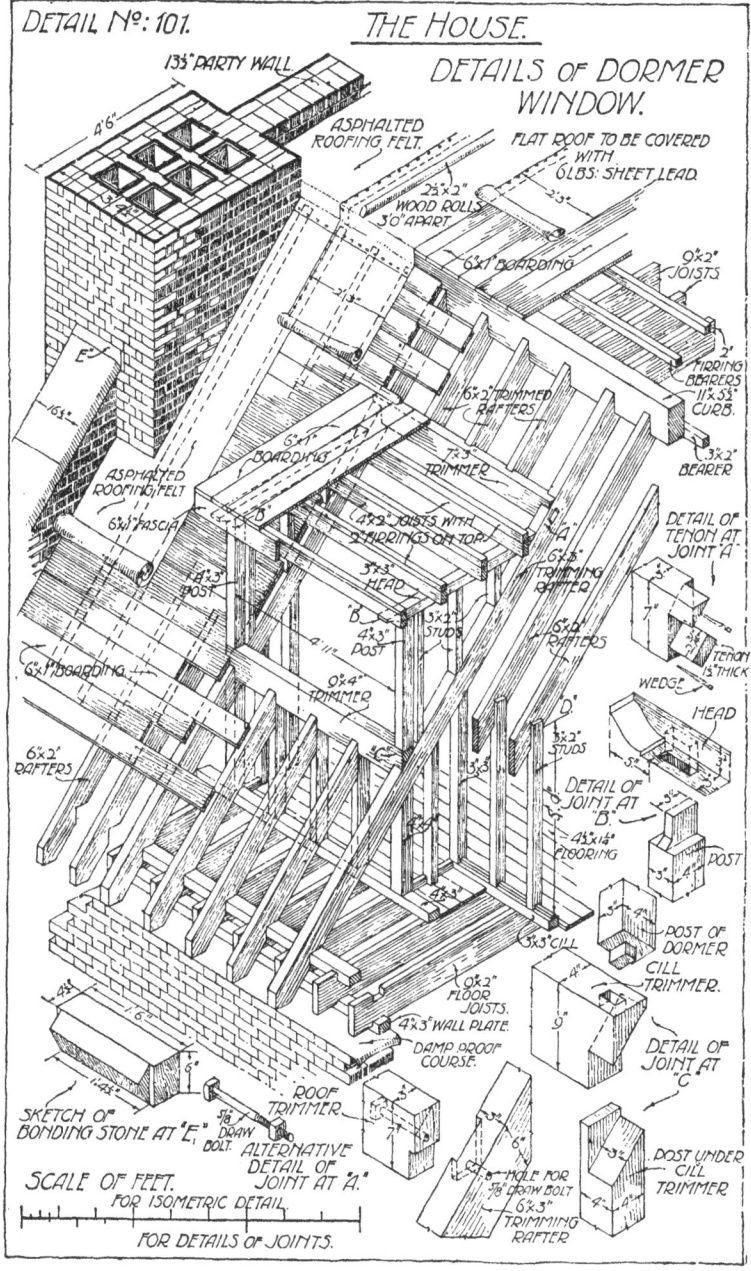

DETAIL Nº: 101.   THE HOUSE.

DETAILS OF DORMER WINDOW.

13½" PARTY WALL.

ASPHALTED ROOFING FELT.

FLAT ROOF TO BE COVERED WITH 6 LBS: SHEET LEAD.

4'6"

2½×2" WOOD ROLLS 3'0" APART.

6×1" BOARDING

9×2" JOISTS

2" FIRRING BEARERS 11×5½" CURB.

6×2" TRIMMED RAFTERS

3×2" BEARER

E'

16½"

6×1" BOARDING

7×3" TRIMMER

DETAIL OF TENON AT JOINT A

ASPHALTED ROOFING FELT

6×1" FASCIA

4×2" JOISTS WITH 2" FIRRINGS ON TOP

5"

7"

6×6" TRIMMING RAFTER

5½×3" POST

3×3" HEAD

'A'

TENON 1⅜" THICK

'B'

4×3" POST

3×2" STUDS

5×4" RAFTERS

WEDGE

HEAD

6×1" BOARDING

9×4" TRIMMER

'D'

3×2" STUDS

DETAIL OF JOINT AT B.

6×2" RAFTERS

'C'

3×3"

4½×1¼" FLOORING

5"

POST

4½×3"

POST OF DORMER CILL TRIMMER.

5×5" CILL

4½"

DETAIL OF JOINT AT C

9×2" FLOOR JOISTS.

4×3" WALL PLATE.

9"

4½"

DAMP PROOF COURSE.

POST UNDER CILL TRIMMER.

SKETCH OF BONDING STONE AT "E."

⅝" DRAW BOLT.

ROOF TRIMMER.

ALTERNATIVE DETAIL OF JOINT AT 'A.'

6"

5"

6"

3"

4"

SCALE OF FEET.

⅞" HOLE FOR ⅞" DRAW BOLT.

6×3" TRIMMING RAFTER

FOR ISOMETRIC DETAIL.

FOR DETAILS OF JOINTS.

312. Trimming for dormer window opening. In the same detail and in detail No. 102, the trimmers required to form the opening in the rafters for the dormer window are shown. Two $6'' \times 3''$ trimming rafters are substituted for the $6'' \times 2''$ common rafters at the sides, and a $7'' \times 3''$ trimmer is set vertically between them at the head of the opening and flush at the bottom angle, while a $9'' \times 4''$ trimmer is similarly placed at the foot of the opening, except that it stands about $4''$ above the rafters to form a seating for the window cill of the casement frame.

The top trimmer, at A, may be jointed in two ways: (a) by a bevelled tenon squared at the edges and wedged to the rafter, as shown in the dissociated detail to the right of No. 101, or (b) stop housed into the rafter and draw-bolted, as shown in the lower part of the same detail.

The bottom trimmer, at C, is notched over the $6'' \times 3''$ rafter and rests squarely upon the series of studs beneath the dormer front. This trimmer may be either spiked or draw-bolted in position, the former method being quite satisfactory if the angle post—shown in the separate detail at C—is left with a square portion to receive the end of the trimmer.

313. Spandril sides to dormer. The spandril sides to the dormer are formed by posts, rails and studding as follows:

A $4'' \times 3''$ angle post is stub-tenoned to the cill, as shown at C, and haunch-tenoned at the top into a $3'' \times 3''$ head which passes over the post to form a $5''$ moulded projection, as at B. At the back end the head is splayed upon the rafter and spiked firmly to it, or in a very large dormer draw-bolted from the under side, or again plain bolted through the two pieces at right angles to the rafter.

Vertical $3'' \times 2''$ studs are then placed between head and rafter to fill the spandril, one piece being near to the angle post for securing the plasterers' laths.

314. Flat roof to dormer. The dormer roof is formed by $4'' \times 2''$ joists placed parallel to the front; the joists give support to firrings, which are varied in depth to give a fall towards the front edge; these firrings receive $6'' \times 1''$ boarding upon which the lead covering is laid.

A plain fascia board is mitred round the face and ends of the joists and firrings, to form a finish and to provide a ground for fixing the gutter, etc.; a $\frac{1}{2}''$ soffit board is also required to close the opening between the joists at the overhanging part, as shown in section through the head in detail No. 102.

315. Roof over cistern room. Within the main roof at the hipped end is a room set apart for the cold water storage cistern. This room

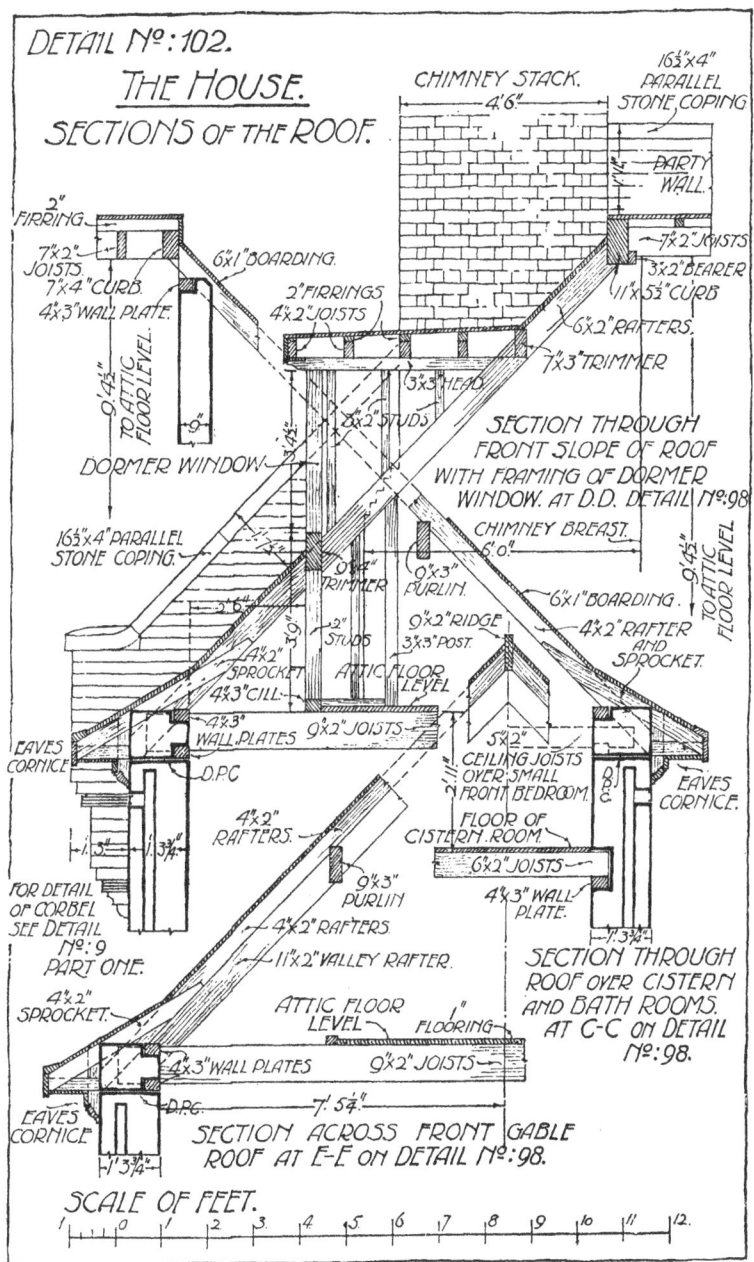

DETAIL Nº: 102.

THE HOUSE.

SECTIONS OF THE ROOF.

CHIMNEY STACK.
4'6"

16½"x4" PARALLEL STONE COPING

PARTY WALL.

2" FIRRING

7x2" JOISTS
7x4" CURB.
4x3" WALL PLATE.

6"x1" BOARDING.

2" FIRRINGS
4x2" JOISTS

7x2" JOISTS
3x2" BEARER
11"x5½" CURB
6"x2" RAFTERS.
7x3" TRIMMER

0'45" TO ATTIC FLOOR LEVEL.

3x3" HEAD

8x2" STUDS

SECTION THROUGH FRONT SLOPE OF ROOF WITH FRAMING OF DORMER WINDOW. AT D.D. DETAIL Nº:98

DORMER WINDOW

CHIMNEY BREAST.
6'0"

16½"x4" PARALLEL STONE COPING.

9"x1" TRIMMER

9x3" PURLIN.

6"x1" BOARDING.

4x2" RAFTER AND SPROCKET.

0'45" TO ATTIC FLOOR LEVEL

2" STUDS
3'9"

9x2" RIDGE
3x3" POST.

4x2" SPROCKET
4x3" CILL.

ATTIC FLOOR LEVEL

4x3" WALL PLATES
9x2" JOISTS

5"x2" CEILING JOISTS OVER SMALL FRONT BEDROOM.

EAVES CORNICE

D.P.C

FLOOR OF CISTERN ROOM.

6x2" JOISTS

EAVES CORNICE.

4x2" RAFTERS.

4x3" WALL PLATE.

FOR DETAIL OF CORBEL SEE DETAIL Nº:9 PART ONE.

9x3" PURLIN

4x2" RAFTERS

11"x2" VALLEY RAFTER.

4x3" WALL PLATE.

SECTION THROUGH ROOF OVER CISTERN AND BATH ROOMS. AT C-C ON DETAIL Nº:98.

4x2" SPROCKET.

ATTIC FLOOR LEVEL

FLOORING

4x3" WALL PLATES
9x2" JOISTS

EAVES CORNICE

D.P.C

7' 5⅜"

SECTION ACROSS FRONT GABLE ROOF AT E-E ON DETAIL Nº:98.

SCALE OF FEET.

1   0   1   2   3   4   5   6   7   8   9   10   11   12.

is entered from the intermediate landing of the attic stair—see general plans—and has a level 2′ 11″ below the general level of the attic floor.

Detail No. 102 includes a section through the cistern room which shows this difference in level and also the position of the ceiling joists over the smaller bedroom—No. 4 on the general plan.

The eaves detail is similar to that of the main front eaves, except that the wall plate is differently placed to suit the position of the floor joists.

In the left hand upper corner of the detail is a section through the head of this roof showing its junction with the flat roof beyond the 9″ division wall. Support for the trimmer at the curb is obtained from the heads of the rafters, which are birds-mouth jointed upon the 4″ × 3″ wall plate for their own bearing and then pass forward 4″ within the attic space and again birds-mouth notched to receive the 7″ × 4″ curb.

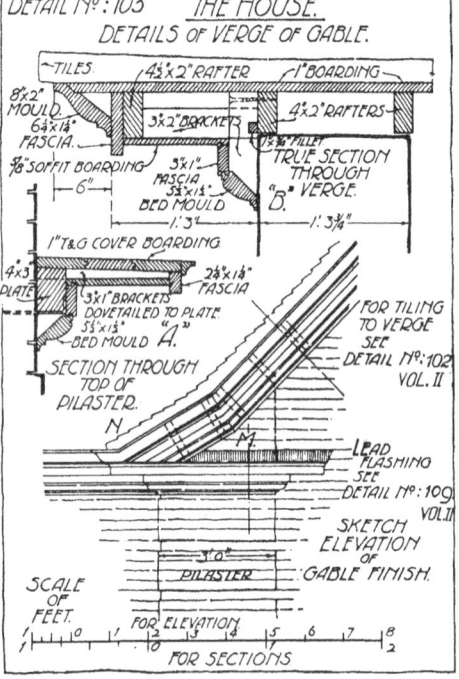

316. Front gable. The lower part of detail No. 102 shows the cross section through the roof of the front gable; 4″ × 2″ common rafters are supported upon 4″ × 3″ wall plates, 9″ × 3″ purlins and a 9″ × 2″ ridge, and the detailed construction of the eaves is identical with that of the main front eaves.

In this section may be seen the two 11″ × 2″ valley rafters in elevation behind the section plane, and the 9″ × 3″ purlins abutting thereon. The valley rafters really recede from the lower to the upper ends, hence the purlins meet them in vertical planes and have plain bevelled joints because the depth of the purlin only slightly exceeds the depth of the valley rafter.

In detail No. 103 the method of finishing the angle of the front gable is illustrated. It consists of a raking wooden cornice of a similar character to the one adopted for the main eaves, with the raking moulds terminating over the pilasters formed by the projecting quoins, which are themselves crowned by a short return portion of the eaves cornice continued across the face to surmount the pilasters.

317. Construction of capitals to the pilasters and verges to the gable. The eaves cornice, with the exception of the gutter, is mitred round the front external angle, carried across the face of the pilaster and returned at 90° to the wall.

At the section A, which is taken through the capital of the pilaster, the method of supporting and covering this feature is shown. (The sheet lead protection to the weathered top has already been shown in Vol. II.) To support the bed mould a $4'' \times 3''$ plate is secured to the wall with $2'' \times \frac{1}{4}''$ wrought iron brackets built-in in cement mortar. $3'' \times 1''$ wood brackets are dovetailed flatwise to the plate, and the mouldings and cover secured as shown.

The raking cornice under the verge of the gable has a profile similar to the eaves cornice and is constructed as shown in section at B in the same detail; the crowning mould of this portion corresponds to the outline of the eaves gutter, while the cradling is rearranged to suit the rafters which run parallel to the cornice. The rafters near the verge are arranged one on each side of the gable wall, and also an overhanging rafter which is 14" from face of brickwork, and without the usual supports except at the ridge. Near the foot it may be secured to the wall plate by using a $2'' \times \frac{3}{8}''$ wrought iron angle plate, which is screwed to the wall plate and behind the overhanging rafter.

For intermediate support, wooden brackets of $3'' \times 2''$ material are secured by screws to the first wall rafter and the gable rafter is spiked to the projecting ends. These brackets also serve to receive the wood boxing and moulds for the cornice. A fillet, $1'' \times \frac{3}{4}''$, fixed to the rafter enables accurate setting up of the parts to be done and gives some support to the load.

At the foot of the cornice the pitch of the sprockets must be followed, and a mitre, to change the direction, is therefore necessary to the moulding at M; a bracket is required at each side of this joint to allow the ends of the two pieces to be fixed. The bed mould, soffit, and fascia of the cornice are carried forward at the sprocket pitch to within $\frac{1}{4}''$ of the lead covering over the pilaster cap, while the cyma recta crowning mould is continued and again mitred at N to stop on the level behind the short return piece of the cast iron gutter, as shown in the elevation.

318. Support to feet of hip rafters. Hip rafters bearing upon the quoin walling of a building require some means of spreading the

load over an adequate area about the angle and some safeguard
against disturbance of the walling which might be caused by the
inclined thrust from the rafter.

The most common method is to halve the wall plates together at
the angle, allowing the ends to continue as far as possible to provide
resistance to thrust. An angle tie is placed at 45° across the two
wall plates and a short beam for receiving and supporting the hip
is carried from the centre of the angle tie to the intersection of the
wall plates.

Detail No. 104 shows the method. It may be varied to suit
circumstances, but the detail may be taken as typical of the general
principles of construction. (*The method shown is adopted only in the
best construction. For ordinary house property, notched and nailed
joints are usual.*)

The wall plates are $4'' \times 3''$, halved and overlapped to a distance
of $4''$ beyond the joint. A $6'' \times 3''$ angle tie is placed at a distance of
$1' \, 10\frac{1}{2}''$ from the angle—this depending upon the position and size
of the hip rafter—with the ends dovetail notched across the plates.
A $6'' \times 3''$ short "dragon beam" is tusk-tenoned to the angle tie
and secured by a wedge, the opposite end being notched over the
wall plates.

To provide for load bearing and the thrust of the hip rafter upon
the dragon beam, a single abutment and tenon joint is employed
similar to that used for the junction of the principal rafter and tie
beam of a roof truss. The purpose of this rather intricate construction is to relieve the angle of the walls entirely from side thrust.

319. Junction of curbs and hips. At the angle of intersection of
the $11'' \times 5\frac{1}{2}''$ main curb, the $7'' \times 4''$ curb trimmer and the $11'' \times 2''$
hip rafter, the junction may be made as shown in detail No. 105.

The larger curb passes forward upon the $9''$ wall and is cut to
a bevel to avoid interference with the slope of the side roof. This
arrangement provides for the curb trimmer to be supported by the
main curb and is thus independent of the support from the rafter
heads; these latter, if necessary, may be cut off thus saving the
labour of forming the second series of notches.

320. Purlin joints at hip and valley rafters. When purlins intersect at a hip or valley, they are usually splayed against the
vertical face of the rafter and spiked or bolted thereto. A satisfactory finish for an open roof is not obtained by this treatment unless
the hip or valley rafter is deep enough to receive the whole of the
splayed end of the joint and in average cases it will be found that
the purlins project below the rafter and produce an ugly finish.
This condition may be immaterial in places hidden from view, so

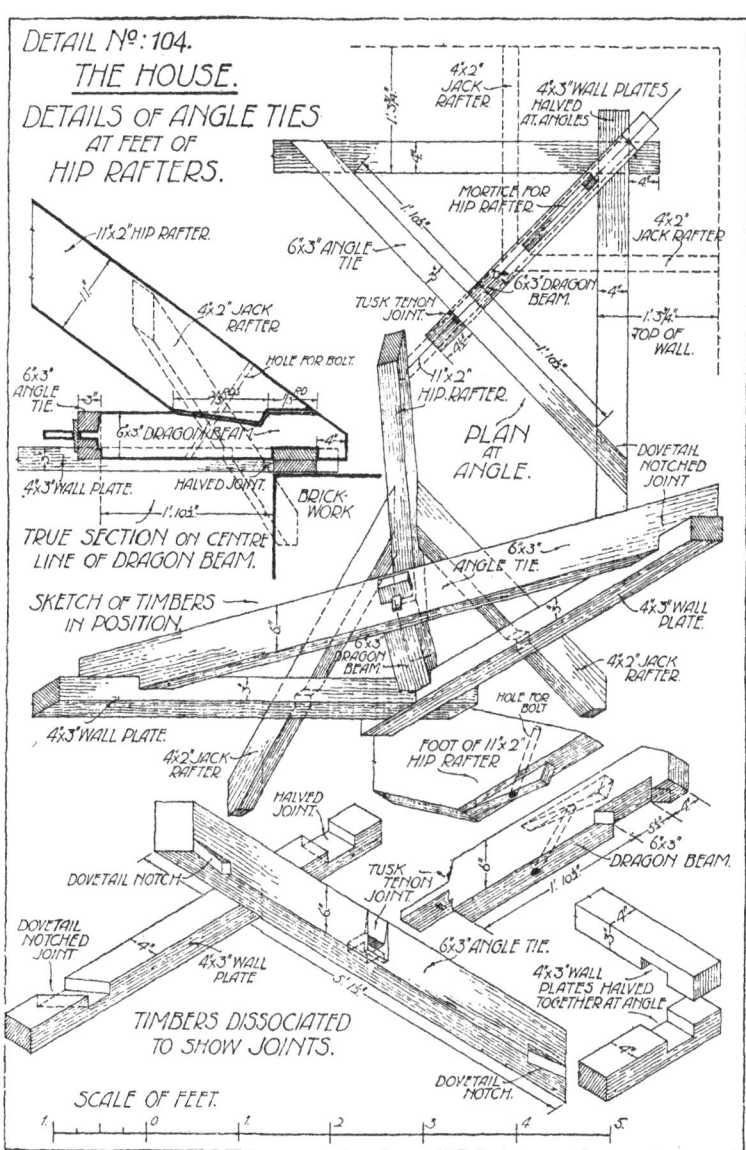

DETAIL Nº: 104.

THE HOUSE.

DETAILS of ANGLE TIES
AT FEET OF
HIP RAFTERS.

11"x2" HIP RAFTER.

4"x2" JACK RAFTER

6"x3" ANGLE TIE

6"x3" DRAGON BEAM

4"x3" WALL PLATE.   HALVED JOINT.   BRICK-WORK

TRUE SECTION on CENTRE
LINE of DRAGON BEAM.

SKETCH of TIMBERS
IN POSITION.

4"x3" WALL PLATE.

4"x2" JACK RAFTER

HALVED JOINT.

DOVETAIL NOTCH.

DOVETAIL NOTCHED JOINT

4"x3" WALL PLATE

TIMBERS DISSOCIATED
TO SHOW JOINTS.

4"x2" JACK RAFTER

4"x3" WALL PLATES
HALVED AT ANGLES

MORTICE FOR HIP RAFTER

TUSK TENON JOINT.

6"x3" DRAGON BEAM.

4"x2" JACK RAFTER

11"x2" HIP RAFTER.

PLAN
AT
ANGLE.

TOP OF WALL.

DOVETAIL NOTCHED JOINT

6"x3" ANGLE TIE.

4"x3" WALL PLATE.

4"x2" JACK RAFTER.

HOLE FOR BOLT

FOOT OF 11"x2"
HIP RAFTER

TUSK TENON JOINT.

6"x3" DRAGON BEAM.

6"x3" ANGLE TIE.

4"x3" WALL PLATES HALVED
TOGETHER AT ANGLE

SCALE OF FEET.

long as the joint is strong enough, but in many cases a better arrangement is desirable.

Detail No. 105 shows the better method. A splayed notch is taken out of each purlin on the top edge, to the correct bevel and half the thickness of the hip or valley rafter, the remaining projection on the lower part of the joint being passed beneath the rafter and mitred to the corresponding purlin on the opposite side. When the purlins are deep enough, the mitred parts may be bolted through below the rafter and the upper splay joints spiked to the face of the rafter from each side. In the example there is insufficient rafter projection of the purlins to allow of the bolt being placed below the rafter. In most hip and valley work the upper surfaces of the angle rafters are placed approximately flush with the upper surfaces of the spars. Hip rafters should, strictly, have their top edges worked to an inverted **V** section in order that half of their upper surface should lie in each roof plane; but with

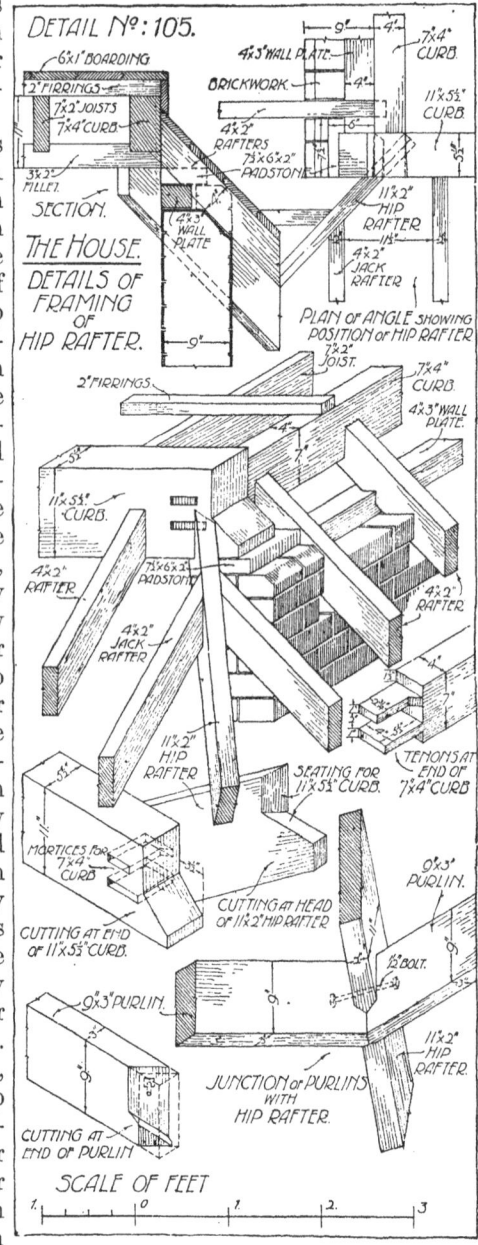

DETAIL Nº: 105.

SECTION.
THE HOUSE.
DETAILS OF FRAMING OF HIP RAFTER.

PLAN OF ANGLE SHOWING POSITION OF HIP RAFTER.

JUNCTION OF PURLINS WITH HIP RAFTER.

SCALE OF FEET

thin hip or valley rafters, this is usually dispensed with and the outer edges of the square top made to lie in the plane of the spars. In a similar way valleys need not be sunk to the V section, but have usually the centre line of their square top edge in the same plane as the two roof surfaces, thus forming their line of intersection. If a purlin is jointed to two angle rafters and bears upon one or more intermediate walls, as at X in detail No. 98 A, the purlin acts as a double cantilever and provides assistance in stiffening and supporting the rafters. Such assistance should be secured wherever the opportunity is provided, as for example by continuing a purlin beyond a freely supported bay of the roof at one or both ends to intersect at a hip or valley, if the length required for the purpose is not excessive nor impracticable.

321. Substitutes for purlins in roof construction. In Vol. i a purlin has been defined as a horizontal beam intended to provide intermediate support to common rafters. It may derive its own support from walls, roof trusses or angle rafters, as seen in previous studies.

It sometimes occurs that the ordinary purlin may be omitted, as shown in detail No. 101 at D, by utilising a row of vertical studs where such happen to be required to form a partition, as occurs in the attic bedroom of the example. In this instance the studding is primarily required to cut off the useless eaves angle of the roof space and its position allows it to be employed as a substitute for a purlin.

The selection of the mode of jointing depends upon the stiffness of the floor which is, or should be, made to provide support for the loaded studding.

If the floor is rigid the joints illustrated at D are very satisfactory, but if the floor is somewhat lacking in stiffness, the studs are better nailed to the sides of rafters and joists, so that they resist any tension which may occur when the joists sag.

CHAPTER THIRTEEN

TEMPORARY CARPENTRY

TIMBERING, CENTERING AND SHORING

322. While details of the temporary support of structures must, in some measure, be left to the discretion of the builder in order to avoid unnecessary expense, many conditions occur in constructional work which render such temporary support of vital importance. It is therefore felt that the person responsible for the design of the structure should be capable of supervising and, if necessary, designing such work. The temporary carpentry work included is suited to the purpose of this volume, and more especially for the use of the general building student.

TIMBERING OF EXCAVATIONS

In Vol. I were given methods of timbering shallow trenches in firm, loose and soft soils. The methods there discussed are only suitable for trenches not exceeding 3 ft. in depth and no difficulty occurred in throwing the excavated material out of the cutting.

323. Deep trenches. When the depth of an excavation for the foundations or drainage of a building exceeds 6 ft., some arrangement must be made for the removal of the earth, as a man can only lift and throw—with difficulty—to this height. The earth must be either hoisted directly from the bottom of the trench or platforms must be provided at suitable levels, not more than 5′ 6″ apart, so that the earth may be spade lifted at two or more "throws".

324. Drain trench in fairly firm ground. If all the earth is dry and fairly firm, the example of detail No. 106 will be a suitable method of timbering. The ground is assumed to consist of 1′ 6″ of top soil, 5 ft. of gravel and sand, and the rest of the depth of fairly dry clay. Let the trench be 13 ft. deep and splayed from 4 ft. wide at the top to 3 ft. at the base, this splay or "batter" being a safeguard against struts working loose as the surrounding soil loses its moisture and contracts on being exposed to the air, and at the same time giving greater freedom to the excavator.

In this detail the timbering is erected in three heights of approximately 4 ft. each, using poling boards, walings and struts as shown in the example in Vol. I; each height is treated in the same manner but with the poling boards placed closer at the edges at each lower stage to meet the greater pressure due to the height of earth supported.

One difference only occurs between this example and the more elementary case, viz. two rows of walings and struts are employed,

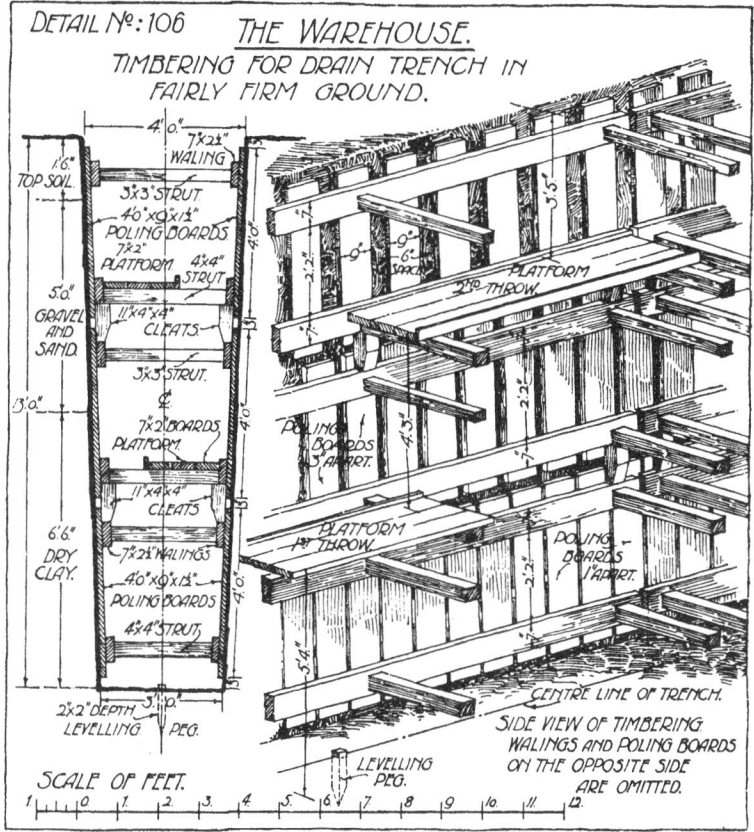

placed near the top and bottom ends of each set of poling boards, because the latter are too long for one series; two such rows are required where the poling boards exceed 2′ 6″ in length.

325. Platforms. Between the throws, or lifts—of which there are necessarily three—a platform is required, on which the material is deposited from the lower stage and re-thrown a stage higher by another workman, and so on until the surface is reached. The last throw should be less than the normal, to allow the earth to be cast well away from the edge of the excavation.

The platforms are constructed of 2″ battens, laid across the struts, which are tightly driven and further supported by wooden cleats nailed beneath the end of the strut and the edge of the waling, and lapping across the ends of two sets of poling boards, as shown on the section. Platforms are arranged in various ways to suit the conditions of the excavation and may be either the full breadth of the trench or a suitable part of it, but generally alternated in position in the direction of the length of the trench.

This is shown in the same detail, where the platforms are not the full width of the excavation and are placed in alternate positions both in the width and length of the cutting, thus ensuring greater ease for the excavators. Where the edges of the platforms are exposed they should have a guard piece at each free edge to prevent the falling of material during the process of shovelling.

326. Dimensions of timbers and methods of placing. The timbers employed are: poling boards, $9″ \times 1\frac{1}{2}″$; waling pieces, $7″ \times 2\frac{1}{2}″$; struts, from $3″ \times 3″$ in the upper part of the trench to $4″ \times 4″$ in the lower parts where the thrust is greater and platforms have also to be supported.

These timbers are placed in position gradually; at first, single struts are used to each pair of poling boards, being placed at the centre of their length, until sufficient have been inserted to allow a pair of waling pieces and the necessary struts to be placed at the upper and lower ends of the series. These are inserted and in the process of tightening the temporary struts fall away and are removed.

327. Loose soil. When the earth is deep and loose, but not wet, it may be necessary to arrange the timbering in another way, which is illustrated in detail No. 107. In this instance we have assumed 12″ of top soil, 4 ft. of loamy soil and 8 ft. of dry, loose sand. Such a case would serve with the same support as that of detail No. 106 at the top of the excavation, but lower down would require more continuous support. Thus, for the second stage, horizontal sheeting boards are adopted, of $9″ \times 1\frac{1}{2}″$ material, which would be inserted row by row, in pairs, and strutted lightly until the series of five boards could be clasped by one $7″ \times 3″$ vertical poling board having $4″ \times 4″$ struts at top and bottom, as shown.

To prevent the upper timber slipping, a different method from that previously described is employed, the poling boards extending upwards to the edges of the lower walings of the top stage.

The lower part of the excavation is 6″ narrower than the upper one, to allow the lower and middle series of poling boards to overlap. By this means and a few spikes, very effective tie between the stages, and support from stage to stage, can be obtained.

Although not emphasised in the sketch, the lower stage would be slightly battered inwards.

Platforms are arranged as stated in paragraph 325.

328. Levelling pegs. To determine the exact levels or falls for the bottoms of trenches, 2″ wooden pegs are driven in at intervals

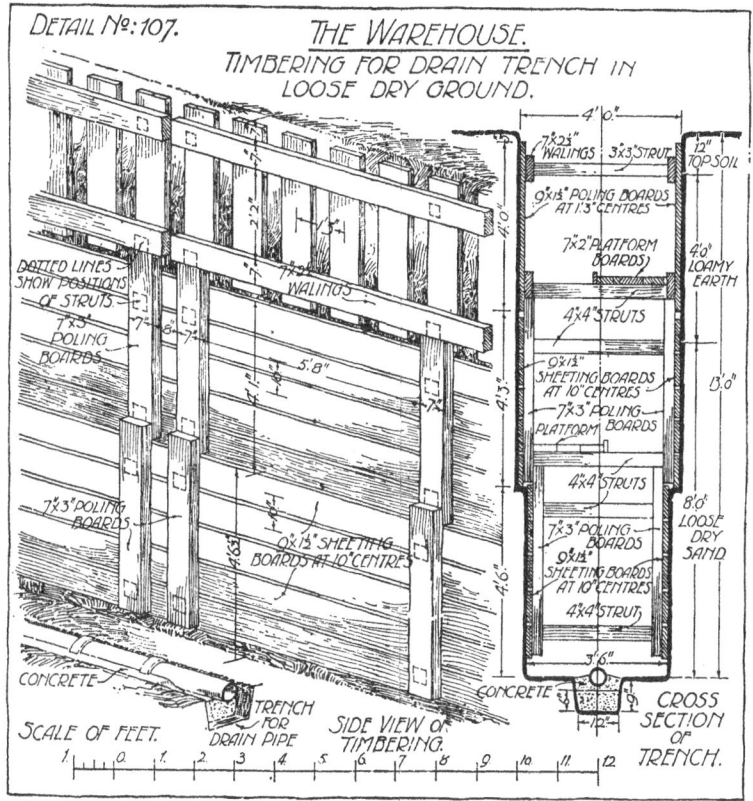

of 10 ft., or less, the tops being carefully adjusted to the required position and then utilised to determine heights of foundation details. A similar method is used to adjust the inclination of the drains; see paragraph 495 in the chapter on Drainage.

329. Stepped excavations. Excavations often have to be arranged with stepped bottoms; they may be either shallow or deep, and the general method of timbering would apply as previously illustrated and described. One example of a stepped excavation is shown suitable for the party wall of the warehouse which is situated upon falling ground.

The section and isometric view of the timbering shown in detail No. 108 explain the disposition of the supports at one of the steps in the foundation, this step being approximately 2′ 9″ deep.

To avoid interference with the supports when the concrete is being placed, a space clear of timbers is left, 2 ft. deep. This would only be possible in fairly stiff soils. In other cases rows of horizontal sheeting in short lengths would be employed and single lengths removed as the filling reaches them in length and in height. In some extreme cases the side supports have to be left in position.

Poling boards 3 to 4 ft. long are employed, these varying in length as required so that not more than 6″ of unsupported earth is left above them. Two rows of 7″ × 3″ waling pieces, in 13 ft. lengths, are provided, with three struts to each row.

At the greater depth of excavation two tiers of poling boards are required. The maximum depth in this case is about 8 ft. and a throwing platform would be necessary, as described in paragraph 324, until the depth diminished to 5 or 6 ft.

330. Arrangement of poling boards. It is wise to utilise short lengths of poling boards wherever possible, 3 ft. being quite suitable for the general run of work, though in some cases greater lengths cannot be avoided. The object of using the short lengths is to enable support to be adequately provided at close stages as the cutting proceeds, and similarly to enable small sections of the timbering to be removed as the trench is filled by walling or earth, and thus avoid any large surfaces being left without support.

331. Spacing of poling boards and sheeting. The spacing of boards and sheeting would be determined by the nature of the soil and adjusted by the person in charge of the work. Pockets of soft material may call for close spacing at some points, while stiff clay would stand with little or no support.

332. Use of waste material. Provided that the timber is sufficiently strong for its purpose short lengths of scaffold poles or putlogs may be made suitable for struts, waste ends of flooring will serve for poling boards and short or waney pieces of joisting will make satisfactory walings. The dimensions in the details are quite arbitrary and are given only for the purpose of guiding the student in the selection of sizes.

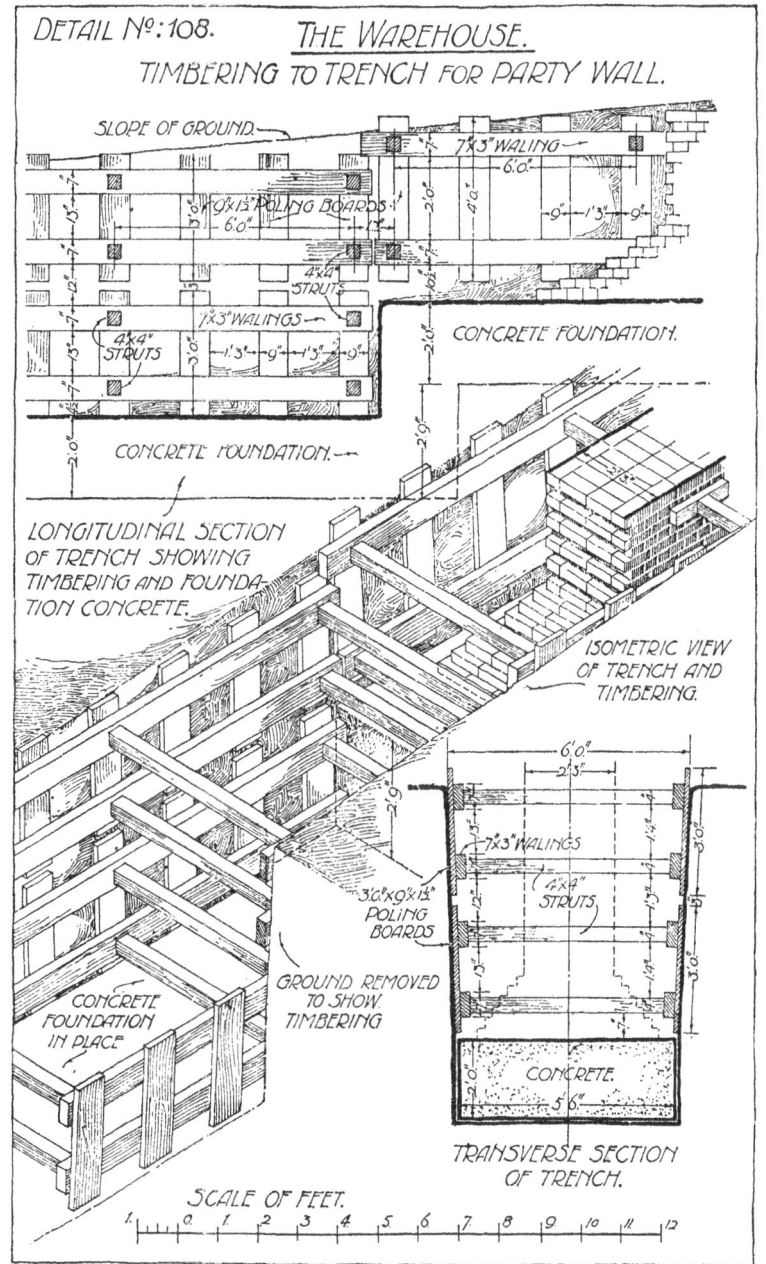

DETAIL Nº:108.   THE WAREHOUSE.
TIMBERING TO TRENCH FOR PARTY WALL.

SLOPE OF GROUND.

7×3" WALING
6'.0".

3'.0"×9×1½" POLING BOARDS
6'.0".

4×4"
STRUTS

7×3" WALINGS

4×4"
STRUTS

CONCRETE FOUNDATION.

CONCRETE FOUNDATION.

LONGITUDINAL SECTION
OF TRENCH SHOWING
TIMBERING AND FOUNDA-
TION CONCRETE.

ISOMETRIC VIEW
OF TRENCH AND
TIMBERING.

6'.0"
2'.3"

7×3" WALINGS

4×4"
STRUTS

3'.0"×9×1½"
POLING
BOARDS

GROUND REMOVED
TO SHOW.
TIMBERING

CONCRETE
FOUNDATION
IN PLACE.

CONCRETE.
5'.6"

TRANSVERSE SECTION
OF TRENCH.

SCALE OF FEET.
1. 0 1 2 3 4 5 6 7 8 9 10 11 12

CENTERING

Centering for arches of simple form and small span were illustrated in Vols. I and II. The following paragraphs illustrate the construction and principles of designing centers for elliptical arches and other arches of larger span.

333. Semi-elliptical center. The setting out and the construction of a semi-elliptical arch have already been shown in Vol. II. These were applied to the opening from the ground floor to the loading shed of the warehouse.

Detail No. 109 shows a method of constructing the center for this arch. The span is 9′ 9″ and the wall is 18″ thick, requiring two built-up ribs, spaced at about 12″ centres, connected with $2'' \times 1\frac{1}{4}''$ laggings 18″ long and placed approximately one to each brick.

To hold the ribs in position while lagging, a $7'' \times 2''$ seating piece is nailed across the ties at each end and a $4'' \times 1''$ cross stay near the heads of the centre posts. Seating pieces are similarly required for all centers, being laid immediately upon the folding wedges in pairs upon the cross bearers, which in turn are supported by the posts.

The arc, or bow, is constructed of two-ply material, $1\frac{1}{4}''$ thick and 6″ to 9″ wide, the segments being overlapped and bolted with two $\frac{1}{2}''$ bolts at every joint; segments should be at least 2′ 6″ long and 4″ minimum width at the ends.

A double tie is employed as before, but in this case no notching or other special jointing of the tie is necessary.

Examination of the detail will show that all notching is performed upon the struts, posts and braces, bare-faced[1] tenons being formed at the joints and where possible passed between double material spaced to avoid mortising.

At the head of the braces a $6'' \times 1''$ head tie or cover piece encloses the oblique bare-faced tenons.

334. Principle of trussing ribs for centers. There are many ways of arranging the rib framing to ensure the transmission of the load to the supports without distorting the shape of the frame. Detail No. 110 illustrates the chief methods of framing for spans not exceeding 10 ft. in the first example and 15 ft. in the second. If the second type is used for spans under 10 ft. the two dotted struts may be omitted.

The former type (marked A in the detail) and employed for several examples, acts exactly like a king-post roof truss, the bow replacing the rafters. When loaded gradually from the springing

---

[1] Bare-faced tenons are only cut and shouldered on one side of the material.

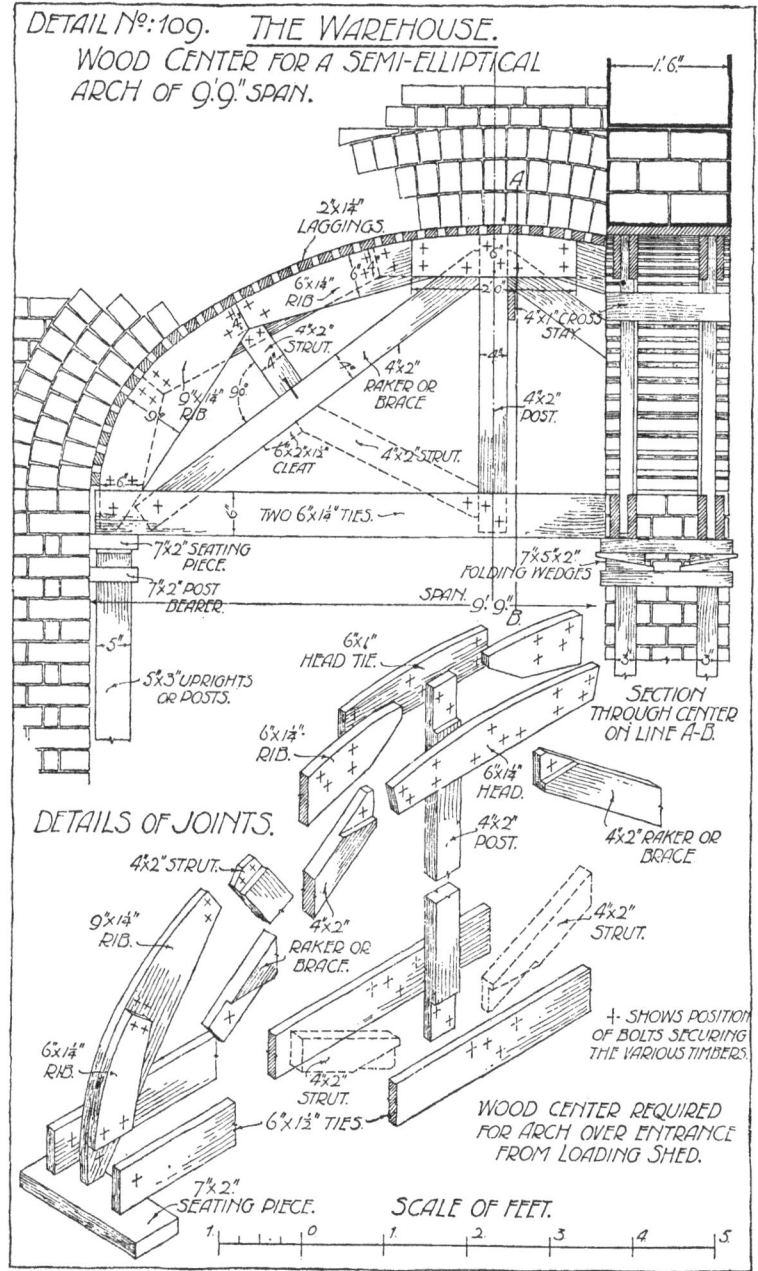

DETAIL Nº: 109.   THE WAREHOUSE.
WOOD CENTER FOR A SEMI-ELLIPTICAL
ARCH OF 9'.9" SPAN.

1'.6"

A

2"x1¼"
LAGGINGS.

6"x1¼"
RIB

6"x1¼"

2'.0"

4"x1"CROSS
STAY

4"x2"
STRUT.

4"x2"
RAKER OR
BRACE

4½"

4"x2"
POST.

9¾"x1¼"
RIB.

90°

6"x2½"
CLEAT

4"x2"STRUT.

TWO 6"x1¼" TIES.

7"x2" SEATING
PIECE.

7"x2" POST
BEARER.

7"x5"x2"
FOLDING WEDGES

SPAN   9'.9".
B.

5".

5"x3"UPRIGHTS
OR POSTS.

6"x1"
HEAD TIE.

SECTION
THROUGH CENTER
ON LINE A-B.

6"x1¼"
RIB.

6"x1¼"
HEAD.

4"x2"
POST.

4"x2"RAKER OR
BRACE

DETAILS OF JOINTS.

4"x2" STRUT.

4"x2"
STRUT.

9"x1¼"
RIB.

4"x2"
RAKER OR
BRACE.

+. SHOWS POSITION
OF BOLTS SECURING
THE VARIOUS TIMBERS.

6"x1¼"
RIB.

4"x2"
STRUT.

WOOD CENTER REQUIRED
FOR ARCH OVER ENTRANCE
FROM LOADING SHED.

6"x1½" TIES.

7"x2"
SEATING PIECE.

SCALE OF FEET.

1   0   1   2   3   4   5

upwards, which always occurs in construction, the bow tends to distort and conveys load through the struts to the foot of the post; as the bow flattens under the load at the haunches, it tends to rise at the crown but is restrained by the post if properly jointed; as the erection of the arch proceeds load is accumulated over the crown and the post is somewhat relieved of tension but is always subject to tension from the downward thrust of the struts of a greater amount than the compression caused by the load on the head of the post. For this reason a third support with folding wedges is often placed at the centre of the span should the latter exceed 10 ft.

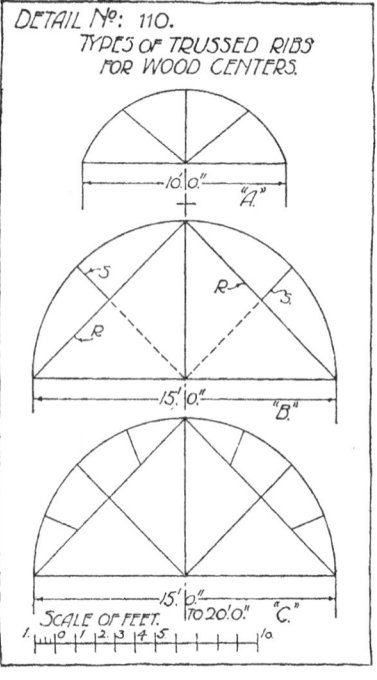

DETAIL No: 110.
TYPES OF TRUSSED RIBS
FOR WOOD CENTERS.

The second type (marked B in the detail) and employed in the example of detail No. 109, is a combination of raking beam and trussed frame. Omitting the dotted members it may be considered as a triangulated frame like a simple roof truss with the rafters acting as beams to carry load transmitted to them from the short struts, SS, which in their turn receive it from the bow. The rakers (or braces), RR, thus tend to deflect and must be rigid enough to resist appreciable deflection. A better arrangement is to use lighter rakers and to insert the inclined strut, shown dotted, on each side, which makes the transmission of load direct and produces a perfect frame.

The principle may be extended to centers of larger span, as illustrated at C in the detail, by merely adding short struts between the rakers and the bow as the unsupported length of the latter increases with the span.

335. Necessity for sound jointing of centers. A consideration of the stresses in the supporting ribs of a center, especially if verified by experiment, will show the necessity for making the joints capable of resisting either tension or compression, because both kinds of stress may act at different times during the process of erecting an arch as the load accumulates from the springing, past the haunches to the crown. Bolted joints, as shown in the above details, accomplish this object, but in some larger centers iron plates may become necessary.

SHUTTERING FOR CONCRETE FLOORS

336. For many types of concrete and other fire-resisting floors, temporary platforms are required to support the concrete, or the units of construction, until set.

There are two general methods available:

(*a*) to suspend this shuttering from the floor beams;

(*b*) to support the shuttering by posts and bearers from the floor below.

The selection depends upon the conditions and nature of the work.

337. Shuttering for concrete floors of lavatories. In Vol. II the construction of these floors was studied. Reference may be made to these details but the present purpose will be served if it is noted that two alternative forms of beam are employed.

In the upper detail of No. 111, in which steel **I** beams are used, $\frac{1}{2}''$ wrought iron hangers or dogs are suspended by wrought iron plates and nuts from the beams. These form a loop for the reception of $3'' \times 2''$ sheeting bearers upon which the boards of the platform are supported.

In the lower detail, in which steel double-channels are used,

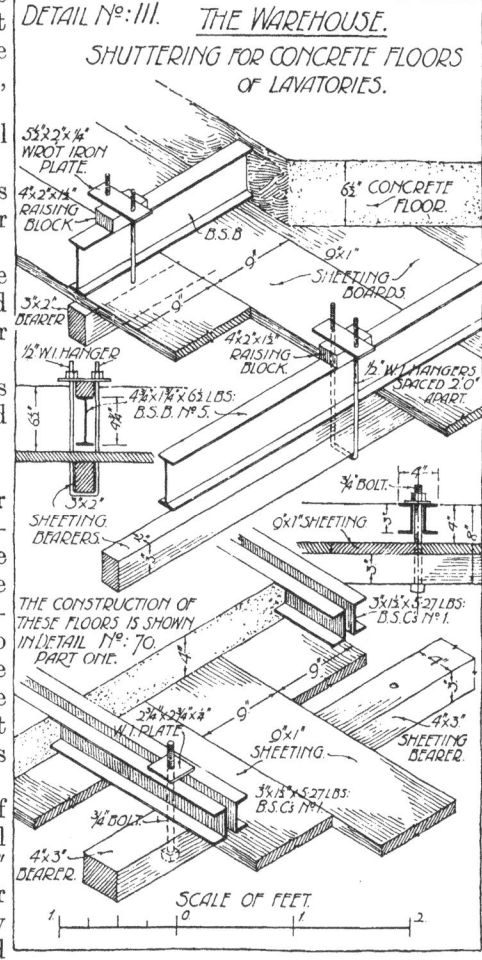

DETAIL N°: III.   THE WAREHOUSE.
SHUTTERING FOR CONCRETE FLOORS
OF LAVATORIES.

SCALE OF FEET.

ordinary $\frac{3}{4}''$ bolts are passed through $4'' \times 3''$ wood bearers and between the beams; they bear upon large wrought iron plates or washers and are secured with nuts. The bearers support the $9'' \times 1''$ sheeting as in the previous example.

This form of shuttering leaves the floor below free for other activities.

338. Shuttering to hall floor of house. Detail No. 112 shows the second type of shuttering applied to the floor over the basement of the house.

It is a simple construction consisting of a platform to each of the bays of the floor, supported upon $4'' \times 2''$ bearers, which in turn rest upon $4'' \times 3''$ cross heads and posts, these being braced diagonally with $6'' \times 1''$ boards to form frames or standards. These standards are placed in series as at A, B and C, and connected by diagonal bracing to obtain a rigid structure.

The posts rest upon sole plates except when bearing upon a hard surface.

The cross heads are made adjustable to levels by pairs of folding wedges placed between heads and posts.

On the outer faces of the bearers in each bay $2'' \times 1\frac{1}{2}''$ fillets are nailed at the top edge to form a casing, which provides a recess for the formation of the small concrete haunch at each side of the steel beam.

The key plan shows the position of the standards for the several bays of the floor.

339. Shuttering for composition walling. When walls have to be raised in a semi-plastic material such as wet chalk, clay compounds, and concrete, some kind of adjustable forms or shuttering becomes necessary.

Detail No. 113 shows a type of shuttering successfully employed in the erection of composition walling for cottages at Amesbury in Wiltshire.

The sides consist of $1\frac{1}{2}''$ sheeting boards, fixed together by $6'' \times 1\frac{1}{2}''$ battens; the forms are clamped to the top edge of the wall by straining wires bearing upon iron plates and twisted together at the loose ends to obtain a grip. The top edges are kept at the correct thickness spacing by rebated struts, dropped loosely into notches at the heads of the vertical battens. These forms were made in lengths suited to the work and connected end to end by a tongued overlap as shown at A. The joint formed by the projecting ends of the sheeting was secured by long hardwood wedges passed through iron staples, which ensured the correct lining of the boards to obtain a continuous plane face.

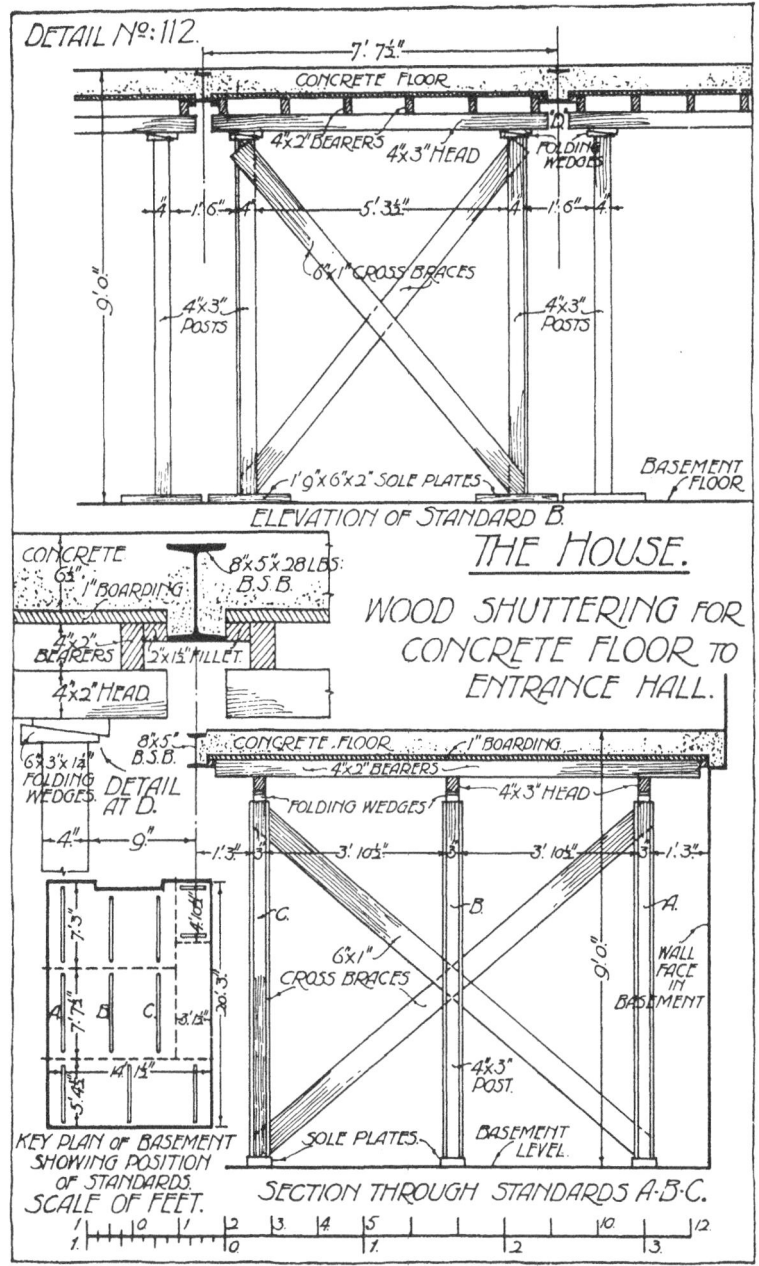

DETAIL Nº: 112.

7' 7½"

CONCRETE FLOOR

4"x2" BEARERS   4"x3" HEAD   FOLDING WEDGES

4"   1' 6"   4"   5' 3½"   4"   1' 6"   4"

9' 0"

4"x3" POSTS   6"x1" CROSS BRACES   4"x3" POSTS

1' 9"x6"x2" SOLE PLATES   BASEMENT FLOOR

ELEVATION OF STANDARD B.

CONCRETE 6½"   1" BOARDING   8"x5"x28 LBS. B.S.B.

4"x2" BEARERS   2"x1½" FILLET

4"x2" HEAD.

8"x5" B.S.B.   6"x3"x1½" FOLDING WEDGES   DETAIL AT D.

THE HOUSE.

WOOD SHUTTERING FOR CONCRETE FLOOR TO ENTRANCE HALL.

CONCRETE FLOOR   1" BOARDING
4"x2" BEARERS
FOLDING WEDGES   4"x3" HEAD

4"   9"   1' 3"   3"   3' 10½"   3"   3' 10½"   3"   1' 3"

C.   B.   A.

6"x1" CROSS BRACES   9' 0"   WALL FACE IN BASEMENT

4"x3" POST.

SOLE PLATES   BASEMENT LEVEL.

KEY PLAN OF BASEMENT SHOWING POSITION OF STANDARDS.
SCALE OF FEET.

7' 3"   20' 3"   18' 1½"

A   7' 7½"   B   C   13' 1½"

5' 4½"   14' 1½"

SECTION THROUGH STANDARDS A·B·C.

To remove the forms, the straining wires were severed, the sheeting removed and the ends of the wires cut flush with the face of the work and the strand left embedded in the wall.

A special arrangement of the forms was adopted for the construction of angles, as shown to the left of detail No. 113, and providing for a rounded external angle.

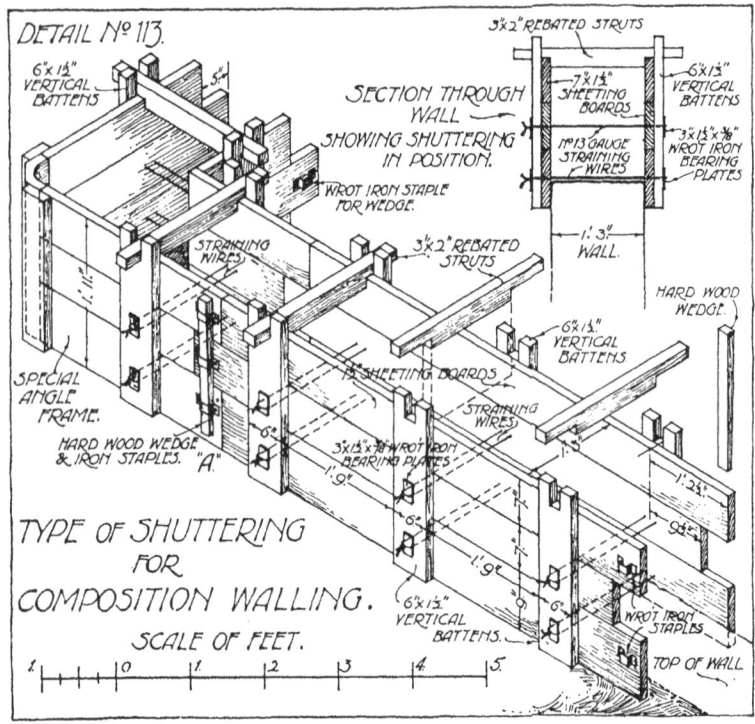

By the employment of one angle frame and a few lengths of plain shuttering, the walls were kept continuously in progress, the earlier deposits being sufficiently set to allow removal of the forms by the time the last deposits were made.

SHORING

340. Shoring is temporary framing providing support to a structure which shows some signs of instability, or which is liable to disturbance in consequence of alterations to surrounding and attached property, or to the removal of some of its support for repairs. Most shoring is framed in large timber.

341. Classification of shores. Shores in ordinary use may be divided into three classes, viz. dead, raking and flying shores. They may be applied singly or in combination according to the nature of the support required.

Dead shores are vertical posts supporting horizontal beams upon which the load is borne and transmitted to the shores. A dead shore carries vertical load only and is commonly employed to support the walls of a building, in addition to their floor loads, while foundations are improved or openings formed in the walls.

Raking shores are large timbers set in an inclined position against a structure and act as struts to prevent overturning, but should not be employed with the intention of forcing a wall back to its original position.

Flying shores are struts or framing placed between two structures to provide mutual support against outward movement; such shores receive no support from the ground.

342. Raking shores. When a building shows some sign of movement by an external wall heeling over, due to unequal settlement of the foundations, or bulging due to eccentric loads from floors or roof, the defect may be prevented from increasing by placing inclined struts, called raking shores, in suitable positions, from the ground to the face of the wall, and securing them to prevent movement. This method is suitable for property in wide streets or where a sufficient spread of the shores can be obtained without undue interference with street traffic.

If two or more shores are required a "system" is built up on a general principle, which, with slight modifications due to the height of the building and the ground space available, is the same for any number of raking timbers.

343. System of shores. Detail No. 114 shows a system of three shores placed against the walls of a four-storey building, an old building having to be removed for the erection of a new structure, as suggested by the sketch in detail No. 115. Such shores are particularly necessary near the angles of buildings where side support is to be withdrawn.

It is obvious that a shore which is nearly vertical would effectively support dead weight, but would have very little effect by its own weight in retaining the damaged wall, while on the other hand a shore approaching the horizontal would sag unduly, become impracticably long, and encroach uselessly upon ground space. If a single raker be required it is found that an angle of 60° is suitable but the practical conditions to be met are so variable that set inclinations can scarcely be entertained.

Probably the best practical arrangement is to spread the foot of the shore as far outwards as the footpath space allows, unless excessive; or, if the available width is insufficient, as far into the roadway as may be necessary or as approved by the Local Authority.

To provide for each shore effecting its purpose as a strut, resistance to thrust must be efficiently met at both ends. This is done by abutting the head of each shore upon a wall piece and against a needle or stop, which passes through the wall piece into the wall, while the foot is supported and secured upon a sole piece partially embedded in the ground.

The functions of the parts referred to are:

Wall piece; (a) to form a seating for the head of the shore and thus distribute the thrust over a considerable area of wall; (b) to receive the needle which, by its shape and the thrust against it, binds the wall piece to the wall; (c) to connect the heads of the shores in series where two or more rakers are employed.

Needle; to resist the thrust of the head of the shore and prevent slipping, while at the same time transmitting the thrust, through the wall piece, to the wall.

Sole piece; to receive the

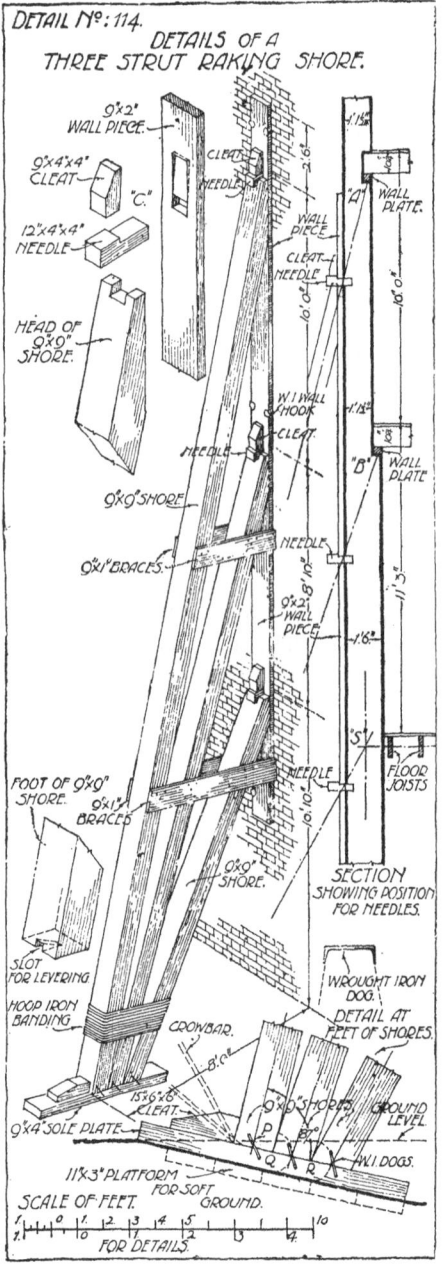

DETAIL Nº: 114.

DETAILS OF A THREE STRUT RAKING SHORE.

group of shores at the foot, provide for their support and prevent slipping, while allowing the shores to be tightened by levering them forward on its surface.

344. Position of shores. Considering the structure in detail it is necessary first to determine the most effective position for the shores. Having settled their maximum projection determine the position of the centre point P of the outer raker on the ground level, and direct the centre line so that it passes through the centre of base of the wall plate supporting the floor joists, as shown at A. Then place the sole piece in position as shown, with its top face making an internal angle less than 90° and sufficiently below the ground level to limit the projection of the top corner while giving ample room for the cleat. Complete the outline of the top raker about the centre line.

DETAIL Nº: 115.   SKETCH SHOWING A SUITABLE APPLICATION OF RAKING AND FLYING SHORES.

To place the second raker describe an arc, centre Q and radius equal to half its breadth; direct its centre line to support the wall plate at B and at the same time tangential to the arc. Complete the second raker.

To place the third raker perform the same operation with centre R, but directing the centre line tangential to the curve through a point S at the centre of depth of the floor and in the middle of the wall. The reason for this is, that in the upper floors the joists run at 90° to the front wall and derive their support from the *wall plate*, but in the first floor the joists are parallel to the wall and strutted tightly against it. The resultant of the thrust and dead weight of wall passes approximately through the point S, hence the position of the shore.

345. Construction of series of raking shores. Holes are first cut in the walls to receive the needles, care being taken not to shake the surrounding work; wall pieces are then set out, mortised for the needles and notched for cleats. The wall piece is initially secured in position by wrought iron wall hooks driven into the joints and clasping the edges of the plate. Wall pieces vary in size from $7'' \times 1\frac{1}{2}''$ to $11'' \times 2''$.

Needles are prepared from $4'' \times 4''$ material, notched down to $3''$ thick at the back, as shown in detail No. 114 at C, and placed in position through the mortices to a depth of $4\frac{1}{2}''$ within the wall. A bevelled cleat above the needle is fitted into the housing and spiked to the wall piece, thus providing a solid abutment for the shore.

The head of the shore is cut to the angle and notched so as to clasp the needle and prevent side movement during assemblage, or at a later stage should the shore accidentally become loose. The shoulder bearing upon the needle should be at least $3''$ wide.

The foot of the shore is cut to the required length and bevel so as to tighten up by sliding forward on the sole piece which is solidly bedded in position, as shown in the lower portion of the detail.

346. Tightening the raking shores. The rakers are tightened by levering them forward with a crowbar, as indicated in the detail, for which purpose a notch is cut on the outside at the foot of the shore to receive the bar. It is clear that the outside angle at the foot should be less than 90° so that forward movement tightens the raker.

When there are several rakers the base angle of the inner one often becomes too acute to prevent slipping. It is therefore levered into position and fixed by wrought iron dogs driven into the sides of sole piece and raker; it is wise to secure each raker in the same way as the outer ones are erected, and tightened upon the inner raker, and finally to secure the series by a rough cleat spiked to the sole piece. Violent wedging should be avoided when tightening, gentle levering being much more effective and attended with less danger of increasing existing defects.

347. Sole pieces and bearings. Sole pieces should be the same width as the shores in order to allow dogs to be driven in their sides. If the ground is soft and an ordinary sole piece—say $3' \times 9'' \times 4''$—is insufficient to provide solid bearing, a foundation of $3''$ planks may be placed transversely beneath the sole piece to form a bed for its reception, as shown in the detail by dotted lines.

The function of the sole piece is to distribute the thrust over sufficient ground to prevent penetration or collapse under the worst conditions in which the ground may be expected to exist, hence any steps appearing necessary for securing a good foundation bed should be taken, *e.g.* ramming to consolidate loose soil, or cutting out unsatisfactory material and replacing with dry hard rubbish.

348. Bracing of raking shores. After erecting and tightening the shores they should be braced as shown in the isometric detail of No. 114, the objects being to make the system into a unit, to stiffen the longer rakers by shortening the unsupported length and to reduce the practical effect of any one member becoming loose.

The braces consist of one or more 1″ boards, 6″ to 11″ wide, nailed across the series of rakers and to the sides of the wall pieces and approximately at 90° to the outer member. It is therefore convenient to make the wall piece of the same width as the shore.

Near the foot, the shores are braced or banded by similar boards or by bands of hoop-iron wound round the series and secured with clout nails. This assists in preventing disturbance and damage by traffic.

349. Flying or horizontal shores. Where two buildings in a narrow street are opposite each other, or a building has been removed from between two others, the opposite walls may be mutually supported by shores framed between the buildings. These are called flying shores because they have no ground support, or horizontal shores when their main members are horizontal.

Such shores, when applied in a street, leave a clear road for traffic, or a free ground space for new construction, where a building in a terrace has been removed for replacement by another building.

This latter case has been assumed in the example, hence the purpose of the shores is to prevent disturbance or damage to the party walls of the adjacent property while preparing new and perhaps deeper foundations for the new buildings.

Detail No. 115 shows the general position of the flying shores, which, as a rule, should not exceed 36 to 40 ft. span for any type of framing, nor more than 30 ft. for the simple form illustrated in the sketch, and fully detailed in No. 116.

Three points of support are provided to each building, by using a horizontal member at the intermediate level opposite a floor, and placing pairs of struts inclined upwards and downwards towards the centre of pressure from the floors above and below. These struts are virtually raking shores receiving support from the horizontal member instead of from the ground, though they are placed at a flatter angle and made to thrust more directly than an ordinary raking shore. Lighter timbers may therefore be employed in these positions as the strut exerts less direct force to produce the same horizontal resistance against overturning.

350. Construction of flying shores. A series of flying shores may be required to support a building, the spacing in vertical planes averaging 10 to 15 ft., while shores may be required in special

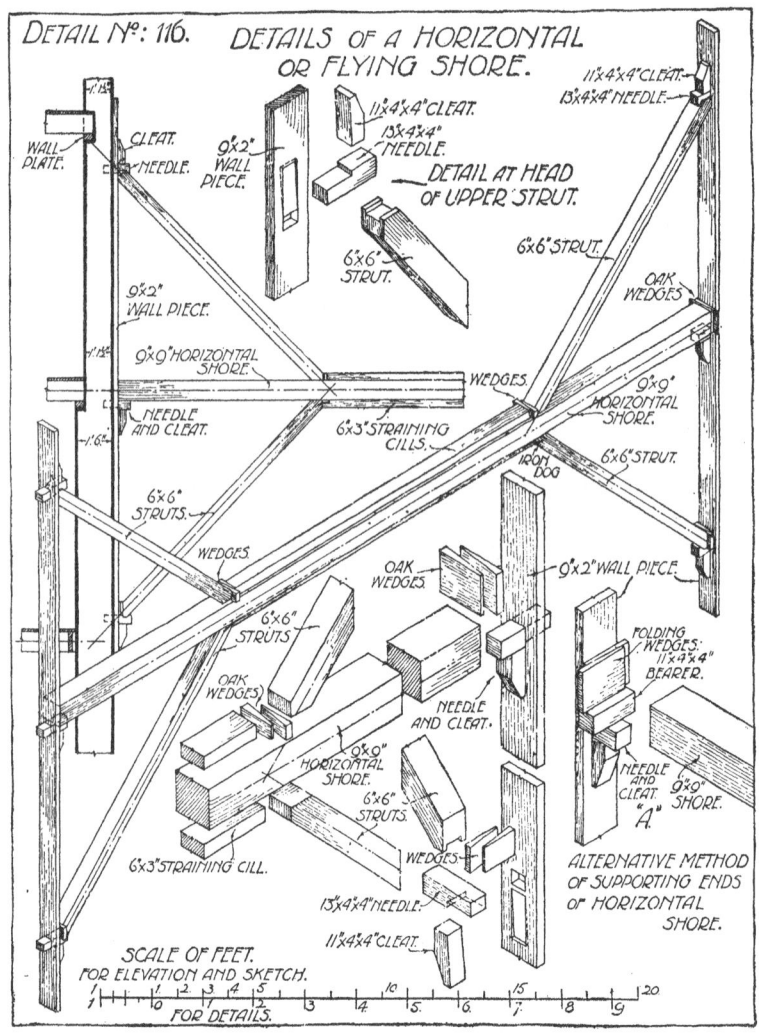

positions opposite division walls, to prevent disturbance of the party wall at the junction. Each system of shores is constructed as shown in the illustrations.

Holes for needles are cut, wall pieces prepared and fixed, needles and cleats inserted all as employed for raking shores. The horizontal member, $9'' \times 9''$, is then placed in position with the $6'' \times 3''$ straining cills nailed on the upper and lower surfaces; the shore is carefully wedged in position between the wall pieces while resting upon the needles or cross bearers shown at A.

The lower struts, $6'' \times 6''$, are then placed, notched over the top of the needles and secured by wrought iron dogs at the upper end, where they should fit closely. The upper struts are next fitted with folding wedges intervening between their lower ends and the straining cill; these wedges are carefully and gradually driven until the whole system is rigid, the lower struts becoming tight through the tendency of the horizontal member to deflect as the pressure is applied to the upper struts.

In some cases the lower struts are also wedged in order to lift the horizontal member and remove the natural sag due to its own weight. It is considered preferable in this case to place the wedges at the foot of the strut against the wall piece because easier to manipulate and less liable to fall out, though they are not so effective as when placed at the opposite end of the strut.

351. Wedging of shores. It should be observed here that wedging of shores is a delicate operation. The vibration due to sudden and violent blows may cause serious damage to a structure by increasing the disturbance which the shore was intended to arrest. The function of a shore is not to force a distributed part of a building back to its original position but to prevent *further* movement, or to improve stability where danger of an unstable condition may arise through alterations to a structure. Hence, if wedging is resorted to, it should be done under expert supervision, the wedges being accurately prepared and carefully driven until the object of wedging has been fulfilled, but not exceeded.

352. Timber used for shoring. The timber employed in shoring is chiefly northern pine or pitch pine, the latter being more particularly suitable for large and heavy shores because it is obtainable in large sizes and great lengths. For further particulars of timber see the chapter on Materials, Vol. II.

OTHER ITEMS OF TEMPORARY CARPENTRY

It is obvious that other items of temporary carpentry are indispensable in the erection of an important building and these include such features as storage sheds, mess rooms, temporary sanitary conveniences, hoardings, gantries over footpaths, derrick towers, temporary offices, scaffolding, etc.

The builder must understand thoroughly the most expeditious

ways of providing or erecting these items, especially those which must have considerable stability, while the architect should at least be capable of judging whether the more vital erections are safe and satisfactory for their intended purpose.

In connection with the warehouse the only items needing special reference are hoardings and scaffolding.

353. Hoardings. Hoardings are merely required to keep the ground immediately surrounding the building free from trespass during the building operations and to prevent accidents to persons using the footpaths; they consist of posts, rails and boarding, the former being embedded in the ground.

Before erecting a hoarding, permission or license must be obtained from the Local Authority.

Advantage is often taken of a hoarding to sublet it for advertising purposes, or it may be treated in some suitable form of decoration and used to advertise the reconstruction of old property or the erection of new buildings.

354. Scaffolding. Scaffolding is a temporary erection of round timber and planks, forming platforms from which building can be carried on in stages as the work rises above the ground.

In some parts of the country scaffolding is erected *within* the building taking advantage of the floors for the main support and erecting planks upon rectangular trestles which may be used lengthwise or edgewise to lift the platform to different heights above a floor. This provides for overhand face work which cannot be clearly seen nor inspected; see also Vol. I, chapter on Brickwork.

In the best practice scaffolding for the external walls is erected outside the building and the work satisfactorily accomplished from the face.

There are two forms of scaffolding known as (a) the bricklayers', (b) the masons'.

These scaffolds differ only in the method of supporting the platform bearers or "putlogs".

355. Bricklayers' scaffold. The bricklayers' scaffold consists of a vertical frame of round deal poles about 30 ft. long and 5″ diameter at the base, comprising standards about 8 ft. apart, ledgers placed horizontally at about 5 ft. centres inside the standards, lashed together at the junctions with manilla rope, and tightened with large rounded wedges.

Special "scaffixer" chains, and other patent appliances, are also used in place of ropes.

To obtain a rigid support for the bases of the standards they may be (*a*) buried two feet in the ground and bearing on a block of stone, (*b*) placed upon flagstones and spiked to and guarded by a fender beam to prevent them from being displaced, (*c*) buried in a tub of solidly rammed earth.

The purpose of the ledgers is to receive the ends of short square birch timbers, $3'' \times 3''$, split from the log to ensure continuity of fibres. At the wall end these putlogs rest in holes provided by the occasional omission of face bricks.

The working platform, consisting of $9'' \times 2''$ boards 12 ft. long, rests upon the putlogs, the ends of the boards being bound with hoop iron to prevent splitting.

It is important to place a putlog at each end of the boards so that they cannot tilt as would be the case if the boards overhung. As an alternative two boards are often overlapped on one putlog, which is bad practice and may cause a worker to stumble.

A guard board and guard rail, fixed to the inside of the standards, are necessary at the outer edge of the platform and are required by many regulations.

The height of a scaffold is increased by lashing additional poles to the lower standards, and security and rigidity is obtained by lashing diagonal bracing across a series of standards upon the outside face.

For average work a convenient width of working platform is about 3 ft., and in the case of thick, high walls a scaffold would be employed on each face.

356. Masons' scaffold. This is similar to the bricklayers' scaffold except that it receives no support from the wall and therefore needs a second series of standards placed as close to the wall as may be convenient. Because such a scaffold would lack transverse rigidity, it is cross braced transversely where possible and tied to its position by poles passed through any convenient openings in the wall, secured to internal members such as stanchions or floor beams, and lashed to the inner and outer standards of the scaffold frame.

Where single scaffolds are employed the poles are lashed to floor beams, stanchions or stumps driven into the ground.

Since the materials used by the mason are in larger and heavier pieces than those used by the bricklayer, heavier timbers are often used throughout the entire scaffold.

In some cases square timbers are employed instead of poles, the standards being built up of two or more thicknesses which are lapped and bolted together. The mason's scaffold often requires to be more spacious and also stronger than a bricklayer's scaffold.

For many large and important buildings special forms of scaffold have been designed, both in timber and in steel, in order to facilitate the execution of the work. All-steel scaffold frames are now in common use. The photograph printed below shows a typical use of a steel scaffold for a large block of flats.

Students should take every opportunity of gaining information regarding temporary as well as permanent constructions, by making notes and sketches of any unfamiliar items and observing the special conditions for which they were designed.

Example of Modern Steel Scaffolding, by Scaffolding (Great Britain) Ltd.

CHAPTER FOURTEEN

ROOF COVERINGS

Plain and Pan Tiling

Plain tiling, including the treatment of eaves, verges, ridges and minor details were considered and illustrated in Vol. ii.
The succeeding paragraphs consider the finishings to hips and valleys.

357. Formation of hip coverings in plain tiled roofs. Hips may be covered in several ways. One method is to cut the tiles to mitre over the hip and to render the joint weathertight by interleaving soakers, in a manner similar to that shown for valleys in detail No. 118, or, as an alternative, to form a narrow secret gutter of lead beneath the joint.
Neither of these methods are considered to produce first-class results since the hip is not sufficiently safeguarded against the effects of wind, and both systems are costly in maintenance.

358. Plain hips. A better method is to use one of the several forms of special hip tile which are available. Three of these are shown in detail No. 117.
The first is a plain purpose made tile having two wings of the correct form and bevel to allow it to bed upon the hip and bond with the ordinary courses, as shown at A. The double overlap is thus continued up the hip exactly as in the plain courses and the tile is secured upon the hip rafter by a single nail. In order to avoid the possibility of the hip tiles failing to bed down properly at the tail where they join against the plain tiles, the internal angle is moulded distinctly less than the actual angle required. This allows for a possible flattening of the angle during drying and burning, but if no alteration occurs the hip tiles are easier to lay and the whole hip line is slightly lifted, and thus emphasised. A common disadvantage with these hip tiles is the difference in colour and texture due to the plain tiles being machine made while the hip tiles are hand made.

359. Bonnet hips. The second form of covering is known as a "bonnet hip". It consists of a series of tiles, each of which is a curved slab suggesting the form of a sun-bonnet at the tail edge, as shown at B.
The centre of the tile is lifted considerably in cross section forming

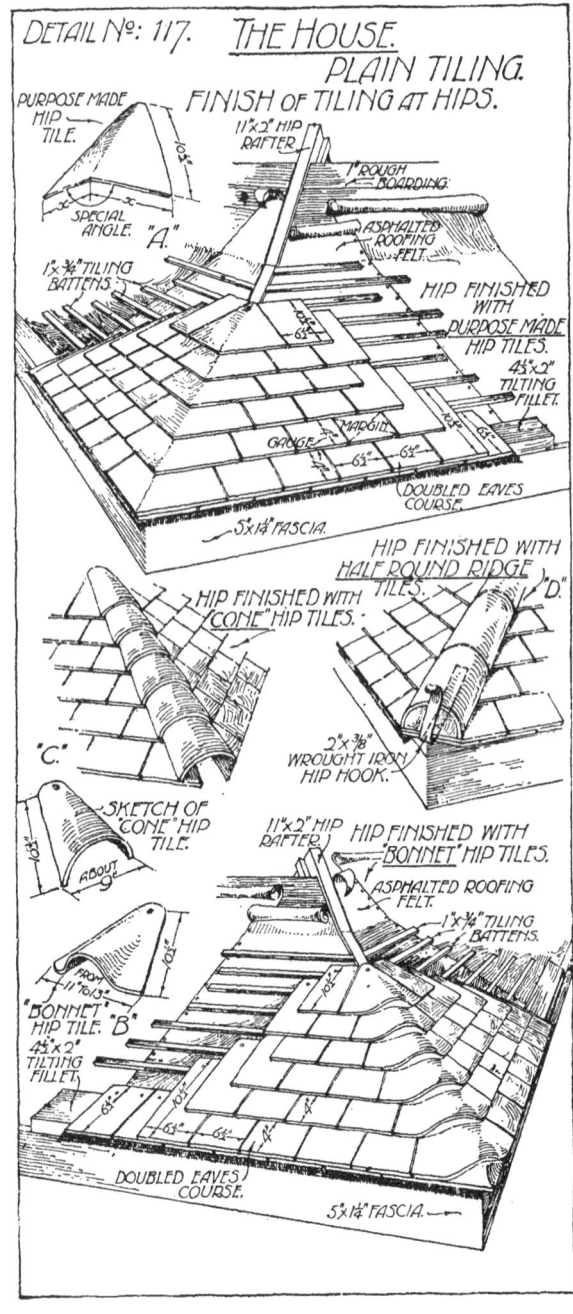

a double ogee or "wave" curve, which dies out to a flat concave outline at the head of the tile, where it is nailed to the hip rafter.

The bonnet edges do not bed upon the back of the tile below, but each tile leaves a clear space at the centre and beds down at the side wings only; the space may be filled by bedding the tiles in cement mortar or by pointing the edges on completion. Better results are obtained by bedding, but care must be taken to avoid marking the exposed surfaces of the tiles by contact with the mortar; the exposed mortar should be cut off clean and smooth to avoid the necessity of "pointing" at a later stage of the work.

Bonnet hips produce a very good architectural effect if the tiling is of a satisfactory texture and colour.

360. Cone hips and half round hips. In the same detail at C is shown a sketch of a cone hip tile, which is sometimes employed for covering hips by lapping and bedding in mortar over the mitred angle of the plain tiling.

The finished appearance somewhat resembles a line of half flower pots and is not altogether pleasing.

Hips are also often covered in a similar way to ridges, with half round tiles, as illustrated at D, especially where the latter have been adopted for the ridge in the same structure. This finish often presents a very clumsy appearance, and the tile at the foot of the hip in both the above methods should be held in position by a wrought iron hip hook screwed to the foot of the hip rafter.

361. Formation of valley coverings in plain tiled roofs. Use of lead soakers. Valley coverings may be arranged by using mitred plain tiles and lead soakers, as shown in detail No. 118, where the soakers act as flexible tiles interleaved between the separate courses and bent across the angle of the valley; they are kept out of sight by making them shorter than the tiles by the length of their margins.

A sketch of a separate valley soaker is shown in the detail and a series of these are also shown in position by dotted lines, the one upon the top course being clearly visible.

The soakers are long enough at the head to dress over the thickness of the tile and extend about $\frac{3}{4}''$ upon the roof boarding. They will therefore hold in position without nails, though there is no objection to open copper nailing along the top edges.

This method of trimming a valley is open to some of the same objections as were mentioned in paragraph 357, in relation to mitred hip tiles.

362. Open lead valley gutters in tiled roofs. A method which has been much used in some parts of the country but not altogether pleasing is to form an open lead gutter at the valley, as shown by section and isometric view in the lower part of the same detail. The preparation for the lead consists of two $1\frac{1}{2}'' \times \frac{3}{4}''$ triangular

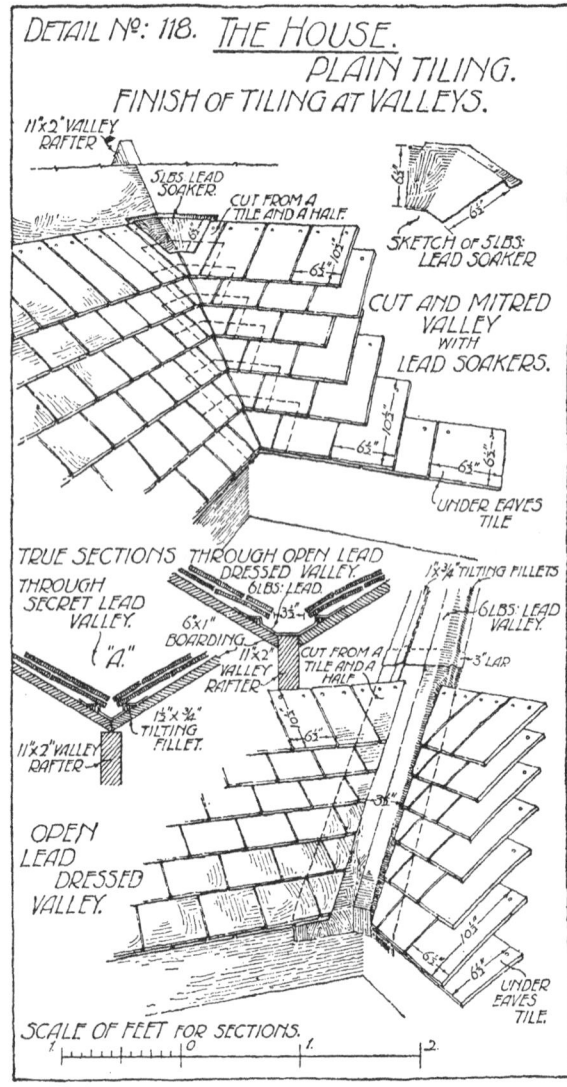

DETAIL Nº: 118. THE HOUSE.
PLAIN TILING.
FINISH OF TILING AT VALLEYS.

11"x2" VALLEY RAFTER

5LBS. LEAD SOAKER.

CUT FROM A TILE AND A HALF.

SKETCH OF 5LBS: LEAD SOAKER

CUT AND MITRED VALLEY WITH LEAD SOAKERS.

UNDER EAVES TILE

TRUE SECTIONS THROUGH OPEN LEAD DRESSED VALLEY.
6LBS: LEAD.
THROUGH SECRET LEAD VALLEY.
"A."
6x1" BOARDING
11"x2" VALLEY RAFTER
CUT FROM A TILE AND A HALF
13"x3/4" TILTING FILLETS
6LBS: LEAD VALLEY.
3" LAP
13"x3/4" TILTING FILLET.
11"x2" VALLEY RAFTER

OPEN LEAD DRESSED VALLEY.

UNDER EAVES TILE.

SCALE OF FEET FOR SECTIONS.

gutter fillets fixed upon the boarding, one on each side of the centre, and forming a shallow gutter 5″ or 6″ wide.

Sheets of lead are dressed up the slope, overlapped 3″ in length, and turned over the fillets with a side lap of about 1″ beyond them.

The two lines of tiles are thus lifted throughout the length of the gutter by resting on the fillets, but the break in the roof surface is too small to be noticeable. The termination of this gutter at the foot is shown in the same detail while the head is shown in detail No. 116.

In order to screen the lead work of the open valley from view, the tiles are sometimes drawn closer together, resulting in a secret valley, as shown in section at A. The arrangement of the lead work is the same as in the open valley, but secret valleys, like other secret gutters, are liable to hold dirt and leaves and are difficult to clean.

363. Purpose made valley tiles. The most common modern method is to weatherproof valleys by purpose made plain valley tiles, of the form shown at A in detail No. 119, being exactly the inverse of the purpose made hip tile, except that the top edge is often cut back at the mitre, thus reducing the weight of the tile without destroying the overlap.

The edges of the valley tiles are usually made to lie in lines parallel to the main sloping joints between the common tiles, but in this example they are shown bevelled, which necessitates the bevel cutting of the plain tiles joining up to them. This bevel depends largely upon the slope of the roof, but should in no case be excessive, and is better made, wherever possible, to give parallel joints. The same bevelled joints will be noticed in connection with the junctions of the bonnet hips and plain tiles in detail No. 117.

364. Special methods of covering valleys. Laced valley. In detail No. 119 at B is shown one of the oldest methods of tiling across a valley with plain tiles and which is known as a "laced" valley. "Ordinary" tiles and "tile and a half" tiles are employed, being lifted up and laced in opposite directions upon an $11'' \times 1''$ valley board which bridges the angle between the intersecting surfaces and affords support to the "tile and half" units. These are alternately placed, as shown, immediately upon the valley board and nailed thereto. The plain tiles in each course are made to conform to a gentle curve and to fit snugly against each other by packing up the units with bits of tile and rough mortar. The extent of the lift may be seen at the edges of the bottom course.

The method is artistic and practically effective but requires special skill in arranging the units.

365. Swept valley. Another method for covering valleys in plain and stone tiling, but involving considerable cutting, is shown at C in the same detail, and is known as a "swept" valley.

In this arrangement each course of tiles is swept round a quadrant curve in plan by cutting the units to a wedge form, and lifting

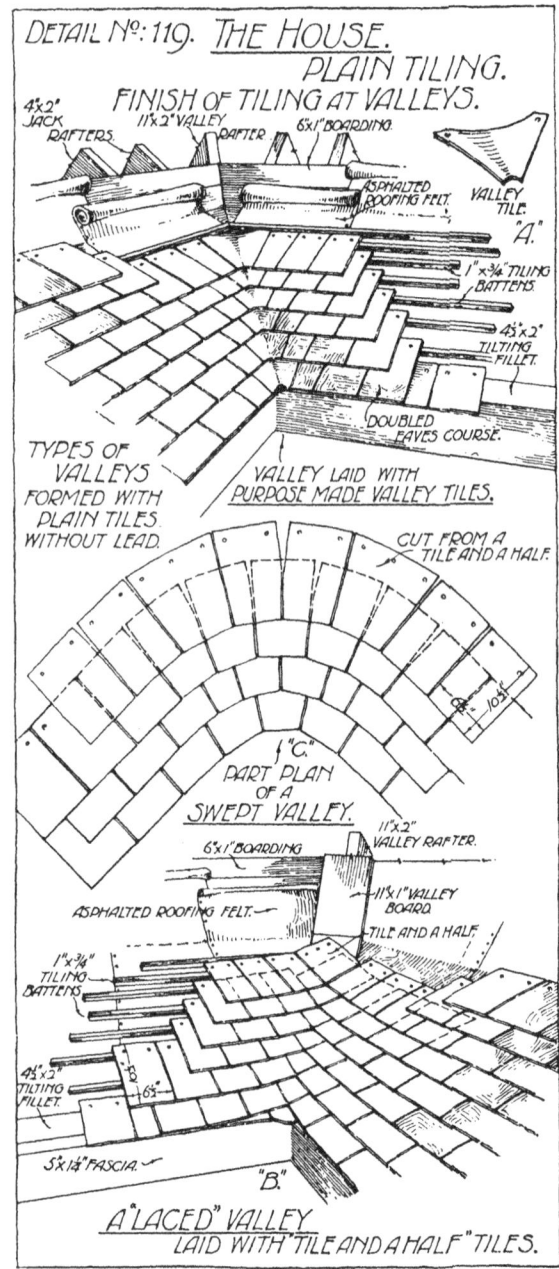

DETAIL Nº: 119. THE HOUSE.
PLAIN TILING.
FINISH OF TILING AT VALLEYS.

4"x2" JACK RAFTERS.
11"x2" VALLEY RAFTER.
6"x1" BOARDING.
ASPHALTED ROOFING FELT.
VALLEY TILE. "A."
1"x¾" TILING BATTENS.
4½"x2" TILTING FILLET.
DOUBLED EAVES COURSE.

TYPES OF VALLEYS FORMED WITH PLAIN TILES WITHOUT LEAD.

VALLEY LAID WITH PURPOSE MADE VALLEY TILES.

CUT FROM A TILE AND A HALF.

10.5"

"C."
PART PLAN OF A SWEPT VALLEY.

6"x1" BOARDING.
11"x2" VALLEY RAFTER.
11"x1" VALLEY BOARD.
ASPHALTED ROOFING FELT.
TILE AND A HALF.
1"x¾" TILING BATTENS.
4½"x2" TILTING FILLET.
"B" 6½"
5"x1½" FASCIA.
"B."

A "LACED" VALLEY LAID WITH "TILE AND A HALF" TILES.

them up from the angle of the valley by a board or other packing. The lower courses have a small radius, but this is increased course by course until a satisfactory curve is obtained at, say, the fourth course.

The swept angle of tiling has at this stage become part of a cone and this shape is maintained at the same size until a termination is required at the ridge.

Swept valleys are particularly suited to stone tiling because valley tiles in stone are not feasible.

Laced valleys could also be suitably employed with thin stone tiles of medium size.

366. Tiling of vertical surfaces. Vertical surfaces are very often hung with tiles, which could have been applied to the sides of the dormer window in detail No. 98.

This is accomplished by fixing horizontal tiling battens in the same manner as adopted for sloping roofs and hanging the tiles by their nibs over these; in vertical applications every tile should also be nailed, because there is great risk of lifting and rattling in a wind, due to the difficulty of fitting the tiles solidly in position.

In situations not much exposed, alternate tiles, or courses, may be nailed.

PANTILING

367. Pantiles are slabs of baked clay made with a flat wave-shaped cross section, and if correctly used may make a pleasant and economical roof covering. The tiles are usually about 14″ long × 9″ or 10″ wide × $\frac{1}{2}$″ to $\frac{5}{8}$″ thick.

Their common form is shown in detail No. 120, and is adopted because it allows the tiles to be laid in single thicknesses with laps at the head and one side. There is no double overlap as in slating and plain tiling, because the wave allows sufficient protection of the side joints at the overlapped edges.

The side lap is $1\frac{1}{2}$″ to 2″ and the longitudinal lap 3″ to 4″. The tiles are hung by a lug or nib, over 2″ × 1″ tiling battens.

At diagonally opposite corners of each tile the sharp angles are removed, as shown in the detail, this being necessary to facilitate the laying of the tiles at the angle A, where the side lap causes a double thickness and would interfere with the head lap if the sharp angle were retained.

It will be seen that by removing these opposite angles there is no interference with the bedding, and the courses can thus continue up the slope with the long edges in an unbroken line.

The single thickness of the tiling thus obtained considerably reduces the weight of the covering, but does not produce so warm an interior as the plain tiled covering with its double lap.

368. Bedding of pantiles. Pantiles do not bed accurately one upon another at the overlap, because two curved slabs of identical sections and parallel thickness cannot fit together exactly; for this reason the tiles are usually bedded in mortar at the overlap. If used on a domestic building the least that can be done to make the surface wind and weather tight is to point or torch the side and back joints at the overlaps.

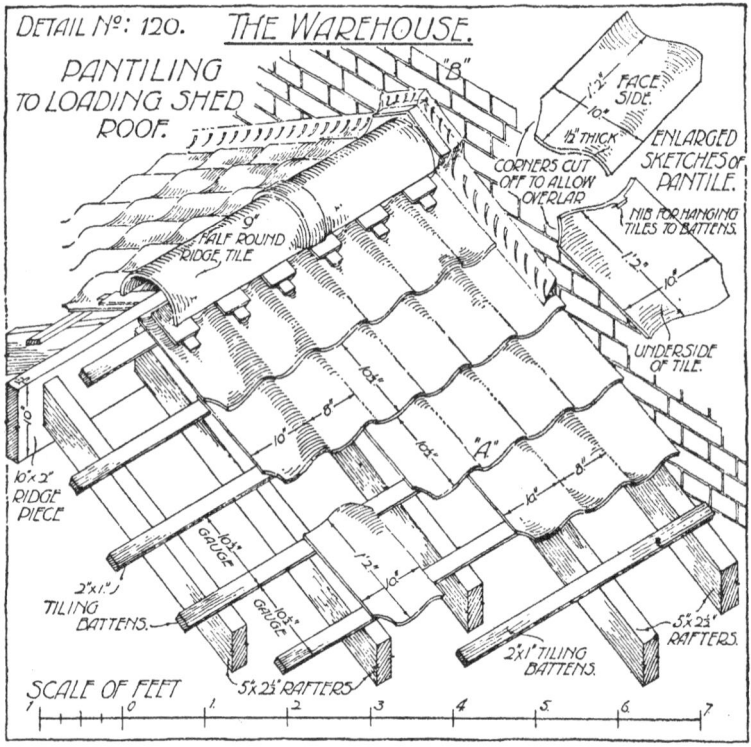

Another good method is to provide close lathing upon the rafters, spread mortar over the laths and bed every pantile solidly upon it. The overlaps should also be bedded to prevent the edges being caught by the wind.

In foundries, sheds, farm buildings and other places where open joints are an advantage for ventilation and cooling, the tiles may be laid dry with an overlap of at least 4″, on a minimum pitch of 26°.

369. Stops and ridges. Hips and valleys. When a pantiled roof intersects with and stops against a side wall, it may be finished at the angle in the manner shown in the detail. A large triangular fillet of "one to two" cement mortar is trowelled into the angle, upon thoroughly wetted surfaces of the tiles and brickwork. The fillet is finished with curved **V** cuts or depressions which compress the mortar and tend to localise any contraction cracks which may develop as the cement sets.

Lead flashings may suitably be employed in this case, as in all roof coverings, to obtain a watertight joint.

The ridge in this example is covered with 9″ half-round tiles with socketed joints. Bearing for the bottom edges of the ridge covering is obtained by packing the hollows of the tiles with pieces of plain tile set in cement mortar; the ridge tiles are placed and bedded upon these and the bedding mortar cut off neatly to the line of junction.

Ridge coverings may be stopped against walls by the method illustrated at B, where the junction is protected by pieces of plain tile solidly bedded to the ridge and finished above by a fillet of cement.

Hips are mitred and covered with half-round tiles (similar to the ridge tiles) bedded solid in cement mortar. Valleys are mitred and flashed with lead soakers.

Pantiles are not well suited to broken roofs involving valleys and hips.

370. Pitch of pantiled roofs. The ordinary pitch of a roof suitable for pantile covering is usually accepted as 24° to 26°, but there is an advantage in employing steeper slopes (up to 35°) for unbedded tiles, in order to ensure adequate resistance to the weather.

When tiles are satisfactorily bedded the minimum pitch of 24° is quite reliable.

371. Suitable uses of plain tiles and pantiles. Plain tiles, being small, are suitable for roofs containing many intersections involving cutting of the units, while pantiles are only suitable for plain straight slopes, valleys and hips being difficult to arrange and clumsy in appearance.

Generally, it may be taken that roof coverings with small units lend themselves to cutting and intricate shaping, while large units must be treated simply and with restraint.

CHAPTER FIFTEEN

EXTERNAL PLUMBERS' WORK

Gutters to Main Roof of Warehouse

Lead work in general roof flashing and in the construction of flat roofs has been considered in Vols. I and II.

This chapter deals with tapering gutters, skylights, dormers, etc.

372. Parallel gutters. The main roof is drained on the party wall side by a parallel box gutter of the same type as those previously illustrated and described in Vols. I and II. It is longer and deeper than in the previous cases, with more falls and drips and may be either:

(a) drained from the centre to the ends, necessitating a very deep and expensive gutter of practical inconvenience, but allowing of discharge into external down pipes on the front and back walls;

(b) drained to points about one-quarter of the length from each end, with falls both ways toward them; this would allow of a normal depth of gutter with 11″ pole plates but would require internal down pipes.

The student should work out the details of this gutter from the knowledge gained in studying previous examples.

373. Combined eaves treatment. It often happens that one type of eaves treatment cannot be continued throughout a building. This occurs in the side of the warehouse overlooking the yard. In this case the staircase and lift enclosure are carried to a higher level than the rest of the building, hence it is convenient to use an open overhanging eaves above the loading shed and a closed eaves with a parallel parapet gutter behind the staircase wall.

This gutter falls towards the back of the building and discharges into the same down pipe as the eaves gutter.

An examination of the general plans and of the main roof plan, detailed in Vol. II, will make the arrangement clear.

374. Tapering gutters to hipped ends of main roof. As both ends of the main roof of the warehouse are hipped and parapet walls rise above the line of intersection of the roof at each end, lead-lined gutters are employed to convey the water collected on the hipped surfaces to the main slopes.

It is convenient to employ tapering gutters in these positions in the manner illustrated in detail No. 121, which shows a portion of the gutter behind the parapet wall of the front gable.

The gutter bottom is formed of $1''$ boards nailed upon $3'' \times 1\frac{1}{4}''$ bearers placed across from the rafters to a $3'' \times 1\frac{1}{4}''$ firring plate which in turn rests upon a $9'' \times 3''$ rafter footplate or purlin.

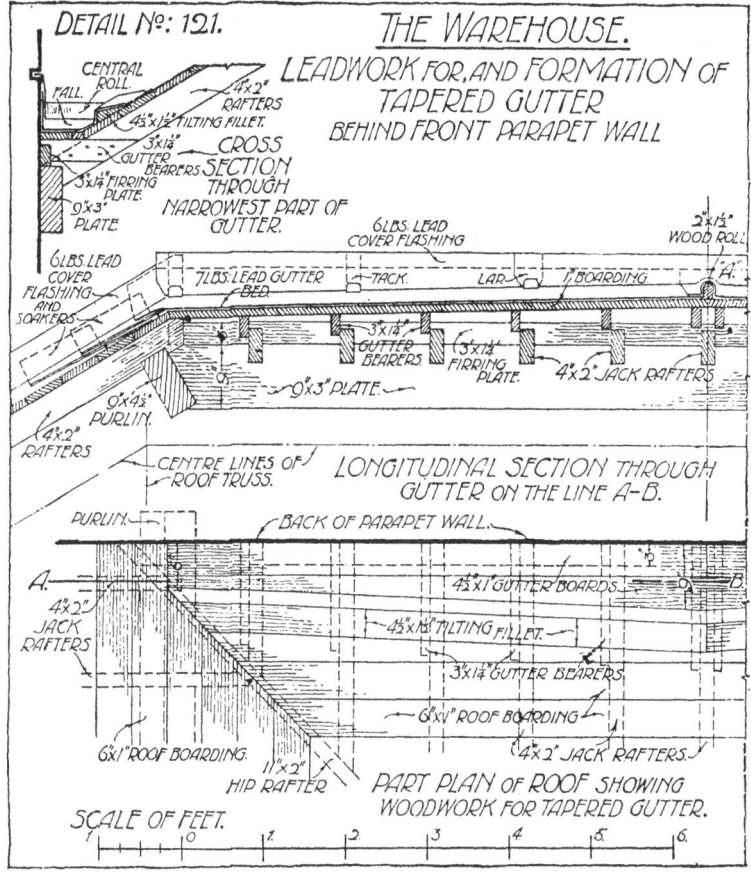

Owing to the fall the gutter widens towards the highest point, as shown in the detail, tapering boards being required at the intersection of gutter and roof slope.

In this instance no drip is required, one sheet of lead being continued from the crown roll at A, to the curb, over which it extends to provide a $4''$ overlap to the under-course of slates at the head of the slope.

The section shows the use of soakers on the inclined gable and the passing of the cover flashings of the gutter and raking portion.

375. Use of tapering gutters. Tapered parapet gutters are commonly employed in continuous lengths behind parapet walls but are not so convenient to the slater or tiler as parallel box gutters, because the latter do not interfere with the normal run of the courses of slating. In the former case the eaves of the slates generally follow parallel to the sloping inner edge of the gutter, as shown by the position of the tilting fillet in plan, and a further interference occurs where drips are employed due to the sudden widening of the gutter at these points at the change of level.

Tapering gutters may, however, often be usefully employed, especially between the slopes of V and M roofs, and in positions where the projection of a box gutter below the wall plate would be an objection.

SKYLIGHTS

376. Opening wood skylight. Upon the lead-covered flat of the house, and close to the chimney, is formed an opening skylight to give direct light and provide ventilation to the staircase and landing at the attic floor level.

The curb in this case rises 5″ above the flat at the lower edge and 18″ at the upper edge and is covered with 1″ boarding. The construction of the necessary woodwork is shown in detail No. 124, Vol. II. Upon the curb and boarding is laid a covering of 6 lbs. lead carried fully across the top edges, and down the sides to a level line overlapping the upturn and to within 1″ of the level of the flat.

Behind the skylight and between it and the chimney, a lead gutter is formed, being 3″ above the level of the flat at the centre of the gutter and 2½″ at the ends, where a drip is formed, as shown at A in detail No. 122.

The general arrangement of the covering, including lead sheeting, upturns, overlaps at angles, use of clips and copper nailing should be followed from the smaller detail, while the turns and dressings of both sheets at the gutter end, with the necessary lead tacks and provision for stopping the covering of the joint roll, should be understood from the details at B and C.

At D and E the shaping of the upturned leadwork is illustrated for the lower edge and angle of the curb, and at F is shown a bossed end for the side cover flashing where it envelopes the angle.

In this latter detail the surplus lead, which would occur at the overlap, is beaten and drawn into a solid clothing of thicker material, in preference to folding and lapping into a dog's ear, as previously illustrated in Vol. I, where the folding is done internally, instead of externally as would be required in the present example.

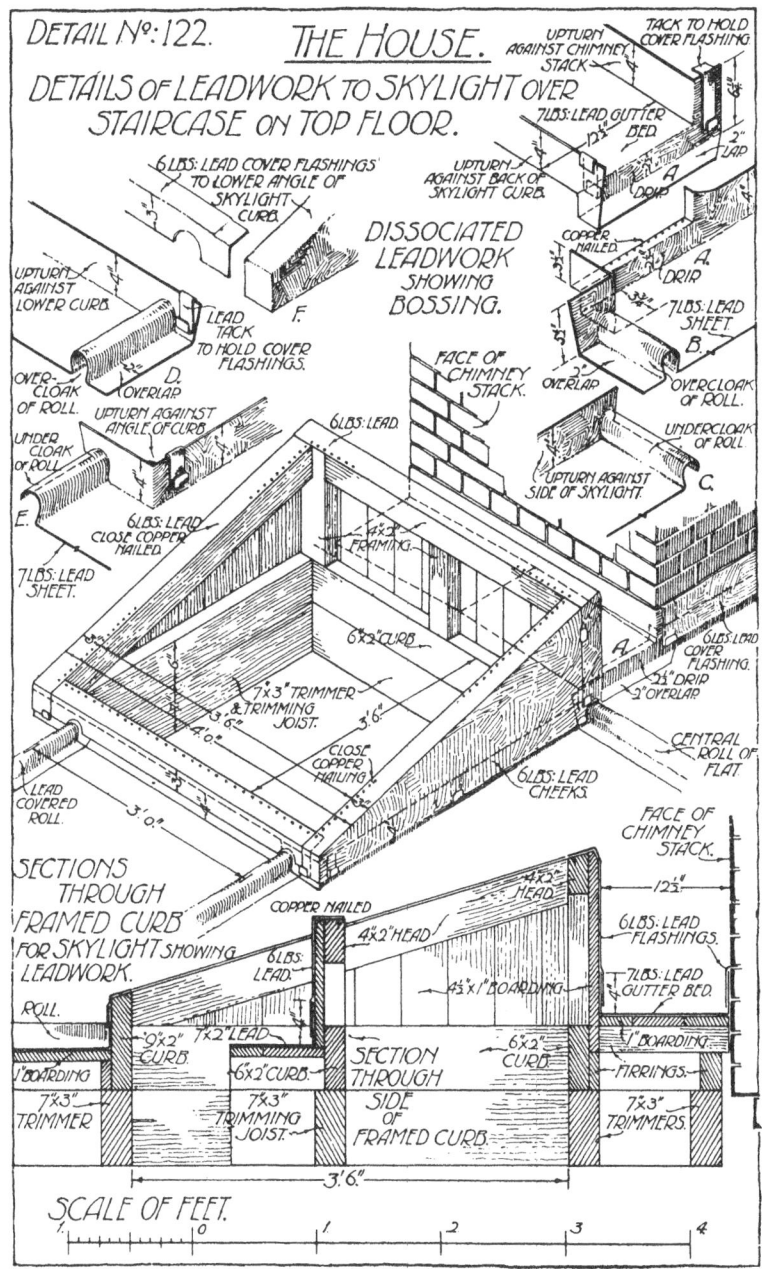

DETAIL Nº: 122. THE HOUSE.

DETAILS of LEADWORK to SKYLIGHT OVER STAIRCASE ON TOP FLOOR.

TACK TO HOLD COVER FLASHING

UPTURN AGAINST CHIMNEY STACK

7 LBS: LEAD GUTTER BED.

6 LBS: LEAD COVER FLASHINGS TO LOWER ANGLE OF SKYLIGHT CURB.

UPTURN AGAINST BACK OF SKYLIGHT CURB.

DISSOCIATED LEADWORK SHOWING BOSSING.

LAP

A

DRIP

UPTURN AGAINST LOWER CURB.

OVER-CLOAK OF ROLL.

UPTURN AGAINST ANGLE OF CURB.

UNDER CLOAK OF ROLL.

LEAD TACK TO HOLD COVER FLASHINGS.

D.

OVERLAP

E.

7 LBS: LEAD SHEET.

6 LBS: LEAD CLOSE COPPER NAILED.

F.

COPPER NAILED

A DRIP

7 LBS: LEAD SHEET. B.

FACE OF CHIMNEY STACK.

OVERLAP

OVERCLOAK OF ROLL.

6 LBS: LEAD.

UPTURN AGAINST SIDE OF SKYLIGHT.

UNDERCLOAK OF ROLL. C.

4x2" FRAMING.

6x2" CURB.

7x3" TRIMMER & TRIMMING JOIST.

CLOSE COPPER NAILING.

A

6 LBS: LEAD COVER FLASHING.

DRIP

OVERLAP

CENTRAL ROLL OF FLAT.

LEAD COVERED ROLL.

6 LBS: LEAD CHEEKS.

FACE OF CHIMNEY STACK.

SECTIONS THROUGH FRAMED CURB FOR SKYLIGHT SHOWING LEADWORK.

COPPER NAILED

6 LBS: LEAD.

4x2" HEAD.

4x2" HEAD

4½"x1" BOARDING

12½"

6 LBS: LEAD FLASHINGS.

7 LBS: LEAD GUTTER BED.

1" BOARDING.

ROLL.

9x2" 7x2" LEAD CURB.

6x2" CURB.

SECTION THROUGH SIDE OF FRAMED CURB.

6x2" CURB.

FIRRINGS.

1" BOARDING

7x3" TRIMMER

7x3" TRIMMING JOIST.

7x3" TRIMMERS.

3' 6".

SCALE OF FEET.

0   1   2   3   4

A similar solid bossed angle would be made at the higher end of the curb.

377. Dormer window covering. The dormer windows upon the slopes of the house roof are framed in timber and covered on the sides and top with 6 lbs. lead.

The perspective view of detail No. 123 gives general information as to the arrangement of the framework, roof and covering, and the larger sectional details show the method of lapping and securing the lead.

All the sheet lead in large pieces is laid upon 1" square edged boarding firmly nailed close to both edges, and for this purpose accurately thicknessed boards should be employed.

The protection of the woodwork commences at the cill where a lead flashing is laid across the upper edge of the curb and is subsequently turned up 1" behind the window cill and copper nailed thereto; the outer edge folds down the face of the curb and is laid upon the tiles so that a double overlap is obtained at least equal to that of the tiles, the edge being secured by lead or copper clips. This flashing is carried at least 2" round the angles and is dressed down upon the boarding before the tiling is continued up the slope on each side of the dormer.

378. Lead-covered dormer cheeks. The spandril side of a dormer window is termed a "dormer cheek"; this surface may be boarded and covered with lead, when the covering is termed a "lead cheek".

Lead cheeks may be in one or more pieces according to the size of the dormer side; joints, where required, are made in vertical lines, by plain welts.

To secure the top edge it may be turned at right angles over the edge of the top board and copper nailed between the joists before the roof boards are fixed. At the lower edge the lead cheek acts as a cover flashing to the upturn of a secret gutter, or to lead soakers placed at the intersection of the tiling with the dormer, as previously described.

The front vertical edge of the lead cheek is lapped 3" along the front face of the angle post, close copper nailed and folded back 2" to cover the latter.

379. Necessity for support to lead cheeks. In large dormers and similar structures where lead cheeks are used, they may become so heavy that their own weight drags them downwards and so continually and permanently stretches the lead, which is not elastic. This action is accentuated during rises of temperature; the lead expands and moves downwards, but is not elastic enough to regain its original position when the material cools and attempts to

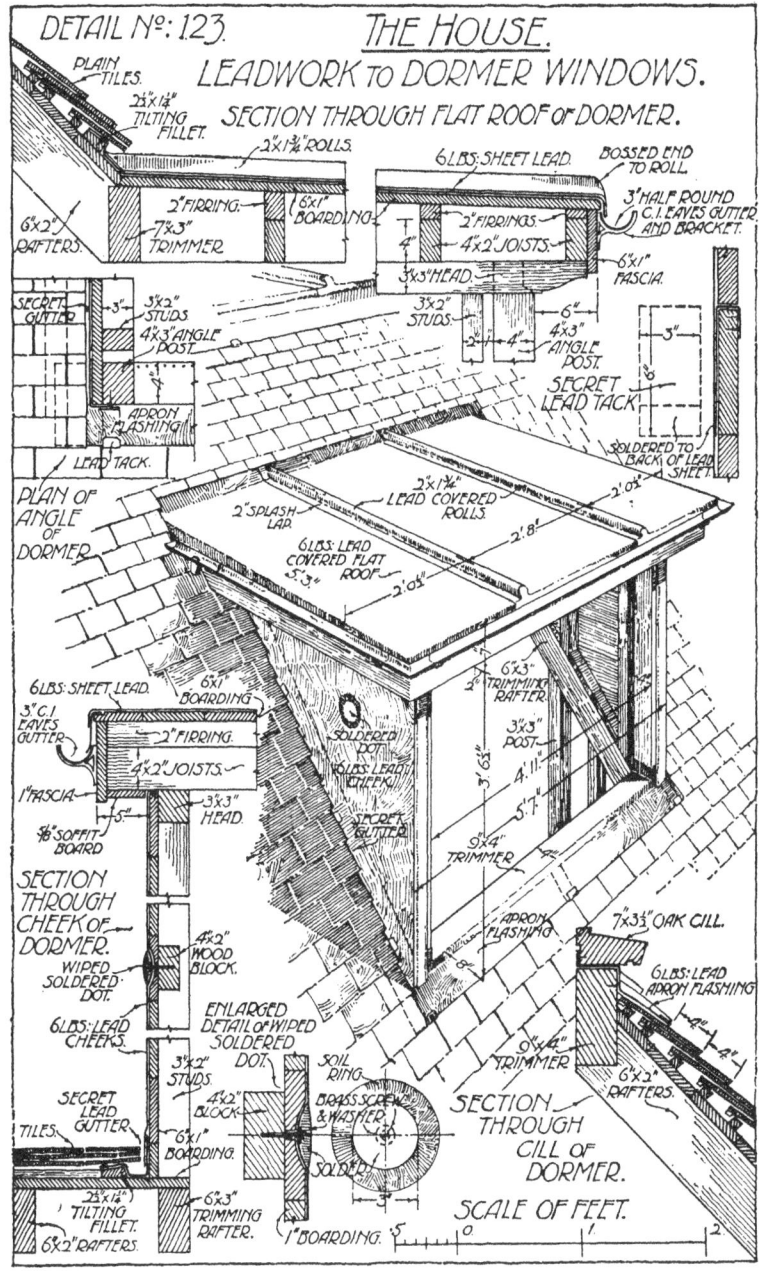

DETAIL N°: 123.   THE HOUSE.
            LEADWORK TO DORMER WINDOWS.
        SECTION THROUGH FLAT ROOF OF DORMER.

PLAIN TILES.
2½"×1¼" TILTING FILLET.
2"×1¾" ROLLS.
6 LBS. SHEET LEAD.
BOSSED END TO ROLL.
2" FIRRING.
6"×1" BOARDING.
6"×2" RAFTERS.
7"×3" TRIMMER.
2" FIRRINGS.
4"×2" JOISTS.
4"
3½"×3" HEAD.
3" HALF ROUND C.I. EAVES GUTTER AND BRACKET.
6"×1" FASCIA.
SECRET GUTTER.
3"
3½"×2" STUDS.
4"×3" ANGLE POST.
3½"×2" STUDS.
2½" 4"
6"
4"×3" ANGLE POST.
3"
6"
SECRET LEAD TACK.
SOLDERED TO BACK OF LEAD SHEET.
APRON FLASHING.
LEAD TACK.
PLAN OF ANGLE OF DORMER.
2"×1¾" LEAD COVERED ROLLS.
2" SPLASH LAP.
6 LBS. LEAD COVERED FLAT ROOF.
5'·3"
2'·0½"
2'·8"
2'·0½"
2'·0½"
6"×3" TRIMMING RAFTER.
1"
6 LBS. SHEET LEAD.
6"×1" BOARDING.
3" C.I. EAVES GUTTER.
2" FIRRING.
4"×2" JOISTS.
SOLDERED DOT.
6 LBS. LEAD CHEEKS.
3½"×3" POST.
3'·6½"
4'·11"
5'·7"
1" FASCIA.
5"
3"×3" HEAD.
⅞" SOFFIT BOARD.
SECTION THROUGH CHEEK OF DORMER.
SECRET GUTTER.
9"×4" TRIMMER.
WIPED SOLDERED DOT.
4"×2" WOOD BLOCK.
APRON FLASHING.
7"×3½" OAK CILL.
6 LBS. LEAD APRON FLASHING.
6 LBS. LEAD CHEEKS.
ENLARGED DETAIL OF WIPED SOLDERED DOT.
4"×2" BLOCK.
SOIL RING.
9"×4" TRIMMER.
4"
4"
SECRET LEAD GUTTER.
3½"×2" STUDS.
BRASS SCREW & WASHER.
6"×2" RAFTERS.
TILES.
6"×1" BOARDING.
SOLDERED.
SECTION THROUGH CILL OF DORMER.
2½"×1¼" TILTING FILLET.
6"×2" RAFTERS.
6"×3" TRIMMING RAFTER.
1" BOARDING.
3"
SCALE OF FEET.
5   0   1   2

contract. It may partially succeed in contracting with light pieces, but some of the expansion becomes permanent.

If supported at intermediate points, instead of the top edge only, there would be less stretch because there is less dead weight to maintain it between points of support, and in addition such stretch as did occur would be spread over short portions and consequently more uniform.

Two methods of support are illustrated:

(a) By the use of soldered dots.

(b) By employing secret tacks.

380. Soldered dots. In this method a circular sinking is scooped out of the covering board about 3″ diameter and ⅜″ deep. The surface sheet is dressed into this and screwed down at the centre of the depression by an ordinary brass screw bearing upon a brass washer. The screw head should project as much as possible to ensure adhesion of the solder which is used to fill up the depression. To prevent adhesion of the solder to the portion outside the depression a ring of black "soil"—a compound of lamp black and size—is painted upon it, the lead surface within scraped bright and clean and then wiped flush with plumbers' solder.

The circle of solder with its enclosing black ring may be considered of some architectural value to the surface to which it is applied.

381. Secret lead tacks. These are strips of stout lead 2″ to 3″ wide and 6″ to 9″ long, soldered to the back of the covering which is to be suspended, and having one end laced through a notch in the joint of the boarding; the end should be turned down 1½″ or more inside, and copper nailed.

Both the above methods are effective, but the latter is invisible and chiefly employed where it is desired to hide the means of securing, which is not always a commendable object.

382. Roof covering to dormer. The roof covering of the example is a simple application of roll jointing on a lead flat.

To collect the rain-water, a half-round cast iron gutter is employed and is supported on wrought iron brackets. The gutter is returned at each angle and made to discharge through a nozzle upon the main roof surface as near the head of the dormer as possible.

In some cases the nozzle outlet is omitted and the end of the gutter left open for free discharge upon the tiles.

In many instances, especially for small dormers, gutters are not employed, water being allowed to drip freely from the flat and clear of the vertical faces of the dormer.

CHAPTER SIXTEEN

INTERNAL PLUMBERS' WORK

WATER SERVICES AND SANITARY FITTINGS

383. Cold water supply. An adequate and constant supply of cold water is necessary to every dwelling and to almost every building.

In urban districts the supply is generally provided and controlled by the local authority. The water is transported under its own head through cast iron pipes, the size of which depends upon the magnitude of the supply demanded along their route; their diameter will depend upon the nature and quantity of the required supply.

384. House or private supply. The private building is supplied by withdrawal from the street water main, through a main supply pipe, and for a dwelling of average accommodation (as, for example, the semi-detached house of our studies) may be a lead pipe of $\frac{3}{4}''$ bore and weighing about 9 lbs. per yard run. This pipe is connected to the cast iron street main by a wiped soldered joint to a brass elbow which is screwed into a hole tapped into the main; it is carried across the intervening space to the building at a depth of at least 2 to 3 ft. below the ground level for protection against frost and traffic.

Main supply pipes must be provided with a stop cock or valve[1] which may be at a short distance from the building, and operated through a cast iron box with a hinged lid, built above the fitting, or the stop cock may be fixed within the building, as shown in detail No. 124, at a point below the lowest draw-off branch.

In the best arrangement there would be an external stop cock close to the boundary of the site and a second fitting (often called a stop tap) within the building. In normal circumstances for interior repairs the latter only would be manipulated, the former being useful if a burst occurs in the length of pipe between the two stop valves.

385. Rising main. The horizontal length of main is usually carried to a point where it may conveniently rise in a vertical line to supply the fittings en route. The vertical length of pipe is called the rising main, and should be maintained at $\frac{3}{4}''$ bore until it reaches the highest required level, or beyond this where branches are numerous.

[1] See paragraph 423.

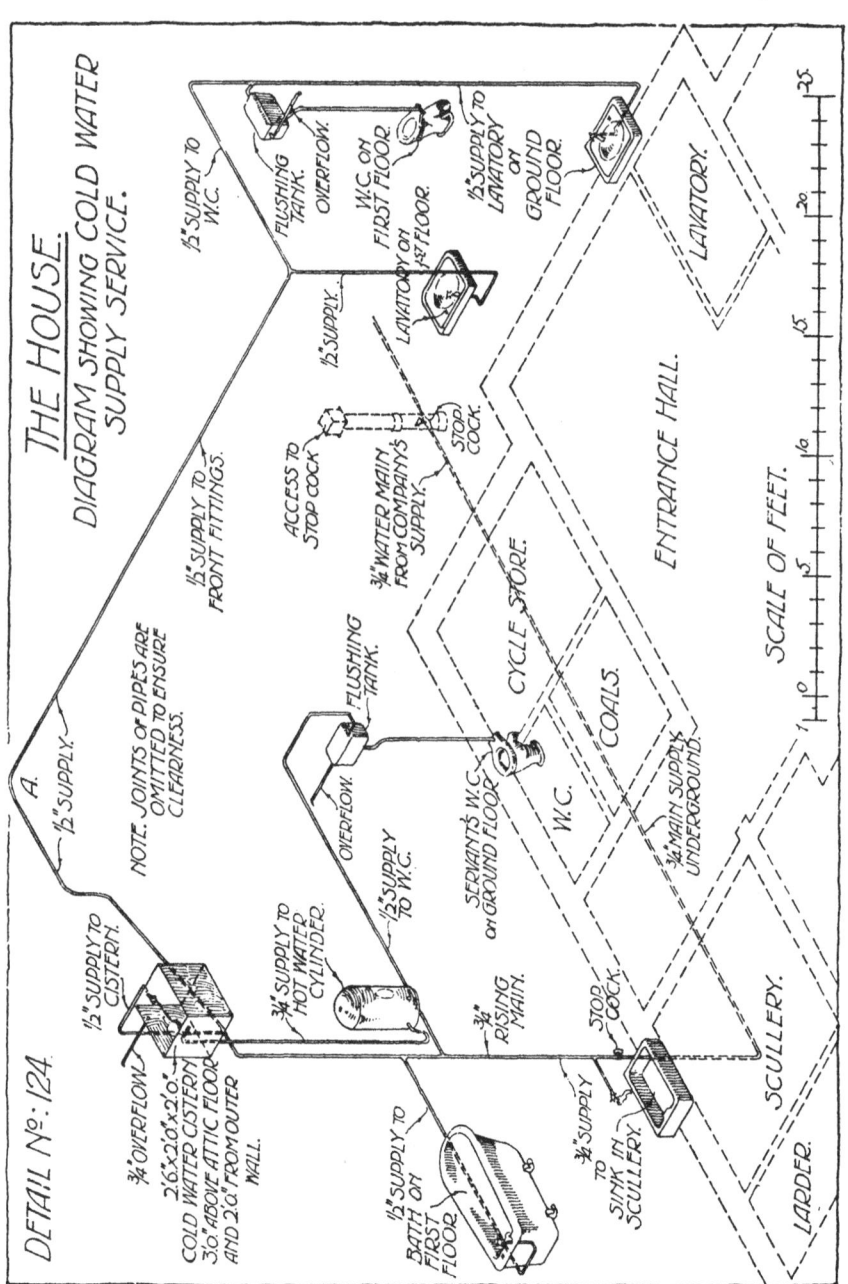

THE HOUSE.

DIAGRAM SHOWING COLD WATER SUPPLY SERVICE.

DETAIL Nº 124.

SCALE OF FEET.

386. Cold water supply to house. In the scheme of supply to the house—detail No. 124—the rising main supplies a scullery sink, bath, and W.C. by draw-off branches $\frac{3}{4}''$ and $\frac{1}{2}''$ bore.

Other fittings requiring cold water service in different parts are: two lavatories, W.C. and a cold storage cistern which holds a reserve for the hot water supply; pipes of $\frac{1}{2}''$ bore weighing 6 lbs. per yard run are provided for all these fittings and the service pipes supplying them may be traced from the top of the rising main following the lines of wall or flooring.

The service pipe to the servants' W.C. on the ground floor is taken from the rising main in preference to dropping it from the $\frac{1}{2}''$ supply above, which, though shorter, might cause restriction of the supply to other fittings at the front of the house if these happened to be in simultaneous use. It would have been necessary to carry the $\frac{3}{4}''$ pipe from the rising main to the angle A to secure an adequate supply under the alternative arrangement.

387. Cold supply reserve for the hot water system. This is met by a $\frac{1}{2}''$ branch service fed through a ball valve[1] to the cold storage cistern, which is a galvanised iron tank $2'\ 6''\times2'\ 0''\times2'\ 0''$; the function of this tank is to hold a sufficient reserve to supply the hot water system with additional water as the heated liquid is withdrawn. The store should be sufficient to meet all needs for several hours, in the event of a discontinuance of the supply during repairs.

In some districts the supply from the street main is intermittent and some such reserve becomes imperative both for the special purpose named and for the general cold supply.

388. Hot water supply. In every dwelling there should be an economical means of providing hot water, and at present it is left to each house to provide its own supply, although it is possible that in the future attempts will be made to establish cooperative supplies.

There are many methods of providing a supply of hot water for domestic purposes, some of which differ only in the details of their arrangement and fittings and, as some of these are rapidly becoming obsolete while others are unsuitable for the small supply required for a house of the size under consideration, it was decided to deal with the well established system which is still considered most generally applicable to small modern installations. (But see also paragraph 390.)

The main purpose of all such installations is to provide a store of hot water which may be withdrawn at certain sanitary fittings or other points of convenience; the supply must be also automatically renewed within a short time.

[1] See paragraph 428.

Any means of heating may be adopted, using coal, coke, or gas as fuel, and the fittings may be adapted for any of these.

389. Method of providing store of hot water. To provide a store of hot water two fittings are primarily required: (*a*) a boiler, which may be either independent or set within the most convenient fire range; (*b*) a container for receiving and storing the heated water, set at a higher level, usually a rectangular steel tank or a copper cylinder.

These two fittings are connected by two lines of pipe; one, called the flow pipe, rises from the top of the boiler and terminates in the upper half of the storage vessel, and a second, called the return pipe, is·connected to the container near its base and terminates at a height of about 3″ from the base of the boiler, being generally (in the case of fire back boilers) passed through the top and continued within.

The store of hot water is accumulated as follows: Suppose the boiler, storage vessel and connecting pipes to be full of water. When the boiler is heated, the water within it is raised in temperature, expands in bulk and therefore reduces in density.

As this change goes on, the colder and therefore denser water in the system tends to fall to the lowest level and the heated water to rise. This unbalanced condition of the liquid causes a movement which is encouraged and guided to develop a circulation of the water in a constant direction. By placing the inlet to the flow pipe at the top of the boiler the heated water is free to rise and pass upwards and to traverse the flow pipe, while its place is taken by the cooler water forced down the return pipe by the column of water reaching to the storage vessel above.

Thus a constant circulation is ensured, the hottest water constantly accumulating in the top of the storage vessel.

390. Methods of supplying heat to the water. Heat may be supplied by the kitchen fire, by a boiler placed at the back of the range. This might be supplemented by a system of gas heating to an additional and independent boiler, so that hot water could be obtained when the kitchen fire was not required for other purposes. Or, an independent boilder may be used of the Ideal or similar type, such as is now commonly fitted in the kitchens of modern dwellings. In these circumstances there is no fireplace of the usual open type, but the independent boiler—usually set in a tiled recess—supplies sufficient heat to warm a small kitchen thoroughly although the primary reason for its installation is to supply hot water by circulation to a storage cistern (or cylinder), from which the house supply— to sink, bath and lavatory—is drawn in exactly the same way as when the heat is supplied from a fire back boiler of the standard type.

Independent gas boilers and geysers are also available. They are efficient but more costly to operate than a coal or coke fire.

Electric immersion heaters have also been well developed. The heating units are protected and immersed in the storage tank and heat is directly communicated to the water in the tank.

For a reference to modern storage tanks see paragraph 394.

391. Fire back boilers. There are several types of boiler suitable for placing at the back of a kitchen fire or other range. The commonest is shown at A in the detail and is set upon a bed of fire bricks at each side with a $4'' \times 3''$ flue passing beneath and behind it. To control the draught through this flue a damper is placed in the vertical arm of it, by means of which the flue may be opened, closed, or regulated.

Thus the front of the boiler is exposed to heat, and also a part of the base and back when the damper is open, and the flames and hot gases are drawn through the flue.

An improvement on this boiler is the arched flue type shown at B. A greater area of flue surface is brought into effect for heating the contents of the boiler.

The type of boiler shown at C is called a boot boiler, owing to its shape, which has an upright portion extending into a broad foot. These boilers, designed for use with closed kitchen ranges, are very efficient, because they expose a large surface of the boiler to the heat, due to their shape and manner of setting.

Boilers are made of cast iron, welded wrought iron and cast copper; the latter are the most efficient.

392. Connection of pipes to boiler. The flow and return pipes are connected to the boiler by pieces of $\frac{3}{4}''$ iron or copper tube tapped in and fixed with screw nuts. These tubes avoid the danger of melting which might occur if lead circulation pipes were brought very close to the boiler. To connect the lead flow and return pipes to the copper tubes they must be tinned and secured by a wiped soldered joint (see paragraph 407).

For hard waters wrought iron pipes may be used throughout the system, but for soft waters they are unsuitable and lead or copper pipes are required.

393. Safety valve and hand hole. Every boiler should be provided with a dead weight safety valve placed on the top surface, as shown at D. A sketch detail shows the action of the valve. It consists of a short vertical tube having the top end shaped and ground to fit a spherical ball, which is kept in position by a weighted cap. The weights may be adjusted to suit the pressure developed by the head—or vertical height—of water in the system; if any appreciable excess of pressure is developed in the boiler, above the normal pressure for a full system, the valve should relieve it by lifting the weighted cap.

In districts where hard water is used, deposits, often called furring, accumulate in the boiler and need to be periodically cleared. To allow this to be conveniently done a hand hole is provided, marked E.

394. Hot water storage cylinder. Where a bath is to be supplied, in addition to other fittings, *e.g.* sink and lavatory, the storage required is at least 30 and preferably 40 gallons. The storage vessel is either (*a*) a cylinder of sheet copper 1′ 3″ to 1′ 6″ diameter, and height according to capacity; its base and top are both dome-shaped, the better to withstand internal pressure, or (*b*) a rectangular galvanised steel tank of similar capacity, made of stout metal and particularly suited to hard water districts. Detail No. 125 shows the complete arrangement of what is known as the cylinder system.

In some cases it is found advisable to provide a hand hole in the cylinder or tank as well as in the boiler, to allow of cleansing and to screw up back nuts at the pipe connections.

Storage cylinders and tanks are often placed in cupboards, with a view to providing a means of airing household linen and wearing apparel.

The most modern cylinders—used with the latest developments of gas and electric heating—are beautifully finished insulated containers which are scarcely warm on the outside when the water has reached full heat. The heat is conserved instead of being wasted by radiation. The source of heat for such fittings (and also for the electric immersion heaters which can be fitted into any ordinary storage tank) is thermostatically controlled so that when a pre-arranged heat has been attained, the source of heat is automatically cut off. As the temperature of the water falls, from whatever reason, the heat is renewed and so the tendency is to maintain a standard temperature.

These cylinders are not so efficient in providing heat for linen airing, but this is only an incidental use in most cases. They are efficient however for their primary purpose of supplying hot water, conveniently, without personal labour and at fairly reasonable cost.

395. Flow and return pipes. These may be of lead, $\frac{3}{4}$″ bore × 9 lbs. per yard, or iron pipes for hard water. They should take the shortest convenient route between the boiler and storage cylinder with as few bends as practicable, and where these must occur they should be of large radius to avoid obstructions to the flow.

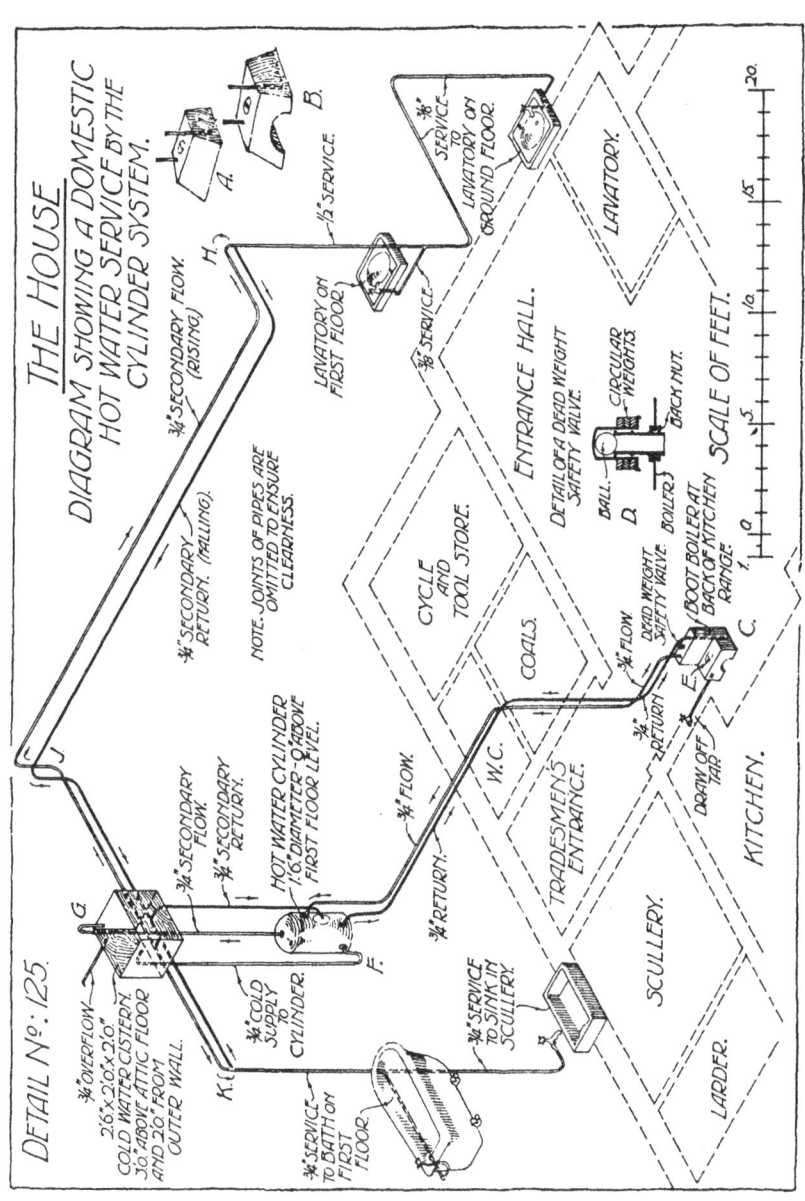

THE HOUSE

DIAGRAM SHOWING A DOMESTIC HOT WATER SERVICE BY THE CYLINDER SYSTEM.

DETAIL Nº: 125.

In some cases it might be convenient to place the storage cylinder almost directly above the kitchen range and hence increase the efficiency by avoiding loss of heat in transit. In this case the cylinder is placed in the most convenient and available position, as will be realised by an inspection of the general plan, although it necessitates a somewhat long run of flow and return pipes. It would therefore be wise to enclose the vertical pipes in wooden casings and to pack round the horizontal lengths with slag wool at their passage across the floor. An alternative is to wrap the pipes spirally with hair felt before securing in position.

396. Supply of cold water to system. To fill the heating system and to replace water on withdrawal from any part of the system, a $\frac{3}{4}''$ cold supply pipe is carried from the cold storage tank to the bottom of the hot water cylinder. It may enter at either the side or the base, but must be dipped as shown at F, which assists in preventing a back flow of hot water up the pipe. A stop valve is often placed just above the dip, so that repairs to the cylinder may be carried out without running off the water in the cold store above, which would become a nuisance if this tank supplies the cold taps in various parts of the house.

By means of a draw-off tap at the boiler the water may be emptied from the whole system.

397. Provision for expansion of water. When water expands in a confined vessel or system of pipes, great internal pressure is developed which would burst the pipes or storage vessels unless relieved, hence provision must be made for such expansion, which is usually done by carrying an open ended pipe from the top of the cylinder and bending it over the cold storage tank as shown at G—detail No. 125. In the case of excessive expansion or formation of steam when the water reaches boiling point there is thus a free escape. Air which is expelled from the water as it is raised in temperature may also find an exit here, and this is constantly occurring because cold water contains large quantities of fine air bubbles which become enclosed during its exposure and conveyance through the pipes.

398. Arrangement of draw-off pipes. In small closely grouped installations the hot water is usually drawn off directly from the top of the cylinder, by branching the draw-off pipes from the expansion pipe and carrying sub-branches to the sanitary fittings. These must be at a lower level than that of the cold water in the storage cistern in order to obtain pressure to give a reasonably quick delivery.

It is advisable to have the cistern at least 6 ft. above the highest draw-off tap, and where possible it should be 10 ft. above.

It will be seen that by using a direct draw-off to every fitting along single branch pipes, a defect occurs in the supply, for, as the drawing is intermittent and water cannot circulate in a single pipe, the contents of the branch quickly become cold, and thus when the next supply of hot water is required the cold contents have to be withdrawn and probably wasted.

For long lengths of hot water service pipe this defect is annoying, but it has a compensation, in that the water is kept hotter in the cylinder, since there is a more rapid loss of heat from the pipe surface than from that of the cylinder.

399. Secondary or auxiliary circulation. To remedy the above defect it is wise to arrange an auxiliary circulation of the hot water from the cylinder through secondary flow and return pipes and thus carry the water as close as may be convenient to the more important points of withdrawal, or, to ensure that only short lengths of pipe have to be emptied of cold water.

In the illustration the secondary circulation is branched from the expansion pipe above the floor level of the attic store room and below the cistern. The flow pipe is made to branch to the right and left, the former rising constantly and following the floor and ceiling until it reaches its highest point over the first floor lavatory at H— detail No. 125; it then drops and returns at a lower level with a constant fall, until it enters a vertical return pipe connected to the cylinder near the centre of its height. From the rise at J to the highest point the pipes pass within the depth of the floor.

This branch circulation feeds the lavatories on the first and ground floors and conveys the hottest water constantly to within a reasonable distance of the fittings.

The left hand branch is very short and could almost be dispensed with in this case. It is included to show the treatment where the vertical drop feeding the bath and scullery sink is at a greater distance from the cylinder. This branch circulation reaches its highest point at K, then falls back to the branch connection and enters the same return pipe as the longer branch.

It is obvious that no part of these auxiliary circuits may be higher than the water level of the cistern and it is advisable to keep them entirely below the level of the base of the cistern.

400. Loss of heat in secondary circulation. The loss of heat which occurs in secondary circuits must be compensated for (a) by using a larger range boiler, or (b) by encasing the pipes with slag wool or felt. So far as possible all pipes should be kept clear of external walls in order to avoid the effects of frost. Where they must be fixed to such walls the pipes should be placed on wood grounds backed with thick felt or asbestos, surrounded with slag wool and enclosed by a face board.

401. Grouping of service and circulating pipes. On comparison of details Nos. 124 and 125 it will be seen that both the cold and hot water service pipes are grouped along similar routes so that they may be economically protected and easily accessible. Except where risks of freezing are exceptional modern fitters advocate the placing of all pipes so that they are clearly visible except where they must pass through floors and ceilings. Where this plan is adopted care should be particularly exercised in selecting the route of such pipes so that they do not interfere with architectural features.

402. Points of danger in use of hot water systems. In connection with hot water systems the following points of danger arise through (a) stoppage of water supply, and (b) frost.

(a) The failure of the water supply can only be fruitful of danger in those cases where a system is fitted up in such a manner as will admit of the system being emptied down to the boiler level.

Some of the older systems have this drawback, but in the cylinder system the water cannot be withdrawn below the top of the cylinder, even in ordinary cases where the draw-off is branched immediately above it. The only danger lies in the use of a draw-off cock at the boiler, and the continued withdrawal of water therefrom when the main supply is stopped and the fire kept burning. Should this occur, and the heated boiler be suddenly filled with water from the supply, steam will quickly form and the boiler is liable to burst, if the draw-off cock has been closed in the meantime. A safety valve, if properly fitted, is the correct provision to meet this danger.

(b) During periods of frost it is always possible for the water in a heating system to become frozen and especially so if disused during such periods. It is well known that water which has been heated, freed from air and then allowed to cool freezes quicker and more solidly than water which has not been heated and the risks of certain hot water draw-off pipes being frozen during the night are thus greater than in the case of cold water pipes.

The freezing of single branches has no effect on the working of the system so long as free expansion is available, but if some portions of the flow and return pipes between the boiler and the cylinder become frozen and a fire is lighted in the grate the expansion will probably cause a burst unless a safety valve has been fitted. These pipes often freeze near the boiler when the latter is fixed in a range to a fireplace having a thin back in an external wall. Such an arrangement invites trouble.

403. Collapse of cylinders. Another trouble with frost is the ossible freezing of the expansion pipe where this has unwisely been terminated above a roof, or the pipe and cistern have been placed in an exposed position. Should this occur while the system is full of hot water, as the latter cools it contracts largely—or a quantity of hot water may be withdrawn—then, with all taps closed a partial vacuum is formed generally resulting in the collapse of the weakest vessel in the system, viz. the cylinder. The cause of this collapse is the external pressure of the atmosphere acting upon the surface of the cylinder without being met by a balancing pressure from within, nor resisted by the strength of the plate, which is generally designed to resist internal water pressure only.

Expansion pipes should be turned over the cold water cistern and the latter situated immediately above the hot water cylinder enclosure, communicating therewith by a latticed floor, so that in cold weather the water in the cistern does not easily become frozen.

404. Protection of cistern. The cold storage cistern should be encased with boarding and covered with a dust proof lid having wide overlapping edges. If any doubt exists of its vulnerability to frost, it should be packed round with slag wool, 2″ or 3″ thick, within the timber casing and the lid made double and likewise packed.

405. Lead safe. Accidental leakages from cisterns are often provided against by placing a "lead safe" underneath. With an overflow of adequate diameter the safe should be unnecessary.

The safe is usually a sheet lead tray fitted inside wooden bearers and dressed over their top edges. A free outlet is provided in direct communication with the atmosphere, by means of a lead pipe, $1\frac{1}{4}″$ diameter, fitted at the open end with a hinged flap which opens under water pressure and closes when not in use. When disused for lengthy periods these pipes are liable to be blocked with dust and dirt and may therefore become insanitary.

406. Overflow. Every cistern requires an overflow pipe for the removal of excess water, should the ball valve become inoperative. Where a safe is provided with a discharge pipe as described above, the overflow may be turned down the side of the cistern, within the casing, and terminated immediately above its outlet.

Where a safe is not provided the overflow would be taken directly through an external wall, and in any case the discharge from safe or overflow should be delivered where it cannot fail to be observed.

JOINTING AND SUPPORTING OF PIPES
FOR WATER SUPPLY AND WASTES

407. Wiped joints in service pipes. Lead pipes transmitting supplies of water under pressure must have joints at least as strong as the pipes; this is especially necessary for cold water services which in towns may be subjected to an internal pressure of from 25 to 75 lbs. per square inch or even more.

It is the usual practice to make all junctions in hot and cold service pipes by "wiped joints". The ends of the pipes for a longitudinal joint are prepared by scraping them clean for a length of at least one diameter from the ends, tapering one pipe and widening and shaping the other to receive it, as shown in detail No. 126 at A. A ring of black soil is then painted round each pipe to define the extent of the joint and prevent adhesion; the pieces are then assembled and a coating of semi-molten plumbers' solder wiped round to the required shape with a tallowed moleskin pad.

Branch joints may be made in a similar way and may be either at right angles or curved in the direction of the flow. The method of fitting the branch pipe into the continuous one is shown at B, and consists of cutting a hole in the side of the latter and expanding it to fit the splayed end of the former. When assembled the angle is coated with solder and wiped to an outline similar to the detail, the extent of which is determined by the soil margins.

Especial care should be taken to avoid obstruction at the joints from protruding edges or by inward leakage of the solder.

408. Taft or blown joints. Where lengths of pipe which will not be subjected to much internal pressure nor to considerable stress due to their own weight or to great changes of temperature are to be connected, the joints are made as shown at C—detail No. 126. One end of the pipe is expanded to a bell shape while the other is rasped to fit conically within it. The expanded end is then filled with solder, using a hot copper bit for small pipes or a blow pipe flame for larger joints.

This type of joint is only suitable for overflows, waste pipes and gas tubing.

409. Flange joints and block joints. Soil and other lead pipes, 3″ diameter and over, require, because of their weight, a very solid support which is often provided at a joint. A chase is formed in the wall, 9″ wide × 4½″ deep, and at intervals across this chase blocks of hard stone or wood 3″ thick are inserted, with holes cut through as shown at D—detail No. 126.

The supporting blocks will be placed at a little under 10 ft. centres because the pipes are supplied in 10 ft. lengths.

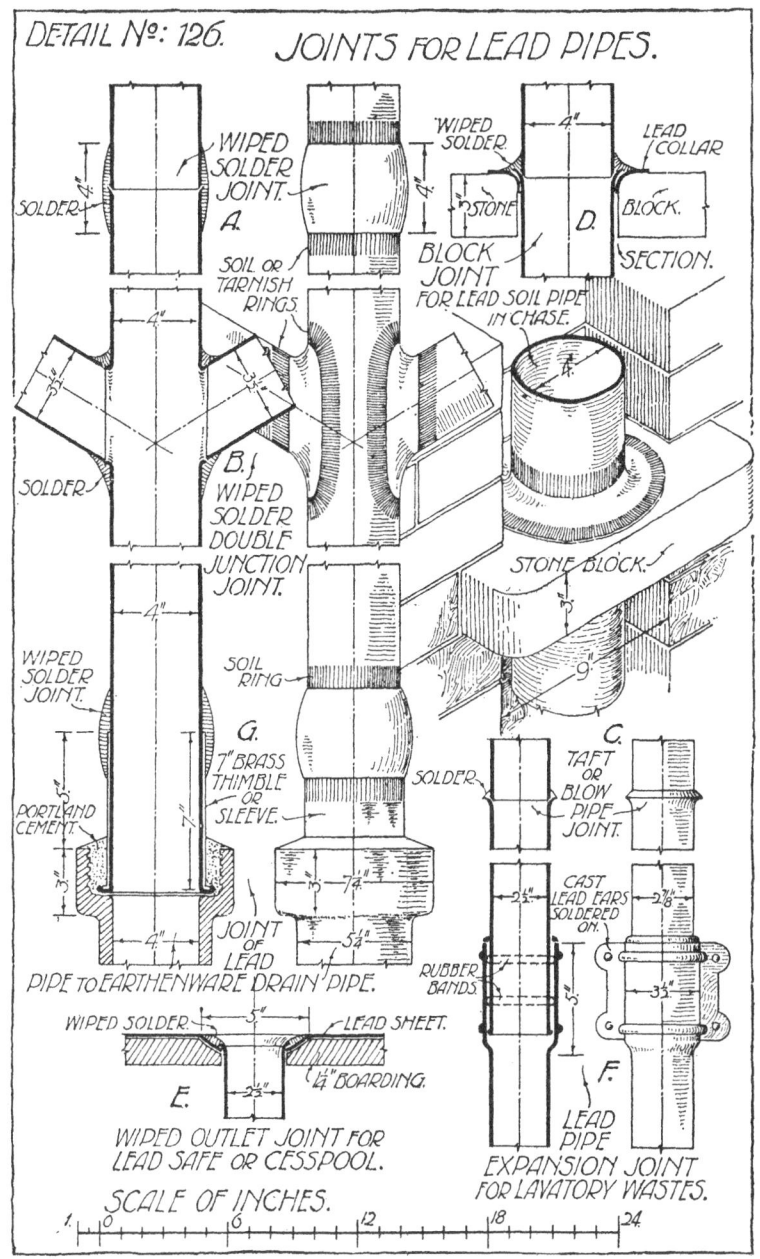

DETAIL Nº: 126.   JOINTS FOR LEAD PIPES.

To provide support the upper edges of the block are rounded and a sheet lead collar, $1\frac{1}{2}''$ to $2''$ wide, dressed down upon it. The lower length of pipe is inserted from below the block and its edges tafted out to prevent withdrawal. The upper length of pipe is then fitted into the expanded end by slightly tapering and lastly the joint surfaces "soiled", scraped clean and wiped to the form shown, leaving narrow soil margins to define the joint. The three pieces of lead at the joint are well united and the solder forms a firm ring of metal to support the pipe upon the block. This form of joint is more often used where pipes pass through floors.

410. Expansion joints to large lavatory wastes. Lead wastes to lavatories suffer by considerable expansion and contraction of the pipes where quantities of hot waste water are passed through them. The large changes of temperature so caused drag the joints, cause bulging of the lengths and so disfigure and damage the wastes.

Attempts to improve these conditions have been made by introducing expansion joints at the connection of the lengths. For this purpose special socketed pipes are required, as illustrated at F—detail No. 126—with strong cast lead ears soldered on. The socket and beaded rings can be prepared by the plumber from ordinary lengths of pipe.

A clearance of $\frac{1}{4}''$ is given within the socket and two pure rubber rings are forced to position as shown, the top end of the socket being loosely closed by a cast lead ring which moves up and down with the pipe as it changes in length. If properly fitted these joints avoid all bucklings of pipes and are very satisfactory until the rings perish. These must be overhauled and renewed at intervals, which is a drawback to the use of the joint.

Wiped outlet joint for safe or cesspool. The wiped soldered joint at E is repeated here as in Vol. I, so that it may be compared with other forms of joint.

411. Supporting and fixing small lead pipes. When run in vertical lengths small pipes are usually supported and secured to brick and stone walls by pipe hooks of the form shown at A in detail No. 127; these hooks are driven into the joints of the wall.

For horizontal or raking pipes the use of pipe hooks alone is most unsatisfactory, soon allowing of a distinct sag between the supports, as illustrated by the sketch at the top of the detail. In all such cases continuous supporting fillets should be plugged to the wall or nailed to timber, as illustrated at B, and the pipe clipped to the wall at intervals by hooks. At changes of direction an easy bend is much preferable to a quick one, as the latter causes excessive friction of the water and restricts the flow; wide bends may be

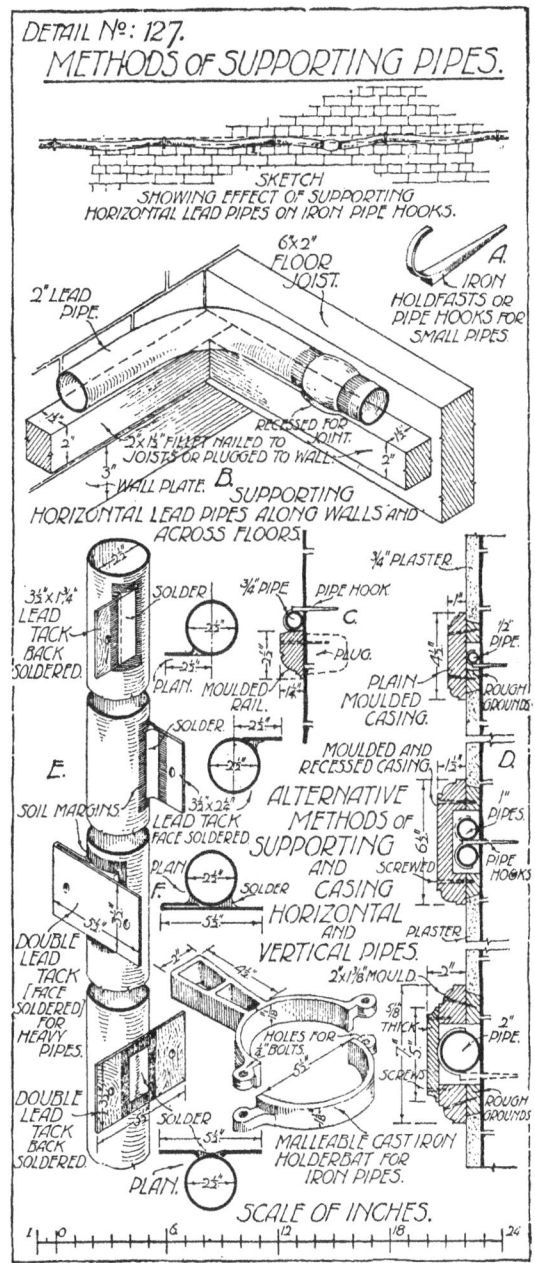

DETAIL Nº: 127.

METHODS of SUPPORTING PIPES.

SKETCH
SHOWING EFFECT OF SUPPORTING
HORIZONTAL LEAD PIPES ON IRON PIPE HOOKS.

6"x2"
FLOOR
JOIST.

A.
IRON
HOLDFASTS OR
PIPE HOOKS FOR
SMALL PIPES.

2" LEAD
PIPE.

RECESSED FOR
JOINT.
3"x1½" FILLET NAILED TO
JOISTS OR PLUGGED TO WALL.

WALL PLATE. B. SUPPORTING
HORIZONTAL LEAD PIPES ALONG WALLS AND
ACROSS FLOORS.

¾" PLASTER.

3½"x1¾"
LEAD
TACK
BACK
SOLDERED.

SOLDER.

¾" PIPE   PIPE HOOK

C.

PLUG.

½"
PIPE.

ROUGH
GROUNDS.

PLAN. MOULDED
RAIL.

PLAIN
MOULDED
CASING.

SOLDER.

MOULDED AND
RECESSED CASING.

D.

E.

SOIL MARGINS.

3½"x2½"
LEAD TACK
FACE SOLDERED

ALTERNATIVE
METHODS of
SUPPORTING
AND
CASING
HORIZONTAL
AND
VERTICAL PIPES.

1"
PIPES.

PIPE
HOOKS.

PLAN.

SOLDER

SCREWED

PLASTER.

DOUBLE
LEAD
TACK
[FACE
SOLDERED]
FOR
HEAVY
PIPES.

2x1⅛"MOULD.

⅝
THICK

2"
PIPE.

HOLES FOR
⅜"BOLTS.

SCREWS

DOUBLE
LEAD
TACK
BACK
SOLDERED.

SOLDER.

MALLEABLE CAST IRON
HOLDERBAT FOR
IRON PIPES.

ROUGH
GROUNDS

PLAN.

SCALE OF INCHES.

supported on angle blocks bevelled and nailed against the supporting fillets.

In exposed positions the fillet may be suitably moulded on the under edge and bevelled to prevent the accumulation of dust on the upper edge, as at C, or on plastered walls, as variously indicated at D.

When pipes are to be secured to visible woodwork, clips are preferable to hooks. These are supplied in galvanised iron and brass for fixing with screws, and are illustrated as applied to a W.C. flush pipe at detail No. 138.

412. Supporting and fixing large diameter lead pipes. Vertical lead pipes are secured and supported by heavy lead tacks $2\frac{1}{2}''$ or more in width and soldered to the pipe. Four methods of forming and attaching the tacks are shown at E.

The tacks may be single or double eared, of single or double thickness and soldered on the face or back.

Back soldering is easier to perform but face soldering is much the stronger method as may be seen by comparing the two upper sketches.

Single ears are suitably used for horizontal or raking pipes and double ears for vertical pipes. For heavy soil pipes and the like, the double thickness—or better, a cast lead ear—should be employed and the face soldered as shown at F.

Soil margins are indicated to each joint. The portion to be soldered is scraped bright to the correct size and shape for the soldered connection, leaving the margin in view; this margin remains when the work is completed and has a legitimate ornamental effect.

The ears are secured to walls by pipe nails or to woodwork by round headed brass screws and washers.

JOINTS IN IRON PIPES

Wrought iron pipes are often used for domestic water services and are more particularly suited to hard water.[1] They are usually $\frac{3}{4}''$ to $1''$ bore and supplied in 12 ft. lengths.

413. Straight joints. Joints in straight pipes are made by screwed jointing ferrules called "unions". The ends of the pipes are screwed for a length a little more than half that of the union and the joint made by coating the thread with red lead or special cement, wrapped with hemp fibre and screwed home.

Detail No. 128 at A shows a section of the joint.

414. Bends and branches. Bends and branches are made by special malleable castings or forgings, except in the case of very flat bends which may be formed in the lengths of pipe.

[1] See Chapter on Materials, Vol. II.

Right angle bends are most commonly required, as shown at B, the ends of the pipes being screwed in as before described and care being taken to maintain a uniform bore.

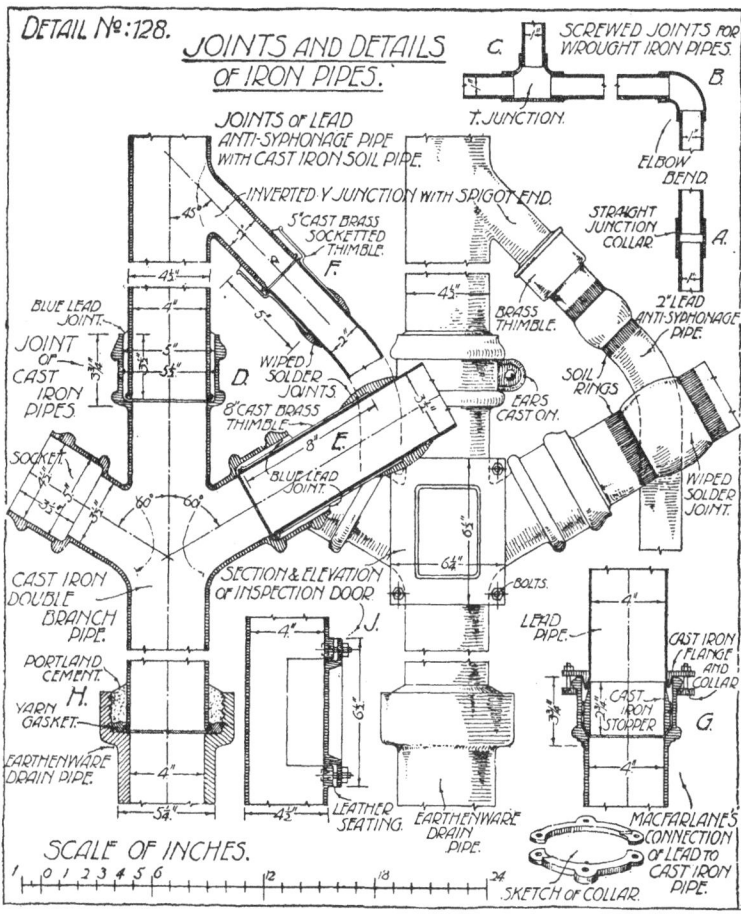

Right angled branches are made by inserting a tee piece as indicated at C, the continuous pipe being assembled first and the branch pipe screwed into the right angled arm of the junction afterwards.

415. Galvanised pipes. If used for soft water wrought iron pipes should be heavily galvanised. Even this protection is not sufficiently permanent, as many soft waters attack and dissolve the zinc coating and corrosion then takes place which discolours the water and is very objectionable.

416. Waste pipes of cast iron. Wastes from W.C.'s, slop sinks, and other groups of fittings are now usually conveyed through vertical cast iron pipes to the drains. Such pipes can be cast with branches in any desired position, as required for grouping discharges or attaching vent pipes, etc.

These pipes may be obtained from 2″ to 6″ internal diameter (or bore) and 6 ft. to 9 ft. long and are jointed by spigot and socket joints, as shown at D. The spigot end is cast with a projecting bead and the socket is large enough to leave an annular space round the entering pipe from $\frac{3}{16}$″ upwards; the larger pipes require $\frac{3}{8}$″ clearance. Sockets are usually moulded externally to strengthen the metal and emphasise the joint since it cannot be hidden; supporting lugs or ears are either cast on the socketed end or separately provided.

The jointing material is molten lead poured upon a ring of yarn gasket which is previously stuffed tightly into the bottom of the socket to prevent the jointing material entering the bore of the pipe. Lead filling of this kind is lightly caulked to consolidate the joint and the top edge splayed off to avoid rain being retained on the rim.

A material called lead wool may very suitably be employed.[1] It is a loose mass of lead fibre which can be hammered into the space and easily consolidated, thus avoiding the troublesome operation of melting the lead.

JUNCTION OF LEAD AND IRON PIPES

A special joint is required between lead and cast iron pipes, due to one metal being much softer than the other, which prevents a plain spigot and socket joint being used.

417. Use of brass thimble in "lead to iron" joints. The joint is shown at E, where the end of the lead pipe is first stiffened by securing a sleeve or thimble of brass about 8″ long to the outgo end. The thimble is prepared with a bead or flange at the lower end and the lead pipe passed through it, turned over to prevent withdrawal and then soldered by a wiped joint at its top end.

The end, so stiffened, can then be placed in the branch socket, packed with gasket and run with lead as before described.

At F, a variation of this joint is shown in which the thimble is provided with a socket to receive the spigot end of an iron pipe.

Inspection doors. An access or inspection door is shown at the junction of these cast iron pipes; its object is to provide for easy

---

[1] See Chapter on Materials, Vol. II.

clearance of obstructions and is further referred to in a later paragraph.

418. Patent connections, lead to iron. There are several methods, chiefly patented ones, which are employed for this purpose. An example is shown at G in the same detail.

A cast iron stopper is dropped into the socket of the iron pipe; its top end is tapered on the outer rim and its height is less than the depth of the socket.

The end of the lead pipe is slightly expanded, turned back and outwards to envelop a loose cast iron flange. This is, in turn, fitted to the top of the socket bearing upon the upper rim of the cast iron stopper and is held in position by a two-piece collar bolted to the loose flange.

Such joints are easily dissembled for repairs and renewals.

JUNCTION OF LEAD, IRON AND EARTHENWARE PIPES

419. Lead to earthenware. If a lead waste be employed, it will require to be jointed to the earthenware drain in one of two ways: (a) it may first be jointed to a cast iron intermediate piece, as previously described, and the latter connected to the drain, or (b) it may be stiffened by a brass thimble and the latter inserted into the earthenware socket, and jointed by gasket and cement in a similar manner to that employed in the detail No. 126 at G.

420. Iron to earthenware. The detail No. 128 at H shows a cast iron waste connected to an earthenware pipe. Because of the difference in the thicknesses of two such pipes, the annular space is excessive and care is required in making the joint and obtaining true alignment of the pipes. Yarn gasket is employed to pack the base of the joint and well cooled neat portland cement is inserted and rammed home. If the cement is not mixed too wet, a reliable joint is obtainable.

Other and special cements are also employed for these joints.

421. Lead to earthenware fittings. Joints of this kind occur at the outgo arm of a W.C., as in detail No. 129 at A, where the lead arm or bend receives the discharge and conveys it through the wall to a waste pipe. Stiffness is acquired on the lead pipe by jointing a socketed brass thimble to the former, the thimble passing within the lead to avoid any obstruction to the discharge. A wiped soldered joint is employed at this junction.

The brass thimble then receives the spigot end of the W.C. outgo and the joint is made by gasket and cement.

The detail also shows two special methods of jointing a lead pipe to the earthenware flushing arm at the back of a W.C. fitting.

In the sketch marked B the flush pipe is bent into a horizontal direction and a cast lead "octopus" connector slipped along it, followed by a rubber ring. The position of the latter limits the entry of the pipe into the socket and the joint is made by setting the connector in the correct position, soldering it to the pipe on the beaded edge, then drawing the pipe and connector close to bear against the rubber ring and binding the octopus rim in position with soft copper wire, as shown in section.

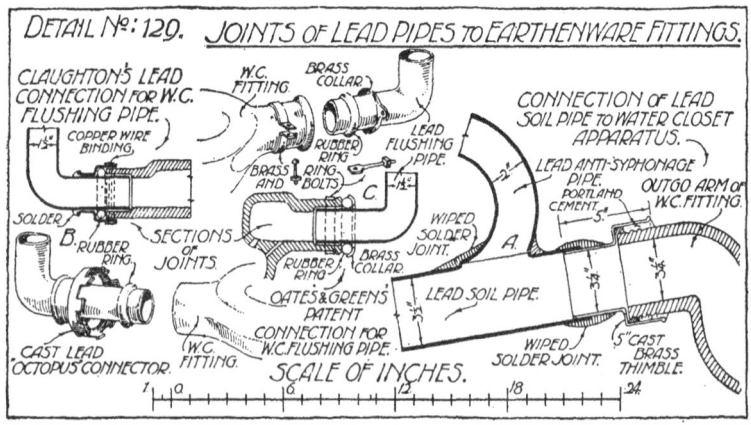

DETAIL Nº: 129.   JOINTS OF LEAD PIPES TO EARTHENWARE FITTINGS.

The example C shows another method in which a brass ring in two segments is clasped round the porcelain arm and attached with bolts; between the wings of this ring a horizontal draw bolt is placed and held by the eyelet end.

A brass collar and rubber ring are slipped on the flush pipe, the latter fitted into the socket, and a watertight joint formed by pulling the collar close against the ring of rubber as illustrated in the section.

These joints allow free movement of the pipe under changes of temperature, without damage to the connection, and only become defective when the rubber ring perishes. It is a simple matter to renew such parts.

SANITARY FITTINGS.

THEIR WATER SUPPLY, ACCESSORIES AND WASTES

Having considered the general provision for water supply to internal sanitary fittings and matters incidental thereto, this section is devoted to a study of the forms of fittings, methods of connecting and controlling their water supply, and the disposal of their wastes.

422. Taps, valves or cocks. Water is admitted to every fitting by some form of tap, generally of the screw down type. Those in common use include stop taps, bib taps (for sinks, baths and lavatories) and ball taps,[1] for automatically controlling supplies to cisterns. Most taps are of brass or gun metal, and are chromium plated. The best taps are of white metal throughout.

423. Stop cocks are placed on a line of pipe to shut off the supply. They are generally employed outside the building and actuated by either (a) a loose key fitting over a square headed spindle, or (b) an ordinary tap head. The former keeps the control in the hands of a responsible person, but the latter allows immediate use by anyone, in the case of a burst or other emergency requiring immediate discontinuance of the supply.

The body of a stop cock is formed with two tapered ends for convenience in jointing to the lead pipe by wiped soldered joints, or prepared to receive union connections for wiping to lead pipes, as at A in detail No. 130.

424. Bib taps are those forms designed to discharge water vertically over a sink, bath, or lavatory. They vary in shape according to their particular use and thus have distinguishing names.

Besides the screw down tap there are lever and spring fittings in use for special purposes. The latter are intended to prevent waste of water in schools and public institutions, because they close automatically when pressure is released. Screw down taps are often accidentally left open by careless people, but they do not so easily get out of order as other forms, hence their general use.

425. Screw down taps. All screw down taps are constructed on similar principles.

As an example of the type we may refer to the section and elevation of the bib tap at B. The fitting essentially consists of a swelled tube prepared for connection to the pipe at one end and having a pocket for the insertion of the mechanical portions of the tap. Across the body of the fitting is a diaphragm with a hole and ground seating to receive a plug or jumper which carries a washer of leather or india-rubber; the latter is forced down against the seating by the screw down spindle working in a tapped cap or casing, screwed into the body.

The spindle must work freely, but if left with clear play the water would leak out around it, hence the upper part of the cap is formed into a stuffing box, being recessed and tapped to receive a milled headed screw bush and the lower part of the stuffing box filled with greased gasket. On screwing the bush into position the gasket is

---

[1] See paragraphs 428 and 447.

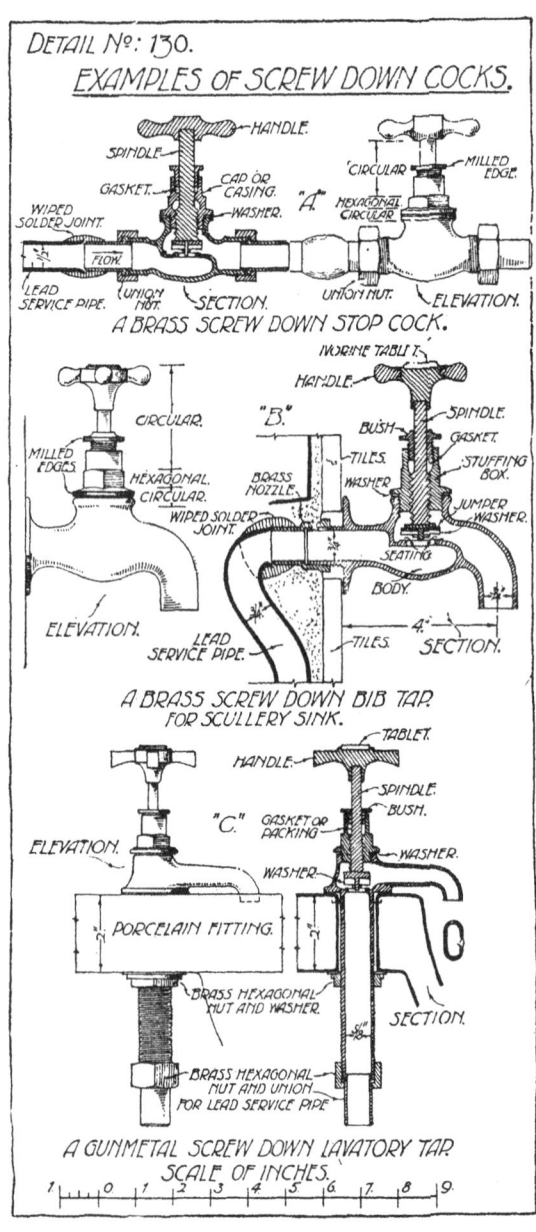

DETAIL N°: 130.

EXAMPLES OF SCREW DOWN COCKS.

A BRASS SCREW DOWN STOP COCK.

A BRASS SCREW DOWN BIB TAP.
FOR SCULLERY SINK.

A GUNMETAL SCREW DOWN LAVATORY TAP.

SCALE OF INCHES.

spread diametrically and pinches against the spindle, forming a close joint without restricting the necessary freedom. A separate cross handle or starred head is pinned or riveted to the spindle, providing leverage and convenience for actuating.

All the screw down taps in the detail have similar working parts, and are all made of brass or gun metal.[1]

Bib tap to sink. This is prepared for fixing against the tiled wall over the scullery sink and has a $\frac{3}{4}''$ bore at inlet and oultet, though somewhat restricted below the size at the valve. It is provided with a square shoulder bearing upon the tiles and is screwed at the inlet end to a brass nozzle which is wiped to the $\frac{3}{4}''$ lead service pipe.

To obtain secure fixing the wall is recessed to a depth of $2\frac{1}{2}''$, to form a dovetailed chase for the pipe, and the latter, with its nozzle end, is bent and sunk into the recess to present the nozzle at the correct depth for attaching the tap and the recess then filled with neat cement containing a small proportion of plaster of Paris. Iron holdfasts or pipe hooks may be driven into plugs to clasp the recessed portion of the pipe.

426. Lavatory bib tap. Lavatory taps in modern fittings are attached to the flat rim of the fitting surrounding the basin and are of simple construction. Detail at C shows the section and elevation.

The body of the tap is a screwed pillar with a flat arm at right angles thereto, and a flat seating for bearing upon the fitting.

To attach the tap to the rim it is screwed firmly from below by a brass nut bearing upon a leather washer. The extra length of the screwed pillar is intended to meet variations in the thickness of the rim of the lavatory basin.

In this case the supply pipe is assumed to be of wrought iron tube, $\frac{5}{8}''$ bore, with a shouldered end, enabling the connection to be made by a union nut which slips over the iron pipe, engages with the shoulder and draws it tightly against the end of the pillar; the surfaces in contact should be a ground fit or a washer of lead or leather should intervene.

If the connection is to be made to a lead pipe, a brass union piece takes the place of the wrought iron pipe and the lead pipe is wiped to the union.

427. Bath taps. The taps for a bath may be either connected through the sides or through the rim surrounding the top. If side connections are employed the taps are practically identical with the bib tap but with a vertical discharge as at A in detail No. 131.

The connection is made by a brass elbow with back flange and male thread and is in turn wiped to the lead service pipe.

At B is illustrated a pillar tap for attachment to the rim of the

---

[1] See Chapter on Materials, Vol. II.

bath. Its construction is similar to a lavatory bib tap, but the stuffing box and bush are cased in by a conical cover screwed down to a level shoulder on the body of the tap, the latter being of such a shape as to leave an unbroken surface which is easy to clean and has no corners for the accumulation of dirt.

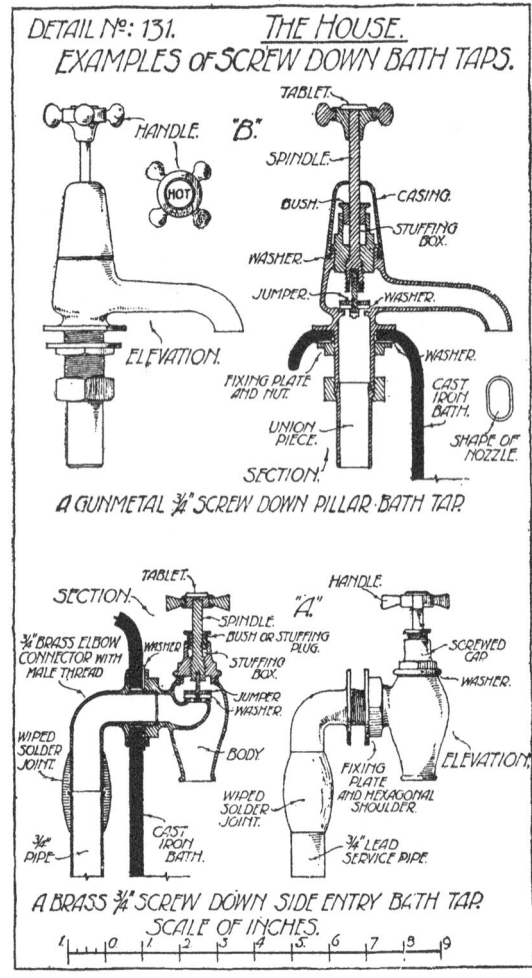

DETAIL No: 131.   THE HOUSE.
EXAMPLES of SCREW DOWN BATH TAPS.

A GUNMETAL ¾" SCREW DOWN PILLAR BATH TAP.

A BRASS ¾" SCREW DOWN SIDE ENTRY BATH TAP.
SCALE OF INCHES.

428. Ball taps or valves. There are several types of ball tap, three of which have been selected for illustration in detail No. 132.

The Portsmouth valve, marked A, is a low pressure tap and has a direct horizontal action. This tap has a cylindrical jumper or plunger carrying a rubber washer which impinges against an angular seating and is actuated by a cranked lever working in a slot and about a pin situated below the plunger. At the free end of the lever arm is a spherical copper float which rises or falls with the level of the water. When approximately horizontal the head of the lever

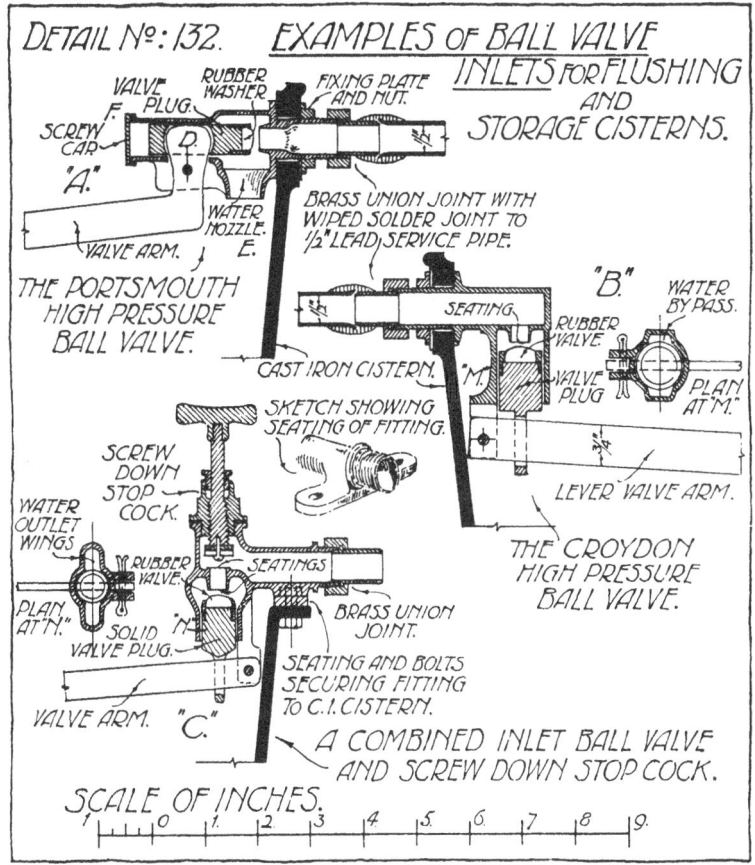

DETAIL № : 132. EXAMPLES OF BALL VALVE INLETS FOR FLUSHING AND STORAGE CISTERNS.

at D should press the plunger against its seating. If the lever falls the pressure is relieved and the water forces the plunger back, allowing it to stream through the orifice at E.

With high pressure supplies the water often leaks through the horizontal arm around the plunger and streams out horizontally; this is provided for by a screw cap at F which compels all water to pass downwards at E.

The tap is fixed to cast iron flushing cisterns by the usual nut and washer connection, and the lead pipe requires an intervening brass nozzle and union as in the case of bib taps.

The example at B is specially designed for high pressure services and is known as the Croydon valve; its plunger acts vertically in a cylindrical socket with two rectangular wing-spaces providing a by-pass for the water when the rubber washer is not in contact with the seating. A plan of the by-pass is given at M. It will be seen that the plunger has an eyelet at the base which threads loosely on the lever and moves vertically with it.

A further example at C shows a combined valve and stop tap which is largely used in some northern districts.

The ball valve has a vertical discharge and a vertical movement of the valve. Adjustment of flow can be made by the tap and the discharge stopped altogether for re-washering or in emergencies.

At the point of admission of water through the valve, it is shown in section that the plug is fixed to the base of the main spindle and screws up and down with it; a leather washer is screwed to the plug and bears upon the seating surrounding the small orifice of primary discharge. On passing this point the water travels round the plug to the outlet wings shown by the plan at N.

429. Use of stop taps to branches for cisterns, etc. Instead of using the combination valve and tap described above it is better to place a separate stop tap near the junction of the branch and main service pipe. It is then possible to cut off the whole of the branch service and repair any portion of the pipe or fittings. Should a frost burst occur on such a pipe the stop tap combination is of no use, because it only controls the supply to the fitting itself.

TRAPS IN SANITARY FITTINGS

430. It is stipulated in most sanitary regulations that every pipe conveying waste from a sanitary fitting must be provided with some form of trap.

Traps are bends or dips in pipes and channels, the centre lines of which are contained in vertical planes. Their function is to trap or enclose the air contents of the drain by confining such air within certain limits or compelling movement to occur along prescribed channels and thus prevent its exit at objectionable points.

There is considerable doubt as to the wisdom of employing traps at every waste outlet and especially to sinks, lavatories and baths. While traps are formed with a view to being self-cleansing every trap slowly accumulates waste solid matter; this is recognised by the provision of an access screw in order to remove obstructions when they interfere with the discharge.

These screws are often the direct cause of such accumulations, by destroying the continuity of the bore; they are seldom used until the pipe is blocked or objectionable smells arise. Since it is not a common practice to flush the fitting with fresh water after use these traps retain quantities of soiled water which may remain for some time, if the fittings are intermittently used.

It is usual to disconnect sink wastes, etc., by discharging them in the open air, over or near other trapped gulleys, and it is suggested that all such traps could be omitted and the pipes directed in as straight a line as possible to the outside air, to discharge over a channel leading to a gully.

Vertical waste pipes from upper storeys may be carried to the eaves as ventilators.

431. Traps in waste pipes to house. Since the present building and sanitary regulations require the insertion of traps to all the internal sanitary fittings, the details which follow necessarily comply, and show usual types of trap.

Traps are formed by bending or casting pipes so that they retain from $1\frac{1}{4}''$ to $2''$ depth of water after each usage, thus forming a "seal" between the waste pipe and the fitting. A trap must be placed immediately beneath and will be of **P** or **S** form according to the position of the discharge end.

432. Scullery sink and waste. A suitable sink for the scullery is shown in detail No. 133 and is a rectangular fireclay dish, $2' \ 4'' \times 1' \ 5'' \times 10''$ outside dimensions, glazed inside and out and provided with a weir overflow and channel, arranged to discharge into the brass waste connection below.

The sink is made to fall to its outlet and is supported on two $3'' \times 3'' \times \frac{3}{8}''$ B.S. Tees built $6\frac{3}{4}''$ into the wall, and projecting $1' \ 4''$. On each side of the sink is a $1\frac{1}{4}''$ teak draining board having $2'' \times \frac{3}{4}''$ margins, and supported by $3'' \times 2''$ wood cantilevers built $4\frac{1}{2}''$ into the wall.

Behind the sink and extending the full width of the draining boards the wall is covered to a height of 2 ft. with $6'' \times 6''$ white glazed tiles, their surfaces being flush with the surrounding plaster. The sink should be placed close against the bare wall so that the tiling finishes upon its top rim and leakage water from taps cannot drain behind it.

The waste pipe from the sink is $2''$ diameter and receives the discharge through a $2''$ solid drawn **P** trap set close below the outlet.

433. Connection of waste to scullery sink. The waste is connected by the use of a flanged brass tube, fitting through the outlet and bedded into a prepared rebate.

This waste is secured by a brass nut bearing upon a lead washer and the trap connected by wiping a shouldered brass union piece to its inlet end and screwing this to the waste pipe by a rebated union nut. The section at A illustrates the arrangement of the weir overflow to carry away excess water.

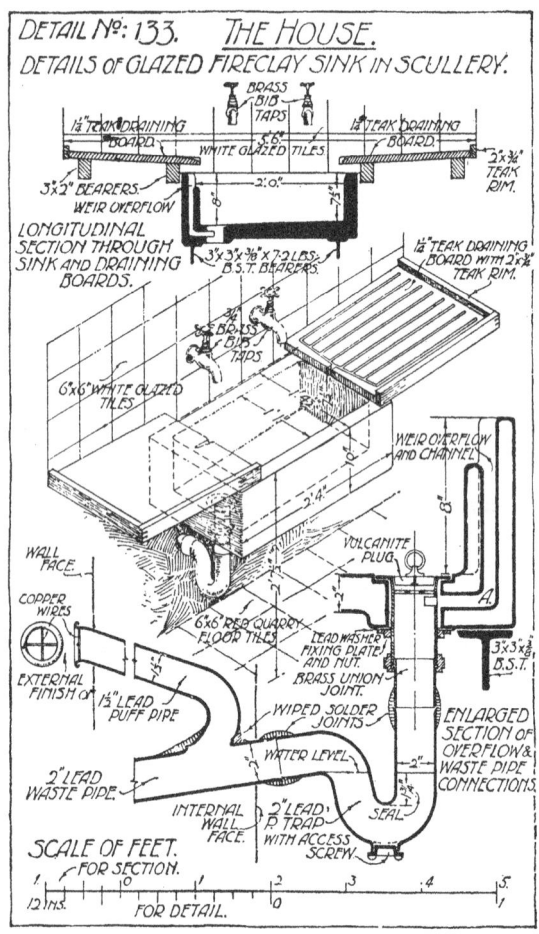

The overflow "weir" and channel are formed in the thickened end of the sink at about 2″ from the top and lead to a rectangular opening cut in the side of the brass waste.

434. Ventilation of waste. The waste pipe passes through the external wall and discharges into a fireclay channel (see Drainage). This pipe would become fouled with soap and grease and circula-

tion of air could not take place along the single pipe which is sealed
at the trap. Hence, a 1½″ vent or "puff" pipe is often attached to
the waste near the trap and the pipe carried through the wall as
shown in the lower section of the detail. The open end is left slightly
projecting from the face of the wall and guarded by cross wires.

435. Lavatory basin and waste. Detail No. 134 illustrates the
form of a lavatory basin. Its section is somewhat irregular in shape
to give a depth of 7″ near the back, gradually diminishing towards
the front margin.

The overflow is of a similar type to that previously described for
the scullery sink, except that surplus water enters the channel
through holes in the back. This channel is accessible for cleaning,
being continued to the sunk top of the fitting.

It should be observed that no ornamentation is included in the
lavatory basin which might tend to accumulate dirt or liquid, and
all surfaces which might appear horizontal are really sufficiently
inclined to drain into the basin.

The fitting is supported at the back by building the earthenware
lugs into the wall and at the front by turned metal legs housed
slightly into the floor and pinned into the rim of the fitting.

The brass waste is of the same form as that employed for the
scullery sink, but the trap has the form of a letter **S** because it is
required to discharge into a vertical waste pipe. As before, the
trap might be of lead, but a 1¼″ cast brass **S** trap is illustrated,
with an access screw on the dip, and a branch opening on the upper
loop to which a puff pipe is attached.

The brass trap is formed in the least possible width by closely
grouping the bends of the channel and the discharge leg may be
carried to the floor in one piece or jointed to any ordinary lead waste
pipe.

An iron service pipe supplies water to the bib taps and a short
length of lead pipe is interpolated to provide some relief under the
effects of expansion and contraction.

436. Bath and waste. Baths are made of cast iron with a stove
enamelled finish or vitreous glazing, of porcelain enamelled earthenware, and sometimes of galvanised zinc. Their top edges are curved
outwards to form a roll, which is often flattened to allow pillar taps
to be employed.

Every bath should either be left open to view and stand sufficiently clear of the walls and floor to allow proper access for
cleaning, or, be completely and efficiently encased.

*The modern parallel bath, completely encased with ornamental or
white enamelled slabs is now in general use and finds great favour.*

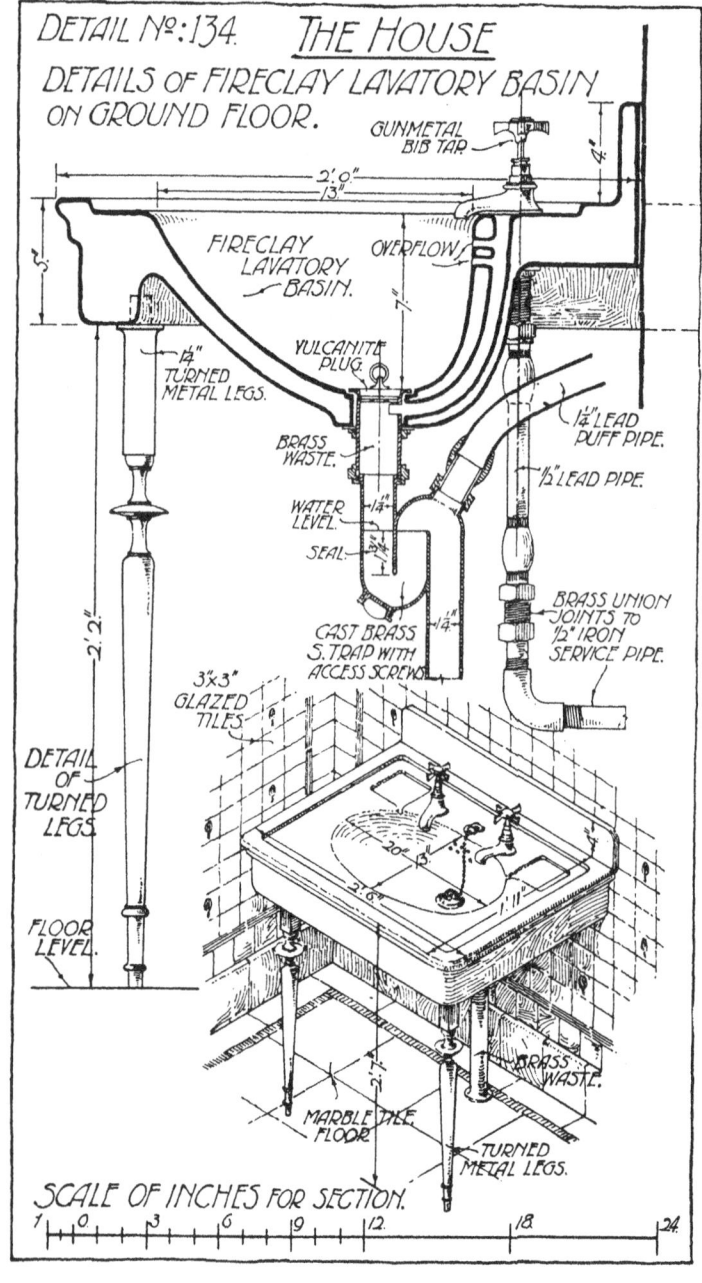

DETAIL Nº:134.   THE HOUSE

DETAILS OF FIRECLAY LAVATORY BASIN ON GROUND FLOOR.

GUNMETAL BIB TAP

2'.0"
13"

4"

FIRECLAY LAVATORY BASIN.

OVERFLOW

5"

7"

14" TURNED METAL LEGS.

VULCANITE PLUG.

1¼" LEAD PUFF PIPE.

½" LEAD PIPE.

BRASS WASTE.

WATER LEVEL.

1¼"

SEAL.

3¼"

BRASS UNION JOINTS TO ½" IRON SERVICE PIPE.

2'.2"

CAST BRASS S. TRAP WITH ACCESS SCREWS.

1¼"

3"×3" GLAZED TILES.

DETAIL OF TURNED LEGS.

20"

2'.0"

1'.11"

FLOOR LEVEL.

2'.1"

BRASS WASTE.

MARBLE TILE FLOOR.

TURNED METAL LEGS.

SCALE OF INCHES FOR SECTION.

1   0   3   6   9   12   18   24

The waste outlet pipe, with vulcanite plug and chain, may be used, the former being connected to an ordinary lead trap.

In detail No. 135 the bath is shown fitted with a "pillar overflow" waste, shown enlarged in detail No. 136.

The waste outlet of the bath is connected to a $1\frac{3}{4}''$ brass tap with deep seal, and wiped to a similar sized waste pipe. Large bore pipes should be used for bath wastes so ensuring speedy discharge and rapid re-use of the fitting on occasions, besides aiding in cleansing the pipe.

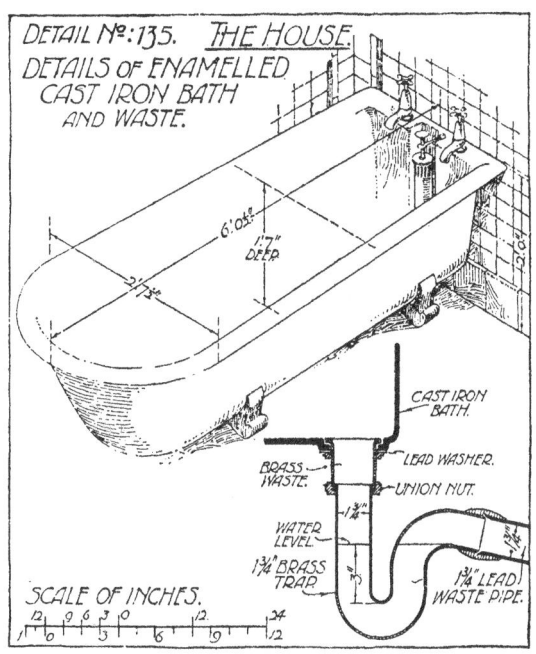

437. Overflow waste pipes. In detail No. 136 are grouped enlarged details of various overflow wastes. That at A is the ordinary type of brass waste with side inlet overflow, vulcanite plug and brass grating; this latter, though often a fixture, should be made removable.

At B is a side outlet overflow for a bath, which consists of a sharp brass elbow, having a back flange and screw nozzle to receive a cast brass grating; the joint is made watertight by a leather washer. The overflow pipe is jointed to the elbow and connected to the inlet arm of the lead trap.

A detail of the pillar waste and overflow is shown at C. It consists of a cylindrical pipe plug with two series of holes, one situated near the top to act as an overflow and the other at the base and enclosed by the outer brass lining of the waste when the plug is dropped.

The plug may be turned to disengage from a pin in the horizontal slot and lifted vertically, then turned backwards so that the lower rim rests on the pin. In this position the lower series of holes is

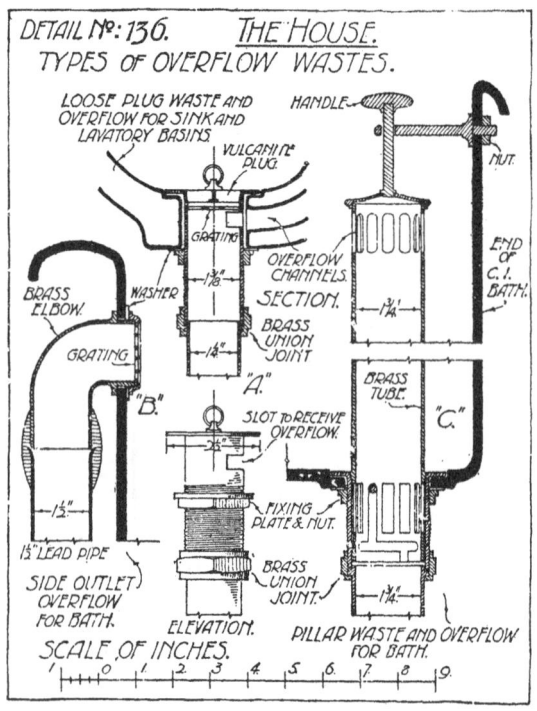

above the base of the bath and the waste water is discharged. When dropped into position the conical surfaces of plug and lining fit accurately and stop the outlet.

438. Lead safe. A lead safe, as described in paragraph 405, is often placed beneath a bath to provide for accidental leakages from joints, burst pipes and careless usage.

It is formed by nailing $2'' \times 1''$ fillets on the floor to enclose a rectangle and sheet lead in one piece is turned up $3''$ against the wall and dressed and copper nailed over the marginal fillets.

The safe is drained as already described. It is, however, better to provide an impermeable floor to a bath room, surfaced with marble tiles, jointless flooring or terrazzo.

WATER CLOSETS

439. A modern water closet generally consists of a glazed earthenware pan and pedestal, the former being of suitable form for the deposit of human waste and allowing it to be flushed away into the drains through closed channels of lead, iron, or stoneware.

An essential part of every W.C. is a trap at the outgo, which, when sealed with water, prevents the ingress of foul air from the drains to the apartment. This trap is *usually* an integral portion of the pedestal and consists of a double bend of **P** or **S** shape leading from the base of the pan to the outgo arm; its purpose is to seal the outlet by interposing a constant depth of about $2\frac{1}{2}''$ of water between the outgo level and the top inner surface at the dip of the trap.

The outlet arm may be made to discharge vertically or almost horizontally, the former requiring an **S** trap and the latter a **P** trap at the outgo.

In all cases the pedestal is provided with a hollow internal rim, communicating with a flushing inlet at the back to which the down pipe from the flushing cistern is connected. A portion of the flush may be directed immediately downwards into the pan, but the greater portion should pass round the rim and thus discharge down the inner surfaces of the pan at the front and sides.

An examination of detail No. 138 should make the form clear and familiarise the student with the names of the parts.

440. Essential features of a W.C. fitting. The essential features of a good W.C. are:

(*a*) The pedestal should be smooth, well glazed and easily cleaned; raised ornament should be avoided.

(*b*) The pan should contain sufficient water to cover the deposit without unduly retarding the flushing action.

(*c*) The shape of the pan should be such as to avoid fouling the surfaces when properly used.

(*d*) The trap should gradually contract to the least cross section at the bottom and retain an even bore from this point to the outgo.

(*e*) The flushing rim should be arranged to wash the entire inner surface of the pan while an adequate flush is retained to act directly on the contents.

(*f*) An adequate and sufficiently sustained flush should be provided to remove the contents and thoroughly cleanse the surfaces of the fitting.

(*g*) The whole of the fitting should be as simple as possible.

Types of W.C. There are considerable differences in the arrangement and sectional form of W.C. fittings, and detail No. 137 assembles a few of the better known types.

441. Wash-out W.C. No. 1 is known as a "wash-out" pedestal closet; the pan is very shallow and in receiving the deposit creates an unnecessary nuisance.

The discharge of flush water carries the contents of the pan against the face of the pedestal before entering the trapped outlet and this surface is very liable to be fouled. Further, much of the cleansing force of the flush is lost before acting on the contents of the trap. For these reasons the wash-out pedestal has been almost entirely replaced in recent years by the "wash-down" and other types.

442. Wash-down W.C. In this form, see detail No. 138, the one in most common use is shown, a considerable body of water being contained in the pan to receive the deposit, which directly enters the mouth of the trap. The flush acts directly upon the contents and with greater effect.

Examples Nos. 2 and 3 in detail No. 137 show alternative arrangements for the wash-down W.C.

The former has a front vertical outlet, while the latter is an improved form from a sanitary standpoint, the pedestal being omitted and the fitting supported by building into a wall; it is known as a "corbel" W.C., and is well adapted for hospitals, etc., where it is necessary to keep the floor space free for rapid cleansing.

443. Syphonic W.C. Example No. 4 represents an improvement on the ordinary wash-down W.C. In the latter form the closet pan is emptied by the gravitational effect of the flush, but the improvement causes a syphonic action in the removal of the contents, by turning the outgo

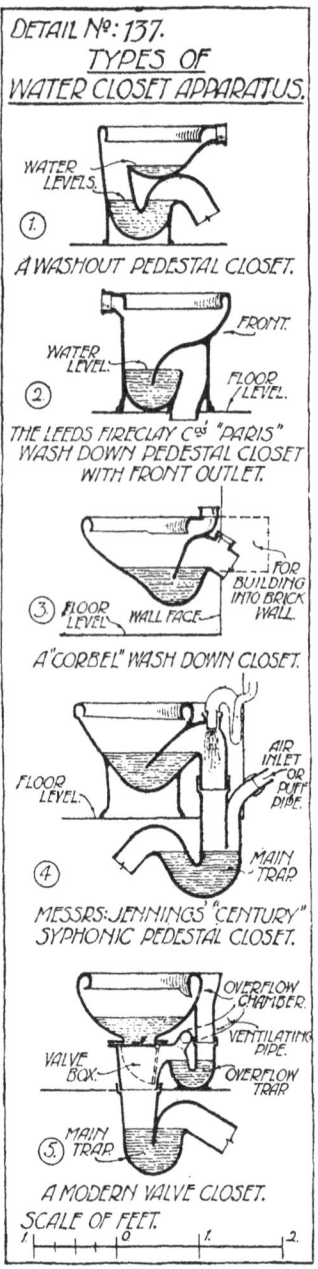

DETAIL Nº: 137.
TYPES OF
WATER CLOSET APPARATUS.

WATER LEVELS.
(1.)
A WASHOUT PEDESTAL CLOSET.

FRONT.
WATER LEVEL.
FLOOR LEVEL.
(2.)
THE LEEDS FIRECLAY Cº "PARIS" WASH DOWN PEDESTAL CLOSET WITH FRONT OUTLET.

FOR BUILDING INTO BRICK WALL.
(3.) FLOOR LEVEL. WALL FACE.
A "CORBEL" WASH DOWN CLOSET.

AIR INLET OR PUFF PIPE.
FLOOR LEVEL.
MAIN TRAP.
(4.)
MESSRS: JENNINGS "CENTURY" SYPHONIC PEDESTAL CLOSET.

OVERFLOW CHAMBER.
VENTILATING PIPE.
VALVE BOX.
OVERFLOW TRAP.
MAIN TRAP.
(5.)
A MODERN VALVE CLOSET.
SCALE OF FEET.

pipe vertically downwards from the trap to a sufficient distance
to ensure the effect.

The object of this W.C. is to empty and cleanse the pan with an
economical volume of water and its efficiency depends upon reduction of air pressure in the outgo leg; there should be a sufficient
flush to charge this leg fully and a 2″ flush pipe is usually necessary.

A well designed syphonic W.C. is probably the ideal modern
fitting. It should be simple in construction, function with certainty
and require little or no attention.

444. Valve W.C. Example No. 5 illustrates a "valve" W.C.
which consists of a large pan having a central outlet controlled by
a hinged valve which is actuated by a lever attached to a lifting
handle at the side of the fitting. There is no flushing cistern required, the contents being carried downwards by the large body of
water and the pan refilled automatically for subsequent use. The
filling is done direct from a service pipe and the control exercised
by an air-checked valve through a weighted system of levers; the
air-check regulates the rate of flow so as to deliver the correct
quantity of water to the pan.

The outlet valve is placed in an iron box, intervening between
the pan and a lead trap. This latter has a considerable vertical drop
at the inlet and is reduced in diameter before reaching the foot of
the bend.

A good valve W.C. is accepted as being amongst the best obtainable fittings, but is expensive to instal.

445. Wash-down pedestal closet to house. Detail No. 138 gives
a perspective sketch of the complete apparatus selected for the
house, together with sectional details of the pedestal and accessories.
The wash-down pedestal is known as the Municipite pattern, made
by Messrs Oates and Green of Halifax.

It provides ample water surface and volume, with an almost
vertical back, a seal 1¾″ deep, a small after flush chamber and
shows alternative forms of outgo arm. A 1¼″ hinged mahogany seat
is provided and secured by bolts to the pedestal. A typical water
waste preventing cistern is shown fitted with a Portsmouth ball
valve, and provided with a ¾″ lead overflow pipe.

The 1⅜″ brass flush pipe is jointed to the cistern by a union,
secured to the wall with cast brass clips and connected to the flushing
rim of the closet with Oates and Green's patent joint, as illustrated
in detail No. 139.

In many cases this joint is made by passing the end of the pipe
into the arm—or widening out and dressing over it—and making
the joint watertight by wrapping with strips of oil painted canvas
or linen, tying with oiled cord and again painting. This is a clumsy

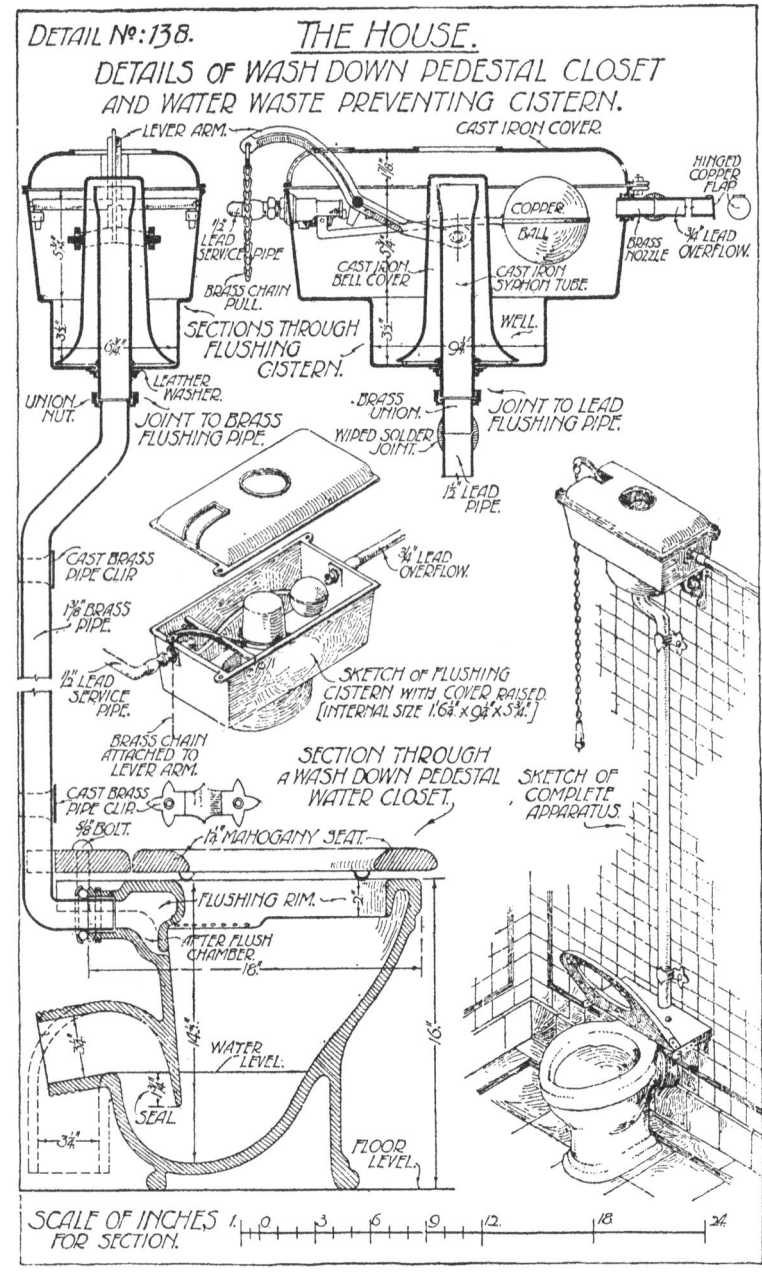

DETAIL Nº:138.   THE HOUSE.
DETAILS OF WASH DOWN PEDESTAL CLOSET
AND WATER WASTE PREVENTING CISTERN.

LEVER ARM.
CAST IRON COVER.
HINGED COPPER FLAP.
COPPER BALL.
BRASS NOZZLE.
¾" LEAD OVERFLOW.
1½ LEAD SERVICE PIPE.
CAST IRON BELL COVER.
CAST IRON SYPHON TUBE.
BRASS CHAIN PULL.
WELL.
SECTIONS THROUGH FLUSHING CISTERN.
LEATHER WASHER.
UNION NUT.
JOINT TO BRASS FLUSHING PIPE.
BRASS UNION.
WIPED SOLDER JOINT.
JOINT TO LEAD FLUSHING PIPE.
1½ LEAD PIPE.

CAST BRASS PIPE CLIP.
1⅛ BRASS PIPE.
¾" LEAD OVERFLOW.
1½" LEAD SERVICE PIPE.
BRASS CHAIN ATTACHED TO LEVER ARM.
CAST BRASS PIPE CLIP.
⅝" BOLT.
SKETCH OF FLUSHING CISTERN WITH COVER RAISED. [INTERNAL SIZE 1'6¼" x 9¼" x 5¾"]
SECTION THROUGH A WASH DOWN PEDESTAL WATER CLOSET.
SKETCH OF COMPLETE APPARATUS.
1¾" MAHOGANY SEAT.
FLUSHING RIM.
AFTER FLUSH CHAMBER.
18"
WATER LEVEL.
SEAL.
3¾"
FLOOR LEVEL.
16"

SCALE OF INCHES FOR SECTION.
0   3   6   9   12.   18   24.

joint and not to be recommended. Another method is to encase the joint by a rubber cone, tightly bound with copper wire; the rubber is likely to perish and the joint thus become defective.

446. Position of outgo of W.C. for various conditions. The outgo of a W.C. may be obtained to pass out horizontally or vertically at the back of the pedestal, vertically at the front or at some inclina-

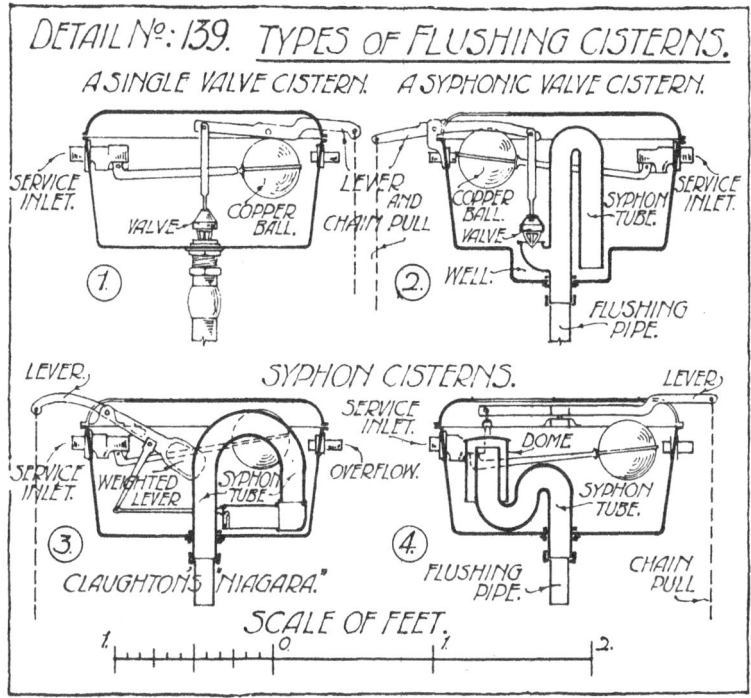

tion to the sides in plan. For isolated ground floor W.C.'s the outgo ends are often vertical and connected directly to the drain pipes at the floor level.

At higher levels, it is usually expedient to discharge the W.C. in a nearly horizontal direction through a branch pipe, which is in turn connected to a vertical soil pipe by bends and branches.

In order to place the outlet in any desired direction for meeting the vertical soil pipe, the pan and trap are sometimes made in two pieces, with a circular spigot and socket joint; the trap can then be rotated into any convenient position and cemented at the connection which should be below the water level of the trap.

447. W.C. flushing cisterns or water waste preventers. Flushing cisterns are necessary accessories to all fittings which receive liquid and solid filth and which require a strong flush of water to remove the contents, float them down the drains and cleanse the surfaces of the fitting and channels in the process.

A W.C. flushing cistern should be designed:

(a) to hold at least two and a half gallons of water to flush and cleanse the pan of the closet;

(b) to prevent waste of water by discharging only the necessary quantity;

(c) to provide for the sudden release of all the contents and for automatic filling of the cistern immediately after it has been emptied;

(d) to be simple in principle and strongly constructed.

448. Types of flushing cistern. Detail No. 139 shows in diagrammatic form some of the common types of flushing cistern in general use.

No. 1 is a single valve type where a conical plug is lifted from its seating by a pull down of the lever and allows the water to flow into the flush pipe. The objection to this form is the necessity for holding down the lever until the whole of the contents have passed into the pipe.

Nos. 2, 3 and 4 show various forms of syphonic action, which depends upon reduction of atmospheric pressure to empty the contents, after the action has been started mechanically by a single pull of the lever chain. The difference between these examples is merely in the detail of the mechanical action initiating the syphonic discharge.

449. Syphons. One form of the common syphon is a bent tube dipping into a liquid at one end and open to the atmosphere at the other. If atmospheric pressure be reduced at the open end by withdrawing the air the pressure on the open surface of the liquid forces some of the liquid up the tube, over the bend and down the open leg. If the quantity of water in the open leg of the syphon is such as to make the column of liquid longer than that in the other leg, then its own "head" or vertical height produces greater *downward* pressure against the atmosphere on this side than on the one connected with the liquid. Equilibrium is thus upset and the water continues to move in the direction of the greater pressure until the whole of the liquid is withdrawn, or until air can enter the feeding end of the tube. The movement is due to the reduction of effective atmospheric pressure on the open end of the tube, caused by the weight of the liquid on that side, which stands below the level of the free water surface in the vessel.

In each application to the flushing cistern, there is a feeding leg and a discharge leg to the syphon. Whatever action is employed to reduce the pressure in the discharge leg, the object is to cause the water to overflow suddenly into the leg from the full cistern, break up into sprays as it descends, and thus carry downwards sufficient air from the discharge leg to reduce temporarily the pressure. If the flush pipe be so charged with water that its level reaches below that of the level in the cistern, movement will continue through the pipe, increase in velocity until the flush pipe is filled and so withdraw the water from the cistern to the level where air is re-admitted to the feeding leg. A good cleansing action to the fitting is thus ensured.

450. Mechanical actions to charge flushing syphons. Detail No. 139, example No. 2, shows a method of starting the syphon by using a bent tube and valve. The flush pipe continues upward within the cistern and is bent over and downwards nearly to the base of the well, which stands below the general body of the cistern. Projecting from the side of the outgo leg is a branch bend, closed by a vertical acting valve which is opened at will by lifting the lever.

When water is admitted by the ball tap it covers the valve and rises within the syphon tube but does not quite reach the head of the bend. Let the lever be pulled and the valve lifted for a sufficient time to charge the flush pipe with a broken stream of water, then air is carried downwards, the internal pressure reduced and the water in the cistern pressed downwards, causing a rise and overflow in the syphon tube.

By this time the valve is closed, but, once charged, the action continues until the cistern is empty; it then automatically refills over the closed valve.

Example No. 3 shows a similar syphon but with an easier bend. It is initially started by a horizontal piston working loosely in the level continuation of the syphon inlet, along the base. Let the cistern be full; if the lever be pulled the weighted end rises and drags the branch lever with it. The latter is fixed at the top end and pivoted freely at the bottom to a horizontal rod having a circular piston; the latter fits rather loosely and has a hinged flap formed in the disc, opening inwards.

As the piston moves forward the flap remains closed, owing to the resistance of the water, and a quantity of liquid is forced up the inlet and over the bend; air is carried downwards and the syphonic action started as before. The out-rush of water holds the piston back until the cistern is emptied, then the weighted lever drags it back into its former position, and as the water again rises its pressure opens the valve and allows the syphon tube to fill ready for the next usage.

Example No. 4 is a further illustration showing a third method of starting the syphonic action.

The syphon tube in this case has a double bend and terminates in an open ended vertical tube. The action is dependent upon the trap at the lower end of the syphon arm being sealed with water—this occurs at the first attempt to use the fitting. Let the trap be full, then, as the in-flowing water rises round the tube and domed cover, it encloses air within the dome and compresses it, because it is locked between two bodies of water; being elastic the air reduces in volume under the pressure of the rising liquid and probably forces a small quantity out of the trap. To be effective the dome must be of thick metal to give adequate weight.

On pulling the lever the dome is quickly raised and the body of water under atmospheric pressure rises in the dome and overflows into the inlet tube of the syphon. The incoming water causes the trap to overflow, carrying air from the bent tube down the outlet pipe, thus relieving the pressure on the outgo side, upsetting equilibrium and starting the syphon. The action continues until air can again enter below the rim of the dome.

451. Flushing cistern and connections as applied to the house fittings. Detail No. 138 shows the construction of a flushing cistern constructed on a different mechanical principle from those shown in detail No. 139. The inlet leg to the flush pipe is a vertical continuation of the latter, with an expanded top. Any water falling into it quickly is guided by the splayed surface towards the centre of the tube, shoots off the slope and drops through the air in a broken stream, carrying some air with it.

The "lift" of the bell or dome is considerable and causes water to overflow the rim of the syphon tube quickly, thus carrying sufficient air to relieve the pressure and start the syphon. This action is accentuated by the falling of the heavy dome.

The detail sections and isometric view show how the bell is lifted and the ball valve placed for clearance of the former.

To connect the valve the screwed end of the tap is passed through a hole in the side of the cistern, packed with bevelled washers and secured with a back nut.

The overflow pipe should pass through an external wall at a convenient point and discharge into the open air, as described in paragraph 406. Overflows should never discharge into gutters, pipes, or other hidden places where the waste would not be observed, or where temptation exists to allow the waste to continue.

Many modern flushing cisterns are of the low-level type, being placed about 18″ above the level of the W.C. seat. They are actuated by a side lever which operates mechanism very similar to other types of cistern. Silent-filling flushing cisterns are also obtainable and should be used where the ordinary cistern would prove an annoyance. The principle usually employed is to allow the water to flow in near the bottom of the cistern.

452. Covers to cisterns. Cisterns are often covered, partly for cleanliness but primarily to stifle the suctional noises caused by flushing and the hissing of the water during filling.

The cisterns shown in the previous details are all of cast iron and covers are easily fitted, as shown in No. 138, where the diagonally opposite angles are provided with lugs for bolting down. In this case the lid is slotted for free play of the operating lever and a central hole provided to ensure a free lift for the dome. Obviously there must always be free admission of air to the cistern.

Cisterns of wood, lined with lead, are commonly used, but are seldom covered.

453. Connection of flush pipe to cistern. Flush pipes are usually of lead, $1\frac{1}{2}″$ diameter × 14 lbs. per yard, though in some cases they are of brass or copper, to avoid denting and to give a better appearance.

These alternatives, with their connections, are shown respectively in the two sections of the cistern in detail No. 138. The lead pipe is soldered to a brass union, which in turn is attached to the projecting syphon tube by a union nut.

454. Supports for cisterns. Cisterns are usually supported by (a) pieces of timber or metal built into the walls across narrow apartments, (b) metal brackets screwed to back boards plugged to the walls.

The former method is ugly but very strong and the latter usually employed where neatness is essential.

455. Soil pipes. The vertical pipes conveying waste from W.C.'s and some other fittings are called soil pipes and may be either of lead or heavy cast iron, and should, wherever possible, be fixed upon the outer faces of external walls.

The joints in such pipes should be perfectly airtight and are arranged as shown in details Nos. 126 and 128.

456. Cast iron soil pipes and connections. Cast iron soil pipes are $3\frac{1}{2}''$ to $6''$ internal diameter and to give reasonable service these pipes should have the following weights:

For $3\frac{1}{2}''$ bore,	24 lbs. per yard	⎫	
„ 4 ″ „	27 „	⎬ Average weight including	
„ 5 ″ „	35 „	⎪ sockets and ears.	
„ 6 ″ „	42 „	⎭	

Soil pipes are usually $4''$ in diameter, the larger sizes only being required where several tiers of W.C.'s are arranged with three or more fittings in each tier.

All soil pipes should be adequately preserved by coating with, or dipping in, some suitable solution. For internal preservation molten bituminous mixtures such as Dr Angus Smith's are best, the pipes being heated and "dipped". The exterior may be painted with oil paint after exposure to the air for some time.

Supports should be strong and the pipes packed $2''$ clear of the wall to allow of subsequent painting at intervals and to give the clear access required for making joints.

Holder bats such as that shown in detail No. 127 are suitable for this purpose.

457. Inspection pockets in cast iron pipes. Stoppages in pipes are most liable to occur within the bend of a branch pipe or near the junction of a branch with the main pipe. This is especially so where two branches join at the same level, because the discharge from one branch pipe may enter the other if the connection is not sufficiently oblique or otherwise well curved in the direction of the flow.

To avoid trouble and expense in clearing stoppages and also to allow of the inspection of junctions, "inspection pockets" are arranged, one of which is shown at J in detail No. 128. This type is approved by the London County Council and consists of a flat panelled cover bolted to a flanged seating cast on the pipe across the junction. Leather packings are used on the seatings to make the joint watertight.

458. Lead soil pipes. Lead soil pipes should be solid drawn, viz. without vertical seams, and of thicknesses which vary with the diameter. These pipes are usually sold in 10 ft. lengths and should weigh at least:

For 3½″ bore,	65 lbs. per 10 ft. length	This is equivalent to being
„ 4 ″ „	74 „	„	made of lead weighing 7 lbs.
„ 5 ″ „	92 „	„	per ft. sup. with a thickness
„ 6 ″ „	110 „	„	of about ⅑″.

Detail No. 140 shows the application of a lead soil pipe to the tier of W.C.'s in the warehouse. The several lengths of pipe are connected by wiped joints and the inlet branches from the W.C.'s are also wiped obliquely, both as described in paragraph 407.

In this case the soil pipe is supported by thick cast lead ears placed at 6 ft. centres and screwed to plugs in the wall.

459. Scheme of ventilation to soil pipes and branches. It is probable that pipes conveying soil wastes will become more or less foul.

To guard against this probability it is necessary to provide for currents of fresh air constantly to traverse the pipes and branches as far as feasible, and thus aerate the channels, and render innocuous any results of decomposition.

Soil pipes are therefore carried above the eaves of roofs and left open at the top, except for a wirework guard; air currents are admitted through one or more openings to the drainage system, usually at an inspection chamber, and are induced to traverse the whole length of the earthenware drains and soil pipes connected to them, see paragraph 493.

In some cases, only the soil pipe is so ventilated, the short branch pipes receiving no attention, and being therefore peculiarly subject to accumulations of soil in all cases where the flush is faulty.

460. Ventilation of soil pipe branches. In detail No. 140 a 2″ lead pipe is connected to the branch as close to the W.C. outgo as may be convenient and sometimes to a special nozzle attached to the outgo arm of the earthenware pedestal, and is known as an anti-syphonage or vent pipe.

Whatever mode of connection be adopted, the vent branch should not present any obstruction to the flow nor create a tendency to accumulate filth in the inlet by having its inclination facing the current. It should bend backwards so that the discharge freely passes the opening, as shown at A in detail No. 129.

For a single W.C. the vent pipe should be passed through the wall and connected to the soil pipe at a height of at least 4 ft. above the soil branch connection. With this arrangement the up-current in the main pipe is partially diverted and traverses the branch pipe and vent, rejoining the up-current in the main pipe at the higher level and being assisted in its circulation by the suctional effect.

461. Ventilation of a tier of soil branches. When several branches enter one soil pipe at different levels as in the case of the warehouse, it would be wasteful and inefficient to make several connections to the main pipe and also difficult to prevent the upper ends of the vents from obstructing the soil discharge. One principal vent pipe is therefore carried from the lower W.C.'s to a height of about 4 ft. above the highest fittings and all the intermediate branch vents connected to it.

In a tier of W.C.'s when a discharge occurs from an upper closet, a suctional effect is set up, tending to withdraw the contents of the traps which communicate with the same pipe, by creating a partial vacuum. The vent pipe then acts as an anti-syphonage pipe by supplying air and thus preventing the formation of such a vacuum.

The main vent pipe may be kept within the building or taken up the external wall; the former arrangement is fairly common and sometimes prevents difficulties of pipes overlapping on the external surface, as well as avoiding the exposure of many pipes.

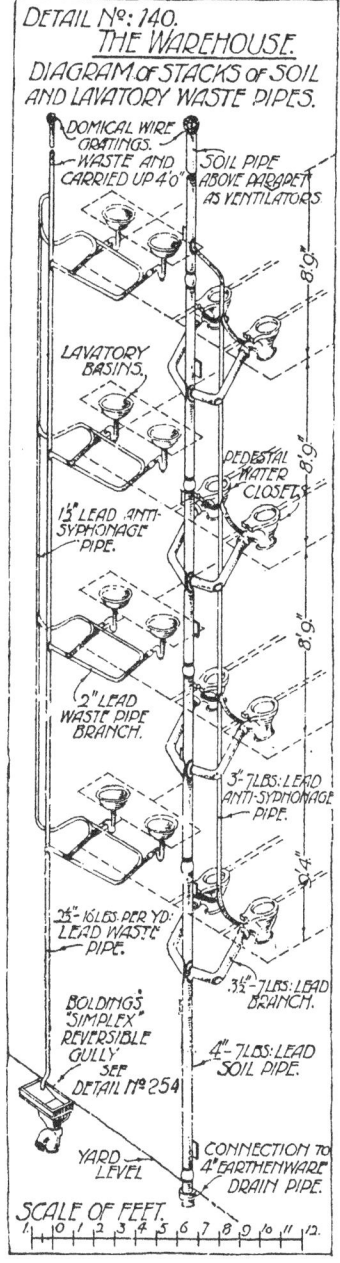

For efficient ventilation the main vent pipe should be larger than the branches if these latter are numerous, though 2″ pipes are commonly employed throughout.

462. Combined wastes to lavatories. When lavatories are required in series of two or more they are fitted on the same principles as the single lavatory, except that the wastes and vents are grouped to economise materials and labour.

A trap is placed under each lavatory basin and the main waste carried from the outer one of the series past the remaining wastes which are jointed into it in the direction of the flow.

The vent pipe is similarly treated. Thus, the lavatories on one floor are discharged through one waste branch and the latter is jointed to a vertical down pipe which receives the discharges from the several tiers, as shown for the warehouse in detail No. 140. Vent or anti-syphonage pipes from each series are connected to a main vertical vent pipe, which terminates in the side of the down pipe above the level of the highest lavatory, in the manner previously applied to vents for W.C. branches.

The down pipe is carried above the eaves to assist ventilation and the foot of the pipe discharges with an open end over a channel leading to a gully trap—see Drainage, paragraph 480.

Suitable sizes for the wastes and vents are: $1\frac{1}{2}''$ branch wastes and traps, 2″ main branches (receiving discharge from two or more lavatories) and a $2\frac{1}{2}''$ or 3″ down pipe, where three or more series of lavatories are connected to this principal waste.

CHAPTER SEVENTEEN

DRAINAGE

463. House drainage. The removal of human waste from dwellings, public buildings and places of employment, is a branch of sanitary science which should be carefully studied by all who are interested in the construction of buildings.

The architect and builder must pay particular attention to this problem, because all buildings must of necessity be drained to the satisfaction of the local sanitary authority in whose district such buildings are erected, and plans of the proposed drainage scheme showing the lines of drain, their connections, adjuncts, means of inspection and cleansing and the final disposition of the discharges, deposited with the authority and approved before work commences.

464. Modern water-carriage system. Modern house drainage consists of the carriage of waste matters from internal and external sanitary fittings, such as sinks, baths, lavatories, W.C.'s, and urinals, through a system of closed channels called drains.

Flushes of clean water are employed where required for the purpose of transporting solid wastes. The resulting flow of sewage from the fittings is directed either (a) to some local receptacle designed for its accumulation or automatic disposal, or (b) to large main drains called "sewers", which direct the discharges to a common centre organised for dealing with the final disposal of accumulated wastes.

The former method is often necessary in rural districts while the latter is adopted for urban or metropolitan areas. The following treatment is confined to the planning of drains within the curtilage[1] of the building, preparing for the collection of the drainage to a point just within the boundary line, from which it may be discharged for either method of final disposal.

465. Drains. Modern drains are closed channels of earthenware or cast iron, generally placed below ground and laid to a sufficient inclination to ensure a quick flow of the contents; they must be air and watertight.

A *main* drain conveys the drainage from one property—or a group of properties—to the sewer and such a drain may have numerous branches for the collection of wastes from the several sanitary fittings.

---

[1] Curtilage: the land adjacent to a dwelling house, etc., and within the accepted boundary thereof.

The ordinary drain consists of socketed cylindrical pipes of glazed earthenware, 4″ and upwards in diameter, in 2 ft. lengths, jointed together to produce an airtight and watertight channel.

466. Function of a drain. A well constructed drain and its accessories should pass soil wastes freely, leaving the interior in a practically unsoiled condition; it should have no sudden or unnecessary changes of direction which might tend to accumulate solid matters, and should also, by scientific arrangement of its inlets and outlets, prevent the escape of foul air from the sewer into the building or within its immediate vicinity.

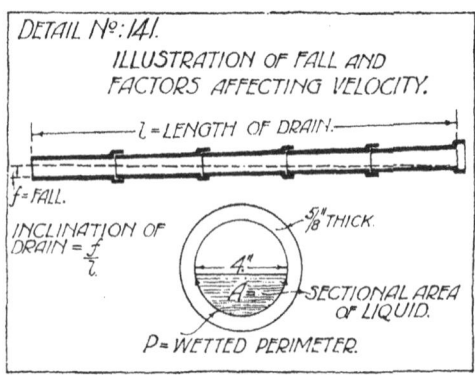

467. Size of drains. A drain is described by the internal diameter or "bore" of the drain pipes, and the sizes in common use are 4″, 6″, 9″ and 12″. The larger sizes are not required in average schemes of house drainage, but are often employed for small sewers.

468. Fall of drains. In order to carry solid wastes through a drain, the flow of the carrying liquid should normally attain a velocity of 4 ft. to 5 ft. per second. This velocity is dependent upon the inclination of the drain to the horizontal and the amount of friction set up between the moving liquid and the inner surface of the pipe.

The common method of referring to the inclination is to state the "fall" or difference in level between two selected points in the bottom or "invert" line of the pipe and to compare this fall with the inclined length between the points as indicated in detail No. 141. Taking the measurements in similar units, say ft.:

let f = the fall between the points determining the length,
and l = the length of straight pipe between these points,

then the inclination may be expressed as $\dfrac{f}{l}$.

XVII] DRAINAGE

If $f = 2\cdot5$ ft. when $l = 125$ ft., the inclination is $\dfrac{f}{l} = \dfrac{2\cdot5}{125} = \dfrac{1}{50}$. This is commonly stated as a fall of one in fifty.

469. Effect of friction. The effect of friction is an uncertain matter depending upon the condition of smoothness or otherwise of the interior of the pipe. It is easily understood, however, that the amount of surface rubbed against by a known quantity of liquid passing through the pipe will be a factor in determining the average rate of flow, and most calculations for computing an approximate velocity contain this factor.

For any but large schemes such calculations are unnecessary, and it is sufficient for our present purpose to quote a rule for guidance in determining the minimum falls for drains.

470. Rule for minimum fall of drains. A velocity of 4 to 5 ft. per second is obtained for the following sizes and falls of drain, with pipes flowing freely and either *full* or *half full*:

Diameter of pipe	Fall		
4″	$\frac{1}{40}$	or 1 in	40
6″	$\frac{1}{60}$	,,	60
9″	$\frac{1}{90}$	,,	90
12″	$\frac{1}{120}$	,,	120

These falls may be stated by the simple formula:

Fractional fall $= 1 \div 10$ times bore of pipe,

or
$$\frac{f}{l} = \frac{1}{10d},$$

hence the actual fall in a given length is

$$f = \frac{l}{10d},$$

where f, l and d are all in the same dimension unit, usually feet.

The fall obtained by the above rule must be looked upon as the *minimum* allowable under average practical conditions.

471. Construction of straight lengths of drain. Straight lengths of drain are usually constructed from earthenware pipes,[1] which should be sound, straight, of regular bore and circular cross section. They should be smooth, well glazed and impervious under considerable pressure of water, and capable of easy laying in straight lines without faulty joints.

472. Laying drain pipes with ordinary joints. Detail No. 142 shows an isometric view of common drain pipes with one joint

[1] See Chapter on Materials, Vol. II.

dissociated and a second completed; at A is a part sectional detail of the common form of joint.

The pipes are of 4″ bore, ⅝″ thick and their length is such that each pipe forms 2 ft. of finished drain. The joint consists of a socket on the head of the pipe, 3″ long externally and 2″ deep, into which

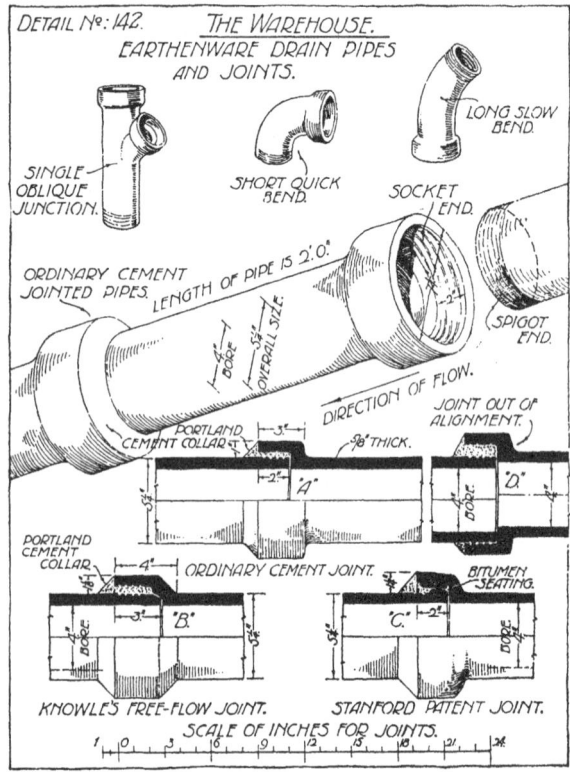

the "spigot" end on the tail of the companion pipe enters, leaving a clear annular space, or ring, ⅜″ thick, which is provided for packing with jointing material.

To obtain a satisfactory adhesion and also to prevent longitudinal movement of the joint, the spigot and socket are both furrowed circumferentially in good pipes and left unglazed. Before insertion the bottom of the socket and the spigot edge are coated with neat cement, then the pipe is pressed home, a double ring of yarn gasket, of hemp or cotton, turned round the pipe and rammed to the bottom of the socket to prevent further cement working through the joint. The pipes are set with their spigot ends in the direction of the flow, in true alignment and with their "inverts" level and supported at the free end; the joint is then completely filled with neat cement solidly packed, and finally terminated at the edge of the socket by a triangular fillet at about 45° to the pipe surface.

The cement for this purpose must not be mixed with any sand and should be fresh yet cool. Hot cements expand much on setting and may break the joints.[1]

On completion of the joint any protrusion of cement paste which has worked through in the initial bedding of the spigot end to the socket, should be scraped clear and a smooth straight invert ensured.

473. Special pipe joints in earthenware. While the form of joint just described is in common use and quite satisfactory when accompanied by good materials and careful workmanship, there are many special forms which are designed to ensure speedier jointing, more reliable connections and easier and better alignment.

474. Knowles' "Free-flow" joint. At B is illustrated Knowles' "Free-flow" joint, a self-centring type, which is prepared with a socket 3″ deep having a conical rebate formed within it, accurately fitting a corresponding taper on the spigot end of the adjoining pipe. The object of this conical joint is to guide the pipes at the junction into true alignment; with the ordinary joint improper setting of the pipes or careless ramming may leave the invert at the spigot end below that of the socket, presenting an obstruction to the flow as at D. The joints are made by covering the spigot end with grease, pushing the spigot home, and cementing as described for ordinary pipes.

475. Stanford joint. The detail at C shows a Stanford joint, the essential feature being a tapered bitumen collar on the spigot and a corresponding lining to the socket. These rings are cast in position $1\frac{1}{4}$″ to 2″ long and turned to an accurate fit. When laid the seatings are brushed over with a bituminous composition, tightly inserted and the remainder of the socket stopped with neat cement and finished to a bevel at the exposed edge.

476 Use of yarn gasket. Some authorities object to the use of yarn gasket for packing drain pipe joints, on the grounds that it rots and eventually becomes filthy if any strands become exposed to the interior. The latter event is quite likely unless the edge of the spigot is properly bedded in cement when laying. It is wise to draw the gasket through cement grout before ramming it home; this preserves it and assists the ramming.

---

[1] See Chapter on Materials, Vol. ii.

477 Bends. While it is desirable to keep lines of drainage perfectly straight between points of access specially provided, it is necessary in many cases to make changes of direction by inserting curved pipes called "bends". The most legitimate use of closed bends is for changes of direction where the axes of the pipes are in vertical planes, as at A and B in detail No. 144.

Changes of direction within the plane of inclination may also be made with similar pipes, but it is preferable to make such changes by half channel bends, or by purpose-made junctions of the kind shown in detail No. 146. These can often be grouped and must be surrounded by small brick chambers having movable covers which permit inspection and cleansing at the point of change, which is also the position of probable stoppage.

Bends are known as right angled, or obtuse, according to the angle enclosed between the axes of the straight pipes connected to them. If the curve of the bend is of small radius, say 6″ or less on the inside, the pipe is named a "quick bend", and if of larger radius a "slow bend", see detail No. 142. Slow bends should be employed wherever the conditions allow.

478. Junctions. These are closed pipes having a branch inlet formed by an oblique or curved arm at the side, and provided with a socket to which a line of branch drain may be connected. There are many forms of junction between pipes of different bore, sufficient to meet all practical needs. A single oblique junction is shown in detail No. 142.

While in the best modern work the frequent use of such junctions would not be approved, they are, however, still in common use in many districts. The objection to them is that if stoppages occur at such positions they are not easily accessible for inspection and clearing.

Open channel junctions are the best, because they expose the point of connection. They must be enclosed in a brick chamber, as described in a subsequent paragraph and shown at details Nos. 146 and 147.

479. Earthenware traps. The function and form of traps in lead wastes have already been discussed and described. They perform the same general function in a drainage scheme, for which they are more justifiable, and are usually separate fittings inserted between, or connected to, other units of the drain; they vary in shape in accordance with their required positions and special functions and those in most common use for drainage are the "gully" and "intercepting" traps.

480. Gully trap. Any domestic waste liquid of an unobjectionable character should be discharged from the sanitary fitting into the open air over a gully trap. This trap is often a deep receptacle holding a considerable body of water, the inlet to which is guarded by a cast iron or earthen-ware grating; it has an outlet arm at a high level for connecting to the drain, while the inlet opening is near the bottom of the trap. General types of gully trap are shown in detail No. 143.

Their purpose is to guard or trap an opening to a drain at or near the ground level and because they may, under certain conditions, be very liable to unsealing by evaporation, the earlier forms were made bulky and with a deep seal to guard against this liability. For drainage inlets in streets and yards the type A is still commonly adopted and without objection, because the liquid contents contain little or no organic waste which may cause foul gases to be emitted should the contents lie unchanged for a lengthy period.

481. Self-cleansing gully trap. If the traps above referred to were employed for domestic buildings where discharges from sinks, baths and lavatories were to be accepted, accumulations of scraps of food and grease would occur in the trap, intermingled with grit and dirt, and this condition, aided by

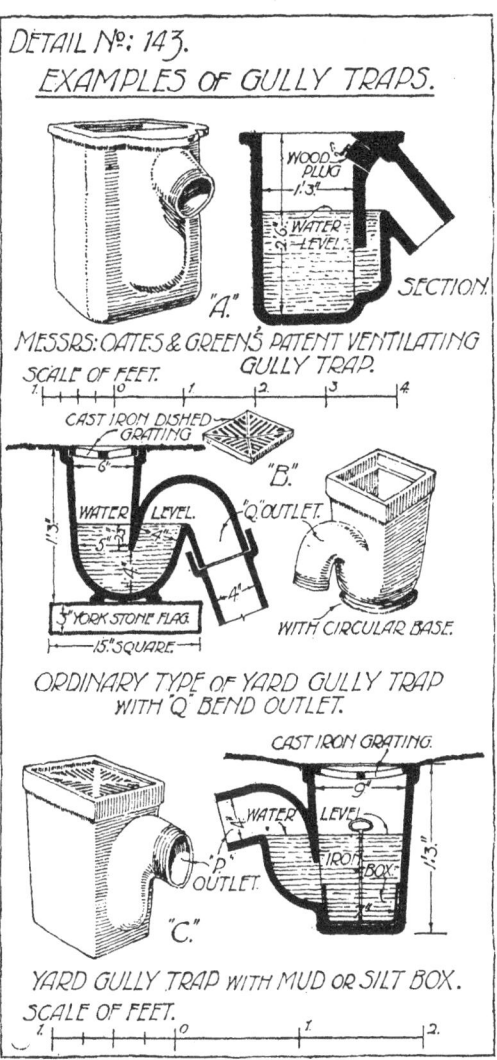

DETAIL № 143.

EXAMPLES OF GULLY TRAPS.

WOOD PLUG. 1'3". WATER LEVEL. SECTION. "A."

MESSRS: OATES & GREEN'S PATENT VENTILATING GULLY TRAP. SCALE OF FEET.

CAST IRON DISHED GRATING. 6". "B." WATER LEVEL. "Q" OUTLET. 4". 5" YORK STONE FLAG. 15" SQUARE. WITH CIRCULAR BASE.

ORDINARY TYPE OF YARD GULLY TRAP WITH 'Q' BEND OUTLET.

CAST IRON GRATING. 9". WATER LEVEL. IRON BOX. 1'3". "P" OUTLET. "C."

YARD GULLY TRAP WITH MUD OR SILT BOX. SCALE OF FEET.

periods of disuse, would cause the trap to become a small cesspool liable to cause a nuisance in the immediate vicinity.

This is especially liable when rain-water is not discharged periodically through the same trap and where the waste is accepted from a wash up sink only. For such purposes gullies of the self-cleansing type should be employed, where the sectional area is constant from the bottom of the receiver to the exit; see detail B.

In order to assist in thoroughly flushing gullies, wastes from sinks, baths and lavatories should be grouped together where possible, and it is a further advantage to include in the group a down pipe from the roof.

482. Gully with silt box. The example at C shows a form of gully trap suitable for use in a gravelled yard where mud and grit are liable to be washed into the gully. A loose iron box, with handle, is placed in the bottom of the gully to receive and facilitate the removal of this accumulation.

483. Intercepting trap. This form of trap is shown at detail No. 146 and is used to seal the private drainage system from the sewer, thus intercepting any sewer gases which might enter the drains.

It is known as an "intercepting" or "disconnecting" trap and sometimes as a "drain sentinel".

Many variations of shape and arrangement occur in the intercepting trap to suit special circumstances and positions.

484. Ventilation of drains. Every part of a drain conveying foul wastes should be ventilated in order to prevent the accumulation of drain air, charged with gases arising from the decomposition of organic waste, which may not, even in a well arranged scheme, be carried away completely by the flush of water. In an imperfect scheme, due either to faulty planning or workmanship, some means of ventilation would be the only way of preventing permanent fouling of the channels. Most bye-laws therefore require such ventilation for all soil pipes and soil drains conveying waste from an upper floor and also for those transporting waste from an external ground floor W.C. if the length of the drain exceeds 10 ft. from the outgo of the fitting to the intercepting trap. If longer than 10 ft. then a vertical pipe must be carried up to a height of at least 10 ft. to act as an up cast shaft and air be admitted to it through the drain, and on the house side of the intercepting trap.

485. Fresh air inlet. To admit fresh air to the drain an untrapped inlet of at least 4″ bore is required to every system of soil drains.

The inlet is usually placed near the top of the disconnecting chamber and turned up against a convenient wall, preferably in an open and little used position.

The vertical pipe is generally of cast iron finished at the top with a box-like fitting containing a sheet of mica, hinged behind a grating having openings equal in area to the bore of the pipe. This mica flap is sensitive to air movements and should freely open to admit air if the outside pressure is greater than the internal pressure. Any reversal of these conditions, due to liquid discharges through the drain producing a momentary change of air current causes the valve to close against the grating and thus prevents the escape of drain air.

If such flap valves were dependable they would be efficient for their intended purpose, but experience has proved that the flap deteriorates, changes shape, and often fails to work freely on its hinges after a time, with the result that the function is no longer fulfilled; if jammed, the valve may actually restrict the inlet.

In modern work it has been proved that perfectly open inlet gratings are satisfactory, because, if the system of drains is properly arranged, no serious accumulations of foul air can occur and an unobstructed current is provided for at all times. Such inlets should be placed in boundary walls or on such walls of the building as will prevent any possible emission of foul air being carried directly into the building through windows, doorways or other openings. Vertical pipes, 6 ft. high, having an air inlet box fixed at the top are often employed in open places where wall support cannot be obtained; they should be strong and well secured.

486. Ventilating outlet. Soil pipes are usually carried up above the eaves of roofs to provide outlets for drain air. The assumption is that fresh air will be admitted through inlets, pass along the intervening drains and be drawn up and discharged through the soil pipes acting as ventilators.

In the usual arrangement of inlets and outlets for drain ventilation the inlet is placed at the lower end of the system, delivering air through the disconnecting chamber, while the outlet is the soil pipe near the head of the drains; for this reason the movement of air is arrested and reversed at each considerable discharge from a sanitary fitting.

Sometimes the reverse arrangement is made, which does not arrest and reverse but only disturbs the current and tends to improve it by suctional and propulsive effects.

If a soil pipe acts as the inlet an equally high pipe should be provided for an outlet. See ventilation of warehouse drains, paragraph 494, for a reference to this type of ventilation.

487. Efficiency of ventilating pipe. There is much doubt as to
the reliability of the average ventilating pipe on a drainage system.
It is presumed by some writers that if an inlet and outlet be provided with the latter at a considerably higher level than the inlet,
then an automatic change of air occurs in the pipe, due to the
weight of the column of air represented by their difference in
height. This is a mistake. Apart from other influences such as
wind pressure, aspiration due to wind, and changes of temperature
of the atmosphere, no movement of air would occur in the pipe
except when discharges of waste occurred, because the *pressure of the
still atmosphere is the same at any horizontal line* and no tendency
to move exists until the agencies mentioned act upon it.

The most certain and general cause of movement of air in the
pipes is the combined effect of direct pressure of wind acting
horizontally and entering the vertical openings in the perforated
inlet and the suctional effect caused by the wind blowing horizontally over the outlet above the eaves. Should the temperature
of the air in the vertical pipe become higher than that of the
surrounding air through the sun beating directly upon the pipe,
the air expands, its density decreases, and an up current is started.
Other conditions, such as humidity, position of pipe and inlet, kind
of materials used, etc., will have some effect, but these are of little
importance.

488. Drains not requiring ventilation. Branch drains conveying
waste water as a rule do not need ventilation, especially if well
flushed from time to time with rain-water.

It is possible for drains conveying soapy water to become foul,
though this seldom occurs where rain-water discharges into the
same gully; it is obviously wise, however, to keep all such branch
drains as short as possible.

489. Summary of principles of drainage. Summarising the general
principle of drainage the following points of first importance have
to be satisfied:

(a) Plan the drainage in straight lines wherever practicable.

(b) Guard every drain inlet by a trap to avoid incursion of foul
drain air to the interior of the building.

(c) Intercept the system of private drains from the public sewer,
or from the main discharge pipe leading to other means of disposal.

(d) When a deviation must be made from the straight line of a
drain in plan, arrange for means of access and inspection at the
point of change and maintain straight lines, where possible, between such points.

(e) Make branch drains as short as possible.

(*f*) Disconnect the waste pipes from sinks, baths and lavatories from the drains by making them discharge in the open air, over a channel leading to a gully trap.

(*g*) As soil drains, from W.C.'s and urinals, cannot be so disconnected, continue the vertical waste pipes from these without break to the drains and provide for their ventilation.

490. Plan of the warehouse drainage scheme. Before examining the scheme of drainage illustrated in detail No. 144, it should be noted that this forms only a part of the complete scheme required for the warehouse as a whole. There are rain-water pipes to the back and front of the building, and in many cases these would have to be taken within the building by iron drain pipes encased in concrete and connected to the inspection and disconnecting chambers shown on the drainage plan.

As the consideration of iron drains does not come within the scope of this volume, these necessary portions of the complete scheme have been omitted.

In the portion of the scheme illustrated in the detail it will be seen that the wastes to be disposed of are: rain-water from roofs and from surface of yard, discharges from lavatories and W.C.'s, and drainage from the base of the lift well which may possibly be of a greasy nature.

The principal drainage work occurs in the yard entered from the front street. Around this area the lavatories and W.C.'s are situated and the lift well is conveniently abutting thereon. Detail No. 144 shows how the wastes may be collected and disposed of in agreement with the general principles previously enunciated.

We are here concerned only with the mode of conveying the wastes from the ground level to the sewer, as the arrangements for trapping inlets to internal fittings have already been dealt with in the Chapter on Plumbers' Work.

Detail No. 140 shows that the series of W.C.'s have their wastes collected into one soil pipe, which enters the drain just below the ground level, its position being shown at C in detail No. 144.

In a similar manner the lavatory wastes are collected into one pipe discharging over the gully at D, while the rain-water pipe from the flat roofs above the staircase and lift is brought to the same point. There is also a rain-water pipe from the loading shed delivering at E, and drainage from the lift well enters at F.

491. Collections of branch drains. As the branch drains from the lift well and loading shed are too far from the soil pipe and lavatory wastes to group them all at one point, it is convenient here to collect the two former as near the head of the drain as possible and to group the remaining branches at a point near the front boundary

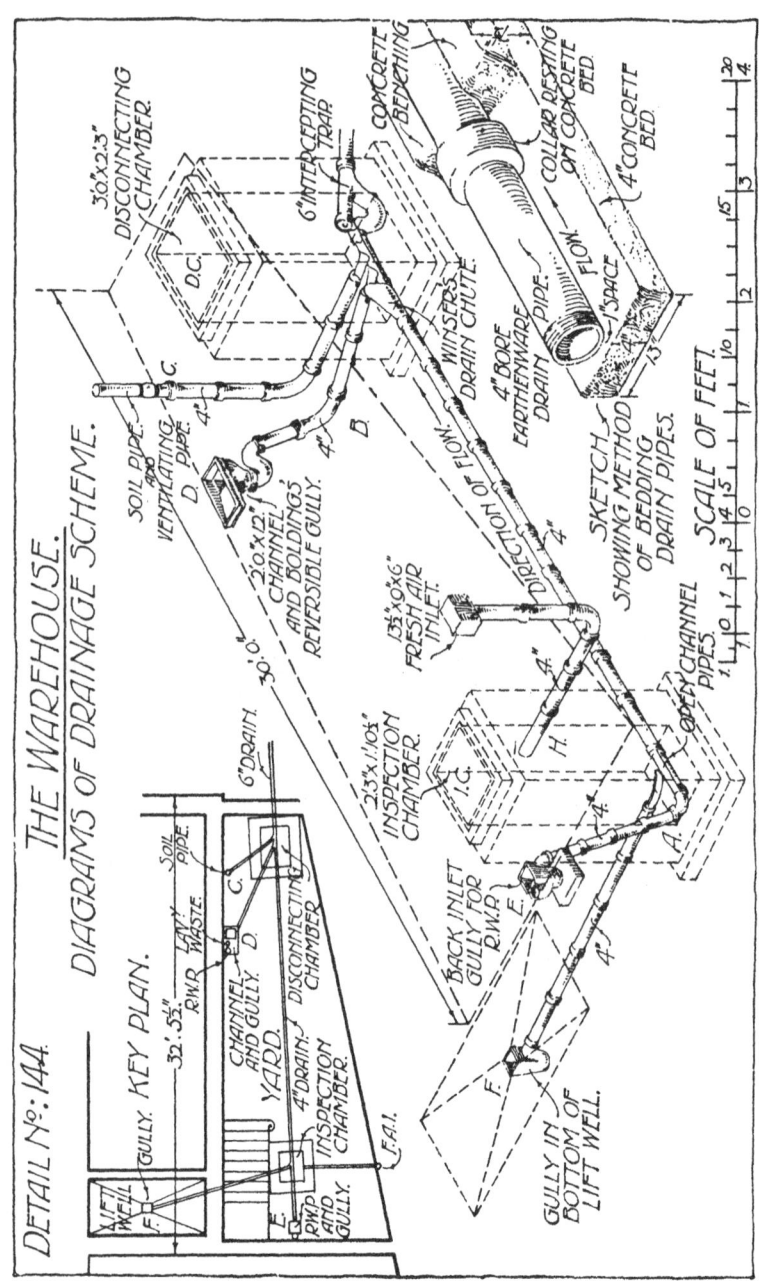

DETAIL No. 144.

THE WAREHOUSE.
DIAGRAMS OF DRAINAGE SCHEME.

KEY PLAN.

SCALE OF FEET.

of the yard. Brick chambers are provided for this purpose and the changes of direction in each case made within the base of the chamber.

If required only for access, as in the chamber marked IC, they are named "inspection chambers" or "inspection pits", but if used to house an intercepting or disconnecting trap they are termed "intercepting" or "disconnecting" chambers, as shown at DC.

These chambers may vary in size from 1′ 6″ square upwards; the smaller sizes are suitable for shallow chambers only, but if required to admit the body of a man they should be at least 2′ 3″ × 1′ 10½″ and may be of such larger size as required to contain the junctions of the branch drains collected within them.

The term "manhole" is commonly applied to all such chambers but they are referred to here by their distinctive names.

It should be observed that the line of drain between the chambers should be perfectly straight since it is then easy to find whether a stoppage occurs between chambers by sighting from one to the other and also renders drain clearing easy. Stoppages in this straight length are, however, very unlikely unless soil pipe branches are made to deliver into them, which is not good practice, but often accepted and carried out.

492. Arrangement of short branches. As the depth of the branch from the lift well determines the depth of other branches at the points of collection of the latter, it only remains to decide whether the branch drains from inlets at the yard level shall fall uniformly from the gully to the chamber or be dropped down to the level as quickly as possible after leaving the gully outlet. The latter method has been selected as being more likely to produce a sound and permanent drain, and the greater depth makes it less liable to fracture by loads travelling over it; in addition the flow into the inspection chamber, though rapid, will not be so liable to splash or oscillate over the main channel as would be the case when a double change of direction occurs within the chamber.

493. Depth of drains. The depth of a system of private drains depends upon (a) the level of the lowest point where a waste is to be discharged, and (b) the depth of the existing sewer, proposed sewer, or other means of removal.

Where a sewer already exists, a new building must generally be designed to allow its drainage to be collected thereto, with the necessary falls to secure self-cleansing drains.

In the example it has been assumed that the sewer in the front street is sufficiently deep to allow the lift well to be drained, the latter being slightly below the basement floor level and thus allowing drainage from floor cleansing, etc., to be collected to the well. The level of the lift well has therefore decided the minimum depth of the drains, which must be sufficiently above the level of the sewer to allow the necessary fall to be obtained.

494. Ventilation of warehouse drains. In this scheme, because the soil pipe branch drain enters the disconnecting chamber, while a considerable length of drain exists at a higher level, it is convenient to place the air inlet pipe in the side of the inspection chamber near the head of the drain and allow the air to travel down the straight length to the disconnecting chamber and through this to the soil branch which acts as an up cast shaft.

On discharging any of the W.C.'s the up current in the soil pipe is reversed but the current in the upper part of the drain is strengthened by the suctional effect. When flow stops, the up current is restored and the soil pipe and branches are continuously ventilated between usages of the fittings.

495. Construction of warehouse drains. The construction is commenced by excavating the ground to the necessary depth at the sewer end of the drain where the disconnecting chamber is to be placed. The depth of this excavation is tested by levelling from some unalterable level in the vicinity—called a datum level—and allowance made for the thickness of the foundation in order to ensure that the invert level of the pipe at the outgo of the trap is in agreement with that previously decided and indicated on the plans.

The excavation would then be conducted for the whole length of the drain trench and its branches, and timbered according to the nature of the ground as indicated in details Nos. 106 and 107.

At the positions of the brick chambers the excavation would be widened to allow for the projection of the concrete base.

496. To set out fall of drain and lay pipes truly. The fall of the drain is determined by driving 2″ square pegs into the bottom of the trench and testing the levels of their heads until the difference in height between pairs of pegs is correct; then ramming the trench bottom true to the fall. If the pipes are to be laid on concrete, this would be deposited to the level of the pegs and screeded in straight lines between them, so that the sockets of the pipes may be placed direct upon the concrete.

497. Boning the fall of trenches and drain pipes. Long lengths of drain should have their inclinations tested by employing a boning rod to ensure a straight line gradient between two initial points in the line. The sketch detail No. 145 will help to make this operation clear. It is obvious that the invert of each pipe should lie in the line of gradient of the drain, hence, if two horizontal rails are fixed in planes at right angles to the drain and each of some known height above the invert, then a line joining the two rail levels, and in the direction of the drain, will be at the same constant height above the invert.

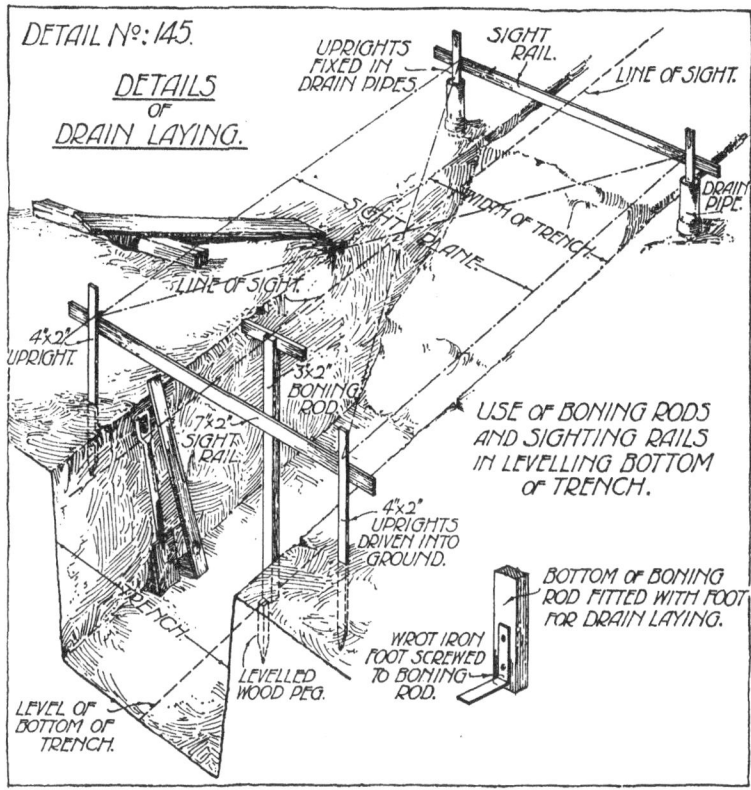

DETAIL Nº: 145.

DETAILS
OF
DRAIN LAYING.

UPRIGHTS FIXED IN DRAIN PIPES.

SIGHT RAIL.

LINE OF SIGHT.

DRAIN PIPE.

WIDTH OF TRENCH.

SIGHT PLANE.

LINE OF SIGHT.

4'x2' UPRIGHT.

3'x2' BONING ROD.

7'x2' SIGHT RAIL.

USE OF BONING RODS AND SIGHTING RAILS IN LEVELLING BOTTOM OF TRENCH.

4'x2' UPRIGHTS DRIVEN INTO GROUND.

TRENCH.

LEVELLED WOOD PEG.

WROT IRON FOOT SCREWED TO BONING ROD.

BOTTOM OF BONING ROD FITTED WITH FOOT FOR DRAIN LAYING.

LEVEL OF BOTTOM OF TRENCH.

If a person looks across the top edge of one sight rail towards the top edge of the other, he establishes a "plane of sight", and any intermediate object can be placed in the same plane to a great degree of accuracy by raising or lowering it.

To ensure that the invert of the drain is parallel to the plane of sight, a wooden rod with a cross head like a large T square is prepared and a foot attached which enters within and lies upon the invert of the socket end of the pipe to be tested. This instrument is called a "boning rod".

The length from the under edge of the foot to the top of the bar agrees exactly with the vertical height between the invert line of the pipes and the plane of sight; thus, by placing the pipe in position

and adjusting its height until the top edge of the boning rod is in
this plane a perfect gradient is obtained. The pipes, so adjusted,
must be solidly packed and jointed.

Some pipes are not quite straight in length, but otherwise good
and sound; these should be laid so that any deviation from the
straight line of the invert is sidewise and not vertical. In first class
work such pipes should be rejected if the curvature is very appreciable.

498. Adjusting of levels. Levels at the principal points in a line
of drain are correctly determined by employing a surveyors' dumpy
level and a staff, and operating from a datum line selected for the
scheme.

499. Adjusting falls by straight-edge and spirit level. Short
lengths of drain are usually laid to the fall by preparing a straight-
edge of wood having its edges out of parallel to the extent of the
inclination of the drain to the horizontal. The disadvantage of this
method is that the inclination is not determined at the invert, but
externally over the sockets; it is not so accurate as the boning
method.

When a drain has ample fall and great accuracy throughout the
entire length is unnecessary, this method serves quite well. In some
cases long straight-edges are prepared and cut to clear the sockets
of some of the intervening pipes; the fall is then tested at every third
or fourth pipe.

500. The "Gradiograph". The gradiograph is an instrument for
testing the inclination of drains and other channels. It is constructed on the principle of the straight-edge and level described
in the last paragraph and is placed *inside* the pipe, thus ensuring
that the two ends of the invert of each pipe are in the correct line of
gradient. The instrument can be adjusted to any required fall and
maintained thereat by locking the adjusting apparatus.

Either long or short lengths of drain can be laid by this instrument
with considerable accuracy, the levels at important points being
determined by the surveyor's level.

501. Details of disconnecting chamber. Detail No. 146 shows the
plan and isometric view of the disconnecting chamber, with the
intercepting trap in position upon its concrete seating. The chamber
is constructed of 9″ brickwork laid in English bond, bedded and
flushed in cement mortar and built upon a 6″ bed of cement concrete. For deep chambers, iron horse-shoe steps are built $4\frac{1}{2}″$ into
the walls at convenient heights, and provide a means of descent.

Impervious bricks are employed in the construction, such as

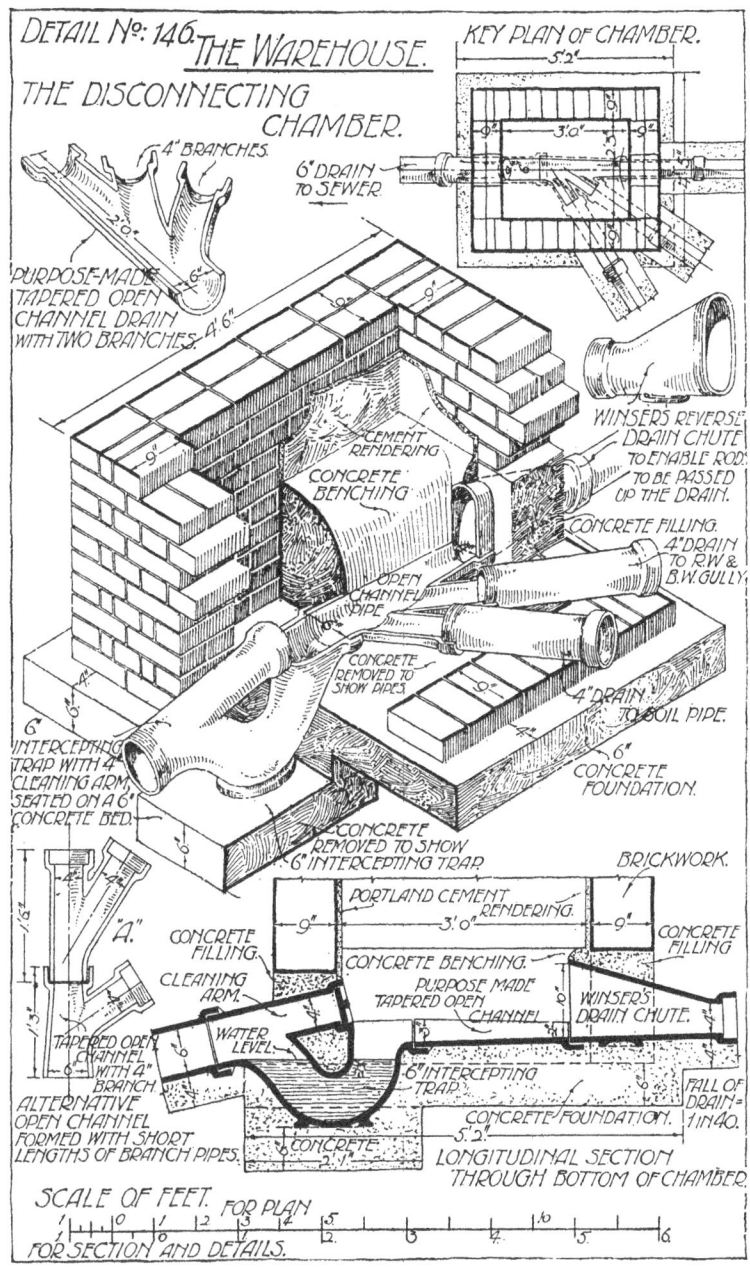

DETAIL No: 146. THE WAREHOUSE.

THE DISCONNECTING CHAMBER.

KEY PLAN OF CHAMBER.

4" BRANCHES.

6" DRAIN TO SEWER.

PURPOSE-MADE TAPERED OPEN CHANNEL DRAIN WITH TWO BRANCHES. 4' 6"

CEMENT RENDERING.

CONCRETE BENCHING.

WINSER'S REVERSE DRAIN CHUTE TO ENABLE RODS TO BE PASSED UP THE DRAIN.

CONCRETE FILLING.

4" DRAIN TO R.W. & B.W. GULLY.

OPEN CHANNEL PIPE.

CONCRETE REMOVED TO SHOW PIPES.

4" DRAIN TO SOIL PIPE.

6" CONCRETE FOUNDATION.

6" INTERCEPTING TRAP WITH 4" CLEANING ARM, SEATED ON A 6" CONCRETE BED.

CONCRETE REMOVED TO SHOW 6" INTERCEPTING TRAP.

BRICKWORK.

"A."

PORTLAND CEMENT RENDERING.

CONCRETE FILLING.

CONCRETE FILLING.

CLEANING ARM.

WATER LEVEL.

CONCRETE BENCHING.

PURPOSE MADE TAPERED OPEN CHANNEL.

WINSER'S DRAIN CHUTE.

TAPERED OPEN CHANNEL WITH 4" BRANCH.

ALTERNATIVE OPEN CHANNEL FORMED WITH SHORT LENGTHS OF BRANCH PIPES.

6" INTERCEPTING TRAP.

CONCRETE.

CONCRETE FOUNDATION.

FALL OF DRAIN = 1 IN 40.

LONGITUDINAL SECTION THROUGH BOTTOM OF CHAMBER.

SCALE OF FEET. FOR PLAN

FOR SECTION AND DETAILS.

plastic engineering bricks, blue bricks, or second quality salt glazed linings backed by hard burnt stocks.

If, for any reason, common stocks are to be employed, the interior surface must be rendered with $\frac{1}{2}''$ of cement mortar to ensure a watertight enclosure; or another method, largely employed in some districts, is to build the brickwork in stretching bond, employing two independent $4\frac{1}{2}''$ walls $\frac{1}{2}''$ apart, breaking joint at the horizontal courses, each wall being set in cement mortar and the space between them grouted solid with cement or run with asphalte.

As a further precaution against leakage which may occur through defects in the walls if the drains become blocked and the sewage rises in the chamber the brickwork—when common bricks are employed—is sometimes surrounded with plastic clay, worked into condition and rammed all round the chamber from the concrete upwards and 6″ to 9″ thick. Clay so worked to a condition of plasticity is called "puddled clay".

502. Formation of base of chamber. In the example a length of main drain passes through the chamber and two important branches connect on one side of it. The method of forming the base and connecting the branches is one of several methods, which vary chiefly in the detail of their arrangement for preventing disturbance of the flow at the point of junction with the main channel.

It consists of a purpose-made open channel laid upon a bed of concrete, with two branches at the required inclinations to the main channel.

These fittings are separately shown in the same detail with an alternative method of arrangement, as indicated at A, consisting of an ordinary channel junction connected to a tapered channel junction.

All the drains are 4″ in diameter at their entry to the channel, but the latter is enlarged at the outlet end to 6″ diameter in order to allow for the increased flow of waste as the number of branches increases, and to suit the inlet arm of the 6″ intercepting trap into which it discharges.

Straight lengths of 4″ branch pipes enter the chamber and rest within the sockets of the purpose-made channel junctions and are jointed in neat cement.

To make the channels deep enough to prevent overflowing and fouling the chamber, concrete is deposited on each side of them, steeply sloped and rounded over at the edges; this is called "benching-up", and confines the flow strictly within the channels.

503. Drain chute. Every branch and main line of drain should be capable of easy clearance, which is done in case of necessity by pushing clearing rods along the branches from the chamber. These rods are provided in 3 ft. sections which joint together to make up the length required.

To facilitate their passage into a branch from a restricted space such as the chamber, a special pipe is sometimes provided at the entrance of the branch having an enlarged end, formed by increasing the depth on the top side; this fitting is called a "drain chute" and is shown applied to the delivery end of the main line of drain between the chambers. The chute is obtainable with either a spigot or socket at the normal end of the fitting, and is generally provided with a foot for standing upon the brickwork or concrete, when bedding in position.

504. Disconnecting trap to warehouse drains. We have already referred to the general form and use of the intercepting or disconnecting trap. In the section at the bottom of the detail, the trap is seen to enter the chamber sufficiently to allow the cleaning arm socket to project slightly from the brickwork; the arm is provided to give access to the drain on the sewer side of the trap and is closed by a removable airtight stopper in the form of a "screw cap".

The trap is an easy bend of the same section as the 6″ drain taken normal to its curve throughout, and thus offers little obstruction to the flow.

The water level in the trap is 2″ below the channel invert at the point of entry and the seal, or dip, about $2\frac{1}{2}$″.

The object of providing the drop in level is to assist in the passage of solids through the trap by increasing the velocity of inlet. This drop produces a cascade effect, but is really unnecessary where the drains have the minimum fall previously recommended.

A concrete bed attached to the foundation of the chamber, but at a lower level, should be provided to support the intercepting trap.

All such traps should have a flat base to facilitate level bedding upon the concrete.

505. Details of inspection chamber. In detail No. 147 the method of constructing the inspection chamber at the head of the drain is shown, together with the necessary fittings to the base. These fittings are employed as an alternative to those previously illustrated for the disconnecting chamber; in any practical scheme one type would be selected and employed throughout.

The base of the chamber is formed by placing a course of bricks on the concrete as shown, laying a half channel pipe across the chamber upon the concrete base and levelling up to the edges of the channels with rough concrete on each side.

The bend at the back inlet and the straight pipe at the outlet are fitted and the brickwork carried up until these are enclosed; the pipe ends should be surrounded with concrete in preference to cutting the bricks around them.

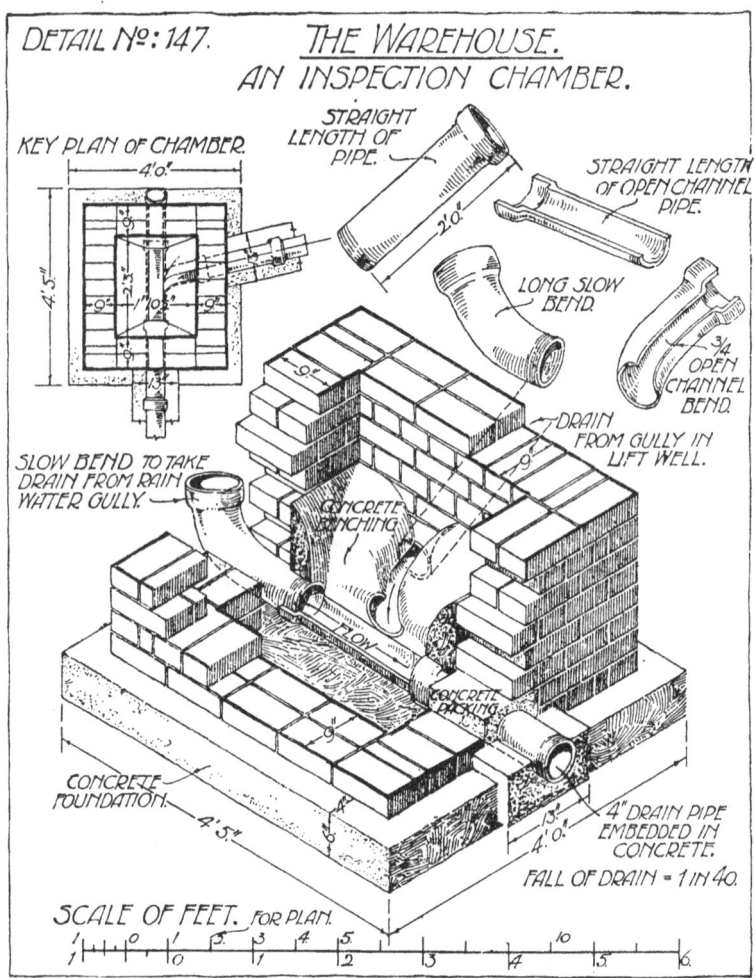

DETAIL Nº: 147. THE WAREHOUSE. AN INSPECTION CHAMBER.

KEY PLAN OF CHAMBER.

STRAIGHT LENGTH OF PIPE.

STRAIGHT LENGTH OF OPEN CHANNEL PIPE.

LONG SLOW BEND.

¾ OPEN CHANNEL BEND.

DRAIN FROM GULLY IN LIFT WELL.

SLOW BEND TO TAKE DRAIN FROM RAIN WATER GULLY.

CONCRETE BENCHING.

CONCRETE PACKING.

FLOW.

CONCRETE FOUNDATION.

4" DRAIN PIPE EMBEDDED IN CONCRETE.

FALL OF DRAIN = 1 IN 40.

SCALE OF FEET. FOR PLAN.

The side branch is connected to the central channel by an "open channel" bend, which is separately shown in the detail, the bend resting upon its slightly flattened base at the edge of the main channel, packed up with concrete for support and the sides benched up as before described.

506. Laying pipes on concrete. In the best drainage work all earthenware pipes are laid upon beds of concrete to ensure continuous support of the line of pipes and to guard against the effects of local settlements of the earth. In roads having heavy traffic concrete is also necessary to prevent fracture of the pipes and joints, which may be caused by vibration or by bending under load. It is not usual in many parts of the country to employ concrete bases except under these special circumstances.

A concrete bed for drain pipes should be 4″ to 6″ thick and at least 12″ wide, laid to accurate falls, and the pipes placed upon it, as shown in detail No. 144. The pipes are laid with their sockets resting upon the concrete base, so that they are accessible for jointing. If the thicker bed be employed, recesses for the sockets may be left and the body of the pipe laid in close contact with the bed.

After the drain has been tested for soundness of jointing the sides of the pipes are supported by benching up, as shown, with finer concrete, packed closely into the spandril angles.

Where carried under a building drain pipes must be entirely encased in concrete.

507. Gullies employed in warehouse drainage scheme. Three forms of gully are illustrated:

Back-inlet gully. The back-inlet gully is employed against the loading shed wall and receives the down pipe from the roof through a socketed back inlet, bending inwards above the water level in the trap, as shown at A in detail No. 148. A free current of air can therefore traverse the rain-water pipe which discharges into it.

The top of the gully is a square dish with a sunk or "dished" grating at the surface inlet to which the yard drainage is conducted. Such a form of grating is not the best, as paper and rubbish collected upon it spread out and obstruct the inlet.

In this example the outlet discharges downwards at a steep inclination, but an easy bend at the outgo can be obtained where required.

Gully with clearing arm. A gully with a clearing arm is applied at the inlet to the drain in the base of the lift well, which is formed in concrete and dished towards the centre. The gully would receive drainage from the whole of the basement floor during cleaning operations and is also liable to contain oil and grease from the lift gearing and slides.

Access for clearing the drain is provided by a screwed stopper placed internally at the head of the outgo arm and beneath the grating. The stopper is sealed with vaseline or grease when screwed home; it is illustrated at B in the detail.

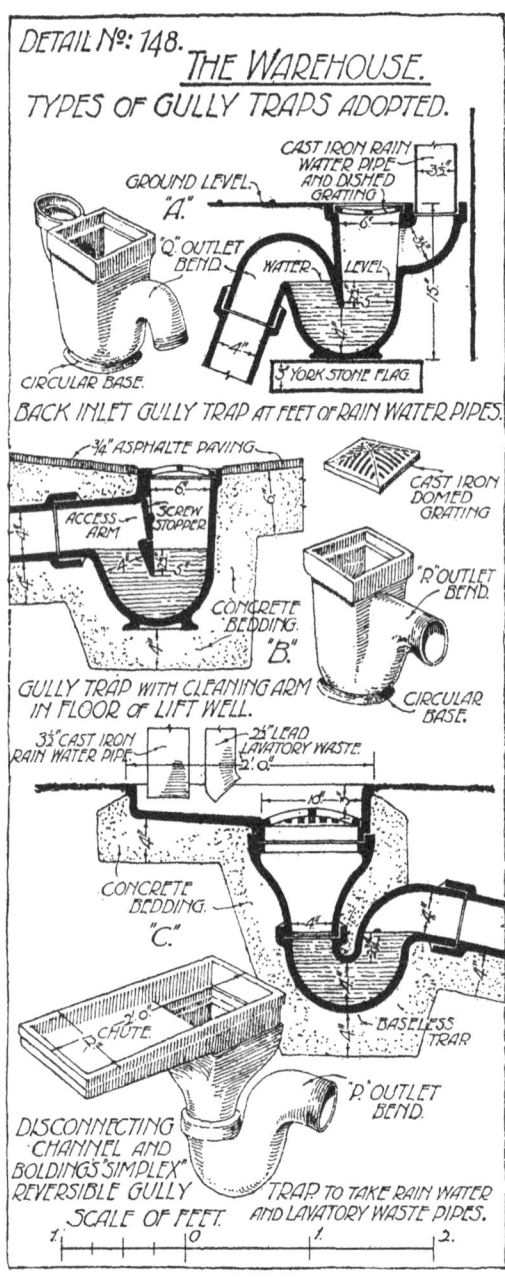

DETAIL Nº: 148.
THE WAREHOUSE.
TYPES OF GULLY TRAPS ADOPTED.

GROUND LEVEL.
"A."
CAST IRON RAIN WATER PIPE AND DISHED GRATING
3½"
6"
"Q" OUTLET BEND.
WATER LEVEL
3½"
15"
4"
CIRCULAR BASE.
3" YORK STONE FLAG.
BACK INLET GULLY TRAP AT FEET OF RAIN WATER PIPES.

¾" ASPHALTE PAVING
6"
ACCESS ARM
SCREW STOPPER
CAST IRON DOMED GRATING
"R" OUTLET BEND.
CONCRETE BEDDING.
"B."
CIRCULAR BASE.
GULLY TRAP WITH CLEANING ARM IN FLOOR OF LIFT WELL.

3½" CAST IRON RAIN WATER PIPE
2½" LEAD LAVATORY WASTE.
2': 0".
10"
CONCRETE BEDDING.
"C."
4"
4"
BASELESS TRAP
2': 0"
CHUTE.
"P" OUTLET BEND.
DISCONNECTING CHANNEL AND BOLDINGS "SIMPLEX" REVERSIBLE GULLY
TRAP TO TAKE RAIN WATER AND LAVATORY WASTE PIPES.
SCALE OF FEET.
1.   0   1.   2.

A domed grating is employed in this example. It is less liable to choke and is therefore better than the dished form.

Disconnecting channel and reversible gully. When wastes from lavatories, baths and sinks—either separately or in addition to rain-water—are to be discharged over a gully, as required by most bye-laws, a channel gully may be usefully employed. There are many and varied forms, but all provide for the liquid to discharge into a chute on one side of the trap.

Bolding's "Simplex" reversible gully is illustrated, fitted with a splayed disconnecting channel and a domed grating, as shown in the detail at C.

The channel allows several wastes to discharge into the chute, which would not be conveniently possible over an ordinary gully grating; there is also less liability for any accidental leakages of foul air from the drains or gases rising from an unclean trap to enter the down pipes.

The form of the trap allows the outlet to be turned in any convenient direction for the branch drain. A square topped receiver rests upon the trap and should be soundly jointed to it in cement. The whole fitting is preferably supported by, and enclosed in, concrete 4″ thick, as illustrated.

508. Seatings and enclosures for gully traps. Gully traps, like all other types of trap, should preferably have a flat base to enable them to be set at the proper level without difficulty. If any trap be tilted, its seal is altered from that originally intended, and if the inclination is towards the outgo may seriously reduce it.

Traps made without level bases require careful setting and should be secured in position by concrete as shown at C in detail No. 148.

The depth of seal in gully traps should not be less than $2\frac{1}{2}″$ for general use and often reaches $3\frac{1}{2}″$ where evaporation is likely to reduce the seal in long droughts.

In stone districts traps are commonly set upon large stone slabs, as at A.

509. Surface finish to yard. Most yard surfaces are paved with stone slabs $2\frac{1}{2}″$ or 3″ thick, laid upon a bed of levelled ashes over rammed earth, or upon dry brick or stone rubbish.

In most good modern work the yard surface would be concreted and this should be made good where it connects to the concrete surrounding traps, etc., to prevent leakage of waste water into the soil below. Such water may cause pollution and may travel to unexpected positions resulting in damage to walls and foundations, or may give rise to foul gases which find their way to the interior.

510. Testing new drains. The whole system of drains to any building should be tested for soundness of jointing before the excavations are filled. If joints are sound they will be both air and watertight, and tests might therefore be made by compressed air or water pressure. Smoke testing is also employed for certain purposes.

Hydrostatic test. The "hydrostatic" or water test is probably the most satisfactory and is insisted upon by most Local Authorities for new drains. It consists of stopping the outlet end of each length of drain to be tested, and filling with water; the weight of the water causes pressure which, at any level, is proportional to the vertical depth from the free surface of the liquid. This vertical height of water pressing upon the containing surface is called the "head" and every foot of head produces a pressure of ·434 lb. to the sq. inch. A reasonable maximum test pressure is about 3 lbs. per sq. inch, which is approximately obtained by a head of 7 feet.

In testing any drainage system for a building some thought should be given to the possible maximum pressure which might be developed in the event of the drains becoming stopped at the point where they discharge into the sewer branch through the disconnecting trap.

Incoming drainage might accumulate to the level of the surface gullies if the stoppage is maintained, although there is a likelihood of the increasing pressure forcing the cause of stoppage through the drain.

As a rule the maximum depth of the disconnecting trap does not exceed 6 ft., so that the test pressure of 3 lbs., or a head of 7 ft., is sufficient for normal cases, especially where ordinary glazed earthenware pipes are employed, because this material is often somewhat porous and will not withstand much greater pressure than 3 lbs. per sq. inch without leakage through the pipes.

It is sometimes objected that in ordinary water tests the pressure, being variable with height of water contained, does not sufficiently test the higher parts of the drains; but it must be remembered that the test conditions are very similar to those of a blocked drain. A more severe test can be given to the higher parts or branches by stopping the gullies after filling to their level and allowing the water to rise a few feet up the soil pipe where such exists; otherwise two lengths of temporary pipe may be fixed above the surface level to obtain the necessary head.

To clear the air from branch drains, so that they may be properly filled with water, a short piece of flexible tube may be inserted through the trap, or the latter may be baled out and a filling hose inserted through the bend of the trap into the branch drain.

When properly filled the test should be extended over a period of at least 30 minutes. There will be some absorption of water by the cement and also by the pipes if the latter are of poor quality.[1] There should be no appreciable fall in the water level if the drains are satisfactory; should faulty joints be discovered the cement must be cut out and the joint re-made.

Tests should be conducted before filling the pipe trenches and again on completion of the filling.

This latter process requires care in depositing the earth round the drain pipes; all shocks and strain must be avoided.

511. Pneumatic or "compressed air" test. This consists of forcing air into stopped portions of a drain. The ends of each portion to be tested are plugged and air pumped through a valve in one of the plugs until it reaches the desired pressure within the pipes. Pressure is registered on an ordinary pressure gauge, or by the head of water or mercury supported in a graduated glass tube; the pressure should not, as a rule, exceed 3 lbs. per sq. inch.

The apparent advantage of air testing is that every portion of the drain receives the same unit pressure; we have already pointed out that all portions do not become subject to equal pressure, nor are the higher portions at any time likely to receive much pressure. The only satisfaction to be derived from pneumatic testing is that all parts of the original drain have stood an equal test and that the air test is more searching than the water test in that leakages are more easily noted, though not easily located. For this reason smoke tests are sometimes applied after a leakage has been proved by the air test, in order to locate it.

512. Smoke tests. Smoke tests are very often conducted, more particularly for vertical pipes such as soil and waste pipes, and also for old drains. The method is to plug the channels and force smoke into the pipes in sufficient quantity to be easily visible at any faulty joint. Very little pressure is obtained in the pipes, hence small faults are overlooked in the inspection.

It is a common practice to test soil pipes by plugging the discharge end, sealing all traps at inlets to the soil branches and leaving the vent open at the top of the pipe. On filling with smoke the latter should be emitted at the vent without any trace of escape at joints and traps.

Smoke testing has little to recommend it beyond ease of location of considerable defects.

513. Manhole covers. A cast iron manhole cover is placed over each chamber on a level with the surrounding ground; there are many types, some suitable for street use and others for private premises where light traffic is to be anticipated.

[1] See Chapter on Materials, Vol. II.

The complete fitting consists of a frame and a lid, the joint between the parts providing an airtight seal.

514. Preparation for frame of manhole cover. In small manholes the brickwork is levelled and floated over with cement mortar, and the frame bedded truly upon it, or, as an alternative, a rebated stone margin is laid upon the brickwork in cement, as shown in detail No. 149, and cramped firmly together at the joints.

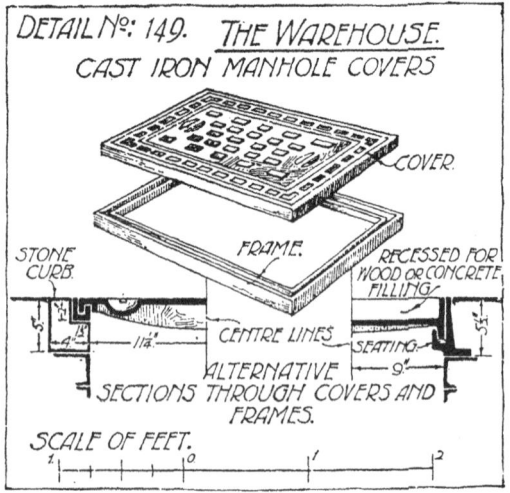

DETAIL Nº: 149. THE WAREHOUSE.
CAST IRON MANHOLE COVERS

COVER.
STONE CURB.
FRAME.
RECESSED FOR WOOD OR CONCRETE FILLING.
CENTRE LINES
SEATING.
ALTERNATIVE
SECTIONS THROUGH COVERS AND FRAMES.
SCALE OF FEET.

515. Sealing the chamber. In order to seal the chamber effectively the frame is provided with a groove, into which a deep rim on the cover is inserted.

The groove may be filled with thin grease, sometimes mixed with sand, to make the joint airtight, or a few strands of soft well-oiled cord may be laid in the bottom of the groove and the edge of the rim bedded firmly upon it.

516. Surfaces of manhole covers. The surfaces of these covers are often studded or indented in patterns, to provide a foothold to persons passing over them, but a better method is to cast the cover with a deep recess and to fill this with concrete having a slight fall towards the edges, or again, with paving blocks of brick, wood or stone, as shown.

The covers are lifted by T-shaped keys engaging in slots, or by cast bars in the covers which provide hand hold for lifting. The better classes of manhole covers are heavily galvanised to prevent oxidation, while the cheaper qualities are painted with several coats of red oxide oil paint, or coated with Dr Angus Smith's bituminous solution.[1]

[1] See Chapter on Materials, Vol. II.

CHAPTER EIGHTEEN

JOINERY

DOORS, FRAMES AND FINISHINGS

This chapter deals with those doors, frames and finishings for the two buildings selected for study, which were not previously dealt with in Vol. II.

517. Folding and swing doors. When a doorway is sufficiently wide to allow two separate doors to be used in the width of the opening it is more convenient to employ them. If hung to each jamb and made to fold together by a rebated joint at their free vertical edges they are known as "folding doors" or "single swing doors hung folding". Such doors are placed in rebated frames and hung on ordinary butt hinges.

For certain purposes it is more convenient to have the doors free to open either way, inwards and outwards; these are known as "double swing" doors and are usually hung with spring hinges which allow rotation either way but withdraw the doors to their closed position when the opening pressure is removed.

Details of swing doors are given in connection with the vestibule at Nos. 169 and 170, and the student should compare these with the details of folding doors in No. 150.

Double doors of either class should only be arranged in openings of sufficient width to allow a person to pass easily when only one of the doors is open.

518. External folding doors to main entrance of warehouse. Masonry detail No. 13 shows the main entrance to the warehouse including the general design of the folding doors, which are four-panelled of 2″ hardwood with stuck moulds on both sides and raised panels on the external face. The proportionate reduction in the height of these panels should be observed and also the general proportions of the door opening. The student should note that fanlights are often adopted for the purpose of gaining height to a door opening and so securing good proportions, even where the glazing is unnecessary for the purpose of admitting natural light.

The plan of detail No. 150 shows how the external doors are intended to fold back across the angle of the vestibule when the warehouse is open for business purposes; it is quite usual to make this provision in modern business premises so that the doors may be locked back to avoid interference by unauthorised persons.

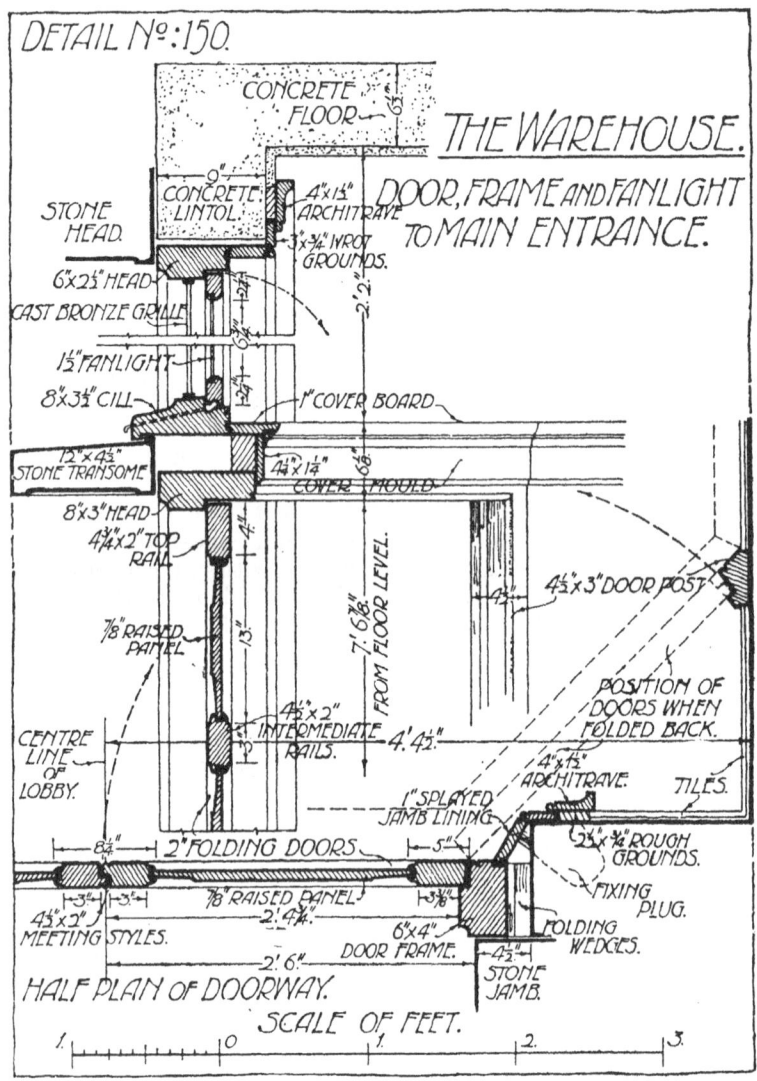

DETAIL Nº:150.

THE WAREHOUSE.

DOOR, FRAME AND FANLIGHT TO MAIN ENTRANCE.

519. Construction of door frame to front entrance of warehouse. The stone opening is divided into fanlight and door openings by a 12″ × 4½″ weathered stone transome. The frame, which is of hardwood, thus becomes divided at this level, the wooden transome being framed in two pieces to obtain a weathertight joint on the top side. The posts of the frame are $6'' \times 4''$ and the fanlight head $6'' \times 2\frac{1}{2}''$, all being rebated and cavetto moulded, as shown in plan in the detail No. 150 and in enlarged section at No. 151. The head of the doorway is $8'' \times 3''$, having the same general shape in section, but with a deeper rebate, thus forming a $2''$ projection internally, which is a part of and a support to the moulded transome shown in the large scale detail.

Above the weathered and sunk stone transome, and rebated over it, is an $8'' \times 3\frac{1}{2}''$ weathered rebated and channelled cill, the overlap being bedded and pointed in oil putty.

The space between the fanlight cill and door head is to be covered by a built-up moulded transome, and to support this a $2\frac{7}{8}'' \times 2''$ wood blocking is inserted, as shown in section.

The frame may be jointed by double tenons at the head of the post and on the members of the wood transome, the joints between cavetto moulds and beads being mitred, though it is easily practicable to scribe the former.

The rough blocking and the projecting door head would pass on the face of the frame as far as the jamb lining, against which they would terminate on the splay; as the blocking cannot be jointed it must be screwed to the posts at the overlap.

520. Finishings to frame and opening. The jambs are covered by $1''$ splayed and tongued linings nailed to bevelled wood plugs in the jambs and covered at the inner margin by a $3'' \times \frac{3}{4}''$ wrought ground placed flush with the $2\frac{1}{2}'' \times \frac{3}{4}''$ rough ground, against which the tiling stops; the architrave is $4'' \times 1\frac{1}{2}''$ and overlaps both wrought ground and tiles.

The lining passes through the full height of the jamb but the architrave is interrupted at the transome by the built-up mould, which is continued across the angle of the vestibule to surmount the doors when folded back and to provide a cornice. The moulding is formed by the projecting head, which is beaded on the lower edge, a $4\frac{1}{4}'' \times 1\frac{1}{4}''$ moulded fascia and a $4\frac{1}{4}'' \times 1''$ moulded cover board. These latter members are tongued together and the cover also tongued to the frame.

521. Retaining frames for open doors. To retain the open doors and allow them to be locked back, splayed rebated posts are fixed to the walls as shown in the plan, and in order to convey a sense of finish and completeness to the angle of the vestibule when the doors are open, a hood or cover of triangular outline is placed across the angle at the transome level and the mould of the latter carried across it and further continued round the vestibule to form a capping to the tiled dado.

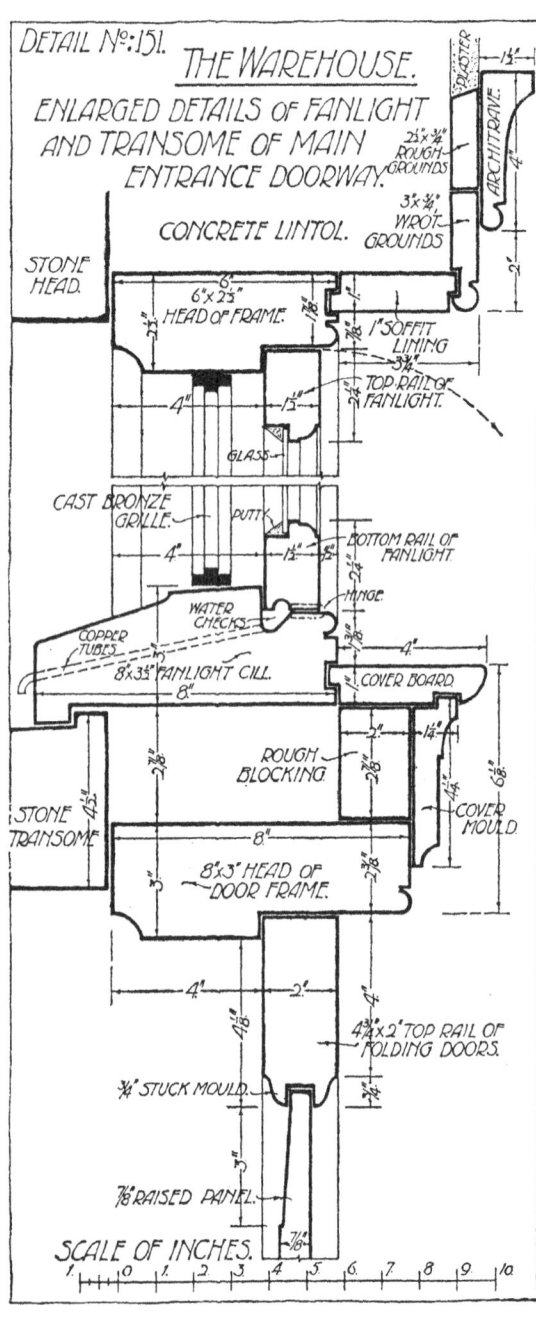

The triangular soffit may be of $1\frac{1}{4}''$ panelled framing, with a close boarded top, or of $2'' \times 1\frac{1}{2}''$ bearers with lath and plaster finish to the soffit and top surface.

522. Construction of doors to main entrance of warehouse. The doors are of hardwood $2''$ thick, each being one panel in width and four panels in height. Each door is about $2' 6''$ wide and $7' 7''$ high and consists of stuck moulded framing of the following sizes: hanging stile $2'' \times 5''$, meeting stile $2'' \times 4\frac{1}{2}''$, bottom rail $2'' \times 9\frac{3}{4}''$, intermediate rails $2'' \times 4''$ and top rail $2'' \times 4\frac{3}{4}''$.

The framing is grooved centrally on the edges, $\frac{5}{8}''$ wide and $\frac{1}{2}''$ deep, to receive the panels which are of $\frac{7}{8}''$ material, raised on one face, with bevelled margins $3''$ wide. The raised or "fielded" portion should not be over emphasised, a projection of $\frac{3}{32}''$ being quite sufficient. It is also unwise to give to the margins too much bevel, which should not exceed $\frac{3}{16}''$ in $2''$. Note the simple outline of stuck mould adopted and the entire absence of fillets, which collect dust and are difficult to clean.

523. Fanlight over doors to warehouse entrance. The fanlight is a hinged hardwood sash $1\frac{1}{2}''$ thick with $2\frac{1}{4}''$ wide stiles and rails. It is hung on the lower edge to open hopper-wise but with a different joint to that adopted for the main entrance fanlight to the house.

Instead of overlapping at the cill, the fanlight sash is set entirely within internal rebates, a method which is only favoured for positions that are shielded, as in this case, by projecting stonework. As there is a possible danger of rain being driven over the cill into the rebate, the cill should be channelled on the inner side of the rebate and the sash rail grooved, as shown in section at detail No. 151, and a pair of copper tubes $\frac{3}{8}''$ bore inserted through the cill to convey any accumulated water to the outside. Such tubes must be bent downwards to prevent winds forcing back the discharging water.

The fanlight is guarded by a metal grille which is shown in section in the details, and in perspective at detail No. 13; this grille is also referred to in paragraph 105.

INTERNAL DOORS TO THE HOUSE

Many of the internal doors, linings and finishings to the more subordinate parts of the ground floor and upper floors could be satisfactorily treated for this dwelling house by adopting any of the suitable details from Vols. I and II.

The ground floor, however, offers facilities for better design and

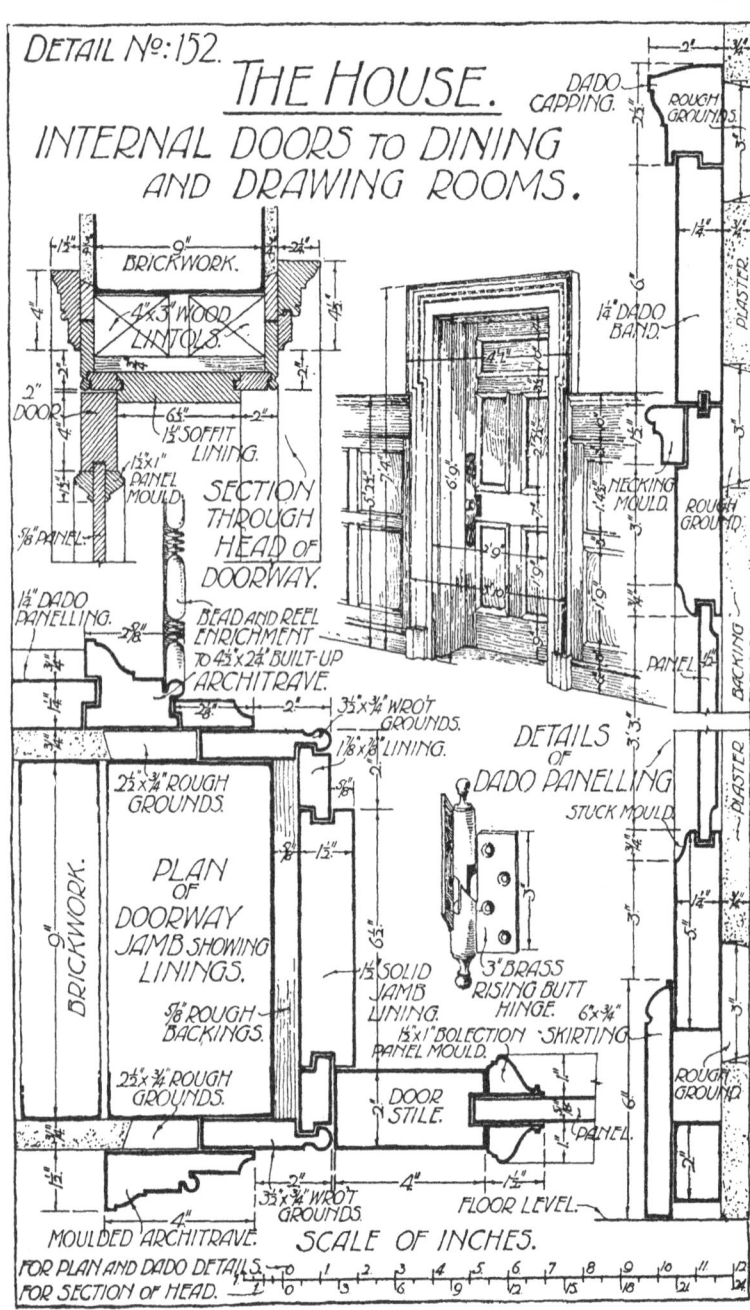

DETAIL N°: 152.

THE HOUSE.

INTERNAL DOORS TO DINING AND DRAWING ROOMS.

DADO CAPPING.

ROUGH GROUNDS.

BRICKWORK.

4"x3" WOOD LINTOLS.

2" DOOR

6½" 1½" SOFFIT LINING.

1½"x1" PANEL MOULD.

⅝" PANEL.

SECTION THROUGH HEAD OF DOORWAY.

1¼" DADO PANELLING.

BEAD AND REEL ENRICHMENT TO 4½"x2¼" BUILT-UP ARCHITRAVE.

3½"x¾" WRO'T GROUNDS.

1⅛"x⅛" LINING.

1¼" DADO BAND.

NECKING MOULD.

ROUGH GROUND.

PANEL

DETAILS OF DADO PANELLING.

STUCK MOULD.

2½"x¾" ROUGH GROUNDS.

PLAN OF DOORWAY JAMB SHOWING LININGS.

BRICKWORK.

9"

⅝" ROUGH BACKINGS.

2½"x¾" ROUGH GROUNDS.

3" BRASS RISING BUTT HINGE.

1½" SOLID JAMB LINING.

1½"x1" BOLECTION PANEL MOULD.

6"x¾" SKIRTING

PANEL

DOOR STILE.

MOULDED ARCHITRAVE.

3½"x¾" WRO'T GROUNDS.

FLOOR LEVEL.

SCALE OF INCHES.

FOR PLAN AND DADO DETAILS

FOR SECTION OF HEAD.

more elaboration of detail in the doorways opening from the hall to the main rooms.

524. Five-panelled doors and finishings to doorways in hall of house. These doorways are formed in 9″ brick walls, are finished with solid linings and tongued rebate pieces, and enclose a five-panelled door. Enriched eared architraves surround the opening on both sides.

The general design is shown in the perspective detail No. 152, which also illustrates wood panelling to the entrance hall and its junction with the finishings of the doorway. The doors are flush on the inside of the room, hence with the projection of the finishings from the wall a good depth of inset to the door from the hall side is obtained which adds to the effect of the woodwork surrounding the opening.

525. Preparation of doorway. The doorway is prepared ready for the plastering and later to receive the joinery work by rough grounding to the jambs and to both faces of the opening.

The $\frac{5}{8}$″ rough backings are stop dovetailed to the wrought grounds subsequently described.

526. Door linings. To form the rebate for the door, which should be $\frac{5}{8}$″ deep, a $2″ \times \frac{7}{8}$″ tongued rebate piece is grooved into the $1\frac{1}{4}$″ solid jamb lining and receives the $\frac{3}{4}$″ framed wrought grounds with beaded edges, as shown on the enlarged plan of the detail.

527. Architraves. Eared architraves are employed on both sides of the doorway, that in the hall being built up of two moulds, $2\frac{5}{8}″ \times 2\frac{3}{8}$″ and $2\frac{1}{8}″ \times \frac{3}{4}$″, tongued together and grooved to receive the wall panelling. Inside the rooms a $4″ \times 1\frac{1}{2}$″ plain moulded architrave is used. A small bead on these architraves can be suitably enriched with a "bead and reel" ornament which is sometimes carved from the solid wood, but if the woodwork is to be painted it can be applied in the form of a cast plaster or papier-maché mould.

The break to form the ears occurs at the level of the dado capping, which is finished against the projection of the normal architrave below the ear.

The architrave terminates upon a moulded foot block $7\frac{1}{2}$″ deep, which receives the housed end of a 6″ skirting.

528. Dado and frieze to hall. The hall is panelled with a wooden dado to a height of $5′ 2\frac{1}{2}$″ including a $6″ \times \frac{3}{4}$″ plinth, a $2\frac{1}{2}″ \times 2$″ capping and subsidiary or necking mould $1\frac{1}{2}″ \times 1$″.

The dado framing is indicated in the perspective detail and a larger scale section is shown on the same drawing.

The framing is $1\frac{1}{4}''$ thick with two heights of panels and the members are selected to show a uniform width of $3''$ on the flat exposed parts of their faces; the panels are $\frac{1}{2}''$ thick with plain faces, bevelled on the back to enter $\frac{3}{8}''$ panel grooves in the frame and the sight edges are finished with cyma recta stuck moulds.

These panels might be replaced with plywood, $\frac{1}{4}''$ thick, either plain for painting or faced with some ornamental veneer if hard wood panelling is adopted.

In order to receive the necking and plinth moulds wider rails are required, the top rail being $5\frac{1}{4}''$ wide and rebated to receive the necking while the bottom rail is about the same width and passes $1\frac{1}{2}''$ or so behind the plinth. The whole piece of framing is $4'\ 6''$ high above the floor and for fixing purposes has a $2'' \times 1\frac{1}{4}''$ rough rail framed in near the floor level; this receives the plinth and is entirely covered by it. The stiles of the framing continue to the floor level.

The upper portion or frieze of the dado is a plain band $6''$ wide, out of $6\frac{1}{2}'' \times 1\frac{1}{4}''$ stuff, tongued at the top edge to receive the capping mould and grooved for a hardwood tongue at the bottom.

The whole of the dado is fixed upon rough grounds, as shown in the sections, and the wall behind is rendered in rough plaster to fill the space to the same plane as the grounds. This and the surrounding plaster must be thoroughly dry before the woodwork is fixed.

529. Methods of fixing finishings to doorway, etc. The whole of this work including the door and dado might be done in American yellow pine prepared for painting in light colours, in which case the fixing of parts would be done by nails placed in such parts as would be most effective, yet hidden where possible, taking advantage of tongued joints to cut their number to a minimum and placing where any subsequent shrinkage would not be liable to cause splitting of the material through being unduly bound by the nails.

Hardwood finishings such as oak, mahogany or walnut would however be more suitable for the hall, and thoroughly seasoned material is necessary for such work.

Very many parts of the fixing can be done by hidden nails driven on the skew through grooves, or directly through surfaces subsequently covered by overlaps; the rest may usually be effected by secret slot screwing of the type shown in detail No. 167. In some cases it may be necessary to use needle points at mitres, viz. steel pins like fine nails without heads, or again to cut and lift the surface fibres, nail through the exposed part and glue the fibres down again. Where very strong fixings are required round holes may be drilled to a depth of $\frac{3}{8}''$, wide enough to receive a screw head and allow it to be turned without damage, and afterwards filled by pellets of the same material with the grain in the same direction as the fixed piece. Difference in colour and grain often makes such filleting objectionable, hence careful selection is necessary.

530. Doors from hall to dining and drawing rooms. These doors are 6′ 9″ × 2′ 9″ by 2″ thick, framed in five panels as shown in perspective at detail No. 152. The mode of framing compares with other doors previously described, and reference need only be made to the proportions.

A common fault is to make the difference in height between the lower and middle panels too great and also to employ rails which are wider than necessary for either construction or design. In this example the stiles and top rail are 4″ wide, muntins and frieze rail 3½″, bottom rail 9″, and lock rail 7″. The upper panel is short in the direction of the height of the door and the full width between the stiles and is known as a frieze panel; if its grain is horizontal, which is the correct way of placing it to avoid much shrinkage, it is called a "lay" panel; this term applies to *all* panels so treated.

If plywood panelling is used for the hall similar panels would be used for the door

Bolection moulds (unrebated and sometimes called risen moulds) are used on both sides of the door, as seen in the enlarged plan of the jamb.

531. Method of hanging door. In order to allow the door to clear thick carpets while fitting reasonably close to the floor when the door is closed, the hinges employed are of the pattern illustrated, and are known as "rising butts". They consist of two wings like an ordinary butt hinge, but instead of moving about the knuckle in a horizontal plane, when one wing is moved away from the other it rises upon a spiral joint and thus lifts the door as it swings from its closed position.

The amount of rise should not be excessive or some difficulty occurs in causing the top hanging edge of the door to clear the upper rebate, which in any case must be bevelled at the joint.

The knuckle of the hinge must stand just clear of the faces of door and lining and the ends are therefore ornamented by moulded terminals. Brass hinges should be employed as they are smaller in the knuckle and of lighter and less objectionable appearance than iron ones. The latter are sometimes employed for painted work but are clumsy in appearance. Hinges are often recognised as ornamental features and are so treated.

MODERN DOORS

The modern tendency is to develop the use of flush doors for all kinds of domestic, commercial and public buildings.

These doors have already been referred to in Vol. I under the headings of hollow framed doors with plywood facings and solid blockwood doors.

Either type of door may be further embellished by veneering the face (or faces) with ornamental hardwood veneers, in matched grain, or by veneering in patterns either in wood, or in enamelled and bright metals.

Some of the simpler treatments of such doors are shown in plates Nos. I to IV which are reproduced from photographs of completed work by permission of Venesta Ltd.

Plates Nos. I and II show single flush doors with hardwood facings of matched veneers, while plate No. II also shows a pair of folding doors of a similar character. It should be noted that special blocks are included in the framing for the insertion of locks and the secure attachment of furniture.

Plate No. III shows a door finished in patterned veneer and plate No. IV shows oak panelling (without framing), a birch plywood floor laid in squares, and birch plywood furniture. The simple but tasteful treatment of modern joinery and cabinet work in these photographs is typical of much modern work. Such treatment, to be effective and permanently attractive must embody perfect preparation of the plywood, an understanding of the appropriate methods of construction and assembly and first class craftsmanship.

FRENCH CASEMENTS

Although the item we are now about to study might strictly be considered as a window to a room, the nature of the details and the intended use make it advisable to include the study under the heading of glazed doors.

532. Casement. The term casement has already been defined in Vol. I as any form of sash hung to the frame on one of its edges. The common form has the sashes hung on the vertical edges and when such sashes are continued to the floor level or thereabouts, so that they may be employed also as doors, they are known as French casements.

Both ordinary and French casements may be arranged to open either inwards or outwards according to the desire of the designer or the special circumstances of the case.

PLATE I

Venesta flush doors. Walnut veneers.

PLATE II

Venesta flush door (walnut). Single and folding doors.

PLATE III

(*opposite*)

Venesta patterned flush door, in hardwood veneer.

PLATE IV

Venesta oak plywood panelling, without framing; Venesta birch plywood floor, in squares;
Venesta birch plywood furniture.

533. Inward and outward opening of casements. There is some diversion of opinion as to whether inward or outward opening of casements is the more justifiable and we may usefully discuss this point before studying the construction of an example.

The relative advantages and disadvantages depend upon position, convenience and the nature of any desired accessories to the window.

Considering weather resistance as a vital point, we find from experience that in exposed positions it is much easier to make an outward opening casement weathertight than one which opens inwards. The most vulnerable portion is usually the joint between the sash and cill and especially at the angles of the latter. If made to open inwards special treatment of the cill and sash joint is necessary, as will be seen in the detailed examples.

Another position causing trouble is the top rail of the sash in a flush frame, where rain can drive in if the casement opens outwards. The obvious remedy for this is a projecting rail in the frame which should overhang sufficiently to allow the soffit to be throated.

Inward opening casements allow of easy access for cleaning both sides of the glass, and are not liable to damage if accidentally left open and unfastened in a strong wind. They do not interfere with the provision of sun-blinds and external Venetian shutters.

Outward opening casements do not interfere with ordinary blinds and curtains and also allow of the use of internal shutters where desired. They take up no space within the room, which is an advantage, especially in thin walls, but unless hung folding and properly proportioned they are difficult to clean.

Selection of type must therefore be made to suit the personal bias of the owner or designer, having regard to all the circumstances of the case.

534. Drawing room windows of the house. These windows are French casements hung folding, with a fanlight over.

Detail No. 153 shows an internal perspective of the window treatment; the casements open inwards and the fanlights are hinged hopper-wise.

The casements are glazed for about three-quarters of their height in small panes and the fanlights are correspondingly divided; the panes in the former are about 10″ high × 5″ wide. It is important to observe the ratio of height to width of the panes, viz. two to one, a ratio which may usually be adopted with architectural advantage, although some designers think it too high.

535. Construction of casement frames. These consist of 4″ × 3″ posts and head, 4″ × 3½″ transome and 5″ × 3¼″ oak cill. As usual with window frames the head and cill run through, and the posts

DETAIL Nº: 153. THE HOUSE.
FRENCH CASEMENT WINDOWS
TO DRAWING ROOM.

ARCHI-
TRAVE.
⅞" FASCIA.
¾" ROUGH
GROUNDS
2×1½" ANGLE
BLOCK.
¾" SPLAYED
JAMB LINING.
2"×¾"
BACKINGS
AT 2'6"
CENTRES
SCALE OF INCHES FOR PLAN.
PART PLAN OF WINDOW.
2' 1⅛"
4"
SASH
BAR.
MEETING
STILES.
4"×3"
FRAME. OAK CILL.
2' 0¾"

are tenoned to them. The posts are deeply rebated and weather-grooved so that the latter will release the pressure of wind-driven rain entering the joint, and allow it freedom to reach the cill. The transome and head are similar to those previously used in window and door frames and the mouldings on all the members of the frame, except the cill, are of the same character as those applied to the outer door frame in Vol. ii. Double tenons are wisely employed for this frame, though only 4″ thick, because the mitred moulding on the inside requires that the joints should be firmly closed and maintained.

A single tenon, if employed, would usually be placed on the square part of the section, so that a full width tenon could be obtained, though a central tenon would be possible with a reduced breadth but would cause more labour in jointing.

536. Section of oak cill. The oak cill is rebated on the inside to allow the casement to open inwards and alternative methods of arranging the junction of sash and cill are shown in detail No. 154.

In the detail at A the cill and sash are both rebated, a hooked and galvanised metal water bar inserted vertically, to stand $\frac{3}{8}$″ above the weathered surface of the cill and the joint guarded by an oak weather board, splay-tongued to the casement. For the latter to be really effective it must be continued at the ends and housed into the frame to the depth of the rebate; otherwise it is no protection at the angles, which are the most vulnerable parts of the cill joint. Owing to the large size of this weather board when made of timber it is often moulded to improve its appearance as shown applied to the fanlight in detail No. 155, but it is sometimes made of metal, as at B in detail No. 154, which allows of housing into the frame at the ends without the excessive cutting and consequent ugly appearance when the casements are open. In the same detail is shown Messrs J. Hill and Co.'s "Jasil" brass water bar; in this case the cill is double sunk and weathered and the water bar of J section is seated upon the upper sinking, standing above it to form a rebate and projecting over the lower sinking to form a throat. If properly fitted and continued the full length between the rebates of the frame, this bar, due to its hooked section, effectively checks entry of water. It should be made of stout material to resist the weight of a person stepping upon the cill or it would be useless for a French casement, though satisfactory for an ordinary casement.

The section at C shows a mechanical form of water bar which has been very successfully used. It consists of a hinged plate, H, which is actuated by a push plate, P, screwed on the bottom rail near the folding side of the casement. When the casement is closed the hinged plate stands vertically against a fixed cover plate, F, closely screwed to the bottom rail of the sash. As the casement opens the hinged plate is drawn downwards by the inward movement of the cover, clearance being provided between the plates F and P to allow the easy rotation of the piece H.

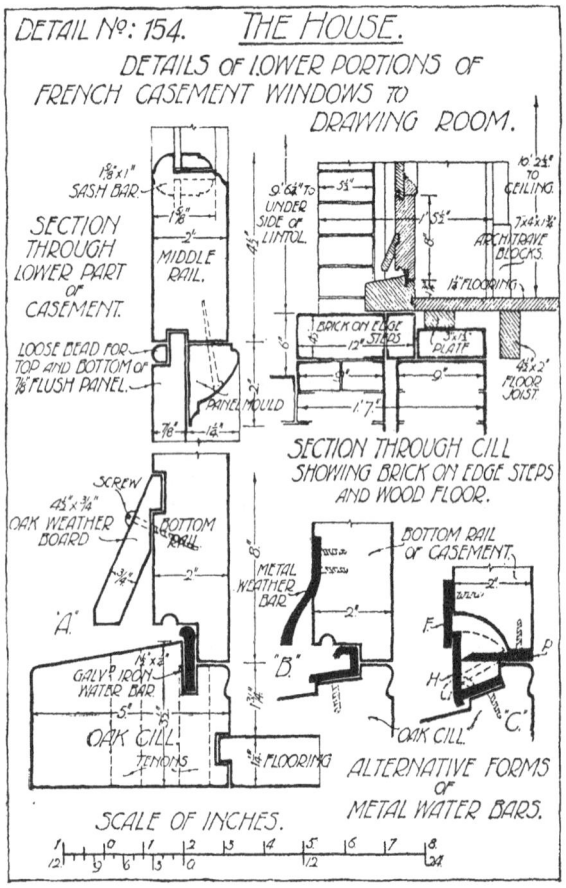

When the casement is opened so that its thickness at the opening edge is clear of the cill the hinged plate H lies upon the wood cill in an inclined position with the edge ready for the push plate to catch and lift up again on closing the casement.

The appliance is long enough to enter the rebates of the frame at each end and is very effective when properly fitted.

For folding casements the plates must overlap at the centre, and for this purpose one set of plates is cranked, or the cover formed by brazing a cover piece to the end of the first closed leaf.

537. Construction of French casement sashes. The sashes in this example are 2″ thick with 4½″ outer stiles and middle rails, 4″ meeting stiles and top rails, and 8″ bottom rails; glazing bars are

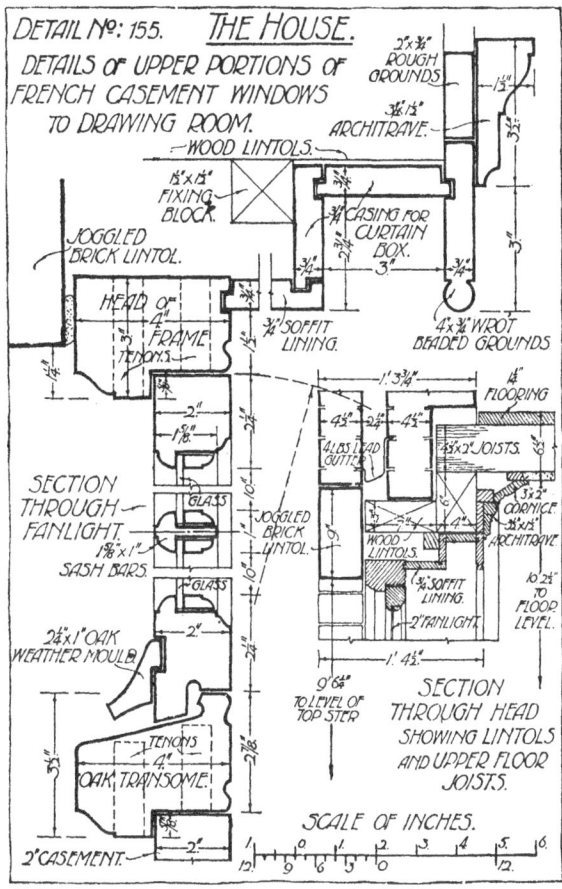

1″ on face and 1⅝″ thick, being flush on the outer surface and inset ⅜″ from the inside face. The lower panel is ⅞″ thick and flush beaded on the outer face—see Vol. II; on the inside the panel is sunk 1⅛″ from the face of the framing and is surrounded by a 2″ × 1¼″ bolection mould; see detail No. 154.

The inner edge of the sash framing has its mould compounded of an ogee and an ovolo, the latter being continued round the sash bars and framing while the ogee forms a rim to the whole, giving the complete sash a panelled appearance and the glazing bars a subordinate effect. The ovolo mould forms a loose glazing bead for securing the glass; see also paragraph 567 for another form of bead. At the folding edges of the casements a bevel-rebated joint is employed and the angle of the rebate weather-grooved, as shown in the plan of detail No. 153.

538. Fanlight sash over casements. The details of this sash are very similar to some fanlights previously considered, but the moulds are of compound form to match those on the casement, while a $2\frac{1}{4}'' \times 1''$ oak weather mould is tongued to the sash.

The bottom rails of fanlights, all weather moulds, cills and transomes should be of oak or teak in good work, and in the best work the entire framing would be of hardwood.

Teak is the best resister of decay due to wet, and is particularly suitable for outside work.

539. Finishings to drawing room windows. The method of finishing the interiors of these windows may be gathered from the perspective detail No. 153. The full height of the room is taken up by the opening, its moulded margin and a light wooden cornice. Linings, $\frac{3}{4}''$ thick, surround the jambs and the marginal mould consists of a $\frac{3}{4}''$ wrought fascia projecting within the lining and forming a curtain recess, and a $3\frac{1}{2}'' \times 1\frac{1}{2}''$ architrave.

To fix these margins $2'' \times \frac{3}{4}''$ rough backings are plugged to the jamb at $2'$ $6''$ intervals and to these backings $\frac{3}{4}''$ tongued and splayed linings are secured. At the front angle a tongued and grooved angle block is fixed, which provides a wide face to receive the grooved and beaded fascia. This fascia is flush with the rough vertical ground and therefore in the same plane as the plaster.

The purpose of the fascia is to allow the architraves to be broken into ears at the transome level by cutting and mitreing, thus showing a wider margin, for which purpose the broken portion of the moulding must be a separate unit from the ground on which it is to be secured.

540. Foot blocks to architraves. Foot blocks to the architraves are provided as previously described in Vol. II, under doorway finishings. The usual methods of jointing the block to the architrave before fixing the latter are shown in detail No. 156. The one marked A is a "dovetail notch" commonly employed in hand work. The example marked B is much more suitable for machine preparation, having no dovetail, but a plain tenon passing completely through and housing the full length of the block and secured by glue and screws.

The housing of the skirting to the side of the foot block is shown in each sketch; its depth should be ⅜".

541. Angle joints of architraves. While joints in architraves for painted work are usually plain mitred, laid upon grounds and nailed through the mitre in both directions, it may be necessary to exclude the use of nails, as for example in polished or high class enamelled work.

For this purpose slot screws are used, these being screwed into the ground and slotted holes drilled and cut in the back of the architrave, exactly as shown in the stair detail No. 167, for the return ends of the nosings. After fixing the side architraves in this way by sliding them downwards into position, the top would be placed similarly, its slots being necessarily cut across the grain. Now this method would not secure a truly assembled mitre and there would be no guarantee of the two pieces keeping flush on the face, hence a neatly fitted dovetail tenon, as shown at C, is employed; this should be tight enough to prevent any movement of the mitre.

542. Blind box and cornice, to head finishings. In many otherwise well-equipped houses, no special provision is made for the fixing of blinds or curtains and the occupant cannot avoid mutilating frames and architraves in making the necessary provision for their support. While the elaboration of drapery about doors and windows is to be condemned, blinds and curtains of some kind are considered necessary and the person responsible for the design of the windows should

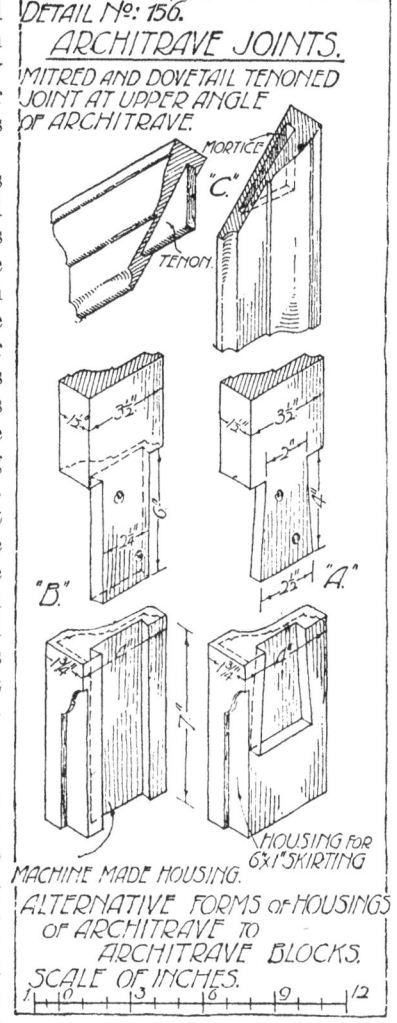

DETAIL Nº: 156.
ARCHITRAVE JOINTS.
MITRED AND DOVETAIL TENONED JOINT AT UPPER ANGLE OF ARCHITRAVE.

MORTICE
"C"
TENON

3½"
3½"
2"

"B."
2½" "A."

HOUSING FOR 6×1"SKIRTING
MACHINE MADE HOUSING.
ALTERNATIVE FORMS of HOUSINGS OF ARCHITRAVE TO ARCHITRAVE BLOCKS.
SCALE OF INCHES.

himself decide a means of supporting blind and curtain fittings in a satisfactory and artistic manner. In detail No. 155 such provision is made. At the head of the window, which is placed as near the ceiling as possible, the linings and architrave are arranged to form a 3″ square casing for a blind or curtain rod, thus covering the heads of the curtains. The box is formed of $\frac{3}{4}$″ tongued head, back pieces and fascia.

To secure the back casing a $1\frac{1}{2}$″ square fixing block is nailed to the lintol; the box is then nailed in position part to part and the tongues ensure correct assemblage.

Above the architrave is a 3″ × 2″ cornice—3″ projection × 2″ deep —obtained from $4\frac{1}{2}$″ × 1″ material and fixed on the splay. This cornice is carried entirely round the room and is broken at the windows round the projection caused by the architrave, as may be seen in perspective detail No. 153.

To secure the cornice 2″ × $\frac{3}{4}$″ rough grounds are fixed on wall and ceiling to which the plaster is levelled. The cornice then covers the junctions by overlapping $\frac{3}{8}$″ upon the plaster. Over the window the ceiling ground is brought forward $1\frac{1}{2}$″ and a packing piece nailed to the wall ground to form a solid back for the cornice.

CHAPTER NINETEEN

JOINERY—WINDOWS AND SKYLIGHTS

THREE-LIGHT WINDOWS WITH CASED FRAMES

Windows and skylights have already received attention in Vols. I and II. This chapter deals only with the construction of three-light cased frames and special windows for ventilation.

When an opening is large, and wide in relation to its height, better proportions may be obtained by dividing the opening into two or three "lights" by forming separate pairs of sashes within the one opening.

An application of this principle, which could be used for the back or side windows of the warehouse, is shown by plan in detail No. 158, and employs wood mullions in place of the stone or metal mullions sometimes adopted.

543. Arrangement of three-light cased frames. The arrangement of these windows depends upon the number and position of the sashes which are desired to slide and on the amount of woodwork it may be considered suitable to show within the opening at the divisions of the lights.

Detail No. 157 indicates three modes of treatment. At A, three pairs of sashes are separately hung, that is to say, the balance weights for each sash are quite independent of the rest. This requires an intermediate weight box to take two weights running side by side, forming a double boxed mullion and taking up 4″ or more in width. On the left of B, the same sashes are shown with one larger weight at the intermediate casing to balance the two adjacent sashes, thus forming a single box 2″ wide and obtaining a much narrower division between the lights. On the right of B a single boxed mullion is shown containing a single balance weight for the centre sash only, the side sashes being fixed.

At C, solid mullions are shown with the arrangement for hanging the centre sash only, by taking the sash lines across the heads of the adjacent lights and attaching to balance weights in the side casings of the frames.

The method of construction adopted is very similar to that of an ordinary cased frame for sliding sashes, the vertical section and plan at the jambs being identical, apart from modifications due to personal preference in the form of cill, nature of mouldings, or to the mode of supporting the walling above the opening.

544. Cast iron window cills. In detail No. 158 the vertical section shows the use of a cast iron window cill in place of stone or brick, a type of construction used in some industrial districts where the method is economical. The cill is 12″ wide × 6″ deep, suited to a 9″ reveal, and will vary in thickness from $\frac{5}{8}$″ to 1″ according to the size of the opening.

To render the joint of the oak cill weathertight a rebate is formed in the metal and is overlapped as shown by the timber, which should be bedded solid in red lead putty when fixing.

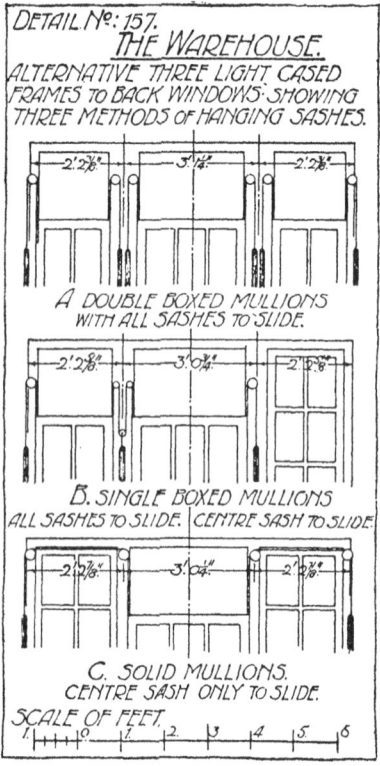

DETAIL Nº: 157.
THE WAREHOUSE.
ALTERNATIVE THREE LIGHT CASED FRAMES TO BACK WINDOWS SHOWING THREE METHODS OF HANGING SASHES.

A DOUBLE BOXED MULLIONS WITH ALL SASHES TO SLIDE.

B. SINGLE BOXED MULLIONS ALL SASHES TO SLIDE. CENTRE SASH TO SLIDE

C. SOLID MULLIONS. CENTRE SASH ONLY TO SLIDE
SCALE OF FEET.

545. Cast iron and steel lintols. At the head, the face walling is supported by a cast iron lintol partly shown in the detail at No. 158, and backed by a 7″ × 3½″ × 20·23 lbs. B.S. channel. The space between the two is filled with concrete *in situ*, leaving a 2″ deep recess to receive the frame.

546. Securing large frames into openings. The frame is secured in position by (*a*) iron holdfasts driven into the jambs (as shown in Vol. 1) and covered by the 2¼″ × 1½″ scribing mould nailed to the frame, or (*b*) plugging the jambs just behind the mould and nailing the scribing mould to them on the skew, or (*c*) where sufficient clearance exists round the frame, by pairs of parallel wedges driven between jambs, head and cill. The two former methods only are satisfactory unless plaster linings are employed, as shown at D in the detail.

547. Construction of weight boxes and divisions in multiple lights. The detail shows the construction of the three arrangements referred to in paragraph 543.

At A a double boxed mullion is shown, being similar to the side box but duplicated to contain two weights, and divided at the centre by a thin lining. With 1¼″ pulley stiles the outer lining becomes 8¼″ wide, which could only be adopted for very large openings to obtain a reasonable proportion.

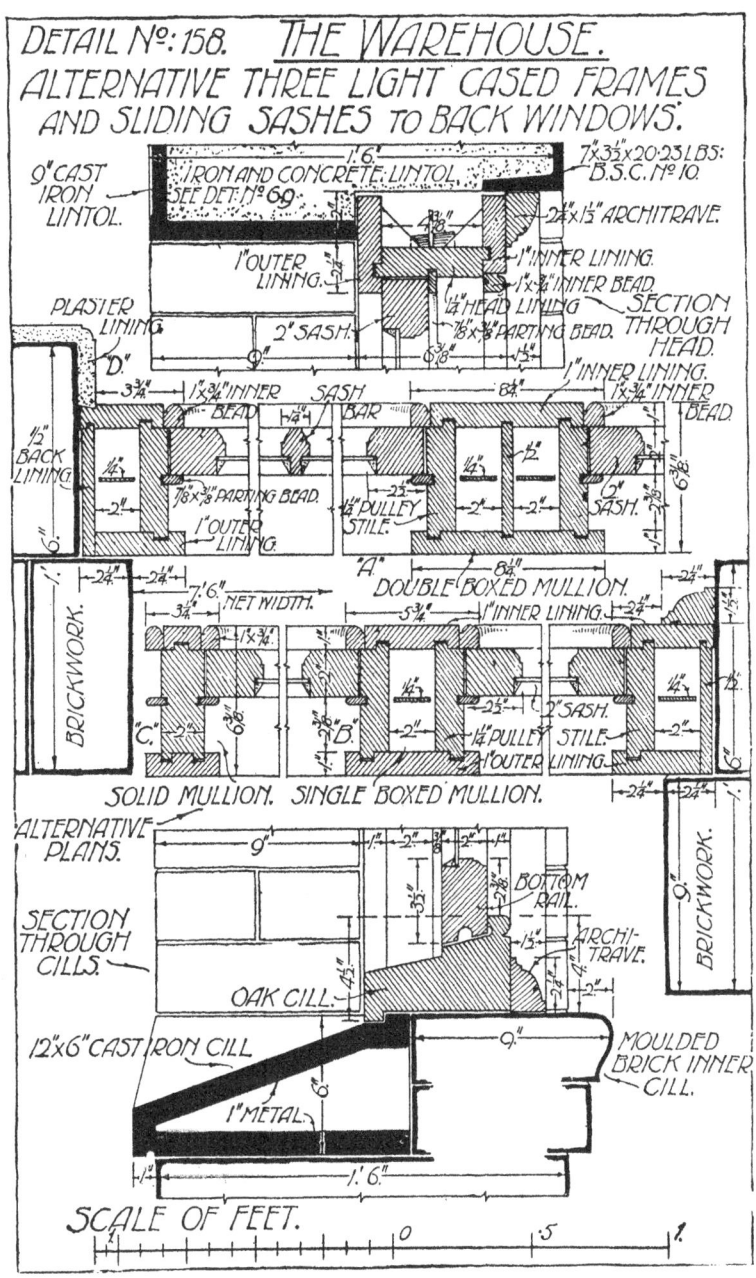

DETAIL Nº: 158. *THE WAREHOUSE.*
ALTERNATIVE THREE LIGHT CASED FRAMES
AND SLIDING SASHES TO BACK WINDOWS.

SCALE OF FEET.

At B the central division is omitted and the pulley stiles brought together to form a box of single width. This reduces the outer lining to $5\frac{3}{4}''$. It should be noted that by the use of the $\frac{7}{8}''$ pulley stile, the width in each case could be reduced by $\frac{3}{4}''$ more.

The box in this example needs to take a larger weight which has a mounted pulley cast into its top end. The suspension cord is in one length, passing under the pulley and continuing to the adjacent sash on each side. As a sash moves vertically the weight only makes half the travel of the sash.

For this purpose the single weight must be equal to half the sum of the weights of the sashes it is to balance: and as a larger weight than the stock size may be required it is often specially cast to a square or rectangular section either in iron or lead. Such weights are troublesome and noisy, as the angles come in contact with the casing, due to twisting of the cord. Round "lead" weights should be used where cast iron would only be possible by adopting a square section.

The section at C shows a $2''$ solid mullion for side hanging of the centre lights. The frame is constructed as before, but the side sashes are screwed fast or held by stops placed in the sash slides. To hang the centre sashes the cords are taken up the face of the mullion, over a pulley of special form fixed at the top and carried over the side lights just below the head, to the side boxes. The horizontal portion of the cord is guarded by the inner sash, which is grooved to enclose the cord, and the other cord is enclosed by a grooved cover which is screwed to the head, and is thus removable for the renewal of cords. Pocket pieces would still be required in the side pulley stiles for the insertion of and access to the weights.

548. General construction of double hung sashes and frames. The example of construction shown in detail No. 158 is typical of the plainest form of good work. There is no unnecessary ornamentation, the parts shown being of the simplest form to suit their purpose. If rigid economy demands the use of less members, or the size of opening allows it, $\frac{7}{8}''$ pulley stiles with $\frac{3}{4}''$ linings may be employed and tongues may be omitted, but the result is seldom satisfactory in the long run, particularly for the larger windows.

549. Windows for efficient ventilation. In schools, public institutions, some office buildings and certain parts of dwelling houses, some form of window which will allow of efficient ventilation without undue draught is a necessity.

A form of window which has been successfully used in school buildings is shown in detail No. 159. It consists of three tiers of sashes, the lower one being hopper hung, hinged at the bottom edge and opening inwards and the two upper sashes being pivot hung.

Air may be admitted in an upward direction, through the hopper opening, either to the full extent or reduced to part opening.

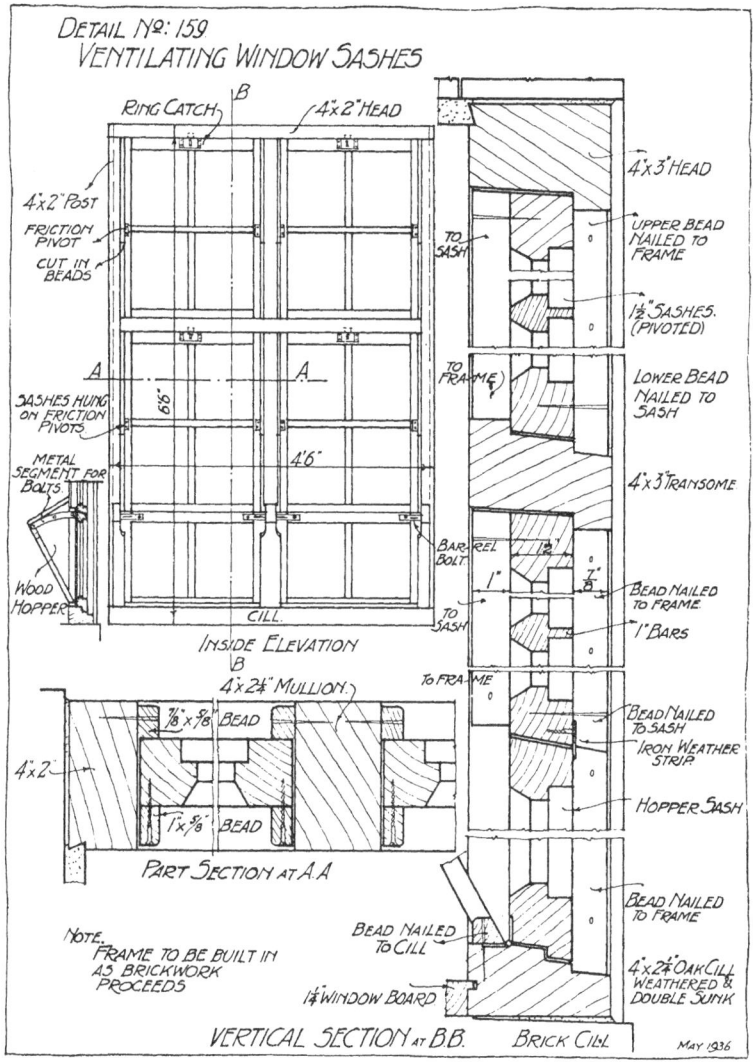

The hopper sash can be secured at one or more intermediate positions by barrel bolts shooting into holes in a metal plate segment screwed to the hopper sides.

The upper sashes are hung on *friction pivots*. The sashes will remain open in any position, the friction on the pivots being sufficient to resist movement even in a strong wind, yet easily movable when pulled from the outer edges of the sash. One or both of the pivoted sashes may be called into use and adjusted for control of air currents. In warm weather practically the whole of the area of the window can be utilised for air movement by opening the pivoted sashes to a horizontal position and pulling the hopper forward to the full extent.

The pivoted sashes are operated by a long rod and hook and are secured by ring handled fanlight catches.

550. Construction of ventilating sashes and frame. The vertical posts of the frame are $4'' \times 2''$ plain members, while the head, transome and cill are rebated, sunk and weathered as shown in the detail.

The pivoted sashes are arranged in the same way as was shown in detail No. 120, Vol. II. The side beads are made separately, cut to allow the sash to pivot clearly, and the beads nailed to frame or sash as required to give freedom for opening and also to offer resistance to the weather.

The sashes are $1\frac{1}{2}''$ thick (or $2''$ if preferred) and the main members—stiles and rails—are at least $2''$ wide, in order to ensure satisfactory jointing at the angles. The introduction of sash bars strengthens the sashes and reduces the sizes of window panes, thus lowering the cost of replacements.

As in all window frames, the cill should be of oak, and the cill and head should run through on the faces with the posts cut between them.

The metal weather strip at the junction of the hopper and centre sash should be heavily galvanised, bedded in thick oil paint and screwed at close intervals to the bottom rail of the centre sash.

For satisfactory working the sashes must be accurately fitted with ample—but not excessive—clearance.

The reader should make reference to Vol. II and sketch out the details of construction in jointing the frame and hanging the sashes.

CHAPTER TWENTY

JOINERY—STAIRS

MAIN STAIR TO HOUSE

551. The main stair from the hall to the first floor of the house is of the open newel type described in the introduction to stairs in the chapter on Masonry.

Its general form may be gathered from the perspective detail No. 160 and the plan in detail No. 161 where the flights of steps are divided by quarter space landings and arranged round a rectangular open well, with newel posts at the foot and at each inner angle of the flights to receive the handrails and support the strings.

552. Type of design. The type of design selected for this stair is based on those built during the Georgian period, and its main feature is the treatment of the handrail and balusters. The former is continued over and around the newels instead of butting against the sides of the newels as in the basement stair.

553. General description of main stair. The stair consists of three flights containing 5, 12 and 3 risers respectively, the short starting and terminal flights being at right angles to the main central flight in plan and divided by square landings.

A lift of 10′ 9″ is required from floor to floor, which is obtained by 20 risers, each $6\frac{9}{20}''$ rise; using a 10″ going, for which there is sufficient space, and comparing the ratio of rise to going on the basis of the rule given in Vol. I, $(2 \times \text{rise}) + \text{tread} = (2 \times 6\frac{9}{20}'') + 10''$ $= 22\frac{9}{10}''$, which is very near the value of 23″ required to give a well-proportioned step.

A handrail is provided on the well side of the stair and continued round the landing, and also on the outside of the short flight at the foot.

Newel posts at the foot and at each inner angle are continued to the ground floor level to obtain support, as shown in details Nos. 161 and 162.

Beneath the outer string and forming an enclosure to the basement staircase is a piece of splayed wood framing called a *spandril frame*, within which a door is placed for access to the basement stair. The panelling of this door and spandril frame is made to harmonise with the panelling of the hall which has been previously detailed in No. 152.

DETAIL Nº: 160.   THE HOUSE.   OPEN NEWEL STAIR FROM GROUND TO FIRST FLOOR LEVELS.

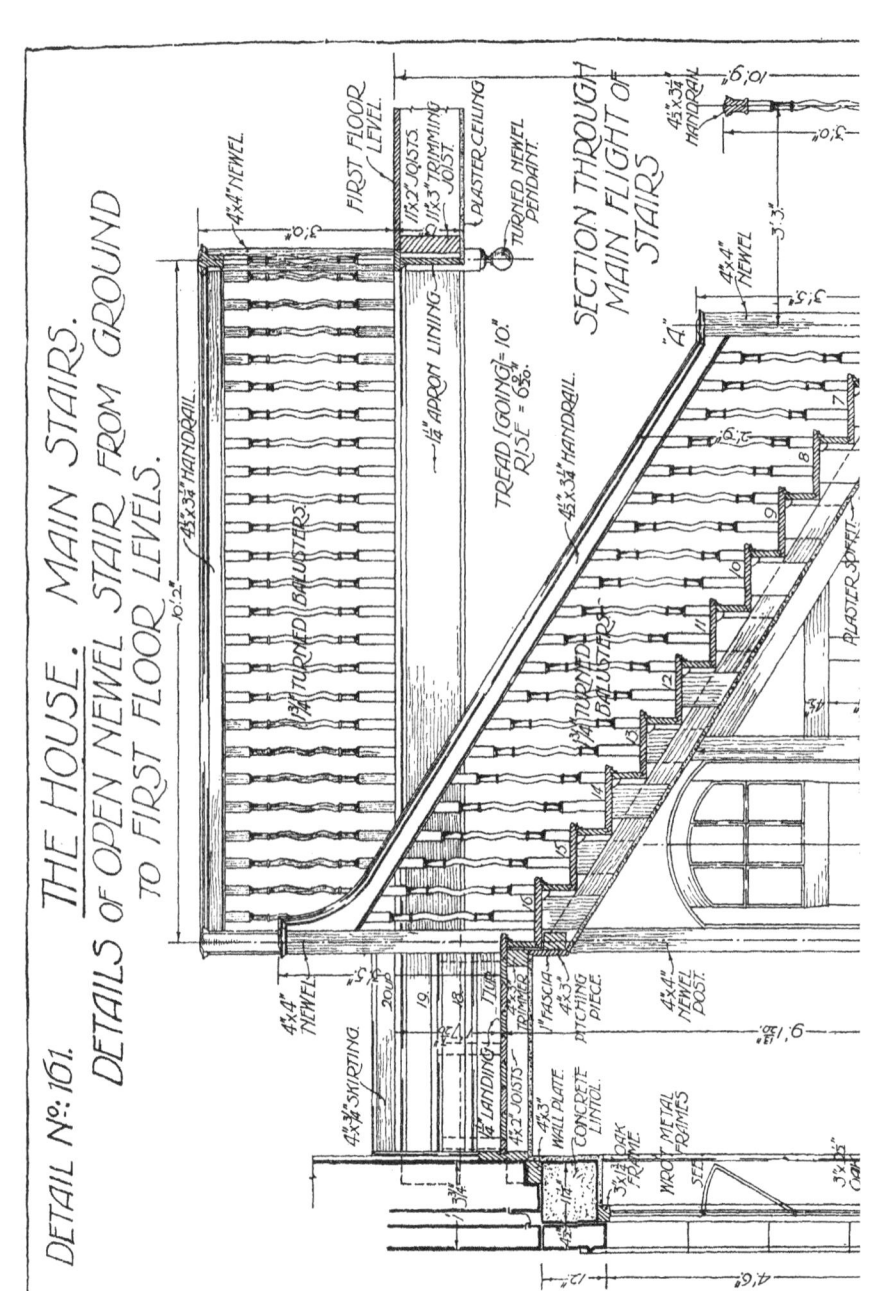

DETAIL Nº. 161.

THE HOUSE. MAIN STAIRS.

DETAILS OF OPEN NEWEL STAIR FROM GROUND
TO FIRST FLOOR LEVELS.

SECTION THROUGH
MAIN FLIGHT OF
STAIRS

FIRST FLOOR
LEVEL.

4"x4" NEWEL.

11"x2" JOISTS
2"x3" TRIMMING
JOIST

PLASTER CEILING

TURNED NEWEL
PENDANT.

1¼" APRON LINING

4½"x3¾" HANDRAIL.

TREAD (GOING)= 10"
RISE = 6⅝".

4½"x3¾" HANDRAIL.

1¾" TURNED BALUSTERS.

1¾" TURNED
BALUSTERS.

4"x4"
NEWEL

4½"x3¾"
HANDRAIL

PLASTER SOFFIT.

4"x4"
NEWEL

4½"x3¾" SKIRTING.

4"x4" TRIMMER

LANDING

4"x2" JOISTS

1" FASCIA.
4"x3"
WALL PLATE.

CONCRETE
LINTOL.

4"x3"
PITCHING
PIECE.

4"x4"
NEWEL
POST.

OAK
FRAME

WRO.T METAL
FRAMES.

3⅝"x⅝"
OAK

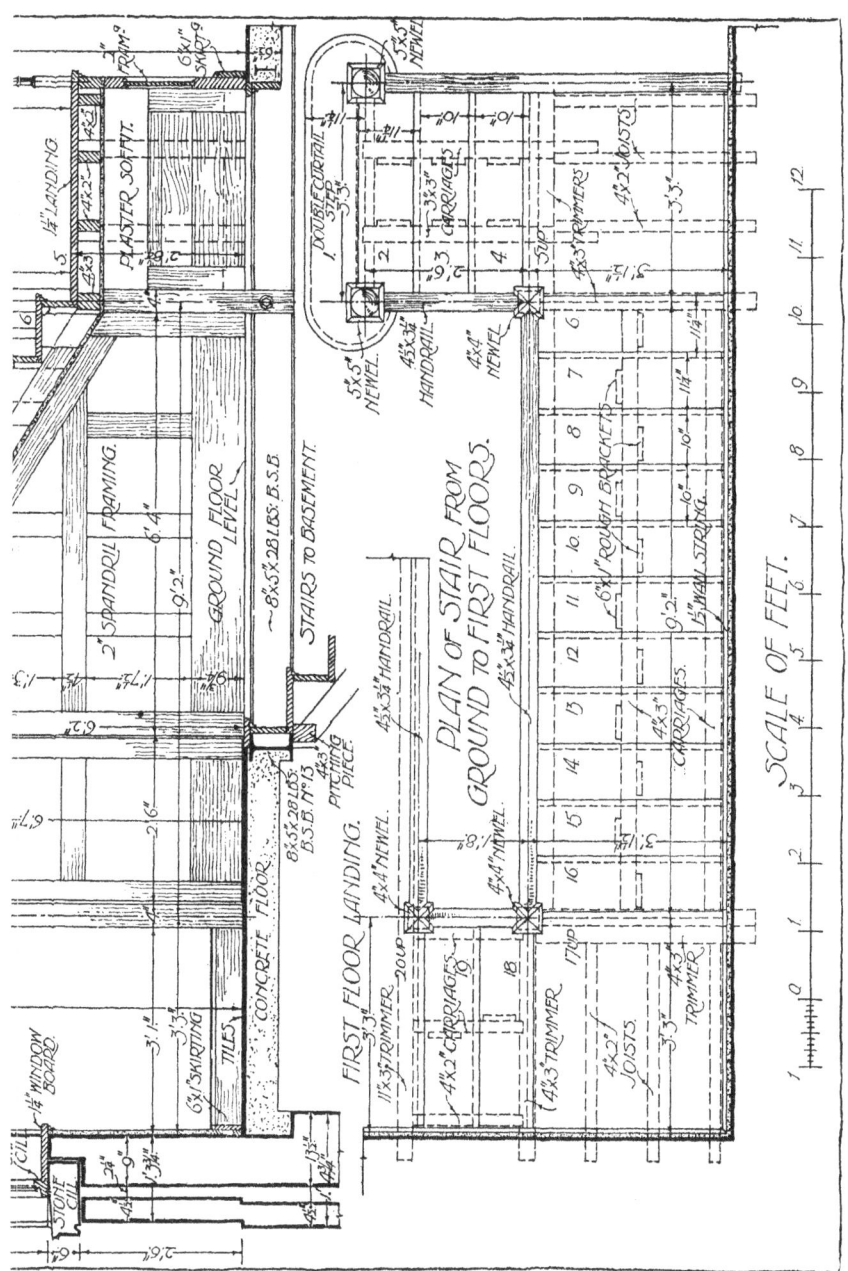

PLAN of STAIR from GROUND to FIRST FLOORS.

FIRST FLOOR LANDING.

SCALE OF FEET.

554. Construction of flights. The flights of steps are constructed in a manner similar to that already described for the basement stair, except that the outer string is of a different type; the top edge being cut to show the outlines of the steps, the nosings of which are continued over it and returned along the ends as shown in the perspective detail. All strings showing the outline of the steps are called "cut strings" and need more depth than closed or parallel strings for equal strength.

The outer strings are tenoned to the newel posts by double pinned tenons, as shown in detail 167, while the wall strings are of the same general character as those detailed in Vols. I and II.

The landings are framed with timber joists tenoned or housed to the newels and to each other, the trimmers being $4'' \times 3''$ and the intermediate joists $4'' \times 2''$; all enter the wall $4\frac{1}{2}''$ at the wall ends.

To provide extra support to the steps and to carry the plasterers' laths $4'' \times 3''$ rough carriages are employed, the centre carriage being rough bracketed on alternate sides. These carriages are bird's-mouth jointed to the trimmers at the head and foot with the exception of the lower series, which are oblique notched and bolted to the sides of the joists, as illustrated in detail No. 162.

555. Newel posts. The smaller newels are $4'' \times 4''$ solid posts, but the two starting newels, which are $5'' \times 5''$, are built up from $\frac{3}{4}''$ finished material with tongued and grooved joints, and with sunk panels on each face. The interior is glue-blocked as far as possible from the ends while the feet and terminals are made from solid material, shaped and enclosed within the casing to a length sufficient to provide solid jointing of string and handrail.

Detail No. 163 at A and B shows the construction of the newels. The smaller newels, A, are solid, and are provided with solid moulded caps having the same profile as the upper part of the handrail. The latter are mitred to the caps, as shown in plan at detail No. 161.

The detail also shows the circular turned pendant for the lower end of the top newel post.

The newel at B is one of the larger ones at the foot of the stair and shows the turned finial with moulded cap and necking, the latter intersecting with the upper portion of the handrail, as at C.

The plinth is mitred round and slightly housed to the post, being scribed to the front and return nosings of the steps, as in detail No. 168.

The large newel post is fixed at the base by housing through the framed curtail step.

556. Handrails. The handrail of this stair is of a form which allows it to be made either solid, or in two pieces, as shown at D in detail No. 163. If made as suggested the labour in forming the

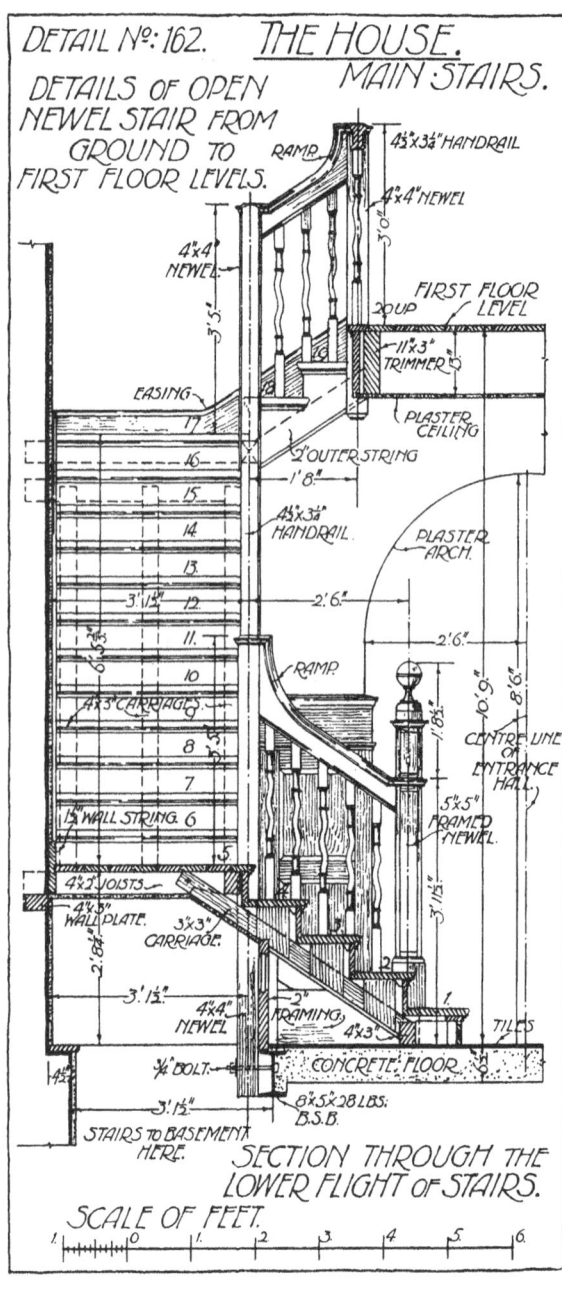

DETAIL Nº: 162.   THE HOUSE.
MAIN STAIRS.

DETAILS OF OPEN NEWEL STAIR FROM GROUND TO FIRST FLOOR LEVELS.

SECTION THROUGH THE LOWER FLIGHT OF STAIRS.

SCALE OF FEET.

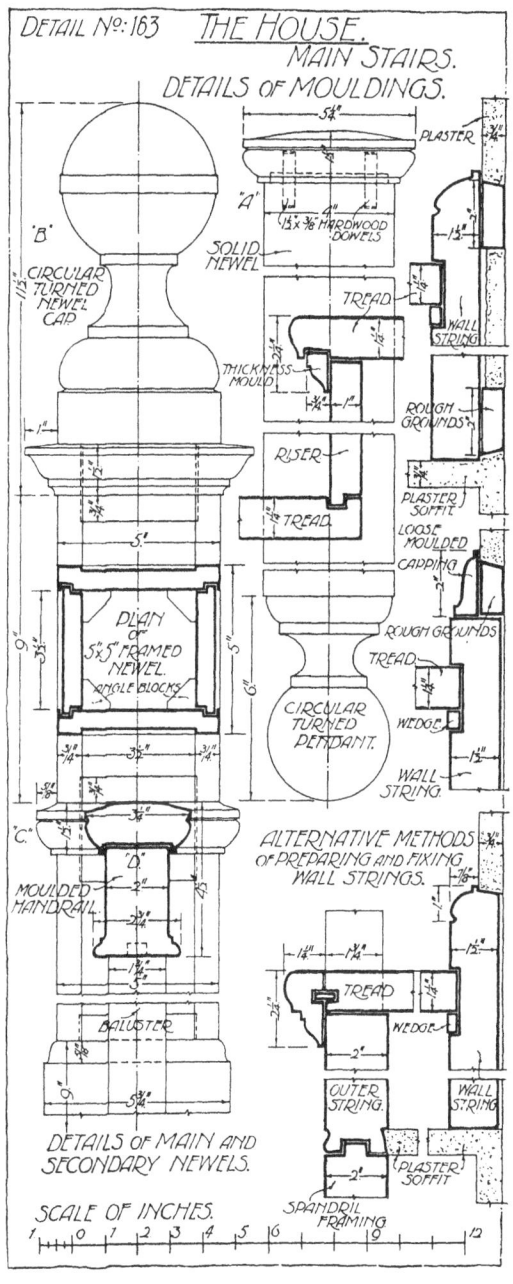

DETAIL Nº: 163   THE HOUSE.
MAIN STAIRS.
DETAILS of MOULDINGS.

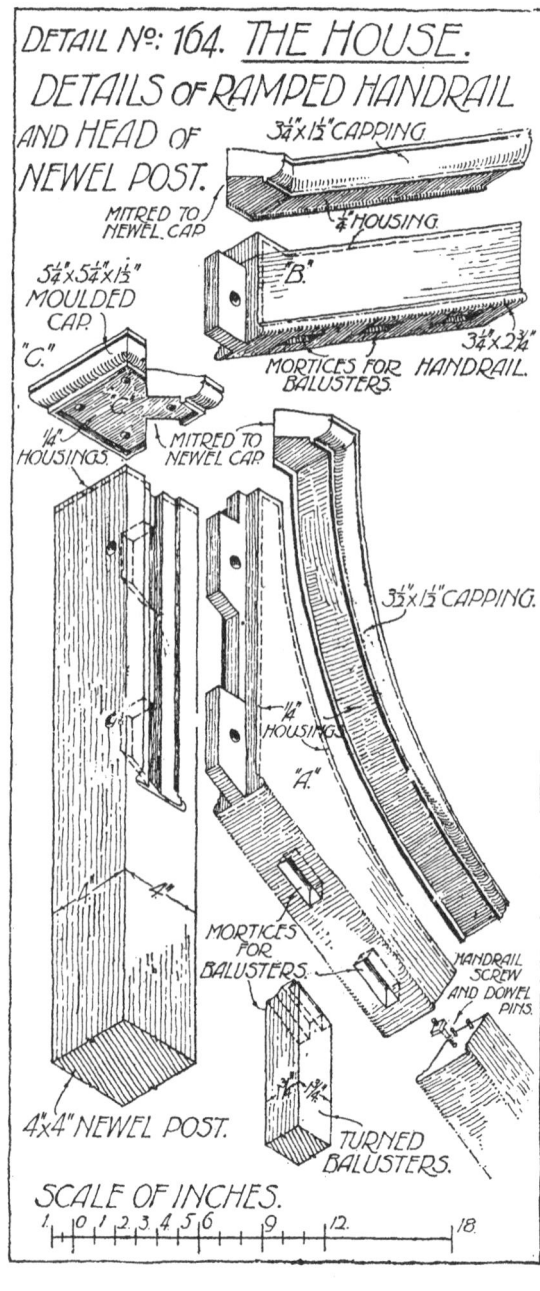

DETAIL Nº: 164. *THE HOUSE.*

DETAILS OF RAMPED HANDRAIL AND HEAD OF NEWEL POST.

$3\frac{1}{4}$"×$1\frac{1}{2}$" CAPPING

MITRED TO NEWEL CAP

$\frac{1}{4}$" HOUSING

"B."

$5\frac{1}{4}$"×$5\frac{1}{4}$"×$1\frac{1}{2}$" MOULDED CAP.

"C."

$3\frac{1}{4}$"×$2\frac{3}{4}$" HANDRAIL.

MORTICES FOR BALUSTERS.

$\frac{1}{4}$" HOUSINGS

MITRED TO NEWEL CAP.

$3\frac{1}{2}$"×$1\frac{1}{2}$" CAPPING.

$\frac{1}{4}$" HOUSINGS

"A."

MORTICES FOR BALUSTERS.

HANDRAIL SCREW AND DOWEL PINS.

4" NEWEL POST.

TURNED BALUSTERS.

SCALE OF INCHES.

moulded portion at the ramps is diminished and a less size of material can be employed which conduces to better seasoning.

The lower portion is cut with the ramp curve on a solid piece, as detailed in No. 164 at A, this piece being connected to the ramp with handrail bolts and dowels and to the newel by oblique tenons. At the foot of each length of rail a level piece, on which the capping

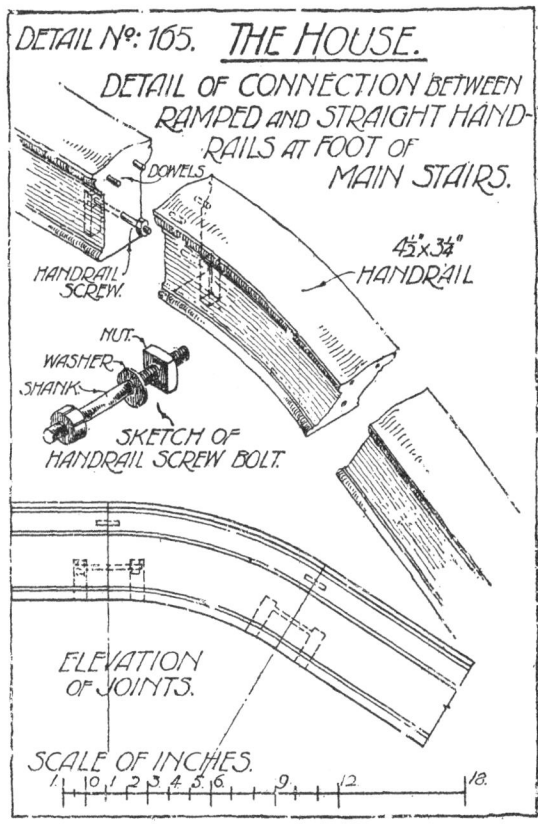

can be mitred and extended, is provided by glueing on a triangular block, as at B. The necessity for this may be seen on examination of the elevation of the newel at A in detail No. 161, where it is clear that the raking capping could not enter the level newel cap except by changing the section of the mould.

557. Newel caps to flat topped newels. The top portion of the handrail is carried round these newels and necessitates a solid cap, as shown at A in detail No. 164 and at C in No. 165.

These caps are mitred to receive the short level pieces on both sides, and are dowelled to the solid posts. The several pieces of the capping are shown in the upper part of the latter detail in their relative positions for fixing. They should be dowelled and glued where possible at end joints and mitres, and also at the longitudinal bed joints of the long pieces.

558. Solid handrails. If the handrails can more conveniently be made in solid pieces of the full section, they are adopted as shown in detail No. 165, where portions of the straight and raking handrails to the side return at the foot of the stair are detailed. The detail shows the joint between a ramp and two straight pieces, including the dowel and handrail screw bolt.

559. Balusters and return nosings. In open stairs with cut strings, balusters form a special feature. A selection is given in detail No. 166 from the collection in the Victoria and Albert Museum, South Kensington. Balusters are usually placed two on each step and set with their outer faces in the same plane as the string if no brackets are employed, as in this case. The front face of the first baluster on each step is in the same plane as the riser.

To obtain a finish to the end of the step, the nosings must be continued on the face of the string and stopped by returning the end at 90° to the string for a distance past the riser face equal to the projection, as seen in detail No. 167. The nosings need to be fitted in position temporarily, then removed for the insertion of the balusters, which are dovetailed from the end, as at A. In ordinary painted work the nosings are nailed in position, but three better methods are shown, viz. secret slot screws, dowels and cross tongues.

In the first of these methods, the nosings are drilled and slotted, so as to pass the heads of the screws in the tread and slide into position, allowing the heads to embed their edges in the sides of the grooves.

The other methods require accurate fitting and well glued joints.

Nosing pieces for the returned ends of steps are generally worked in the solid, the thickness mould and rounded edge being formed in one piece.

560. Junction of riser and cut outer string. The joint between these pieces must not show end grain, hence a mitred joint is necessary.

It is also an advantage to have a square shoulder to ensure the riser being fixed in the correct position, hence the joint usually adopted is that shown at B and known as a rebated and mitred joint.

561. Bracketed stair. In some old examples the outer string was set within the width of the stair and the projecting part of the tread carried upon thick and boldly carved brackets. This has given rise in modern work to the practice of planting a thin sham bracket on the face of the string. Such a bracket, since its constructional value is no longer required, is open to objection.

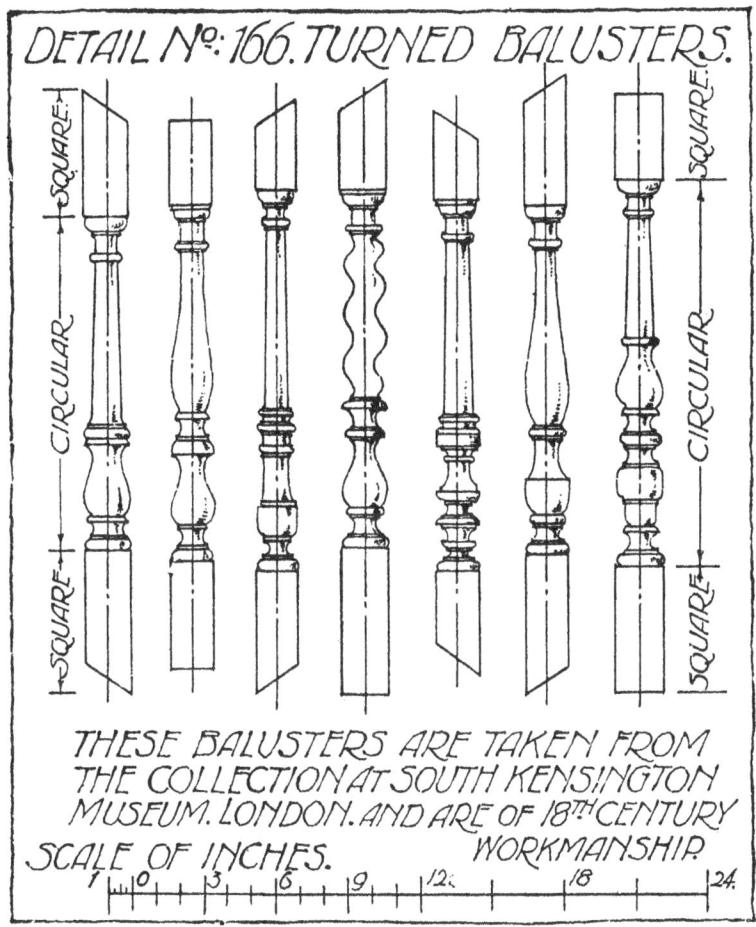

DETAIL Nº: 166. TURNED BALUSTERS.

THESE BALUSTERS ARE TAKEN FROM THE COLLECTION AT SOUTH KENSINGTON MUSEUM. LONDON. AND ARE OF 18TH CENTURY WORKMANSHIP. SCALE OF INCHES.

On the other hand, the use of such brackets simplifies the construction and reduces the large amount of labour necessary in sinking and mitreing the riser and string, as shown at C.

Alternative forms of such brackets are shown in detail No. 167.

The balusters are here arranged so that their outer faces are in the face-plane of the brackets.

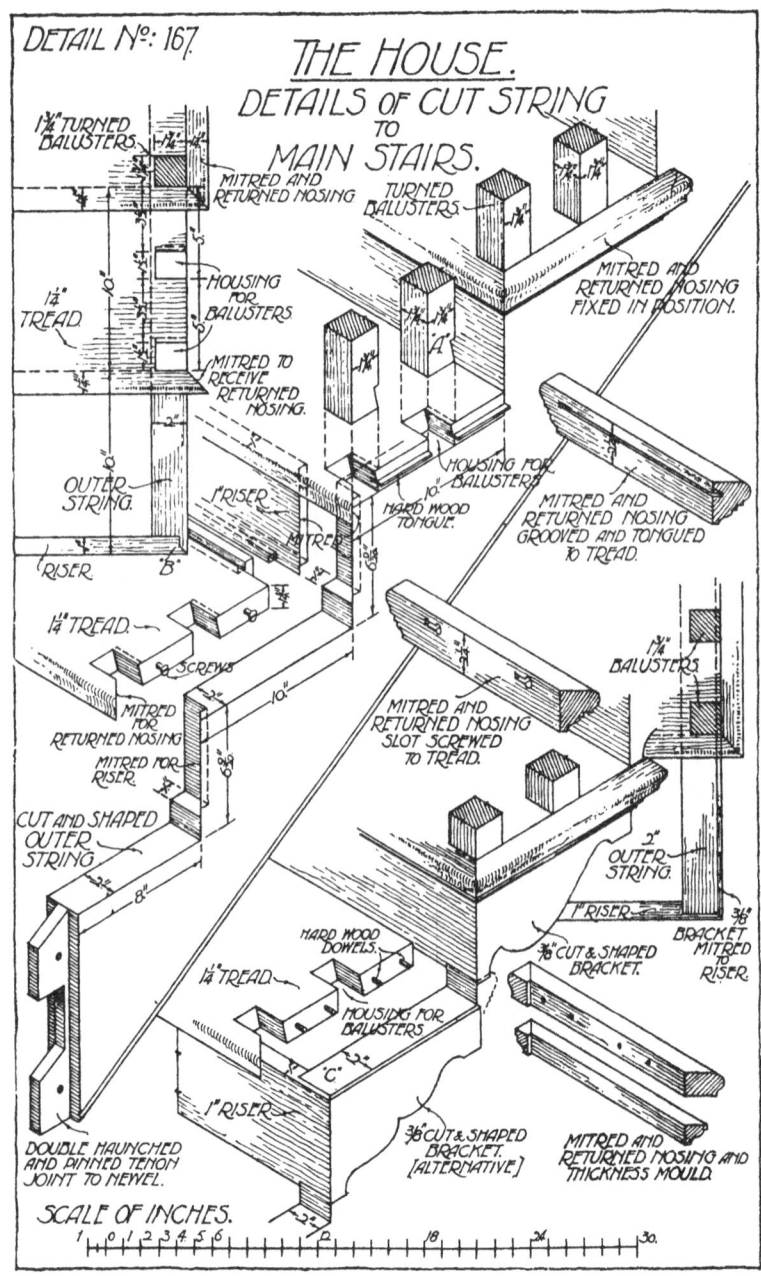

DETAIL Nº: 167.

THE HOUSE.
DETAILS OF CUT STRING
TO
MAIN STAIRS.

1¾ TURNED BALUSTERS

MITRED AND RETURNED NOSING

TURNED BALUSTERS

MITRED AND RETURNED NOSING FIXED IN POSITION.

1¼ TREAD

HOUSING FOR BALUSTERS

MITRED TO RECEIVE RETURNED NOSING.

"A"

OUTER STRING.

1" RISER

HOUSING FOR BALUSTERS

HARD WOOD TONGUE.

MITRED AND RETURNED NOSING GROOVED AND TONGUED TO TREAD.

RISER "B"

MITRED

1¼ TREAD.

SCREWS

MITRED FOR RETURNED NOSING

MITRED FOR RISER.

MITRED AND RETURNED NOSING SLOT SCREWED TO TREAD.

1¾ BALUSTERS

OUTER STRING.

CUT AND SHAPED OUTER STRING

10"

8"

1" RISER

⅜" CUT & SHAPED BRACKET.

BRACKET MITRED TO RISER.

HARD WOOD DOWELS.

1¼ TREAD.

HOUSING FOR BALUSTERS

"C"

1" RISER

DOUBLE HAUNCHED AND PINNED TENON JOINT TO NEWEL.

⅜" CUT & SHAPED BRACKET. [ALTERNATIVE]

MITRED AND RETURNED NOSING AND THICKNESS MOULD.

SCALE OF INCHES.

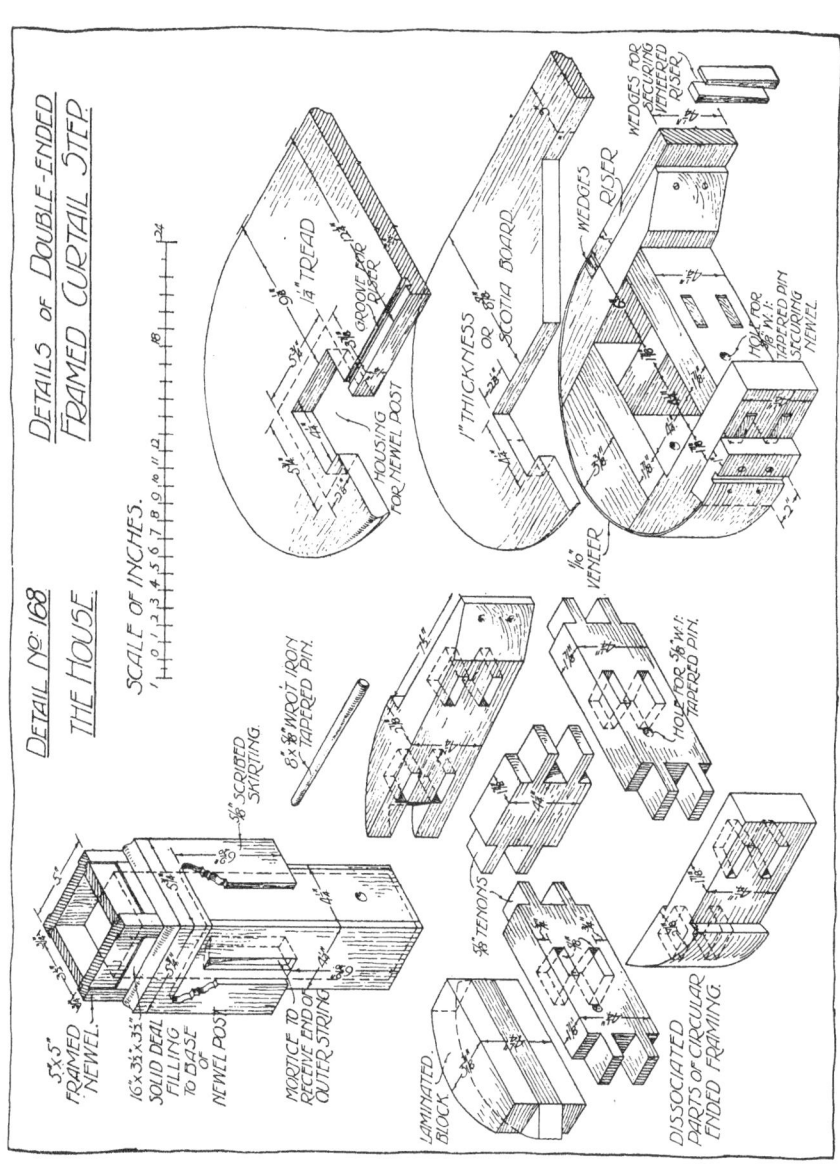

DETAIL Nº 168

DETAILS OF DOUBLE-ENDED
FRAMED CURTAIL STEP

THE HOUSE.

SCALE OF INCHES.

562. Curtail step to main stair. In this case the curtail step is much wider than those of earlier examples in these volumes, being considerably more than the width of a step. A satisfactory method of framing up any similar large steps has therefore been shown, this method being preferable to building up a very large block of laminated material, owing to its greater strength and constancy of form.

Detail No. 168 shows the arrangement of the framing which is out of material not exceeding 3″ thick, double tenoned near the edges and pinned and wedged together as convenient for each joint.

The larger portion of the curved block is made of laminated material which may be in 3 or 5 pieces, splay tenoned and glued between the tips of the outer curved pieces of the main frame. A hole for the foot of the newel is provided in this frame and the outsides are notched to receive the thicker portions of the veneering riser which is reduced to $\frac{1}{8}$″ in thickness, steamed, bent to position, and wedged tight circumferentially.

The scotia board and tread have corresponding notches for receiving the newel and the latter is drawn down tightly upon its shoulders by a steel or wrought iron tapered pin.

CHAPTER TWENTY-ONE

JOINERY—INTERNAL FITTINGS— SCREENS—MANTLEPIECES

VESTIBULE SCREEN TO MAIN ENTRANCE HALL OF WAREHOUSE

The entrance hall of the warehouse is screened from the doorway by a glazed vestibule screen with folding doors intended to be executed in hardwood and illustrated in detail No. 169.

563. Vestibule screen framing. The frame which encloses the filling panels and doors is 4″ thick and is prepared to receive all subsidiary framing at the centre of its thickness, as shown in detail No. 170.

The head, door posts, wall posts and muntins are single members 3″ wide on face, beaded and rebated on one or both edges as required, while the transome is a double member composed of 3″ wide rails, the lower one being left with a plain soffit for the passage of the swing doors at the centre panel, all other parts being rebated and beaded like the rest of the frame.

The two short pieces of cill or dado rail across the side frames are $2\frac{1}{2}$″ wide, each being capped by a projecting mould, as shown at A in detail No. 170; the ends of this are returned for the projecting portion upon each post, while the side beads finish upon its top face.

564. Transome of vestibule screen. Where a heavy moulded transome is required it is usually found best to frame it up from pieces of moderate size. In this case the two rails which form its core are faced by a $5″ \times 2\frac{1}{4}″$ tongued mould which bridges the space between them, as detailed at B.

In the perspective detail this transome mould is shown continued along the sides of the vestibule as a capping to the tiled dado, its height being such that it becomes a frieze mould, cutting off the plastered frieze from the wall tiling.

565. Wall and door posts. The wall posts are prepared with dovetailed rebates to receive plaster or tiles, in which case the frame would require to be fixed first; preferably a $3″ \times \frac{3}{4}″$ rough ground is placed against the wall to receive plaster and tiling so that the woodwork may be inserted when the tilers have finished. The only objection to the latter method is that it is very difficult to place the

DETAIL Nº: 169.   *THE WAREHOUSE.*
GLAZED SCREEN WITH
DOUBLE SWING DOORS TO
VESTIBULE of MAIN ENTRANCE.

4"x3" HEAD.

PLASTER

2'.0⅝"

⅓"

FANLIGHT.

7½"

5"

4½"

3'.2⅜"

9"

7'.6⅛"

3"

2'.5¼"

3"

9¼"

7.6⅛"

8'.7⅜"

4'.9"

3'.0"

1'.0"

10'.1"

PLASTER

4"x3" POSTS.

½" DOUBLE
SWING DOORS.

BRONZE GRILLE.

2'.4½"

3"

4'.4½"

1'.6"

3"

TILES.

SECTION THROUGH
DOORS AND
FANLIGHT.

PART PLAN of SCREEN.
SCALE of FEET.

6
INS.

0

1.

2.

screen in position so that a neat fit is obtained against the face, thus necessitating an allowance for erection clearance, which must be made good by an angle mould. Where such is used it should be a duplicate of the lower member of the transome mould in order to mitre with it.

The door posts have their inner edges hollowed to the radius of curvature from the centre of rotation which is marked "c" in the detail at C. This is arranged to close the opening between door and post when the doors are shut and to keep the distance between edge of floor and post constant during the swing. In any case where centres are employed in preference to double swing butt hinges, the back edge of the door must be rounded, which would leave a wide joint at the face if not recessed as shown.

Wall posts would be secured through the grounds to plugs in the walls, and door posts fixed at the foot by 1″ round iron dowels to the concrete or housed in to a depth of $\frac{1}{2}$″.

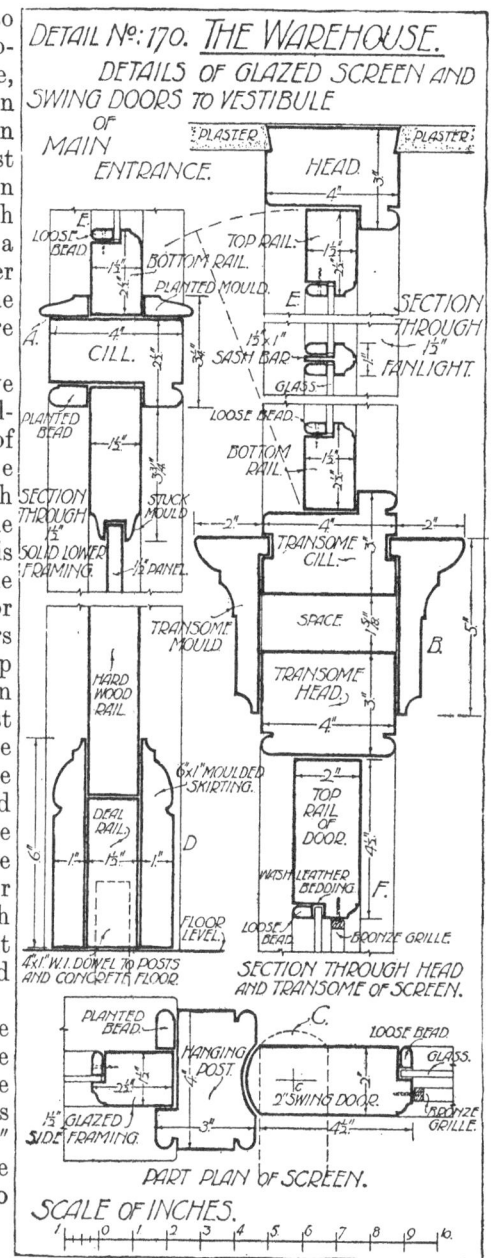

DETAIL No: 170. THE WAREHOUSE.
DETAILS OF GLAZED SCREEN AND SWING DOORS TO VESTIBULE OF MAIN ENTRANCE.

PART PLAN OF SCREEN.

SCALE OF INCHES.

566. Dado panelling to vestibule screen. The side-fillings below the cill on each side of the doorway have panelled wood framing $1\frac{1}{2}''$ thick with cyma recta stuck moulds on the edges and plain $\frac{1}{2}''$ panels. These pieces of framing extend to the floor and have a double bottom rail, the upper half matching the rest of the work and the lower piece being of deal for securing a $6'' \times 1''$ moulded skirting, which overlaps the joint as shown in section at D. The skirting pieces fit between the side beads after the latter have been screwed in position to secure the panelled framing. Secret screws in slotted holes should be employed to fix the skirting, as previously described, and also illustrated at detail No. 167.

567. Sash fillings to vestibule screen. The sashes are $1\frac{1}{2}''$ thick, moulded on the inner face, divided into small panes, and assembled as previously shown.

Plain glazing beads to secure the glass are fixed with small brass screws, as illustrated at E in detail No. 170.

The centre sash is hung to open inwards, being hinged to swing hopper-wise about the lower edge. For this reason beads are omitted on the opening side of the frame.

568. Swing doors to vestibule screen. Swing doors $2''$ thick are fitted to the central opening. The rounded edge of the hanging stile has already been referred to in connection with the frame; in a similar way the free edge of the door must be rounded to the radius of its swing so that the free stiles may clear each other at the centre; this is shown in the plan of detail No. 169.

The doors have lower panels raised both sides, each face being similar to those previously detailed for the main entrance doors, while the glazed portion is in one pane of plate glass protected by a bronze grille, which is fitted to the square edge of the framing within the mould and screwed thereto, as shown at F in detail No. 170.

569. Fittings to vestibule screen doors and fanlight. The doors are pivoted upon centres carried in the floor and frame at a distance of about $1\frac{1}{2}''$ from the door posts. Embedded in the floor is a box-spring hinge, as shown at A in detail No. 171, from which projects a square pin connected to a cam and controlled by a spiral spring. The lower angle of the door is fitted into a metal shoe which bears upon the pin. In the head of the door the lower socket plate, marked B, is housed, and above it in the frame an adjustable top centre is fixed, the centre pin of which is movable, as shown at C. This pin may be lifted by the adjusting screws until it is flush with the plate, thus allowing the heel of the door to be slipped into the shoe and within the opening at right angles to the screen; the top

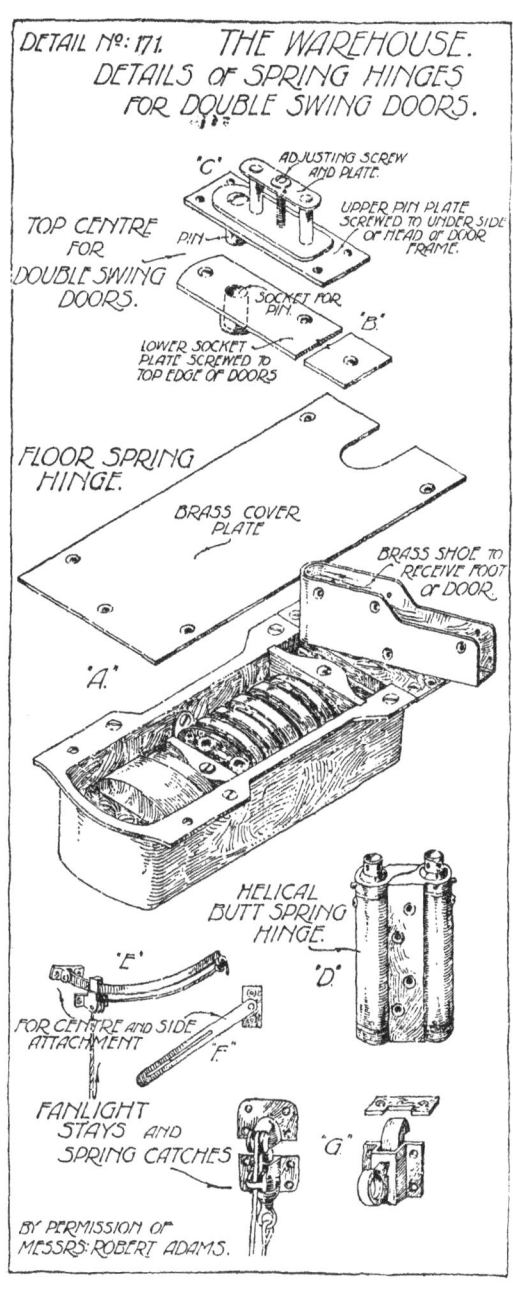

DETAIL Nº: 171. THE WAREHOUSE.
DETAILS OF SPRING HINGES
FOR DOUBLE SWING DOORS.

TOP CENTRE FOR DOUBLE SWING DOORS.

'C'

ADJUSTING SCREW AND PLATE.

UPPER PIN PLATE SCREWED TO UNDER SIDE OF HEAD OF DOOR FRAME.

PIN

SOCKET FOR PIN.

'B.'

LOWER SOCKET PLATE SCREWED TO TOP EDGE OF DOORS

FLOOR SPRING HINGE.

BRASS COVER PLATE

BRASS SHOE TO RECEIVE FOOT OF DOOR

'A.'

HELICAL BUTT SPRING HINGE.

'E'

'D.'

FOR CENTRE AND SIDE ATTACHMENT

'F.'

FANLIGHT STAYS AND SPRING CATCHES

'G.'

BY PERMISSION OF MESSRS: ROBERT ADAMS.

centre is then lowered, the door closed, and the actuating springs in the floor hinge released. Any movement of the door either way causes a compression on the spring, which, by trying to regain its normal position, forces the door towards the centre line of the opening when the pressure is released.

Double swing hinges, like a double butt, with springs in the knuckles are sometimes employed. Their form is illustrated at D; They cannot be recommended, being clumsy, and often inefficient and weak in action.

The sketches at E and F in the detail show common forms of fanlight stay. The former is attached at the centre of the bottom rail of the sash, with the pulley mounted on the frame; then a cord fixed to the eyelet and passed over the pulley allows the sash to be forced outwards.

The fitting at E is suitable for a top hinged sash or fanlight opening outwards. The fitting marked F is secured to each side of the sash and used in conjunction with the spring catches at G. These fittings are suitable for lights hinged at the bottom edge and opening inwards. By the use of a cord and cleat the opening of the sash may be adjusted within the limits of the length of the stay.

Wood Mantels, Overmantels, etc.

The wood panelling to the walls of the entrance hall has already been referred to and its construction described and detailed in connection with the internal doorways opening from the hall (see paragraph 528).

570. Mantelpiece to entrance hall fireplace. A wood mantel harmonising with the hall panelling and suited to the design of the stone fireplace is shown in detail No. 172.

It consists of two pilasters formed by the return of the panelling across a portion of the chimney breast, connected across the latter and above the fireplace by a segmental panelled frame and surmounted by a cornice, which is returned on the level along the sides of the breast to the wall face.

The pilasters are of $1\frac{1}{4}''$ framing, with the skirting, capping and necking mould continued across them to match the hall panelling. The skirting is returned again across the inner edge of the pilaster and butts against the breast, while the capping returns and terminates similarly against the central panel. To form a shelf below the latter the necking mould is brought forward and continued round the edges of the $1\frac{1}{2}''$ board used for this purpose.

A $2'' \times 1\frac{1}{2}''$ bed mould supports the shelf and its edge mould is returned in the solid at the ends, against the faces of the pilasters.

DETAIL Nº: 172. THE HOUSE.
DETAILS OF FIREPLACE IN ENTRANCE HALL.

KEY PLAN OF FIREPLACE.

PLASTER PANEL

6'.1½"

BRICKWORK

SKETCH OF FIREPLACE AND SURROUNDINGS.

SECTION THROUGH OVERMANTEL ON CENTRE LINE.

STONE.

SCALE OF INCHES.

PART PLAN OF JAMB.

ARCHI-TRAVE.

PROJECTION OF CORNICE AT "A"

STONE HEARTH.

STONE JAMB.

ROUGH GROUNDS

FLOOR LEVEL

PROJECTION OF SHELF AT "B".

The plan shows in detail the form of panelling employed and the joints at the internal and external angles. The latter angle is necessarily mitred and tongued, the mitre being as short as possible (say $\frac{3}{8}''$) to avoid warping and opening at the joint. Such a joint is only suitable for longitudinal grain.

It should be noted that all the framing is secured to vertical grounds and that the stone facings at the fireplace jambs have a projection equal to the thickness of the grounds in front of the brickwork. This is shown on the general plan and in the entrance hall fireplace detail in Vol. II.

The centre panel is $\frac{1}{2}''$ thick, framed into $1\frac{1}{4}''$ stiles and rails, averaging 4″ wide, the stiles being necessarily wider to pass behind the pilasters and close the triangular space at the springing of the cornice.

The cornice consists of a $3\frac{3}{4}'' \times 2''$ corona and $2\frac{1}{2}'' \times 1\frac{1}{2}''$ bed mould tongued together as shown in the vertical section and overlapping the tongued edge of the panelling below. A segmental ground, $4'' \times \frac{3}{4}''$, follows the curve of the cornice and serves for fixing the group.

Junctions between the portions of straight and curved cornice, as at A, should be cross tongued, or dowelled and glued. In the larger pieces it is often wise to use handrail bolts to assemble them firmly before fixing to the wall.

571. Chimney-piece to dining room. The dining room is fitted with a wood chimney-piece enclosing a marble mantel, as shown in detail No. 173.

The mantel consists of a heavily moulded surround to the opening enclosing a tiled recess intended to receive a loose fire basket or container called a "dog grate", which, though originally intended for log fires, is now adapted also for the use of coal. Such a fitting is capable of much distinction and care should be exercised in its design or selection.

The jambs of the mantel are solid slabs $6\frac{3}{4}'' \times 7''$ in cross section, having an architrave mould worked on the outer edge, the inner being left square and $4\frac{3}{8}''$ thick; see plan in detail No. 174.

The lintol portion is one piece of marble, $12''$ deep $\times 6\frac{3}{4}''$ thick, having the side and top moulds worked to a solid mitre and displaying a much wider fascia than the sides. A solid rectangular plinth block supports the jambs and all the units are dowelled together with $3'' \times \frac{3}{4}''$ copper dowels and set in fine hard plaster. The marble mantel is placed against the wall, bedded in plaster and secured by iron or copper cramps built into the walls or housed in and cemented, as shown at A in the section of detail No. 174.

572. Woodwork of chimney-piece (in hardwood). The woodwork consists of two pilasters 10″ wide and the full height of the room, each being enriched with carving and the two connected by and enclosing an overmantel frame with a projecting panel. The feature is surmounted by a heavy segmental cornice.

DETAIL Nº: 173. THE HOUSE. OAK AND MARBLE CHIMNEY PIECE IN DINING ROOM.

On each side of the breasts, plain wood linings encase the returns and terminate against similar wood-covered pilasters at the attached piers, which are provided to support the ends of the floor beams and thus avoid building the timber into the party wall; see Vol. II.

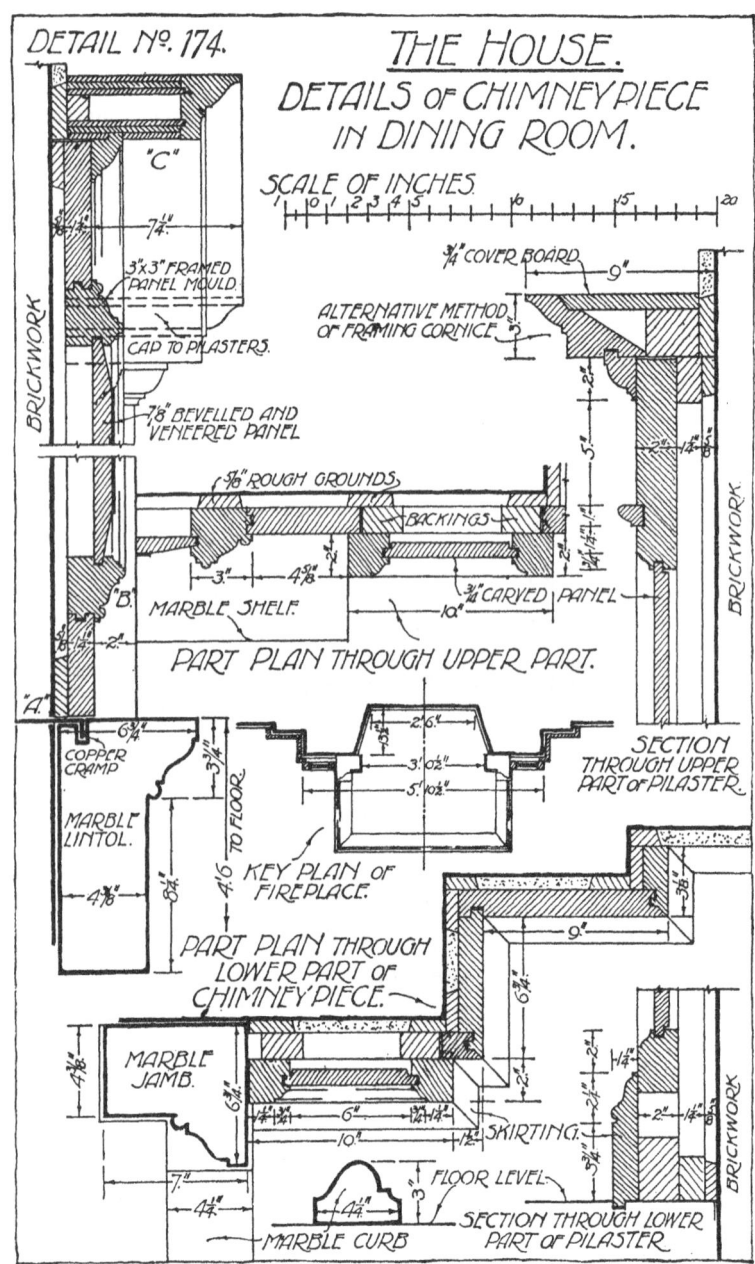

DETAIL Nº. 174.

THE HOUSE.
DETAILS of CHIMNEY PIECE
IN DINING ROOM.

SCALE OF INCHES.

"C"

7¼"

3"x3" FRAMED
PANEL MOULD.

CAP TO PILASTERS.

⅞" BEVELLED AND
VENEERED PANEL

BRICKWORK

¾" COVER BOARD   9"

ALTERNATIVE METHOD
of FRAMING CORNICE

⅝" ROUGH GROUNDS.

BACKINGS.

3"   4⅝"   2"

MARBLE SHELF.

¾" CARVED PANEL

10"

PART PLAN THROUGH UPPER PART.

"B"

2"

"A"

COPPER
CRAMP   6¼"   3¼"

MARBLE
LINTOL.

4⅝"   8¼"   4·6 TO FLOOR

2·6"

3'·0½"

5'·10½"

KEY PLAN of
FIREPLACE.

PART PLAN THROUGH
LOWER PART of
CHIMNEY PIECE.

MARBLE
JAMB.   6¾"

4⅝"

7"

4¼"

MARBLE CURB

4¼"   3

FLOOR LEVEL.

SKIRTING.

4·1¾"   6"   2½·1¼"

10"

1½"

SECTION
THROUGH UPPER
PART of PILASTER.

9"

6¾"

2"

2½"·2"   1¼"

3¼"   2"   1¼"

FLOOR LEVEL.

SECTION THROUGH LOWER
PART of PILASTER.

BRICKWORK.

BRICKWORK.

573. Pilasters and returns. The pilasters are single panelled frames 2″ thick, with 2″ wide stiles, 10″ top rail and 3″ bottom rails, the lower of which is of deal and introduced for fixing the skirting board at the base. A reference to the plans of the pilaster in detail No. 174 will make clear the form and purpose of each part.

The framing is stuck-moulded and the panels are $\frac{3}{4}$″ thick, tongued into $\frac{7}{16}$″ panel grooves and finished with plain faces. A broad top rail is necessary to form a frieze, and also to receive the housed necking mould and the bed mould of the cornice.

Carved pendants are shown in the upper part of the panels; these, in the case of hardwoods such as oak or mahogany, may either be cut in the solid, or carved separately, and secured by screws from behind. In many cases such pendants are carved from a contrasting wood, such as lime or holly, and similarly secured.

If the woodwork is in soft wood the ornament may be moulded in carton-pierre or papier-maché and fixed to the panel, the whole being subsequently painted.

These pilasters are slot screwed to 2″ × 1$\frac{1}{4}$″ framed grounds, while the latter are secured to the $\frac{5}{8}$″ plasterers' grounds, shown in the plan. The latter provide for a coat of rough plaster to be laid behind the woodwork and between the grounds; cement plaster covering the whole surface must be used if there is a flue within the brick-work of the breast nearer than 9″ to the face.

A small break 1$\frac{1}{2}$″ wide is formed at the external angles between the pilasters and the 1$\frac{1}{4}$″ return facing. The latter is therefore jointed to a 2″ × 1$\frac{1}{4}$″ strip by a tongued and mitred joint, the back edge being covered by the pilaster.

The plain returns are broken in the length by a broad necking which is identical with the picture mould.

574. Centre panel of overmantel. A special piece of framing is provided at the centre panel, with framed panel moulds and a projecting and raised field; detail No. 173. The angles of the panel moulds are tenoned and mitred, and the edges tongued and grooved, as shown in section, for enclosure within the main framing.

The latter is in the same plane as the 1$\frac{1}{4}$″ grounds, the stiles being 5$\frac{1}{2}$″ wide and overlapped $\frac{1}{2}$″ by the pilasters and cornice. The framed panel mould is 3″ × 2$\frac{3}{4}$″, having a section as at B, detail No. 174, which brings the centre panel forward 1″ in front of the main framing. This panel is raised or fielded, with bevelled edges, and in the case of hardwoods the fielded portion may be veneered with selected material in four pieces, matched and grouped with the grain set diagonally, thus forming diamond patterned outlines if the grain is strongly marked.

It should be noted that the head of the panel is segmental, and approaching a full semicircle, with short straight pieces at the springing to sharpen the outline.

575. Cornice to chimney-piece. The cornice is of built-up wood sections, surmounting the pilasters at the level of the floor beams and following the curve of the panel mould across the middle. Its rise is equal to the projection of the floor beams below the ceiling and it therefore reaches the ceiling at the crown.

The full projection is $7\frac{1}{4}''$ in front of the principal face of all the framing, but the bed mould is brought forward and broken round the projection of the pilasters. At the returns the projection is reduced to $5\frac{1}{4}''$, thereby corresponding with the overhang beyond the pilasters.

The curved or pediment portion of the cornice suggests a change in the usual form of section, so that the broad soffit can continue without break, hence a solid front to the corona is adopted, with thin cover and soffit boards; these latter may be built up from hardwoods by using broad thin strips, $\frac{1}{4}''$ to $\frac{5}{16}''$ thick, and glueing them together, as shown at C.

Tongued joints should be employed where possible to offset the effects of shrinkage and all mitres of the larger mouldings should be cross tongued or dowelled.

576. Fixing against chimney breasts. Wherever good joinery is to be fixed to a chimney breast over flue walling, the back of it should be kept clear of the wall by rough grounds, and if there is any danger of overheating, strips or sheets of asbestos should be inserted behind the grounds or other woodwork in contact with the wall.

CHAPTER TWENTY-TWO

MISCELLANEOUS

Casing of Pillars—Patent Glazing

577. Casing of steel pillars and beams in steel-framed buildings. The L.C.C. Code of Practice (1932) requires the steelwork of all buildings to have a high grade of fire resistance. For this purpose the steel is to be completely encased and protected by solid void-filled casings of brickwork, terra-cotta, stone, concrete, tiles or other materials having an equal grade of fire resistance.

The component parts of such casings—if built of separate units—must be properly bonded or secured together.

The minimum thickness of the casing must be:

(a) For pillars built into external or party walls: 4″ on all sides.

(b) For pillars built into external or party walls, below ground, 6″ on the external faces where the wall abuts against earth, and 2″ on internal faces.

(c) For pillars *not* built into external or party walls: 2″ on all sides.

(d) For beams: 2″ on vertical sides, 1″ on upper surfaces and 2″ on the soffits. But, if the width of steel exceeds 16″ the covering must be increased on the soffit to $\frac{1}{8}$th of the maximum width of steel.

For pillars and beams in external walls the above casings, if not entirely of concrete, must have at least 2″ of their thickness immediately in contact with the steel member composed of 1 : 2 : 4 cement concrete.

For all other pillars and beams at least 1″ of similar concrete cover must be provided in direct contact with the steel, unless the casing is entirely of concrete.

In all cases the concrete casing must be properly reinforced with steel wire lattice mesh or other reinforcement of not less than 2″ nor more than 6″ square mesh, and providing the equivalent of 16 s.w.g. wires calculated at 6″ square mesh.

Where the casing is at least 3″ thick and of properly reinforced concrete such casings need not be in contact with the steel.

Note. The above requirements as regards casing do not apply to open framed trusses supporting a roof, nor to subsidiary floor beams which are entirely embedded in the thickness of a concrete floor slab, provided that these beams have at least 1″ of concrete cover on their upper and lower surfaces.

Also, they do not apply to industrial workshops, storage sheds and hangars, of one storey height, or to other buildings of one storey not more than 25 ft. high.

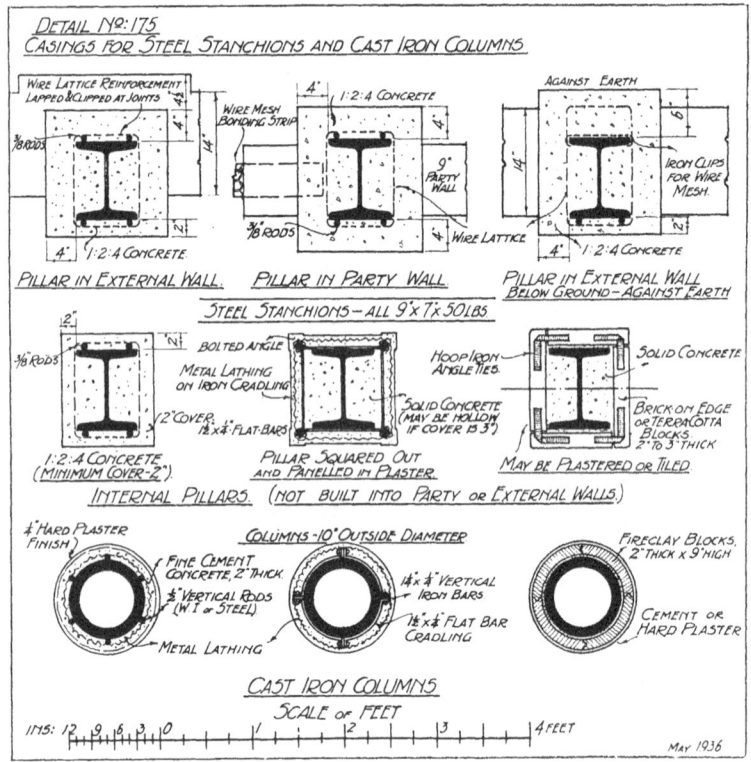

578. Protection to steel stanchions. In detail No. 175, the first six diagrams show arrangements for encasing stanchions. All these comply with the conditions of the L.C.C. Code of Practice for the particular cases selected for illustration. They include (a) a pillar in an external brick wall, (b) a pillar in a party wall, (c) a pillar in an external wall adjacent to the soil, (d) an internal pillar. All these are encased in 1 : 2 : 4 concrete. It must be noted, however, that brick or terra-cotta casings of the required thickness, properly bedded and bonded may be used in lieu of concrete.

All the above examples are filled solid, so that the concrete is in contact with the steel. This is not required if the external casing is at least 3″ thick.

An example is shown of an internal stanchion squared out by iron cradling bolted to vertical bars, filled solid with concrete within metal lathing, and finished in panelled plaster.

An alternative covering is shown in brick on edge or terra-cotta blocks, bonded, tied at the angles with hoop iron and filled solid with concrete. This would be finished in glazed tiles or in hard plaster.

579. Cast-iron column casings. Three examples of independent cast iron columns are shown. The first is encased in 2″ of fine cement concrete bonded by wire lathing packed off the column by vertical rods and clipped at the overlapping joint. The surface is finished with ¼″ of hard plaster.

The second example shows an alternative arrangement for a fine concrete casing. An iron cradling is used at vertical intervals of 18″, bolted to vertical flat bars, wrapped with wire lathing and finished in portland cement.

In the last example terra-cotta segments are employed. They have tongued and grooved vertical joints and are bonded by half-lapping the segments of alternate courses. The final surface is finished in hard plaster or cement.

Types of Patented Roof Glazing

The modern method of glazing roof lights without putty has been referred to and a fully detailed example was given in Vol. ii, but as there are many varieties of glazing bar and many methods of securing the glass and weatherproofing the edge joints, a selection of typical patented methods is given in detail No. 176.

580. Wood glazing bars. Four examples show wood glazing bars to support the glass, and three of these have lead cappings or flashings, with different methods of securing them to the bar. The "Simplex" and "Metacon" bars have plain sheet lead flashings, the former being nailed and folded to receive the edge of the glass, while the latter is secured by a galvanised iron strip, tightly fitted into a groove. The "Three Point Cap" has a special rolled lead capping, with a bulb-edge seating and a three-point contact at the overlapping wing; this arrangement gives a stiffer edge and a more reliable resistance to the penetration of water. Leakage water is conveyed down the channel formed by the bulb.

Water of condensation—and also some leakage—is partially conveyed by the semicircular channels formed in the top edges of all the above-named bars.

The "Drop Dry" bar has a zinc covering and cap which have

elasticity enough to cause a firm grip on the sheets of glass and ample provision is made at the edges for the removal of leakage water.

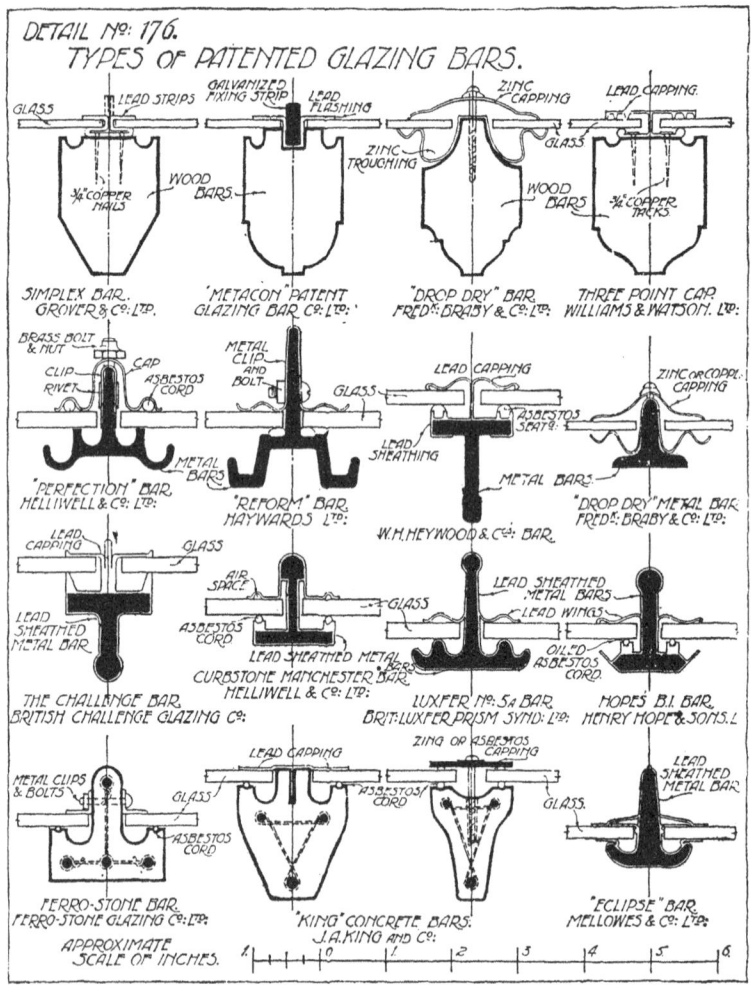

DETAIL Nº: 176.

TYPES OF PATENTED GLAZING BARS.

SIMPLEX BAR. GROVER & Cº: LTD.

"METACON" PATENT GLAZING BAR Cº: LTD:

"DROP DRY" BAR. FRED: BRABY & Cº: LTD:

THREE POINT CAP WILLIAMS & WATSON. LTD:

"PERFECTION" BAR. HELLIWELL & Cº: LTD:

"REFORM" BAR. HAYWARDS LTD:

W. H. HEYWOOD & Cº: BAR.

"DROP DRY" METAL BAR. FRED: BRABY & Cº: LTD:

THE CHALLENGE BAR. BRITISH CHALLENGE GLAZING Cº:

LEAD SHEATHED METAL CURBSTONE MANCHESTER BAR. HELLIWELL & Cº: LTD:

LUXFER Nº: 5A BAR. BRIT: LUXFER PRISM SYND: LTD:

HOPES B.1 BAR. HENRY HOPE & SONS. L

FERRO-STONE BAR. FERRO-STONE GLAZING Cº: LTD:

"KING" CONCRETE BARS. J. A. KING AND Cº:

"ECLIPSE" BAR. MELLOWES & Cº: LTD:

APPROXIMATE SCALE OF INCHES.

581. Metal glazing bars. For supporting roof glazing over long spans, metal bars have the advantage of providing the necessary strength without being so large in section as unduly to obstruct light.

Most bars approach the form of a T in cross section, with a bulb-edge to the web in order to gain strength. They are used with the web or the table uppermost, as may prove convenient for the method of securing the glass to be adopted.

Some bars, *e.g.* the "Perfection," "Reform" and "Drop Dry" sections, have the metal of the bar exposed, the glass being clasped upon the supporting bar by clip caps of zinc or copper.

Leakage channels and in some cases condensation channels, are provided, and in the "Reform" bar the glass is bedded upon asbestos cord with a view of preventing leakage and draught.

582. Lead-covered bars. Most of the systems employing metal bars include the sheathing of these members with lead; examples of these are the "Challenge", "Curbstone", "Luxfer", "Hope's B 1" and the "Eclipse" bars. The lead covering is very effective if perfectly done, giving protection to the metal, but, with defective coverings, exposed ends of iron or steel bars are decomposed by galvanic action between the metals, if water can accumulate at any point. Usually this liability can easily be avoided by careful workmanship and few complaints arise in modern work.

The covering forms a convenient medium in which to fashion bearings and flashings for the edges of the glazing and to provide channels for the removal of leakages. Mellowes' "Eclipse" bar has a double flashing formed by overlapping wings on the lead covering. The rest have single covers, some of which are curved to give two lines of contact, while the "Curbstone" has a small hollow bead leaving an air space near the margin. Heywood's bar may be had either with a complete lead covering or with the glazing section mounted on the steel bar, as shown.

583. Reinforced bars. Glazing bars are also made in artificial stone and fine concrete with steel reinforcing bars. "Ferro-stone" and "King" bars are examples of these. The upper and lower lines of reinforcement are connected by light wire shear stirrups and considerable weight can be borne by members of this type. In each case the glass is embedded on asbestos cord and leakage channels are provided.

Clips of metal, caps of asbestos and flashings of lead are employed to obtain watertight joints.

The great advantage of these bars is their permanency and freedom from atmospheric effects; they are particularly suitable for factory buildings, chemical works and the like. There is no maintenance cost as in cases where the metal is exposed. The same advantage is possessed by lead-covered bars though the permanence of the protection is not so great.

584. Removal and replacement of glass. Repairs to broken sheets of glass are easily carried out; the flashings are bent back or the covers removed, the broken sheets cleared away, new panes inserted, and the flashings or caps replaced. The trouble and expense of cutting out hard putty and the consequent damage to the bars is entirely avoided and the life of the structure thereby prolonged.

585. Reinforced precast window frames. In public buildings, schools and factories large windows are often required to light staircases in which the flights are superposed storey after storey. This is often done by one or more large windows carried in one width and two or more storeys in height. A good medium for this purpose and applied in many modern buildings—especially abroad —is reinforced concrete. The window frame is precast in one or more units of large size with rebated or grooved framing $2''$ to $3''$ wide and $3''$ or more in thickness, the frames being built in as the work proceeds. The units of glazing, from $9''$ to $12''$ in width and $12''$ to $18''$ high, are fitted directly into the concrete bars. The bars usually have one or two $\frac{1}{4}''$ to $\frac{3}{8}''$ reinforcing rods running through them and linked to the rods in the outer members of the frame.

The glazing may be of an ornamental character and some fine effects have been obtained by a wise choice of the glazing material.

Maintenance is reduced to a minimum but these frames do not readily admit of opening parts.

CHAPTER TWENTY-THREE

INSULATION OF BUILDING STRUCTURES

586. Need for insulation. Modern buildings, particularly flats, schools and offices adjoining noisy thoroughfares need some form of insulation from sound and in many cases insulation from temperature changes.

Partitions, and floors most commonly call for sound insulation and roofs may demand insulation from both heat and sound.

There are many insulating materials available usually in the form of lightly compressed fibre boards. The structure is irregularly cellular and the air contained in the cells or voids acts as a buffer against the transmission of sound.

The same material offers a safeguard against heat transmission.

Lloyd insulation board is a well known material for this purpose. A special Lloyd preparation is a bituminous insulation board which has the same properties of heat and sound insulation as the standard insulation board. The bitumen is mixed with the fibre before felting so that the board has absorbent qualities.

The board is impervious to moisture and is particularly suitable for exterior work where subject to damp.

It is claimed to be immune from dry rot and can therefore be used internally as linings to floors and walls where unventilated spaces occur.

Its absorbent and heat insulating properties prevent condensation.

Paint, distemper and plaster is not affected by the bitumen if a thoroughly satisfactory undercoating of paint or distemper is applied.

It provides a good attachment for plaster laid directly upon its surface, but for rough cast and for a thick coat of plaster it is wise to use galvanised wire mesh or expanded metal as a reinforced backing. The mesh can be fixed with galvanised steel staples upon brandering laths as shown in detail No. 177.

Amongst other well-known insulating boards which are usable in a similar manner are Insulite, and Masonite, Celotex, and Insulwood.

587. Uses of insulation board for roofs and walls. A few of the uses of insulation board are illustrated in detail No. 177.

In this detail, at A, is shown the application to the studded wall and roof of part of a timber structure.

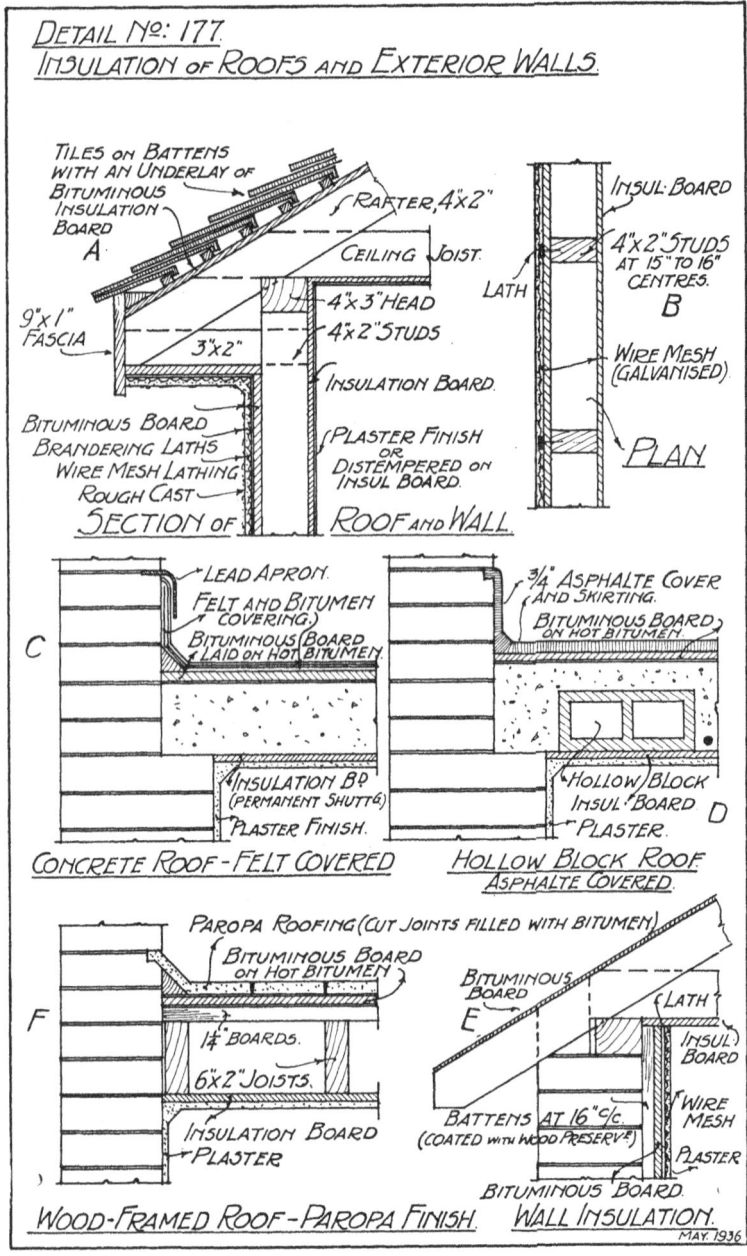

DETAIL Nº: 177.
INSULATION OF ROOFS AND EXTERIOR WALLS.

TILES ON BATTENS
WITH AN UNDERLAY OF
BITUMINOUS
INSULATION
BOARD
A

RAFTER, 4"x 2"

CEILING JOIST.

INSUL. BOARD

LATH

4"x 2" STUDS
AT 15" TO 16"
CENTRES.
B

9"x1"
FASCIA

3"x 2"

4"x 3" HEAD
4"x 2" STUDS

INSULATION BOARD.

WIRE MESH
(GALVANISED)

BITUMINOUS BOARD
BRANDERING LATHS
WIRE MESH LATHING
ROUGH CAST

PLASTER FINISH
OR
DISTEMPERED ON
INSUL BOARD.

PLAN

SECTION OF   ROOF AND WALL.

C

LEAD APRON.
FELT AND BITUMEN
COVERING.
BITUMINOUS BOARD
LAID ON HOT BITUMEN.

INSULATION Bᴰ
(PERMANENT SHUTTᵍ)
PLASTER FINISH.

3/4" ASPHALTE COVER
AND SKIRTING.
BITUMINOUS BOARD
ON HOT BITUMEN

HOLLOW BLOCK
INSUL. BOARD
PLASTER.

D

CONCRETE ROOF - FELT COVERED   HOLLOW BLOCK ROOF.
ASPHALTE COVERED.

F

PAROPA ROOFING (CUT JOINTS FILLED WITH BITUMEN)
BITUMINOUS BOARD
ON HOT BITUMEN

1¼" BOARDS.

6"x 2" JOISTS.

INSULATION BOARD
PLASTER

BITUMINOUS
BOARD

E

LATH

INSUL.
BOARD

WIRE
MESH

BATTENS AT 16" ᶜ/ᶜ.
(COATED WITH WOOD PRESERVᵉ)

PLASTER

BITUMINOUS BOARD.

WOOD-FRAMED ROOF - PAROPA FINISH.   WALL INSULATION.

MAY. 1936.

The inner surfaces of the wall and ceiling are covered with $\frac{5}{16}''$ ordinary insulating board; or, if external insulation is deemed to be sufficient $\frac{5}{32}''$ hardwood board may be used with either the smooth or granulated surface exposed according to the finish desired.

The roof is covered with $\frac{5}{16}''$ or $\frac{1}{2}''$ bituminous board on which the tiling laths are laid.

The external surface of the studding is covered with $\frac{1}{2}''$ or $\frac{5}{16}''$ bituminous board and finished with rough cast cement plaster on galvanised wire mesh or expanded metal. To pack the mesh from the board, slater's laths are nailed vertically on to the studs and the mesh fixed with staples (as shown for the partition at B). This provides an adequate key for the plaster. In some cases the mesh is fixed direct to the board but the efficiency of the mesh is then in doubt as the plaster cannot key so adequately and satisfactorily; adhesion of the plaster to the bituminous board is, however, claimed to be good.

The detail at B shows the application to a stud partition. One side is finished with plaster and the other with ordinary board either insulating or hard as may be required.

In the same detail at C and D are shown applications to flat roofs. In each case thick insulation board is used to form a permanent shuttering (laid on the timber formwork) and again to sheet the completed structure of the roof deck.

At C is shown an application to solid concrete filled roof, with an asphalted felt outer covering and at D a hollow tile construction similarly treated as to insulation but with an asphalte finish to the external surface.

When used as permanent shuttering the adhesion to the floor structure is dependent upon (a) the connection between concrete and board in the case of solid floors, and (b) the connection between the narrow bands of concrete filling between the lines of hollow blocks and the board, unless the hollow blocks are bedded in cement mortar. It is practicable to insert wire hangers through the boards and to carry them up and bend them over the top edges of the blocks within the spaces between lines of blocks, before filling in the concrete.

Unless the ceiling sheets are sufficiently well attached to the floor structure, the addition of plaster to ceiling board (shown in the details) may add sufficient weight to cause sagging and detachment of the boards.

The detail at E is comparable with A and shows the treatment for the insulation of a 9" wall adjoining the roof, where the ceiling and roof are insulated as before. If a plastered finish is required, the order of procedure is (a) fix creosoted vertical battens, 2" × 1" at 15" centres, (b) cover with insulating board, (c) fix slater's laths vertically at the same positions as the battens, (d) secure the wire mesh with galvanised staples, (e) apply the plaster finish.

In the detail at F is shown a flat roof constructed in timber with an insulated ceiling and covered with Paropa[1] roofing, which has good insulating properties. The covering can be laid on any good sound foundation which must have a minimum fall of 1 in 100 ($1\frac{1}{2}''$ in 10 ft.).

The complete covering is formed as follows:

(a) A coat of Paropa bitumastic priming;

(b) a layer of bitumen;

(c) a layer of hessian cloth to reinforce the foundation;

(d) a second layer of bitumen;

(e) a 1″ thick layer of Paropa cement mortar, well tamped and floated to the required fall. This is cut with special tools while soft to divide it into separate panels—not exceeding 30″ sq., but of any shape.

(f) The dividing joints reach the bituminous base and these joints are filled—when the cement mortar is hard—with an elastic bitumen.

The resulting surface is non-slip, elastic as a whole, easy to repair if damaged, allows of any treatment in design, has a good insulating value, and can be used as a floor if it is decided to add an additional storey to the building.

The weight of the whole covering is about 12 lbs. per sq. ft.

588. Insulation of floors and partitions. Detail No. 178 at A shows a section and plan through a 3″ breeze partition and a 9″ joisted floor, both insulated. Sheets of insulation board are laid over the entire floor surface and ceiling, the joists being spaced to suit the 4 ft. wide sheets, viz. 16″ centres.

The breeze partition is erected upon an extra insulation slip $3\frac{1}{4}''$ wide. The floor boards are then laid up to the edge of the extra slip.

The walls are then battened at 16″ centres with 3″ strips of insulation board, of two thicknesses glued together. The wall surfaces are covered with $\frac{5}{16}''$ or $\frac{1}{2}''$ board running down to the floor boards, skirtings are fixed and the thin plaster coating, where required, trowelled down to the top edge of the skirting.

If two-coat work is desired the first coat is floated to be flush with grounds fixed to receive the skirting, then the latter fixed and the final coat of plaster run down to the top of the skirting.

Note. Where a breeze partition is known to be required a thicker joist (or two joists near together) should be placed to receive the extra load.

Detail No. 178 at B shows a section and plan of a staggered stud partition. The studs are staggered in two independent series, each

---

[1] Special roofing by Frazzi Ltd., London.

at 16″ centres, so that the two surfaces of the finished work and their supporting studs have no connection except at the ceiling and floor levels. This greatly reduces the transmission of sound and makes for an efficient result.

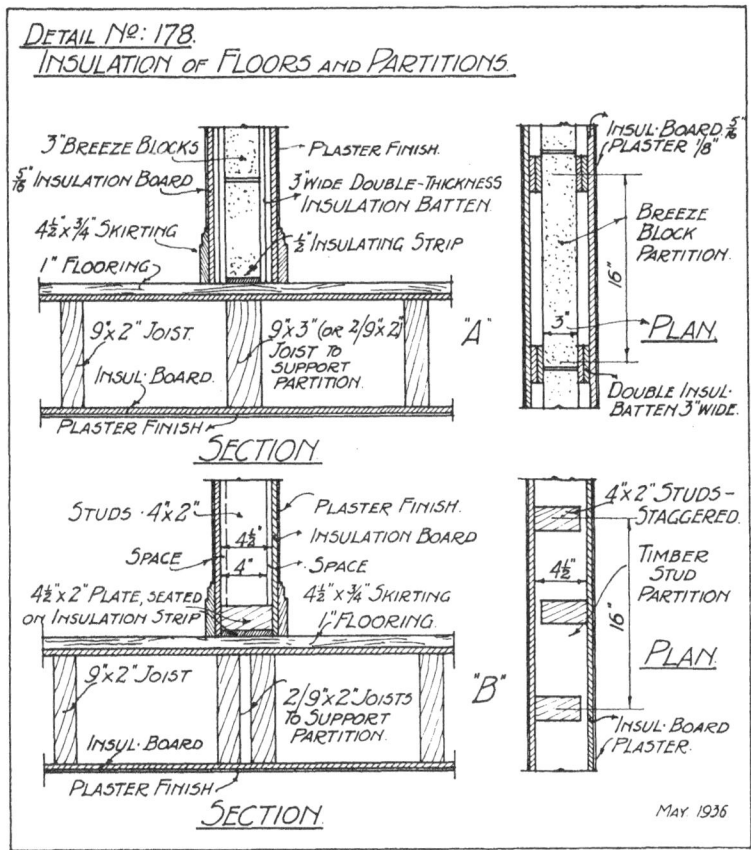

Another method, not illustrated, is to use a double partition having two lines of studs with un-nailed sheets of insulation board between, overlapping at the edges so as to form voids, and merely pushed tightly into position during erection.

Ceilings may be further insulated by using independently supported ceiling joists instead of nailing the insulation board to the underside of the floor joist. This method is good but expensive.

Additional resistance to sound can be obtained by placing sheets of insulation board, supported on fillets, horizontally between the joists at the centre of their depth. This modern arrangement virtually takes the place of the pugging method used originally for sound proofing ordinary wood floors.

589. Sizes of fibre boards and varieties obtainable. Generally fibre boards are obtainable in 4 ft. widths and in lengths from 4 ft. to 16 ft.

The boards are made in thicknesses from $\frac{1}{8}''$ to $\frac{1}{2}''$ and special ceiling sheets up to $1''$ thick.

The finishes may be smooth and glazed for light reflection, matt or granular, or irregular and pulpy in appearance.

The insulating boards are of the latter class $\frac{5}{16}''$ and $\frac{1}{2}''$ thick and the bituminous boards of the same thicknesses with a felted finish. The latter are usually in 10, 12 and 14 ft. lengths.

It is important to avoid waste by not having to cut boards to irregular sizes, and in arranging timber stud work, or strip battens, to place these at centres which will allow of full width boards to be employed as far as possible.

The sizes of the boards to be employed (because they vary with the makers in some cases) should therefore be known before details for the fixing are prepared.

Many makers, especially Edward Lloyd's Wallboards, Ltd., provide adequate technical information as to methods of construction, procedure in working, forms of joint covering, principles of insulation and methods of comparing the efficiency of systems of insulation, and readers interested in this modern form of guarding against noise and serious variations of temperature should consult the information available.

CHAPTER TWENTY-FOUR

MATERIALS

MATERIALS USED BY THE CONCRETOR

590. Concrete is an aggregation of comparatively small units of inert material united by lime or cement.

The coarser inert material may be broken stone or brick, shingle, gravel, clinker, slag, pumice stone or coke breeze, according to the nature of the work; this is commonly called the *aggregate* or *coarse material*.

To bind the coarse material together a cementing material is employed, preferably portland cement;[1] but the interstices or *voids* between the units are large and an uneconomical quantity of cement would be required to fill these voids and unite the pieces, hence a finer inert material is employed, usually sand, to fill these voids, and sufficient cement added to unite the particles of sand to each other and to the coarse material.

The result is to produce a mass of material called *concrete*, which can be conceived as a mortar in which units of coarse material are embedded, though the preparation does not consist of providing the mortar and broken material separately, as it would be impossible to get a thoroughly homogeneous mixture by this process. The mortar is often called the *matrix* though some authorities use this term to denote the actual cementing material apart from the sand.

591. Coarse material. Coarse material is employed in irregular pieces which vary in size from what would be retained on a square mesh screen with $\frac{3}{16}''$ holes to what would pass a similar screen with $1\frac{1}{2}''$ holes, assuming hand screening.

The bulk of the *broken* stuff is, however, prepared by stone crushers and screened mechanically through longitudinal screens with the wires a specified distance apart, but only connected crosswise at intervals of 9″ or more.

When a coarse material is to be employed for reinforced concrete it should pass a $\frac{3}{4}''$ screen and be retained on a $\frac{3}{16}''$ screen, anything finer than $\frac{3}{16}''$ being considered as sand; it is to be screened away and utilised as such. It is important that dust should be screened out and discarded and where the process of breaking up results in fine dust the aggregate should be washed before using.

---

[1] See paragraphs 593 and 394.

For *mass concrete*, where nothing is inserted to reinforce the material, the larger units may sometimes be satisfactorily employed, as in foundations, under portions of basement floors and site concrete, retaining walls, rough backings and similar purposes; provided that the mixture will set properly and attain the required strength any of the aggregates referred to in paragraph 600 may be employed.

For *reinforced concrete* certain materials may not be employed as aggregates because they may contain chemicals injurious to the steel; amongst these are coal residues such as clinker, ashes, breeze and slag; products from blast furnaces and sulphates.[1] Other materials such as limestones and marbles are forbidden because, where exposed, they are liable to be calcined in a fire and become quicklime; their strength would then be so diminished as to endanger the structure.

In all cases the material should be well graded or varied in size between the specified limits; this reduces the voids in the dry mass and requires less matrix to unite the material.

592. Fine material or sand. Sand or stone screenings to be used as fine material should be clean[2] and not too fine. If the material passes a sieve having 2500 meshes to the sq. inch—commonly called a 50^2 sieve—it is too fine and should be discarded; the bulk of it should be *much* larger in the grain, while stone screenings passing a $\frac{3}{16}''$ mesh improve the mixture.

Sand must be clean if strong concrete is required. Clay and vegetable matter reduce the adhesion of the cement to the inert particles; small quantities of clay assist in water-proofing the mixture but at the same time reduce the strength, and anything in excess of $2\frac{1}{2}$ per cent. should be avoided.

Sand from river beds and estuaries is often objected to because of its rounded grain, due to abrasion in the movements of the stream; if large enough in the grain this sand is much better than pit or quarry sand which contains clay and loam.

Sea sand should not be employed for any important purpose and certainly not above ground, as the salt contained in it attracts moisture and dissolves, and on evaporation in dry weather causes efflorescence—a white salt deposit on the face of the work.

593. Cement. For good concrete, only portland cement[3] should be employed, and this should comply with the standard specification of the British Institution Standards (No. 12—1931). In large

---

[1] See L.C.C. regulations for Reinforced Concrete Structures.
[2] Cleanness is much more important than sharpness.
[3] For principles underlying testing see Manson's *Building Science*.

and important works supplies of cement should be periodically tested by a competent and experienced person.

The tests[1] to be conducted under this specification are given under five heads, viz.:

     (a) Fineness.
     (b) Chemical composition.
     (c) Tensile strength (cement and sand). .
     (d) Setting time.
     (e) Soundness.

The results required are, briefly, as follows:

(a) *Fineness.* 100 grammes (4 ozs.) of the cement to be sifted continuously for a period of 15 minutes on a sieve of British Standard mesh No. 170, and the residue for a period of 5 minutes on a sieve of British Standard mesh No. 72, with the following results:

     (1) Residue, by weight, on a sieve of British Standard mesh No. 170, not to exceed 10 per cent.
     (2) Residue, by weight, on a sieve of British Standard mesh No. 72, not to exceed 1 per cent.

Note. Air-set lumps in the samples may be broken down with the fingers but must not be rubbed on the sieve. (Details of the sieves to be employed are given in the complete British Standard Specification.)

(b) *Chemical composition.* The proportion of lime—after deduction of the proportion necessary to combine with the sulphuric anhydride present—to silica and alumina, when calculated in chemical equivalents by the formula

$$\frac{CaO}{SiO_2 + Al_2O_3},$$

shall not be greater than 3·0 nor less than 2·0.

The insoluble residue shall not exceed 1·0 per cent.; magnesia shall not exceed 4 per cent.; total sulphur content calculated as SO_3 shall not exceed 2·75 per cent. The total loss on ignition shall not exceed 3 per cent.

(c) *Tensile strength* (cement and sand). The preparation of the mixture for testing and the filling of moulds is governed by carefully detailed conditions (see full specification) in order to ensure uniformity of preparation of test samples.

Tests to be made on briquettes of standard form, having a section 1″ square at the centre. Newly prepared briquettes are to be kept in an atmosphere with a minimum relative humidity of 90 per cent. for 24 hours after gauging; then immersed in clean fresh water and so remain until required for testing. Temperature of water 58° to 64° F.

Tests to be conducted at 3 days (72 hours) and 7 days (168 hours) after gauging; six briquettes at each period. Standard jaws are required for applying tensile force and steady, uniformly increasing load employed at the rate of 100 lbs. per sq. in. of section in 12 seconds.

At 3 days, breaking strength must be at least 300 lbs. per sq. in.

At 7 days, breaking strength must show an increase on the breaking strength at 3 days and must be at least 375 lbs. per sq. in. of section.

An average of the six tests at each age to be accepted for the above purpose.

The sand used for the test mixture must be standard sand obtained from Leighton Buzzard. It must pass through a sieve of B.S. mesh No. 18 and be retained on a sieve of B.S. mesh No. 25.

---

[1] For full description of tests and apparatus to be employed see British Standard Specification for Portland Cement (No. 12—1931). Price 2s. net.

(*d*) *Setting time.* Tests for setting time on *neat cement* are to be carried out by a standard Vicat apparatus defined in the standard specification.

Unless a quick setting cement is particularly specified, the initial setting time of the cement must be not less than 30 minutes and the final setting time not more than 10 hours.

If a quick setting cement is particularly specified, the initial set shall occur in not less than 5 minutes and the final set in not more than 30 minutes.

The initial set is indicated when the Vicat needle, used as directed, does not completely penetrate the test sample; the final set is indicated when the point of a special collared needle makes an impression but the attached collar fails to do so.

(*e*) *Soundness.* To be tested by the "Le Chatelier" method, using standard moulds in the form of split cylindrical rings with indicator arms attached, which indicate and magnify any expansion as the ring opens out.

Neat cement is gauged and filled between glass plates, the cap plate weighted and the whole submerged in water at a temperature of 58° to 64° Fahr. and left for 24 hours. The distance between the indicator points is then measured and noted, and the mould again submerged in water, which is then brought to boiling point in 25 to 30 minutes and kept boiling for *three hours.*

The mould is then removed, allowed to cool, and the distance between the indicator points measured and noted. The expansion between the indicator points must not exceed 10 millimetres. If the cement fails to comply with this condition, another portion of the same sample may be tested after aeration for 7 days, when the expansion is not to exceed 5 millimetres.

594. Hydraulic limes and special cements. Roman and Medina cements should not be employed in modern work of the first class, though they set both hard and rapidly. Their strength is much less than portland cement and they are not so reliable. They are natural cements prepared by burning, calcining and grinding lime nodules containing from 30 to 40 per cent. clay.

Hydraulic limes have already been referred to in Vol. ii. In the districts where they are plentiful they are used for both mortar and concrete. Concrete made from hydraulic lime is very much weaker than portland cement concrete. It may be usefully employed for foundations of ordinary buildings, and for backings and filling.

Portland cement is, however, now so plentiful, cheap and dependable that it is in general use in preference to lime for mortars and concrete for all purposes.

595. Water for concrete mixing. The water used for mixing concrete or mortar should be clean and fresh and an ample supply should be provided. Sea water is objectionable for general building work because of the salt it contains. It has the same effect as sea sand, described in paragraph 592.

Concrete to be used below ground is sometimes prepared with sea water, which retards the rate of setting and in some cases increases the ultimate strength.

596. Proportioning of concrete. A vital matter in the preparation of concrete is the selection of the quantity of each constituent.

It it usual to express the ratio by volume; thus, for foundations, 1 part of portland cement, 2½ parts of sand and 5 parts of coarse material is a common mixture for ordinary work. If each *part* be thought of as 1 cu. ft. in volume, the meaning is clear.

The greatest difficulty in ensuring correct quantities lies with the cement which can be lightly or heavily filled into the measure, causing a variation of 20 per cent. or more in the actual quantity used. Being so finely ground, it encloses air between the particles when shovelled about and only by shaking the measure is solid filling ensured. In order to guard against variation and fraud it is now common to specify a weight of cement and it is generally required that 90 lbs. of cement shall be considered as the equivalent of 1 cu. ft., so that the cement should be *weighed* and the inert materials *measured*.

For reinforced concrete the ordinary ratio is 1 part cement, 2 parts sand and 4 parts coarse material, but mixtures richer in cement are often desirable.

The L.C.C. Code of Practice (1932) permits the following bearing pressures on foundations formed of cement concrete to specified mixtures:

$$1 : 1 : 2 \text{ cement concrete, 40 tons per sq. ft.}$$
$$1 : 1\tfrac{1}{2} : 3 \quad ,, \qquad ,, \quad 35 \quad ,, \qquad ,,$$
$$1 : 2 : 4 \quad ,, \qquad ,, \quad 30 \quad ,, \qquad ,,$$

If mass concrete is used in which the fine and coarse material are mixed—as in ballast—the following pressures are permitted:

$$1 : 6 \text{ mass concrete 20 tons per sq. ft.}$$
$$1 : 8 \quad ,, \qquad ,, \quad 15 \quad ,, \qquad ,,$$
$$1 : 10 \quad ,, \qquad ,, \quad 10 \quad ,, \qquad ,,$$
$$1 : 12 \quad ,, \qquad ,, \quad 5 \quad ,, \qquad ,,$$

597. Hand preparation of concrete. The three constituents are measured separately—or the cement weighed—in boxes with open bases, heaped *while dry* upon a clean wooden platform or other suitable surface, and turned over several times with a shovel to mix them thoroughly. Water is then added through a rose to obtain uniform distribution and the mass is further turned over until every part of the mixture is wet enough to lay solid. Every part must be moist but without any excess of water.

The concrete should then be deposited immediately, being gently emptied into position from convenient receptacles and lightly rammed or "punned" to solidify it. Tipping concrete from a height is bad practice. It disturbs the composition, making it non-uniform and therefore weaker in parts.

Lime concrete would be prepared in the same way using ground hydraulic lime, but in the ratio of $1:2:5$ or $1:2:4$ for foundation work.

Ratios of constituents are often stated in this concise way: the order is always: cement, sand, coarse material and the figures refer to parts by volume.

598. Machine mixing of concrete. When concrete is required in large quantities it should be mixed in batches by a rotary mixer such as the Ransome and Smith's mixers. Uniformity of quality and consistency is ensured by proper arrangements for measuring the quantities of dry materials and also the quantity of water for each batch and allotting a definite period for the agitation of each filling of the mixer.

599. Quantity of dry materials for concrete. The voids in an average aggregate amount roughly to 44 per cent. of the mass and the voids in sand are often approximately the same, but, as the percentage of voids may vary, tests are desirable in preparing for large quantities of concrete to decide the average percentage of voids in the materials to be employed. Coarse material should be measured, soaked with water and drained off, then filled up carefully to the brim of the measure and the volume noted and compared with the original volume of the material.

Let 10 cu. yds. of coarse material be made into concrete, so that the voids are filled with sand and the voids in the sand filled with cement.

Then the volume of sand required $= \dfrac{44}{100} \times 10 = \mathbf{4\cdot4}$ cu. yds. and the volume of cement $= \dfrac{44}{100} \times 4\cdot4 = \mathbf{1\cdot94}$ cu. yds.

To allow for some shrinkage of sand in mixing and also of the cement use, say, 5 cu. yds. sand and $2\frac{1}{2}$ cu. yds. cement.

The ratio is then $2\cdot5:5:10$ or $1:2:4$, the leanest mixture allowed in reinforced concrete work.

Some aggregates have much less than 44 per cent. of voids, especially if well graded, while a few exceed this value. With reasonable care in measuring and mixing the above proportions may be relied on to give a good concrete. Any change should be in the increase of the cement and this will always increase the strength. Allowances for waste are always necessary and this is done by starting with a larger quantity of aggregate, say 5 to 10 per cent.

It should be noted that if the voids in the aggregate happen to be less than say 40 per cent., there will be sufficient excess in the matrix.

Experience has taught that the average reduction in the bulk of the dry materials when mixed into foundation concrete is somewhat in excess of one-quarter of the original volume in the best cases and fully one-third in others. This forms a rough guide but is not a satisfactory method of determining requirements; tests of actual shrinkage should be made on known quantities, or tests of voids made and calculations worked as above.

The use of concrete for important purposes is constantly increasing, and students should pay special attention to its properties and preparation.

The quantity of water required for mixing is not more than 6 gallons per cwt. of cement employed in the mixture.

According to the nature of the mixture the quantity of water required for a workable degree of plasticity may vary from 4 to 6 gallons, e.g. a $1:1\frac{1}{2}:2\frac{1}{2}$ mix needs about 4 gallons, and a $1:5:7\frac{1}{2}$ mix about 6 gallons per cwt. of cement used.

The strength of the concrete depends very largely on the plasticity or workability of the mixture. Water should not be used in excess, because sloppy mixtures produce porous concretes. A satisfactory and workable concrete mixture will place readily into the forms and can be spaded and tamped into place and produce a dense concrete, free from honeycombing or separation of the constituents.

600. Selection of aggregates. The kind of coarse material to be employed depends upon the nature of the work.

If strength is required granite chippings, hard close grained sandstone or hard crushed limestone are suitable, and some gravels are also satisfactory. For fire-resistance hard broken brick, gravel or shingle, and granite are good. For resistance to water a rich concrete with a non-porous aggregate is desirable—voids must be perfectly filled.

For upper floors where lightness is desired coke-breeze and pumice stone have been employed and also porous brick, but breeze burns out in a strong fire and in damp atmospheres any steel reinforcement employed is insufficiently protected because of the porosity of the concrete. Hard plaster ceilings, $\frac{1}{2}''$ thick, assist in giving protection against damp.

601. Water-proofed concrete. There are many preparations obtainable which are intended to render concrete water-proof even under reasonable pressure. They are variously sold in paste or powder form, the paste being usually employed by adding it to the mixing water and the powder by mixing thoroughly with the dry cement before adding any water.

602. Trus-con water-proofing paste is a good example of the former class. It is added to the mixing water in the ratio of 1 part of paste to 18 parts of water by volume, and is easily mixed to a milk-like consistency and is therefore carried to every part of the cement and to the surfaces of the units of the material in tempering the mixture. The action is colloidal (producing a gelatinous substance) and results in the interstices between the particles—which may be partly filled by colloidal action in the setting of the cement—being completely filled and the mass rendered water-tight.

603. Pudlo is a widely known example of the second class of water-proofers. It is a white powder applied by adding to the dry materials 2 per cent. or more of the bulk of the cement and sand. It should be mixed first with the cement, or with the cement and sand, and then with the coarse material before adding water.

The Pudlo will not mix with water, hence the above procedure. When mixed the slight expansion of the materials causes the pores to be blocked and so prevents the passage of water.

Both the above methods claim not to reduce the strength of the resulting concrete or mortar.

604. Prufit is another excellent paste used in a similar manner to Trus-con paste, using 1 part of paste to 12 parts of water; 1″ rendering so proofed will resist 8 ft. head of water.

605. Ironite cement is a water-proof material which has been specially treated with iron compounds in the manufacture or preparation. This cement is highly water-resisting and will stand a great pressure of water. It is employed for rendering walls and surfacing floors and is also grease-proof, dust-proof and durable under heavy traffic.

606. Surface treatment of concrete to render damp-proof. In cases where ordinary concrete, breeze blocks, brickwork and stonework have failed to resist damp, and some surface treatment is apparently desirable, the face may be either rendered with a water-proofed cement mortar of the kinds described in the above paragraphs, or some indurating substance may be applied to the surface of the porous material.

Induration may be very easily and satisfactorily done by such preparations as Ironite, which are applied with a brush to the surface. Very often one coat will water-proof distinctly porous materials, if the directions of the manufacturers are carried out. The chief objection to the preparation is its colour, which is that of dull iron rust, but it is so effective that its use for commercial and industrial buildings is large. For flat concrete roofs no risk should be taken by depending on ordinary cement rendering with surface treatment; Ironite cement or other integral water-proofer should be employed, or the roof should be covered with mastic asphalte laid to falls, or again with one of the numerous bituminous felts in overlapped layers and with sanded finish.

MATERIAL USED BY THE IRONFOUNDER AND STRUCTURAL ENGINEER

The materials to be studied under this heading are chiefly the kinds of iron in general use, considering their physical and chemical properties and the market forms obtainable for constructional purposes.

607. Kinds of iron. Iron is used in three forms, viz. cast iron, wrought iron and steel; the differences in character are largely due to foreign substances in the composition of the metal.

All iron is obtained from iron ores by smelting. After drying the ore, selecting, mixing and preparing, it is subjected to intense heat generated by coke fuel or charcoal (and sometimes coal) in a blast furnace, assisted by a blast of hot air produced by utilising the waste heat from the furnace to raise the temperature. Fluxes are employed along with the fuel, to combine with the foreign matters mixed with the ore and thus separate the pure metal which, owing to its density, settles to the base of the furnace. When a sufficient accumulation has been made the metal is withdrawn and flows into cooling channels formed in sand, evolving bars of crude iron, called *pig iron*.

608. Cast iron. Cast iron is made from pig iron by selecting the most suitable pigs—having much uncombined carbon—re-melting and running into hollow moulds embodying the form of the required article.

The metal in this condition contains 2 to 6 per cent. of carbon, is hard and crystalline, very brittle and weak in tension, but strong under compressive stress.

Being easily fusible, intricate forms can be run or *cast* in one piece with the molten metal.

Cast iron is improved by repeated melting and old metal or scrap is broken up and fused along with pig iron to improve it. Its density averages 450 lbs. per cu. ft.

Uses. Cast iron is employed for pillar bases, columns and stanchions, lintols and bressummers; gutters; rain water and soil pipes; low pressure heating pipes; pockets for floor beams and roof rafters; external gratings, fire grates and accessories; gates and railings; drain pipes, manhole covers, etc. Also for ornamental work in panelled fillings to masonry façades.

The metal does not oxidise so rapidly as wrought iron and steel but requires constant attention and treatment to avoid undue oxidation if left exposed to a damp atmosphere.

Patterns for enclosing in sand to produce the form of the casting are usually made of American yellow pine and allowance is made for the cost of these patterns in contract work. One pattern only is employed for repetition work.

Pipes are often preserved by dipping, while hot, in Angus Smith's solution.[1]

609. Wrought iron is made from the harder varieties of pig iron by extracting the carbon, and refining the metal by removing other elements. The metal is fused, exposed to a current of air and agitated, so that the carbon combines with oxygen from the air. Good wrought iron contains very little carbon, the limit being ·15 per cent.

Wrought iron is a tenacious malleable substance which is not fusible except at very high temperature, but becomes plastic when heated. If repeatedly rolled while hot, under great pressure, it becomes fibrous and tenacious.

It is easily forged—beaten or pressed—into any shape at a bright red heat, and pieces may be welded together, that is, joined and amalgamated by hammering while white hot.

The density of wrought iron is about 480 lbs. per cu. ft.

Uses. For bolts, straps, large hinges, gutter brackets and beam stirrups; tie rods, connecting plates and holdfasts; low and high pressure heating pipes and steam pipes; heavy door and window furniture. Railings, entrance gates, grilles, etc., provide a fine and artistic field for the use of wrought iron.

Wrought iron oxidises rapidly on exposure to damp. In fittings and furniture it is usually japanned while hot; where this cannot be done wrought iron work should be painted before fixing and immediately on completion.

610. Steel. A form of iron in which carbon varies in quantity from ·15 to 1·5 per cent., or even more, its character being somewhat similar to wrought iron at the lower percentage and increasing in hardness with the increase of carbon. The processes of manufacture have also some effect on the character of the metal, affecting the crystallisation.

Steel is therefore defined by its properties rather than by the quantities of foreign constituents. It is capable of being cast into malleable blocks, and with the higher percentages of carbon can be hardened and tempered by cooling quickly at different temperatures. It is always malleable when hot, being easily forgeable, but only weldable with care except in the milder forms of the metal.

---

[1] See Vol. II for paints and preservatives for iron.

611. Mild or structural steel contains not more than ·5 per cent. of carbon and has a density of about 490 lbs. per cu. ft. Mild steel amalgamates the best properties of cast and wrought iron and is harder, sounder, tougher and stronger than the latter, but possesses its fibrous structure in a large measure; this is due to being rolled under great pressure to the market form of bar or special section.

Structural steel and high tensile steel are both now available for the production of British Standard Sections, which have been frequently referred to in the Chapters on Steelwork.

612. Cast steel is a crystalline or finely granular, hard variety, due to larger percentages of carbon, and is used chiefly for the production of keen edged cutting tools. It is not employed in this refined form for constructional work.

613. Bessemer steel is a mild steel manufactured directly from pig iron or crude cast iron, by removing the carbon as carbon dioxide through an air blast, and then adding a definite quantity of "mirror iron" which is rich in carbon and manganese. The final percentage of carbon is thus controlled. Bessemer steel is used in castings for the bases of large stanchions, bridge bearings and for rolling into structural sections.

614. Siemens or open hearth steel is produced from an ore rich in iron oxide, along with pig iron and steel waste, in an open furnace. When hot, the carbon and oxygen combine and pass off as carbon dioxide. The process is slower and more expensive than the Bessemer process, but this is an advantage for producing a highly uniform quality of metal, because tests can be made and the process arrested at the stage when the desired quantity of carbon is present.

615. Uses of steel. Mild steel is employed for the standard structural sections used in floors, roofs and stanchions, reinforcing bars in concrete, tension members in composite trusses, good bolts and fixings, screws, metal furniture and fittings, hinges, locks and casements.

616. Strengths of cast iron, wrought iron and structural steel. The following are the safe working stresses per sq. inch of section which are usually adopted in modern design. The values are based on steadily applied or dead loads—not moving loads. In the latter case it is necessary to make such allowance for the conditions as will ensure that the equivalent dead load stress does not exceed the given values. The maximum effect of a moving load cannot exceed that of a dead load of twice the magnitude, except there be *impact* instead of smooth motion.

	Tons per sq. inch	
	Mild Steel B.S.S. No. 15	High Tensile Steel. B.S.S. No. 548
For parts in tension		
On nett section. Axial-stress or extreme fibre stress in beams	8	12
On nett section of rivets for axial stress—shop driven rivets	5	7·5
On nett section of rivets for axial stress—site driven rivets	4	6
On nett section of bolts for axial stress	5	7·5
For parts in compression (flanges of beams)		
On gross section for extreme fibre stress of beams embedded in a floor or otherwise laterally secured	8	12
On gross section for extreme fibre stress of uncased beams where the laterally unsupported length "L" is less than twenty times the width "b" of the compression flange	8	12
On gross section for extreme fibre stress of uncased beams where "L" is greater than twenty times "b"	$11 \cdot 0 - 0 \cdot 15 \dfrac{L}{b}$	$16 \cdot 5 - 0 \cdot 25 \dfrac{L}{b}$
Note. There are other limiting conditions for cased beams.		
For parts in shear		
On gross section of webs	5	7·5
On shop rivets and tight fitting turned bolts	6	9
On field rivets	5	7·5
On black bolts (where permissible)	4	6
For parts in bearing		
On shop rivets and tight fitting turned bolts	12	18
On field rivets	10	15
On black bolts (where permissible)	8	12

Under the L.C.C. Code of Practice (1932) all structural steel used in the construction of a building must comply with the British Standard Specification (No. 15) for structural steel, for bridges and general building construction *from time to time in operation*. This specification defines quality and tests.

Such steel used in building construction is governed at present for working stresses by the British Standard Specification for "the Use of Structural Steel in Building" (No. 449—1935).

The working stresses tabulated on page 424 are laid down in the current specification both for standard structural steel and for High Tensile Structural Steel, a steel of higher quality which is governed by British Standard Specification (No. 548—1934).

(The tabulated stresses do not apply to steel grillage and subsidiary floor beams if they are entirely embedded in concrete.)

Where grillage beams are properly enveloped in concrete and spaced at least 3″ apart, with a minimum upper and side cover of 4″, the tabulated stresses may be exceeded by 50 per cent. for structural steel and by 33½ per cent. for high tensile steel.

For filler floor beams which are properly encased in concrete, with a minimum upper flange cover of 1″ of concrete, the extreme fibre stresses in structural steel may be taken at 10 tons per sq. in. and on high tensile steel at 13·5 tons per sq. in. (or the floor may be calculated on the principles of a reinforced concrete floor).

For wrought iron and cast iron the following values are laid down by the L.C.C. General Powers Act, 1909.

	Tons per sq. inch			
	Compression	Tension	Shearing	Bearing
Wrought iron	5	5	4	7
Cast iron	8	1½	1½	10

The above values do not apply to pillars (see Chapter Nine). As a rule the ultimate stress (or breaking stress) for the above materials is at least four times the value given in the table; theoretically, the factor of safety of the structure is 4, viz. the working stress is one-fourth of the stress required to rupture the material.

617. Tests for metals. All metals to be used in structural work should be tested to determine their fitness for the intended purpose.

Cast iron should be struck lightly with a steel hammer, which easily ensures detection of cracks or flaws by the sound produced. It should be inspected for irregularities and air bubbles, and sharp angles should be required. Small beams 1″ × 1″ × 1 ft. clear span or

2″ deep × 1″ broad × 3 ft. clear span, are usually tested for bending; roughly, they should respectively support 1 ton and 1⅓ tons at the centre before breaking and the fracture should be sharp, clean and crystalline.

The Admiralty require cast iron to resist a direct tension of 9 tons per sq. inch before fracture.

Steel is tested mechanically to comply with the British Standard Specification which includes tension, elasticity and both cold and hot bending. The minimum ultimate tension required is from 28 to 33 tons per sq. in. and Young's modulus should be from 13,000 to 13,400 tons per sq. in. The material must show tenacity and ductility by extending 20 to 24 per cent. on test pieces according to their form.

618. Rivet steel is required to be more ductile and is therefore not so strong in tension. The minimum rupture stress allowed is 25 to 30 tons per sq. inch and the material must be capable, when cold, of bending over upon itself without cracking, and when heated the head of a rivet must allow of being flattened to 2½ times its original diameter without cracking at the edges.

619. High tensile steel. In recent years structural steel has been developed which has a high tensile resistance, the range of ultimate stress being from 37 to 43 tons per sq. inch. This high tensile material cannot be obtained by merely varying the carbon content in the steel, but is the result of using small quantities of alloys.

The British Standard Specification (No. 548—1934) does not specify the alloys to be used, but limits the contents of certain substances as follows:

Carbon   ...   maximum 0·30 per cent. for material other than rivet bars.

„   ...   maximum 0·25 per cent. for rivet bars.

Sulphur   ...   „   0·05 per cent.

Phosphorus ...   „   0·05 per cent.

Copper may be present up to 0·6 per cent. as mutually agreed between manufacturer and purchaser.

Tensile tests of high tensile steel to comply with the specification are given for plates, sections and flat bars.

The ultimate stress must lie between 37 to 43 tons per sq. inch. and the yield point must occur at not less than a specified amount (tons per sq. in.) which varies from 19 to 23 tons according to the thickness of the test piece or the diameter of round and square bars.

Elongation before rupture is also specified and varies from 14 to 22 per cent. according to conditions.

Rivet bars must have a tensile breaking stress of 30 to 35 tons per sq. inch.; and an elongation of 22 to 27 per cent. according to the nature of the test piece.

There are six forms of standard test piece specified of varying size and gauge length suited to different commercial forms of the metal and the whole process of testing is definitely controlled.

The range of tests include, tension, cold bending, hot compression, and chemical analysis. Reference to the British Standard Specification should be made for detailed information.

Special forms of Metal

620. Expanded metal. This material is formed by cutting and expanding either plain sheets or ribbed sheets of steel.

The plain sheets are slit alternately in parallel rows, then shorn and expanded to produce a diamond shaped mesh, as shown in detail No. 29 and No. 91, Vol. II. Ribbed sheets are similarly cut along the depressions but with the slits opposite each other in continuous lines, then drawn apart and the alternate ribs forced longitudinally until the ribs and cross strands form rectangular meshes, as in detail No. 30 Vol. II. Both preparations are widely employed.

The prepared sheets are known as *Diamond mesh* and *Rib mesh* expanded steel. Diamond mesh varies from $1\frac{1}{2}''$ to 6″ across the short way of the mesh and is obtainable in sheets from 4 ft. to 8 ft. long × 16 ft. wide or *vice versa*, with great variations in thickness of strands and tensile strength.

Rib mesh is produced in sheets from 16″ to 64″ wide and 10 ft. to 24 ft. long with meshes 2″ to 8″ square.

When used as reinforcement to concrete slabs these expanded sheets should be placed with the rib, or the long way of diagonal, the short way of the slab, because the greater stresses occur in that direction.

621. B.R.C. fabric. This is a steel wire fabric in which tension wires and distributing wires are electrically welded at their crossings and made up in sheets 72″ wide and 34 to 240 ft. long. The longitudinal wires are usually at 3″ centres for floor reinforcement and transverse wires are placed at 12″, 16″ and 18″ centres to prevent cracks due to contraction, changes of temperature or uneven distribution of loads. The fabric is largely employed and has been very successfully applied to the reinforcement of floors, concrete roads, etc. Its particular advantage is that transverse and longitudinal strands cannot be displaced during concreting.

Other well known patent preparations for slab reinforcement are "Johnson's woven wire mesh", "Lock-woven mesh", and for concrete slabs laid on the earth direct "Loop road-reinforcement".

622. Hy-rib. This material is described in the chapter on partitions in Vol. II, and its use for that particular purpose is the more important. Hy-rib is, however, much employed as a floor reinforcement for spanning between main beams like the above-mentioned materials. It may be laid flat upon the lower flanges or upon haunches, or it may be bent and arched across from beam to beam and the concrete deposited upon it. The concrete passes partially through the perforations of the metal and forms a good key for finishing plaster. For sketches of the material see details Nos. 91 and 138, Vol. II.

One great advantage in the use of Hy-rib is that shuttering is not required as in the case of many other slab reinforcements.

EFFECT OF HEAT ON IRON AND STEEL

623. All metals expand considerably when heated. Iron has an average coefficient of expansion of ·000012 per degree Centigrade or ·0000066 per degree Fahrenheit, which means that for a rise of 1° C. an increase in linear dimensions occurs equal to ·000012 of the original size.

Thus, if a girder is 20 ft. (240″) long at normal temperature, say 14° C., and is then raised to 70° C., the increase in length would be $56^{(°)} \times 240^{(″)} \times ·000012 = ·161″$, or about $\frac{5}{32}″$.

In a fire the temperature may be raised many times this amount and the heat may be more intense in one part than another, or one surface of a steel member may be much more exposed than another, with the result that the hotter side expands and forces the girder out of shape.

As the steel stanchions or cast iron columns get hot and expand also, they try to lift their loads, but, assisted by side pressure and twists from the girders, probably buckle in the attempt. Further, if this steelwork attains anything like a red heat, the loads on members twist and bend them out of shape.

With these combined effects the frame-work of the building is twisted into shapeless masses, collapses and probably destroys the walls of the structure.

Cast iron easily cracks under sudden and variable reductions of temperature such as occur in quenching a fire. For this reason it is not a desirable material to employ unless well protected.

624. Protection of iron and steel. Because of the effects described in the last paragraph (and also to preserve against oxidation) iron and steel must be protected if fire-resistance is desired. The L.C.C. require the following protection by fire-resisting material:

Steel skeleton buildings. External pillars to be covered with at least 4″ of material, internal pillars to have at least 2″. Girders require 2″ of protection on faces and soffits, but filler joists spanning between large beams are only required to have 1″ of cover above and 2″ below. For further details of special provisions in the L.C.C. Code of Practice (1932) see paragraphs Nos. 577, 578 and 579.

Portland cement concrete, brick in cement, terra-cotta or fireclay blocks and cement plaster on expanded metal lathing give very satisfactory protection.

APPENDIX I

BRITISH STANDARD BEAMS

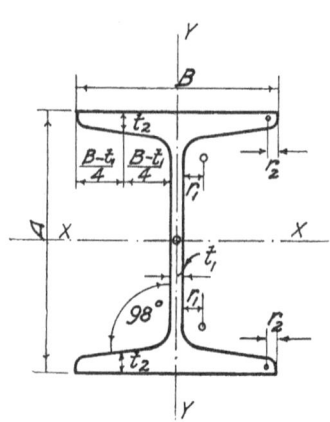

DIMENSIONS
AND
PROPERTIES
OF
BRITISH STANDARD I BEAMS
IN
INCH UNITS

Angle between
web and flange
$= 98°$

Size in Inches A × B	Wt. per Foot in lbs. w	Sectional Area in Square Inches a	Standard thickness Web t_1	Standard thickness Flange t_2	Radii Root r_1	Radii Toe r_2	Moments of Inertia Inch⁴ units Maximum About X–X (J_x)	Moments of Inertia Inch⁴ units Minimum About Y–Y (J_y)	Radii of Gyration Inches Max. About X–X (i_x)	Radii of Gyration Inches Min. About Y–Y (i_y)	Moduli of Section Inch³ units Max. About X–X (Z_x)	Moduli of Section Inch³ units Min. About Y–Y (Z_y)	Centres of Holes C Inches
1	2	3	4	5	6	7	8	9	10	11	12	13	14
3 × 1½	4	1·18	·16	·249	·25	·12	1·66	·13	1·19	·33	1·11	·17	¾
3 × 3	8·5	2·52	·20	·332	·37	·18	3·81	1·25	1·23	·70	2·54	·83	1¾
4 × 1¾	5	1·47	·17	·239	·27	·13	3·66	·19	1·58	·36	1·83	·21	1
4 × 3	10	2·94	·24	·347	·37	·18	7·79	1·33	1·63	·67	3·89	·88	1¾
4¾ × 1¾	6·5	1·91	·18	·325	·27	·13	6·73	·26	1·88	·37	2·83	·30	1
5 × 3	11	3·26	·22	·376	·37	·18	13·68	1·45	2·05	·67	5·47	·97	1¾
5 × 4½	20	5·88	·29	·513	·49	·24	25·03	6·59	2·06	1·06	10·01	2·93	2½
6 × 3	12	3·53	·23	·377	·37	·18	20·99	1·46	2·44	·64	7·00	·97	1¾

6 × 4½	20	5·89	·37	·431	·49	·24	34·71	5·40	2·43	·96	11·57	2·40	2½
6 × 5	25	7·37	·41	·520	·53	·26	43·69	9·10	2·44	1·11	14·56	3·64	2¾
7 × 4	16	4·75	·25	·387	·45	·22	39·51	3·37	2·89	·84	11·29	1·69	2¼
8 × 4	18	5·3	·28	·398	·45	·22	55·63	3·51	3·24	·81	13·91	1·75	2¼
8 × 5	28	8·28	·35	·575	·53	·26	89·69	10·19	3·29	1·11	22·42	4·08	2¾
8 × 6	35	10·30	·35	·648	·61	·30	115·06	19·54	3·34	1·38	28·76	6·51	3½
9 × 4	21	6·18	·30	·457	·45	·22	81·13	4·15	3·62	·82	18·03	2·07	2¼
9 × 7	50	14·71	·40	·825	·69	·34	208·13	40·17	3·76	1·65	46·25	11·48	4
10 × 4½	25	7·35	·30	·505	·49	·24	122·34	6·49	4·08	·94	24·47	2·88	2½
10 × 5	30	8·85	·36	·552	·53	·26	146·23	9·73	4·06	1·05	29·25	3·89	2¾
10 × 6	40	11·77	·36	·709	·61	·30	204·80	21·76	4·17	1·36	40·96	7·25	3½
10 × 8	55	16·18	·40	·783	·77	·38	288·69	54·74	4·22	1·84	57·74	13·69	4¾
12 × 5	32	9·45	·35	·550	·53	·26	221·07	9·69	4·84	1·01	36·84	3·88	2¾
12 × 6	44	13·00	·40	·717	·61	·30	316·76	22·12	4·94	1·30	52·79	7·37	3½
12 × 6	54	15·89	·50	·883	·61	·30	375·77	28·28	4·86	1·33	62·63	9·43	3½
12 × 8	65	19·12	·43	·904	·77	·38	487·77	65·18	5·05	1·85	81·30	16·30	4¾
13 × 5	35	10·30	·35	·604	·53	·26	283·51	10·82	5·25	1·03	43·62	4·33	2¾
14 × 6	46	13·59	·40	·698	·61	·30	442·57	21·45	5·71	1·26	63·22	7·15	3½
14 × 6	57	16·78	·50	·873	·61	·30	533·34	27·94	5·64	1·29	76·19	9·31	3½
14 × 8	70	20·59	·46	·920	·77	·38	705·58	66·67	5·85	1·80	100·80	16·67	4¾
15 × 5	42	12·36	·42	·647	·53	·26	428·49	11·81	5·89	·98	57·13	4·72	2¾
15 × 6	45	13·24	·38	·655	·61	·30	491·91	19·87	6·10	1·23	65·59	6·62	3½
16 × 6	50	14·71	·40	·726	·61	·30	618·09	22·47	6·48	1·24	77·26	7·49	3½
16 × 6	62	18·21	·55	·847	·61	·30	725·05	27·14	6·31	1·22	90·63	9·05	3½
16 × 8	75	22·06	·48	·938	·77	·38	973·91	68·30	6·64	1·76	121·74	17·08	4¾
18 × 6	55	16·18	·42	·757	·61	·30	841·76	23·64	7·21	1·21	93·53	7·88	3½
18 × 7	75	22·09	·55	·928	·69	·34	1151·18	46·56	7·22	1·45	127·91	13·30	4
18 × 8	80	23·53	·50	·950	·77	·38	1292·07	69·43	7·41	1·72	143·56	17·36	4¾
20 × 6½	65	19·12	·45	·820	·65	·32	1226·17	32·56	8·01	1·31	122·62	10·02	3¾
20 × 7½	89	26·19	·60	1·010	·73	·36	1672·85	62·54	7·99	1·55	167·29	16·68	4½
22 × 7	75	22·06	·50	·834	·69	·34	1676·80	41·07	8·72	1·36	152·44	11·73	4
24 × 7½	95	27·94	·57	1·011	·73	·36	2533·04	62·54	9·52	1·50	211·09	16·68	4½

Published by permission of the British Standards Institution, and based upon Report No. 4—1932.

APPENDIX II

BRITISH STANDARD CHANNELS

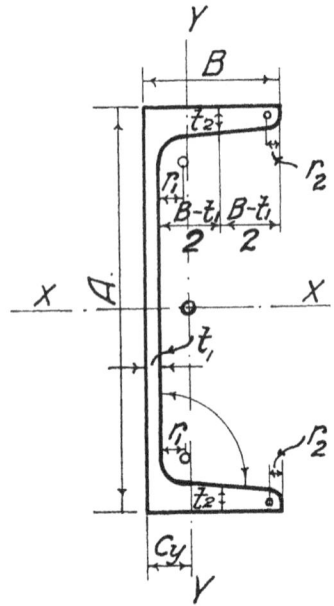

DIMENSIONS
AND
PROPERTIES
OF
BRITISH STANDARD
CHANNELS IN
INCH UNITS

Angle between
web and flange
= 95°

Size in Inches A×B	Weight per Foot in lbs. w	Sectional Area in Square Inches a	Standard thickness Web t₁	Standard thickness Flange t₂	Radii Root r₁	Radii Toe r₂	Moments of Inertia Inch⁴ units Maximum About X-X (J_x)	Moments of Inertia Inch⁴ units Minimum About Y-Y (J_y)	Radii of Gyration Inches Max. About X-X (i_x)	Radii of Gyration Inches Min. About Y-Y (i_y)	Moduli of Section Inch³ units Max. About X-X (Z_x)	Moduli of Section Inch³ units Min. About Y-Y (Z_y)	Centre of Gravity from Back (Cy) Inches
1	2	3	4	5	6	7	8	9	10	11	12	13	14
3 × 1½	4·60	1·35	·20	·28	·30	·15	1·82	·26	1·16	·44	1·21	·26	·48
3 × 1½	5·11	1·50	·25*	·28	·30	·15	1·94	·30	1·14	·44	1·29	·28	·48
4 × 2	7·09	2·09	·24	·31	·36	·18	5·06	·70	1·56	·58	2·53	·50	·60
4 × 2	7·91	2·33	·30*	·31	·36	·18	5·38	·79	1·52	·58	2·69	·54	·59
5 × 2½	10·22	3·01	·25	·38	·42	·21	11·87	1·64	1·99	·74	4·75	·95	·77
5 × 2½	11·24	3·31	·31*	·38	·42	·21	12·50	1·82	1·94	·74	5·00	1·01	·76
6 × 3	12·41	3·65	·25	·38	·48	·24	21·27	2·83	2·41	·88	7·09	1·34	·89
6 × 3	13·64	4·01	·31*	·38	·48	·24	22·35	3·10	2·36	·88	7·45	1·42	·87
6 × 3	16·51	4·86	·38*	·48	·48	·24	26·28	3·70	2·33	·87	8·76	1·77	·91

Size													
6 × 3	17·53	5·16	·43*	·48	·48	·24	27·18	3·95	2·30	·88	9·06	1·84	·90
6 × 3½	16·48	4·85	·28	·48	·54	·27	28·88	5·29	2·44	1·05	9·63	2·25	1·14
6 × 3½	18·52	5·45	·38*	·48	·54	·27	30·68	6·05	2·37	1·05	10·23	2·43	1·11
7 × 3	14·22	4·18	·26	·42	·48	·24	32·75	3·26	2·80	·88	9·36	1·53	·88
7 × 3	17·07	5·02	·38*	·42	·48	·24	36·18	3·87	2·68	·88	10·34	1·70	·84
7 × 3½	18·28	5·38	·30	·50	·54	·27	42·83	5·83	2·82	1·04	12·24	2·42	1·09
7 × 3½	20·18	5·94	·38*	·50	·54	·27	45·12	6·48	2·76	1·05	12·89	2·58	1·07
8 × 3	15·96	4·69	·28	·44	·48	·24	46·72	3·58	3·16	·87	11·68	1·65	·83
8 × 3	18·68	5·49	·38*	·44	·48	·24	50·99	4·11	3·05	·87	12·75	1·79	·81
8 × 3½	20·21	5·94	·32	·52	·54	·27	60·57	6·37	3·19	1·04	15·14	2·60	1·05
8 × 3½	23·20	6·82	·43*	·52	·54	·27	65·27	7·30	3·09	1·03	16·32	2·81	1·01
9 × 3	17·46	5·14	·30	·44	·48	·24	62·52	3·75	3·49	·86	13·89	1·69	·78
9 × 3	19·91	5·86	·38*	·44	·48	·24	67·38	4·18	3·39	·85	14·97	1·80	·76
9 × 3½	22·27	6·55	·34	·54	·54	·27	82·62	6·90	3·55	1·03	18·36	2·76	1·00
9 × 3½	23·49	6·91	·38*	·54	·54	·27	85·05	7·26	3·51	1·03	18·90	2·85	·99
9 × 3½	25·63	7·54	·45*	·54	·54	·27	89·30	7·86	3·44	1·02	19·84	2·98	·97
10 × 3	19·28	5·67	·32	·45	·48	·24	82·66	3·98	3·82	·84	16·53	1·76	·74
10 × 3	21·33	6·27	·38*	·45	·48	·24	87·66	4·31	3·74	·83	17·53	1·85	·73
10 × 3½	24·46	7·19	·36	·56	·54	·27	109·52	7·42	3·90	1·02	21·90	2·93	·97
10 × 3½	28·54	8·39	·48*	·56	·54	·27	119·52	8·50	3·77	1·01	23·90	3·17	·94
11 × 3½	26·78	7·88	·38	·58	·54	·27	141·87	7·93	4·24	1·00	25·80	3·09	·93
11 × 3½	30·52	8·98	·48*	·58	·54	·27	152·96	8·86	4·13	·99	27·81	3·30	·91
12 × 3½	26·37	7·76	·38	·50	·54	·27	159·73	7·15	4·54	·96	26·62	2·68	·83
12 × 3½	30·45	8·96	·48*	·50	·54	·27	174·13	7·96	4·41	·94	29·02	2·86	·81
12 × 4	31·33	9·21	·40	·60	·60	·30	200·09	12·12	4·66	1·15	33·35	4·12	1·06
12 × 4	36·63	10·77	·53*	·62	·60	·30	218·81	13·80	4·51	1·13	36·47	4·44	1·02
13 × 4	33·18	9·76	·40	·62	·60	·30	246·86	12·76	5·03	1·14	37·98	4·31	1·04
13 × 4	38·92	11·45	·53*	·62	·60	·30	270·66	14·51	4·86	1·13	41·64	4·64	1·01
15 × 4	36·37	10·70	·41	·62	·60	·30	349·10	13·34	5·71	1·12	46·55	4·40	·97
15 × 4	42·49	12·50	·53*	·62	·60	·30	382·85	14·97	5·54	1·09	51·05	4·71	·94
17 × 4	44·34	13·04	·48	·68	·60	·30	520·18	15·26	6·32	1·08	61·20	4·96	·92
17 × 4	51·28	15·08	·60*	·68	·60	·30	569·31	16·96	6·14	1·06	66·98	5·28	·91

* Increased thickness of web—and hence increased weight of section—is obtained by raising the rolls in the final shaping of the steel section.

Published by permission of the British Standards Institution, and based upon Report No. 4—1932.

APPENDIX III
BRITISH STANDARD EQUAL ANGLES

DIMENSIONS
AND
PROPERTIES
OF
BRITISH STANDARD
EQUAL ANGLES
IN
INCH UNITS

Size and Thickness $A \times B \times t$ (Inches)	Weight per Foot in lbs. w	Sectional Area in Square Inches a	Radii Root r_1	Radii Toe r_2	Centre of Gravity C_x & C_y	Moments of Inertia J_x & J_y	Moduli of Section Z_x & Z_y	Radii of Gyration i_x, i_y & $i_{(min)}$
1	2	3	4	5	6	7	8	9
$1 \times 1 \times$ ·125	·80	·23	·18	·13	·29	·02	·03	·29 (·19)
,, ,, ·1875	1·15	·34	·18	·13	·31	·03	·04	·29 (·19)
$1\frac{1}{4} \times 1\frac{1}{4} \times$ ·125	1·01	·30	·20	·13	·34	·04	·05	·37 (·24)
,, ,, ·1875	1·47	·43	·20	·14	·37	·06	·07	·37 (·24)
$1\frac{1}{2} \times 1\frac{1}{2} \times$ ·1875	1·79	·53	·21	·14	·43	·10	·10	·45 (·29)
,, ,, ·25	2·34	·69	·21	·15	·46	·13	·13	·44 (·29)
$1\frac{3}{4} \times 1\frac{3}{4} \times$ ·1875	2·11	·62	·23	·15	·49	·17	·14	·52 (·34)
,, ,, ·25	2·76	·81	·23	·16	·52	·22	·18	·52 (·34)
$2 \times 2 \times$ ·1875	2·43	·71	·24	·16	·56	·26	·18	·60 (·39)
,, ,, ·25	3·19	·94	·24	·17	·58	·34	·24	·60 (·39)
,, ,, ·3125	3·92	1·15	·24	·17	·61	·40	·29	·59 (·38)
$2\frac{1}{4} \times 2\frac{1}{4} \times$ ·1875	2·75	·81	·26	·17	·62	·38	·23	·68 (·44)
,, ,, ·25	3·61	1·06	·26	·18	·64	·49	·30	·68 (·44)
,, ,, ·3125	4·45	1·31	·26	·18	·67	·59	·37	·67 (·43)
$2\frac{1}{2} \times 2\frac{1}{2} \times$ ·25	4·04	1·19	·27	·18	·70	·68	·38	·76 (·49)
,, ,, ·3125	4·98	1·46	·27	·19	·73	·83	·47	·75 (·48)
,, ,, ·375	5·90	1·73	·27	·19	·75	·96	·55	·74 (·48)
$3 \times 3 \times$ ·25	4·89	1·44	·30	·19	·83	1·20	·55	·91 (·59)
,, ,, ·375	7·17	2·11	·30	·21	·88	1·72	·81	·90 (·58)
,, ,, ·50	9·35	2·75	·30	·21	·92	2·18	1·05	·89 (·58)
$3\frac{1}{2} \times 3\frac{1}{2} \times$ ·25	5·74	1·69	·33	·21	·95	1·94	·76	1·07 (·69)
,, ,, ·375	8·45	2·49	·33	·23	1·00	2·80	1·12	1·06 (·68)
,, ,, ·50	11·05	3·25	·33	·23	1·05	3·57	1·46	1·05 (·68)

Note. The radius of gyration about any axis is the square root of "the moment of inertia about that axis divided by the area"

or,

$$g = \sqrt{\frac{I}{A}}.$$

The *least* radius of gyration occurs about an axis which cuts the two arms and makes an angle of 45° with **XX** and **YY**. This least value is included in the table (see column 9—in brackets) and is only required for compression members which are long and insufficiently restrained at the connected ends.

Size	1	2	3	4	5	6	7	8	9
4 ×4 ×·375	9·73	2·86	·36	·25	1·12	4·26	1·48	1·22	(·78)
,, ,, ·50	12·75	3·75	·36	·25	1·17	5·46	1·93	1·21	(·78)
,, ,, ·625	15·68	4·61	·36	·25	1·22	6·56	2·36	1·19	(·77)
4½×4½×·375	11·00	3·24	·39	·27	1·24	6·15	1·89	1·38	(·88)
,, ,, ·50	14·45	4·25	·39	·27	1·29	7·92	2·47	1·37	(·88)
,, ,, ·625	17·80	5·24	·39	·27	1·34	9·56	3·03	1·35	(·87)
5 ×5 ×·375	12·28	3·61	·42	·29	1·37	8·53	2·35	1·54	(·98)
,, ,, ·50	16·16	4·75	·42	·29	1·42	11·04	3·08	1·52	(·98)
,, ,, ·625	19·93	5·86	·42	·29	1·47	13·37	3·78	1·51	(·97)
6 ×6 ×·375	14·82	4·36	·48	·34	1·61	14·95	3·40	1·85	(1·18)
,, ,, ·50	19·55	5·75	·48	·34	1·66	19·48	4·49	1·84	(1·18)
,, ,, ·625	24·17	7·11	·48	·34	1·71	23·73	5·54	1·83	(1·17)
7 ×7 ×·50	28·69	8·44	·54	·38	1·76	27·74	6·54	1·81	(1·17)
,, ,, ·625	22·95	6·75	·54	·38	1·91	31·42	6·17	2·16	(1·38)
,, ,, ·75	28·42	8·36	·54	·38	1·96	38·45	7·63	2·14	(1·37)
8 ×8 ×·625	33·79	9·94	·60	·42	2·01	45·12	9·04	2·13	(1·37)
,, ,, ·75	32·68	9·61	·60	·42	2·20	58·26	10·05	2·46	(1·57)
,, ,, ·875	38·89	11·44	·60	·42	2·25	68·58	11·94	2·45	(1·57)
	45·00	13·24			2·30	78·44	13·77	2·43	(1·56)

Thicknesses other than the standard thicknesses for the angles in the above list may be obtained. Angles may be ordered by "width of flanges and thickness", or by "width of flanges and weight per foot", but *not* by both thickness and weight per foot.

Published by permission of the British Standards Institution, and based upon Report No. 4A—1934.

APPENDIX IV

BRITISH STANDARD
UNEQUAL ANGLES

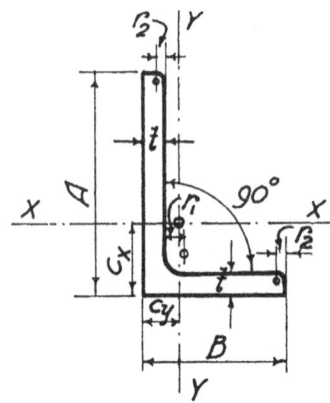

DIMENSIONS
AND
PROPERTIES
OF
BRITISH STANDARD
UNEQUAL ANGLES
IN INCH UNITS

Size and Thickness $A \times B \times t$ (Inches)	Weight per Foot in lbs.	Sectional Area in Square Inches	Radii		Centre of Gravity		Moments of Inertia		Section Moduli		Radii of Gyration
	w	a	Root r_1	Toe r_2	C_x	C_y	J_x	J_y	Z_x	Z_y	i_x, i_y and $i_{(min)}$
1	2	3	4	5	6	7	8	9	10	11	12
2 × 1½ × ·1875	2·11	·62	·23	·16	·63	·38	·24	·11	·17	·10	·62 ·43 (·32)
,, ,, × ·25	2·76	·81	·23	·16	·65	·41	·31	·15	·23	·13	·61 ·42 (·31)
2½ × 1½ × ·1875	2·43	·71	·24	·17	·83	·34	·45	·12	·27	·10	·79 ·41 (·32)
,, ,, × ·25	3·19	·94	·24	·17	·86	·37	·58	·15	·35	·14	·78 ·40 (·32)
2½ × 2 × ·1875	2·75	·81	·26	·18	·75	·50	·49	·28	·28	·18	·78 ·58 (·42)
,, ,, × ·25	3·61	1·06	·26	·18	·77	·53	·63	·36	·37	·24	·77 ·58 (·42)
,, ,, × ·3125	4·45	1·31	·26	·18	·80	·55	·77	·43	·45	·30	·77 ·57 (·42)

Size													
3 × 2 × ·25	4·04	1·19	·27	·19	·98	·48	1·06	·38	·52	·25	·94	·56	(·43)
,, ,, ·3125	4·98	1·46	·27	·19	1·00	·51	1·29	·45	·65	·30	·94	·56	(·43)
,, ,, ·375	5·90	1·73	·27	·19	1·03	·53	1·50	·53	·76	·36	·93	·55	(·42)
3 × 2½ × ·25	4·47	1·31	·29	·20	·89	·65	1·14	·72	·54	·39	·93	·74	(·52)
,, ,, ·3125	5·51	1·62	·29	·20	·92	·67	1·39	·87	·67	·48	·93	·73	(·52)
,, ,, ·375	6·54	1·92	·29	·20	·94	·70	1·62	1·02	·79	·56	·92	·73	(·52)
3½ × 2½ × ·25	4·89	1·44	·30	·21	1·09	·60	1·75	·74	·73	·39	1·10	·72	(·54)
,, ,, ·3125	6·04	1·78	·30	·21	1·12	·63	2·14	·91	·90	·48	1·10	·71	(·53)
,, ,, ·375	7·17	2·11	·30	·21	1·15	·65	2·51	1·06	1·07	·57	1·09	·71	(·53)
3½ × 3 × ·3125	6·58	1·93	·32	·22	1·04	·79	2·27	1·54	·92	·70	1·08	·89	(·62)
,, ,, ·375	7·81	2·30	·32	·22	1·07	·82	2·67	1·80	1·10	·83	1·08	·88	(·62)
,, ,, ·50	10·20	3·00	·32	·22	1·12	·87	3·40	2·28	1·43	1·07	1·06	·87	(·61)
4 × 2½ × ·25	5·32	1·56	·32	·22	1·30	·56	2·54	·77	·94	·40	1·27	·70	(·54)
,, ,, ·3125	6·58	1·93	·32	·22	1·33	·59	3·11	·94	1·17	·49	1·27	·70	(·54)
,, ,, ·375	7·81	2·30	·32	·22	1·36	·61	3·66	1·10	1·38	·58	1·26	·69	(·53)
4 × 3 × ·3125	7·11	2·09	·33	·23	1·24	·75	3·30	1·59	1·20	·71	1·26	·87	(·64)
,, ,, ·375	8·45	2·49	·33	·23	1·27	·77	3·89	1·87	1·42	·84	1·25	·87	(·64)
,, ,, ·50	11·05	3·25	·33	·23	1·32	·82	4·97	2·37	1·85	1·09	1·24	·85	(·63)
4 × 3½ × ·3125	7·64	2·25	·35	·24	1·16	·92	3·47	2·47	1·22	·96	1·24	1·05	(·72)
,, ,, ·375	9·09	2·67	·35	·24	1·19	·94	4·09	2·91	1·45	1·37	1·24	1·04	(·72)
,, ,, ·50	11·91	3·50	·35	·24	1·24	·99	5·24	3·72	1·90	1·48	1·22	1·03	(·72)
4½ × 3 × ·3125	7·64	2·25	·35	·24	1·44	·70	4·59	1·64	1·50	·71	1·43	·85	(·65)
,, ,, ·375	9·09	2·67	·35	·24	1·47	·73	5·41	1·92	1·79	·85	1·42	·85	(·64)
,, ,, ·50	11·91	3·50	·35	·24	1·52	·78	6·94	2·45	2·33	1·10	1·41	·84	(·64)
5 × 3 × ·3125	8·17	2·40	·36	·25	1·66	·67	6·14	1·68	1·84	·72	1·60	·84	(·65)
,, ,, ·375	9·73	2·86	·36	·25	1·68	·69	7·25	1·97	2·18	·85	1·59	·83	(·65)
,, ,, ·50	12·75	3·75	·36	·25	1·73	·74	9·33	2·51	2·86	1·11	1·58	·82	(·64)

Note. For unequal angles the least radius of gyration occurs about an inclined axis, the position of which varies with the ratio of the two arms and the thickness of the angle. The angle is not given but the value of the least radius of gyration is given in column 12.

For further detail refer to British Standard Specifications No. 4 A—1934. See also note on Appendix III.

APPENDIX IV—continued

Size and Thickness A×B×t (Inches)	Weight per Foot in lbs. w	Sectional Area in Square Inches a	Radii Root r₁	Radii Toe r₂	Centre of Gravity Cx	Centre of Gravity Cy	Moments of Inertia Jx	Moments of Inertia Jy	Section Modulii Zx	Section Modulii Zy	Radii of Gyration ix, iy	Radii of Gyration and i(min)
1	2	3	4	5	6	7	8	9	10	11	12	
5 ×3½× ·375	10·37	3·05	·38	·26	1·59	·85	7·63	3·09	2·24	1·16	1·58	1·01 (·75)
„ „ ×·50	13·61	4·00	·38	·26	1·64	·90	9·84	3·96	2·93	1·52	1·57	·99 (·75)
5 ×4 × ·375	11·00	3·24	·39	·27	1·51	1·01	7·97	4·53	2·28	1·52	1·57	1·18 (·85)
„ „ ×·50	14·45	4·25	·39	·27	1·56	1·06	10·29	5·83	2·99	1·98	1·56	1·17 (·84)
6 ×3 × ·375	11·00	3·24	·39	·27	2·12	·63	11·99	2·05	3·09	·87	1·93	·80 (·64)
„ „ ×·50	14·45	4·25	·39	·27	2·17	·68	15·51	2·62	4·05	1·13	1·91	·78 (·63)
6 ×3½× ·375	11·63	3·42	·41	·29	2·01	·77	12·62	3·21	3·16	1·18	1·92	·97 (·76)
„ „ ×·50	15·30	4·50	·41	·29	2·06	·82	16·36	4·13	4·15	1·54	1·91	·96 (·75)
6 ×4 × ·375	12·28	3·61	·42	·29	1·91	·92	13·21	4·74	3·23	1·54	1·91	1·15 (·87)
„ „ ×·50	16·16	4·75	·42	·29	1·97	·97	17·14	6·11	4·25	2·02	1·90	1·13 (·86)
„ „ ×·625	19·93	5·86	·44	·29	2·02	1·02	20·82	7·37	5·23	2·48	1·88	1·12 (·86)
7 ×3½× ·4375	14·97	4·40	·44	·31	2·47	·74	22·22	3·80	4·91	1·38	2·25	·93 (·74)
„ „ ×·50	17·00	5·00	·44	·31	2·50	·76	25·07	4·27	5·57	1·56	2·24	·92 (·74)
7 ×4 × ·50	17·85	5·25	·45	·32	2·39	·90	26·26	6·33	5·70	2·04	2·24	1·10 (·86)
„ „ ×·625	22·05	6·48	·45	·32	2·44	·95	32·00	7·64	7·02	2·51	2·22	1·09 (·86)
8 ×3½× ·4375	16·46	4·84	·47	·33	2·92	·69	32·08	3·89	6·31	1·39	2·57	·90 (·73)
„ „ ×·50	18·70	5·50	·47	·33	2·95	·72	36·24	4·38	7·17	1·57	2·57	·89 (·73)
8 ×4 × ·50	19·55	5·75	·48	·34	2·83	·85	37·95	6·50	7·34	2·06	2·57	1·06 (·85)
„ „ ×·625	24·17	7·11	·48	·34	2·88	·90	46·37	7·87	9·06	2·54	2·55	1·05 (·85)
8 ×6 × ·625	22·95	6·75	·54	·38	2·44	1·45	43·47	21·08	7·82	4·63	2·54	1·77 (1·29)
„ „ ×·75	33·79	9·94	·54	·38	2·54	1·55	62·60	30·14	11·47	6·77	2·51	1·74 (1·29)
9 ×4 × ·50	21·25	6·25	·51	·36	3·27	·80	52·46	6·65	9·16	2·08	2·90	1·03 (·84)
„ „ ×·75	31·24	9·19	·51	·36	3·38	·90	75·45	9·37	13·43	3·02	2·87	1·01 (·83)

Thicknesses other than the standard thicknesses for the angles in the above list may be obtained.
Angles may be ordered by " width of flanges and thickness " or by " width of flanges and weight per foot ", but *not* by both thickness and weight per foot.

Published by permission of the British Standards Institution, and based upon Report No. 4 A—1934.

APPENDIX V

BRITISH STANDARD TEES

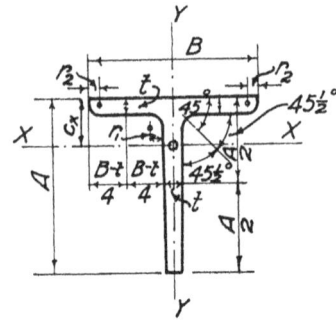

DIMENSIONS
AND
PROPERTIES
OF
BRITISH STANDARD
TEES IN
INCH UNITS

Size and Thickness (Inches) $B \times A \times t$	Weight per Foot in lbs. w	Sectional Area in Square Inches a	Radii Root r_1	Radii Toe r_2	Centre of Gravity C_x	Moments of Inertia J_x	Moments of Inertia J_y	Section Moduli Z_x	Section Moduli Z_y	Radii of Gyration i_x	Radii of Gyration i_y
1	2	3	4	5	6	7	8	9	10	11	12
1½ × 1½ × ·25	2·36	·69	·21	·15	·46	·14	·07	·13	·09	·44	·31
2 × 2 × ·25	3·21	·94	·24	·17	·58	·34	·16	·24	·16	·60	·41
2½ × 2½ × ·25	4·07	1·20	·27	·19	·70	·68	·30	·38	·24	·75	·50
„ „ × ·375	5·92	1·74	·27	·19	·75	·96	·47	·55	·38	·74	·52
3 × 3 × ·375	7·20	2·12	·30	·21	·87	1·71	·81	·80	·54	·90	·62

4 × 3 × ·375	8·49	2·50	·33	·23	·77	1·86	1·91	·83	·96	·86	·87
„ „ × ·50	11·09	3·26	·33	·23	·82	2·37	2·60	1·08	1·30	·85	·89
4 × 4 × ·375	9·77	2·87	·36	·25	1·10	4·19	1·90	1·45	·95	1·21	·81
„ „ × ·50	12·79	3·76	·36	·25	1·16	5·40	2·59	1·90	1·30	1·20	·83
5 × 3 × ·375	9·79	2·88	·36	·25	·69	1·97	3·72	·85	1·49	·83	1·14
„ „ × ·50	12·80	3·77	·36	·25	·74	2·51	5·04	1·11	2·01	·82	1·16
5 × 4 × ·375	11·06	3·25	·39	·27	1·00	4·47	3·70	1·49	1·48	1·17	1·07
„ „ × ·50	14·50	4·27	·39	·27	1·05	5·57	5·02	1·96	2·01	1·16	1·09
6 × 3 × ·375	11·08	3·26	·39	·27	·63	2·06	6·40	·87	2·13	·80	1·40
„ „ × ·50	14·52	4·27	·39	·27	·68	2·63	8·67	1·14	2·89	·78	1·42
6 × 4 × ·50	16·22	4·77	·42	·29	·97	6·07	8·64	2·00	2·88	1·13	1·35
„ „ × ·625	19·99	5·88	·42	·29	1·02	7·33	10·93	2·46	3·64	1·12	1·36
6 × 6 × ·50	19·62	5·77	·48	·34	1·63	19·04	8·56	4·36	2·85	1·82	1·22
„ „ × ·625	24·23	7·13	·48	·34	1·69	23·31	10·87	5·40	3·62	1·81	1·23

Note. Tee sections may be ordered by " width of flange, depth of section and thickness " or by " width of flange, depth of section and weight per foot run ", but *not* by both *thickness* and *weight per foot run.*

Published by permission of the British Standards Institution, and based upon Report No. 4 A—1934.

INDEX